BASIC ENGINEERING FOR BUILDERS

by Max Schwartz

Craftsman Book Company
6058 Corte del Cedro / P.O. Box 6500 / Carlsbad, CA 92018

ACKNOWLEDGEMENTS

The author thanks the following companies and organizations for furnishing materials used in this book.

Concrete Reinforcing Steel Institute
933 North Plum Grove Road
Shawmburg, IL 60173

County of Los Angeles, Department of Public Works
900 South Freemont Avenue
Alhambra, CA 91803

David White Instruments
11711 River Lane
Germantown, WI 53022

Gurley Engineering Instruments
Troy, New York

International Association of Plumbing and Mechanical Officials
20001 Walnut Drive South
Walnut, CA 91789

All portions of the *Uniform Plumbing Code* are reproduced from the 1991 edition, © 1991, with the permission of the publisher, the International Association of Plumbing and Mechanical Officials (IAPMO).

International Conference of Building Officials
5360 South Workman Mill Road
Whittier, CA 90601

All portions of the *Uniform Building Code* are reproduced from the 1991 edition, © 1991, with the permission of the publisher, the International Conference of Building Officials (ICBO).

Nikon Inc.
Melville, New York

Library of Congress Cataloging-in-Publication Data
Schwartz, Max, 1922-
Basic engineering for builders / by Max Schwartz.
p. cm.
Includes index.
ISBN 0-934041-83-0
1. Building -- Handbooks, manuals, etc. I. Title.
TH151.S392 1993 93-34254
690--dc20 CIP

Second printing 1994

C O N T E N T S

6 Masonry 262

7 Plumbing 295

8 Heating, Ventilating & Air Conditioning 343

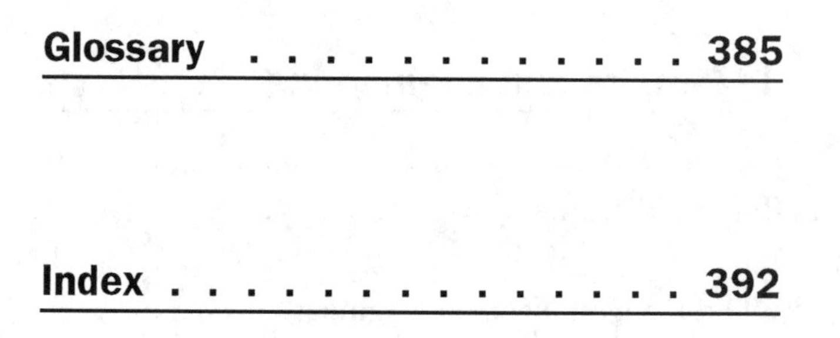

Glossary 385

Index 392

CHAPTER 1

Permits and Engineering

This book is about engineering. But it's not written for licensed professional engineers. It's written for construction contractors, subcontractors, tradesmen, estimators, designers and even building owners. In short, it's for anyone who has to deal with engineering problems in construction — and I suspect that's just about everyone in the construction industry. Engineering is as much a part of construction today as lumber or concrete, steel or wire. In practice, there's no way to avoid engineering issues when building almost anything today. Modern building codes have forced us all to be conscious of good engineering practice, no matter what or where we build.

True, engineering is a specialty that requires specialized knowledge — and lots of it. If you've picked up this book in hopes that it will make you a competent professional engineer, be forewarned. This isn't a training manual for the state license exam. And it certainly isn't a substitute for a four-year college degree. It's something entirely different, but I hope just as valuable to most contractors, builders, and construction professionals.

This book explains in non-technical language the principles engineers use when planning construction projects. You might ask, "Why do I need to know this? Isn't it enough just to hire an engineer when I need one? I'll never be able to stamp the plans with an engineer's seal to get them approved by my building department. So what's the use? Why bother learning anything about engineering?"

The answer is that engineering can be a valuable skill for anyone in the construction industry — like driving a straight nail, reading plans, or estimating costs. The more skills and knowledge you have, the more valuable you are to your customers or your employer. By the time you finish this chapter, I suspect you'll begin to understand how valuable a knowledge of engineering principles can be in your job — even if you're not a licensed professional engineer.

Knowledge of engineering principles will help you:

- build smarter, understanding more quickly what the plans require, why it's required and how you can best meet those requirements
- get building permits issued with a minimum of delay, and often without the help of an engineer
- understand ahead of time when a licensed engineer is required
- size structural members using only preliminary plans
- anticipate requirements for multi-story residential buildings, light commercial and industrial buildings, underground and elevated parking structures
- recognize and follow accepted standards for all concrete, steel, wood, and masonry construction.

But please don't think of this book as a substitute for the building code. It's not. You'll still need a copy of the building code that applies at your construction site. The code is an ordinance adopted by your city or county or state, and enforced by your local building department. The code enforcement office in your community may be called the Department of Inspection Services, Department of Engineering, Department of Public Safety, Housing and Development Administration, Building Department, or some similar name. But the function is the same — to make sure all building done in that community is done according to code.

Another type of code covers construction work in the community's public areas, which usually includes sewer connections, driveways, and curb and street work. It's probably called the *Specifications for Public Work Construction* and *Standard Plans for Public Work Construction*, or some name like that. The Department of Public Works, rather than the Building Department, controls the plan checking, permit issuance and inspections for public work.

Building Codes

Three model building codes are used in the U.S. All three are published by private organizations and have no effect until they're adopted (enacted) by your city or county. Then they become the law and builders have to comply. The three model code organizations within the Council of American Building Officials (CABO) are:

- International Conference of Building Officials (ICBO) located at 5360 South Workman Mill Road, Whittier, CA 90601 (Phone 213-699-0541), who publish the *Uniform Building Code* (UBC).
- Building Officials & Code Administrators International, Inc. who publish the BOCA *National Building Code*. Their address is 4051 West Flossmoor Road, Country Club Hills, IL 60477-5795 (Phone 312-799-2300).
- Southern Building Code Congress International, Inc. who publish the *Standard Building Code*. Their address is 900 Montclair Road, Birmingham, AL 35213 (Phone 205-591-1853).

The ICBO Uniform Building Code is probably the most popular. It's used in most states west of the Mississippi and many Canadian provinces. The BOCA code is used in most communities east of the Mississippi except in the Southeast, where the Standard Building Code is the more common code.

Why, you might ask, are there three model codes? Isn't one enough? The answer is, yes, one model code would be enough. What's good building practice in Florida should be as good in Oregon as it is in New Jersey. But there are three organizations writing codes and all three believe their approach is best for their purposes. In fact, the differences between the codes are growing smaller and smaller as time passes. Eventually we'll probably see a merger of the three model codes into one. But don't hold your breath.

Any city or county government could write their own code rather than adopting one of the model codes. Of course, some do. But most adopt one of the model codes because adopting a code written by someone else is easier than writing your own. And it's a good thing government is a little lazy here. Otherwise every community would have its own code. A builder working in a dozen different cities would have to learn a dozen different building codes. If you've been in construction for a while, you understand how hard it is to learn one code. Imagine having to deal with twelve different codes!

But be aware that every city and county that adopts a model code is free to make all the changes, additions, and deletions they want. In fact, they're free to adopt any code they want, even one that's six years old. And some do. It's not safe to assume that your city and county have adopted the current version of some code without amendment. They probably haven't. The only way to be sure you've got the right code is to buy a copy (with all amendments) and keep it up to date. As an example, the 1990 Los Angeles City Building Code adopted the 1988 edition of the Uniform Building Code, plus city amendments.

Getting a Copy of the Code

So where do you get the code? That's a real problem. Each of the three model code organizations will sell you their current code. Just call and ask for a price list. But that isn't necessarily the same code your community is using. Remember, your city or county probably added, deleted, and made changes when they adopted the code. That complicates the problem. And if you'll let me get up on my soapbox for a minute, I'd like to offer my perspective on this important issue.

The code is a law. True, ignorance of the law is no excuse. But laws are made to be complied with, not as a means of trapping the unwary. Code compliance isn't a guessing game played between you and the inspector. It's a cooperative effort following established rules to protect lives and property.

A building department that's really interested in helping contractors and builders follow the code (rather than just punishing them for failure to comply) should sell the complete, current code at a reasonable cost to anyone who wants it. If this isn't possible, they should advise the public where to get the code and amendments. In many cases, building codes can be purchased from the code's publisher, a technical book store, or from one of the mail order companies that handle construction publications. Amendments to the code, in pamphlet form, are sold by the building departments.

I suspect that some younger, less-experienced inspectors would prefer that contractors and builders didn't have the code. From their perspective, contractors who can read and understand the code just make problems for building inspectors. That misses the point. Here's why.

Ours is a government of laws, not men. No building inspector has the right to demand any more than the building code requires. The only authority inspectors and building departments have comes out of the code. If it isn't in The Book, you don't have to do it. Inspectors know that. Some novice inspectors might prefer that contractors didn't know that. They don't want to argue with builders over code language. They don't want to see contractors studying the code. That just means trouble.

But older, more-experienced hands in the building department usually see it differently. They know the best way to avoid arguments: Help contractors and builders master the code. That way they do it right the first time, before an inspector ever gets involved. That makes an inspector's life lots easier. And what's the best way to help contractors master the code? By putting a copy of The Book in the hands of every builder who wants one. Maybe that's why more and more building departments are selling the code right across the counter. It's something you should encourage.

Remember though, the building official usually has the power to interpret the code based on its intent and purpose. If you still have a disagreement with the inspector which can't be resolved, you can file an appeal with the Board of Appeals of the Building Commission. Treat the inspector in an intelligent business-like manner — and expect the same treatment from the inspector. You should both be courteous and friendly without becoming too personal. And above all, don't argue.

Understanding the Code

Of course, the code isn't easy to understand. It's written to be enforced (like a law) rather than be understood (like a road map). Every building code has exceptions within exclusions within variances. They aren't very well organized. Indexing is poor. But you *can* understand it with a little study. If you're having trouble mastering the Uniform Building Code, I can recommend *Contractor's Guide to the Building Code* by Jack Hageman. It's published by Craftsman, the same company that published this manual. An order form is bound into the last few pages of this book.

As I suggested, the code has grown from a slim little pamphlet (in the 1920s) to a massive volume (in the 1990s). To keep the code from getting even larger, code writers rely on standards published by other authorities. The following references are referred to by building codes because they provide specific details about materials and accepted procedures:

- Aluminum Association (AA)
- American Architectural Manufacturers Association (AAMA)
- American Concrete Institute (ACI)
- American Institute of Steel Construction (AISC)
- American Institute of Timber Construction (AITC)
- American National Standards Institute (ANSI)
- American Plywood Association (APA)
- American Society of Civil Engineers (ASCE)
- American Society of Mechanical Engineers (ASME)
- American Society of Testing and Materials (ASTM)
- American Welding Society (AWS)
- International Association of Plumbing and Mechanical Officials (IAPMO)
- National Fire Protection Association (NFPA)
- Portland Cement Association (PCA)
- Underwriter's Laboratories (UL)

In addition to the three national building codes, you'll see references to other model codes that apply to specialty work. Some of these are:

- CABO Model Energy Code
- National Electrical Code
- National Plumbing Code
- Uniform Administrative Code
- Uniform Building Code Standards
- Uniform Code for Abatement of Dangerous Buildings
- Uniform Code for Building Conservation
- Uniform Disaster Mitigation Plan
- Uniform Fire Code
- Uniform Fire Code Standards
- Uniform Housing Code
- Uniform Mechanical Code
- Uniform Plumbing Code
- Uniform Security Code
- Uniform Sign Code

Remember that building codes set *minimum* standards for safeguarding life, health, and property. You can always exceed code standards. That's fine. Just don't get caught doing less than the code requires.

All building codes are based on experience. As we learn more about construction, construction materials and building hazards, codes change. As new products come on the market, codes change. Any time a building defect results in a major loss of life, codes change. It's safe to assume there will always be changes in the building code. It's safe to assume that the code will become more and more complex. And my guess is that engineering principles will become more and more important to all contractors and builders. Maybe that's the best reason for reading the rest of this book and keeping it handy for future reference.

Building Permits

Building permits are required in nearly all communities now. The penalty for starting work before a permit is issued is usually a doubling of the permit fee. In addition, an investigation fee may be charged. But what's most important is the fact that building without a permit is a violation of the law. The building department may issue a stop order to discontinue use of the building and to vacate the premises.

The permit is an authorization to begin work. It also serves as a public record, a public notice, a plan check list, a statistical record, an inspection record, and a receipt. It provides general information to the permittee (the person or company receiving the permit) and the public. Although the building plans may be destroyed after some time, the permit form is usually kept on file indefinitely.

You apply for a building permit at the building department office. Usually plans are required. Typically, the building official asks you to fill out an application that has space to write in information such as:

- job address
- owner's name, address, and phone number
- contractor's name, address, and active state license number
- architect's or engineer's name, address, and active state license number
- legal description of the property (lot, block, tract, and county reference number)
- brief description of the job, including type of construction, floor area, and number of stories
- a plot plan showing all buildings and the setback from the street
- occupancy classification
- zone
- fire district
- estimated value of the work

The value of the work includes all fixed equipment required to operate and use the proposed building. The evaluation stated on the building permit application may be used by the local governmental appraiser to set the tax value of the property.

Plumbing, electrical, grading, and mechanical permits need other information. For example, a plumbing permit may require a list of all plumbing fixtures, including:

- toilets (water closets) and urinals
- bath tubs and showers
- lavatories or wash basins
- sinks (kitchen, bar, and service)
- automatic clothes washers and dishwashers
- water heaters
- sewer
- cesspool or septic tank
- interceptor and floor drains

The agency granting an electrical permit may want to know about the following items:

- receptacles, lights, and switches
- appliances, such as electric ranges, clothes dryers, water heaters, garbage disposals, clothes washers, and other motors

An application for a grading permit needs topographic plans showing existing and proposed contour lines and drainage devices. The grading department may also require a surety bond in an amount that will cover the entire project. The application will describe how much earth is to be moved, because this determines the permit fee and the amount of the bond. Grading permits are usually required for earthwork cuts that are in excess of 5 feet and fill over 3 feet in vertical depth at the deepest point. Permits are required when earthwork is over 50 cubic yards of cut or fill.

An application for a mechanical or heating, ventilating, air conditioning, and refrigerating (HVAC) permit will also require descriptions of the:

- forced air heating system, floor furnaces, and wall heaters
- gravity heating system
- air conditioning system (evaporative coolers, refrigeration, absorption systems)
- ventilation system
- incinerators, hot water and steam boilers
- automatic clothes dryers

You don't need a permit for any portable ventilating equipment, comfort cooling units, and similar movable equipment.

Make sure the type of construction you plan is legal under the zoning ordinance that applies. Zones are set by the planning or zoning departments which regulate the location of trades and industries. Zoning ordinances limit the ways a building can be used and identify where it can be located on the property.

Zones for residential and commercial property may include:

- Single Family Residence
- Two Family Residence
- Limited Multiple Residence
- Unlimited Residence
- Residential Agricultural
- Neighborhood Business
- Unlimited Commercial
- Commercial Manufacturing

You'll need a building permit for all construction work that will cost over a certain amount, usually $200. You'll need separate permits for each building, even though they're on the same lot. An exception is made for auxiliary buildings, such as a garage.

Building permits are not required for:

- sheds under 120 square feet in area
- fences under 6 feet in height
- walls less than 4 feet high unless supporting a surcharge (used as a retaining wall)
- movable cases, counters, and partitions that are not over 5 feet 9 inches high
- platforms, walks, and driveways not more than 30 inches above grade and not over any basement or story below
- painting, papering, and similar work

Building permits are required for all construction that affects a building's structural stability. The permit process is a means of enforcing the building code to provide better construction and greater safety to the public. The application for the permit requires a legal description of the site, a description of the work, and identification of the parties involved with the work.

Nearly every permit will require building plans drawn with indelible pencil, ink, or reproduced mechanically (such as blueprints). Two sets of plans are required to describe the work, location, and materials to be used. The plans also provide information on emergency rescue and exit from a building, door and window security, and energy conservation. Upon approval of the plans, one set goes to the owner and the other is kept by the inspection department. A simple residential building would probably require the following plans and specifications:

- Plot plan showing lot dimensions, streets, location of all existing and proposed buildings, use, number of stories, and type of construction of each building. Figure 1-1 is a vicinity map showing front and side yard setbacks of several contiguous lots.

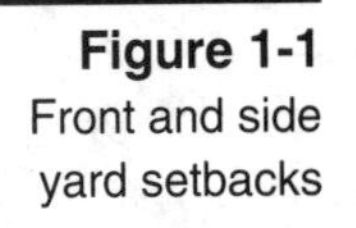

Figure 1-1
Front and side yard setbacks

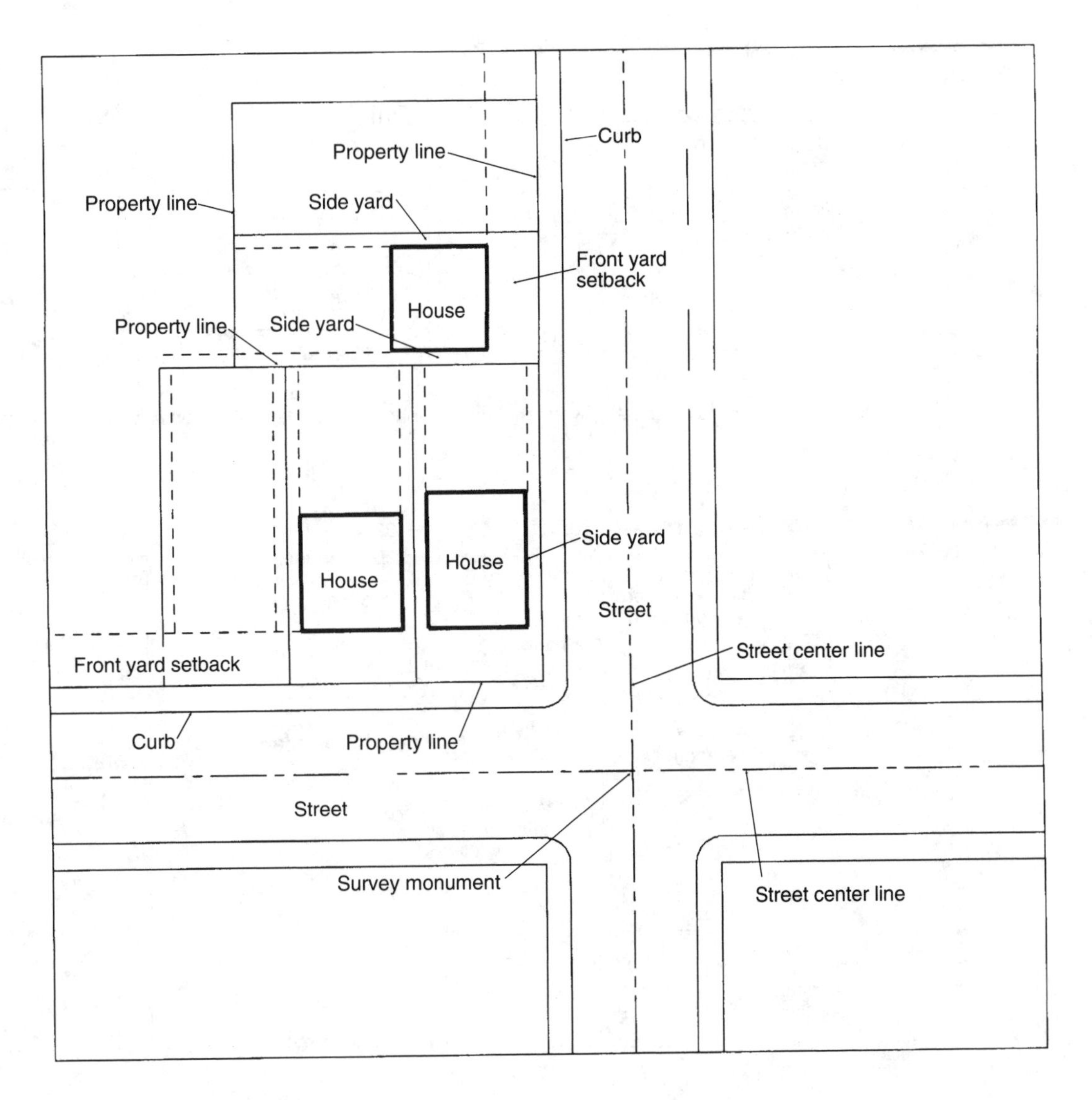

- Foundation plan showing dimensions, locations, and details, pier and crawl space openings, sizes, spans, and spacing of girders and joists for raised floors.
- Floor plan showing all rooms, windows, doors, and building security. Location and sizes of windows, doors, stairways, use of rooms and plumbing fixtures.
- Two or more exterior elevations of the building showing roofing material and slope, location of doors and windows, and attic ventilation. Grade elevations around a building will determine the number of stories. Figure 1-2 shows the difference between a basement and a cellar.
- Construction details which identify the materials you intend to use and the insulation that will be applied to the walls, ceilings, and roof. Figure 1-3 shows simplified construction details for a residential addition.
- Specifications which identify the grade and type of materials to be used for structural parts of the building.

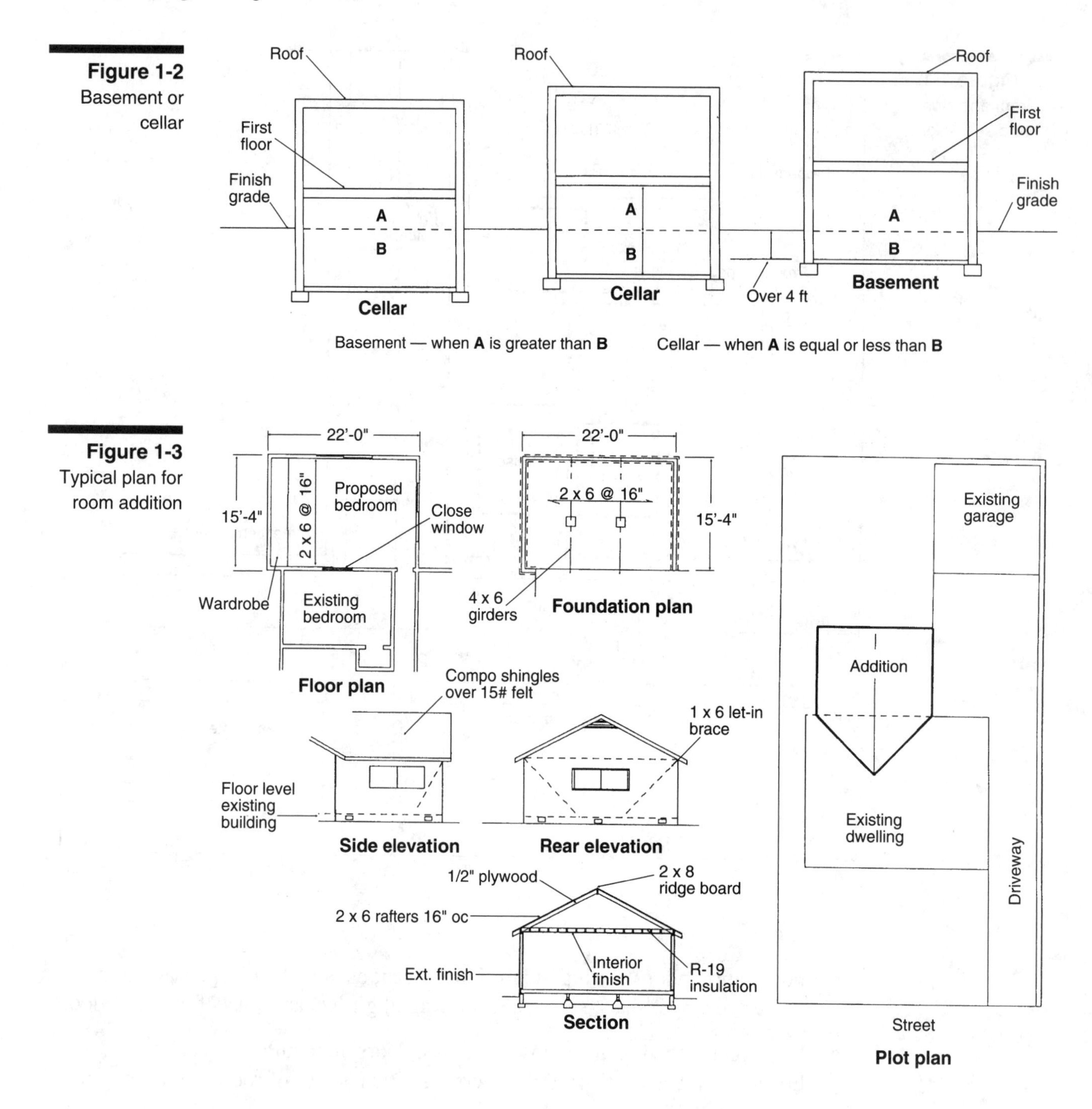

Figure 1-2 Basement or cellar

Figure 1-3 Typical plan for room addition

A building official may waive the requirements for plans and calculations if the work doesn't seem to need them. Some building departments don't require structural calculations when the proposed residential building complies with the code's Type V construction requirements.

If there's doubt about the validity of what your plans show, the building official is going to ask for some help. At your expense, of course. You'll have to submit engineering calculations or the opinion of a responsible person that the plans meet generally-accepted engineering standards. These plans should be prepared under the direct supervision of a licensed engineer and must be signed by that engineer (and usually include a registration seal granted by your state and its expiration date).

Inspections are made periodically during the course of construction. They are made before each major stage of the work is concealed. Inspections are *called* or *special.* Called inspections on wood framed buildings are made on foundations, wood framing, ventilation, insulation, and plaster. Other called inspections are for reinforced concrete or masonry, structural steel, electrical, plumbing, heating, and refrigeration. Final inspection is made at the completion of all work. Special inspection is usually provided by the owner of the building.

Figure 1-4 is a copy of a typical Building Department Inspection Record. It shows the inspections performed during the course of construction.

Projects Needing Separate Permits

Larger projects may need separate permits for grading, plumbing, electrical, and mechanical work. Figure 1-5 shows an elevation view of a six-story building which would usually need separate permits. Contractors doing specialty work usually submit plans covering the work they intend to do. In some communities, you'll need separate plumbing, electrical, and HVAC plans and specifications for an apartment building that's 15,000 square feet or more.

Figure 1-4 Sample inspection record

POST THIS CARD AT OR NEAR FRONT OF BUILDING

JURISDICTION OF ____________________

BUILDING DEPARTMENT
INSPECTION RECORD

Job Address ____________________ Type ________ Occupancy ___

Nature of Work ____________________

Use of Building ____________________ Owner __________

Building Permit No. ____________________ Date Issued ________

Contractor ____________________

Inspector must sign all spaces pertaining to this job

INSPECTION	DATE	INSPECTOR
Foundations:		
Setback		
Footings		
Reinforcing		
Foundation Wall		
Pour no concrete until above has been signed		
Weatherproofing		
No backfill until above has been signed		

Figure 1-4 (cont.) Sample inspection record

INSPECTION	DATE	INSPECTOR
Concrete Slab Floor:		
Electrical (Groundwork)		
Plumbing (Groundwork)		
Gas Piping (Groundwork)		
Do not pour floor until above has been signed		
Rough Electrical		
Rough Plumbing		
Rough Gas Piping		
Rough HVAC (above must be signed before framing inspection)		
Framing		
Cover no work until the above has been signed		
Lath & Plaster (Interior):		
Lath		
Scratch Coat		
Brown Coat		
Finish Coat		
Wallboard		
Lath & Plaster (Exterior):		
Lath		
Scratch Coat		
Brown Coat		
Finish Coat		
Miscellaneous:		
Roofing		
Gypsum Board (before tape and paint)		
Sewer		
Refrigeration		
Electrical Underground		
Final:		
Electrical Fixtures		
Plumbing Fixtures		
Gas Piping		
HVAC		
Job Completed		
All of above must be signed for final		

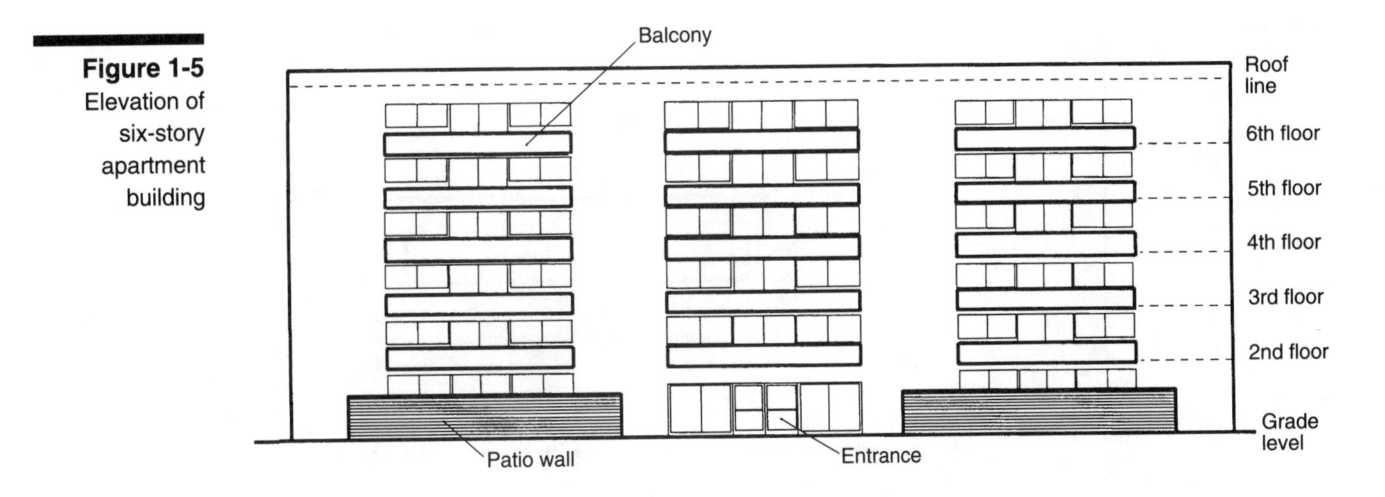

Figure 1-5 Elevation of six-story apartment building

Even if the plans are drawn by the subcontractor who will do the work, the building department may want the plans approved by an engineer licensed to prepare plans for that type of work.

In some states, a licensed contractor who installs electrical or mechanical systems can prepare plans for that work. But the electrical or mechanical contractor may *not* design work that will be installed by another contractor.

Figure 1-6 outlines the procedure that may be required to get a permit issued for a larger building. Obviously, it can be complex. Plan approval can take months or even a year for a larger commercial or industrial buildings in some communities — even when the building department isn't flooded with applications.

The code regulates the occupancy, or intended use, of buildings by their size and type of construction. It also controls the type of building construction based on the size and use. The purpose is to limit the size of buildings that are vulnerable to fire or that have many occupants. When the contents of a building are hazardous, its construction must be more fire resistant.

Building Occupancies

There are many kinds of building occupancies. These can usually be divided into the following general uses:

Group A: Assembly

Group B: Business, including restaurants, stores, and shops

Group E: Education

Group F: Factory

Group H: High hazard

Group I: Institutional

Group M: Mercantile, also private garages and agricultural buildings

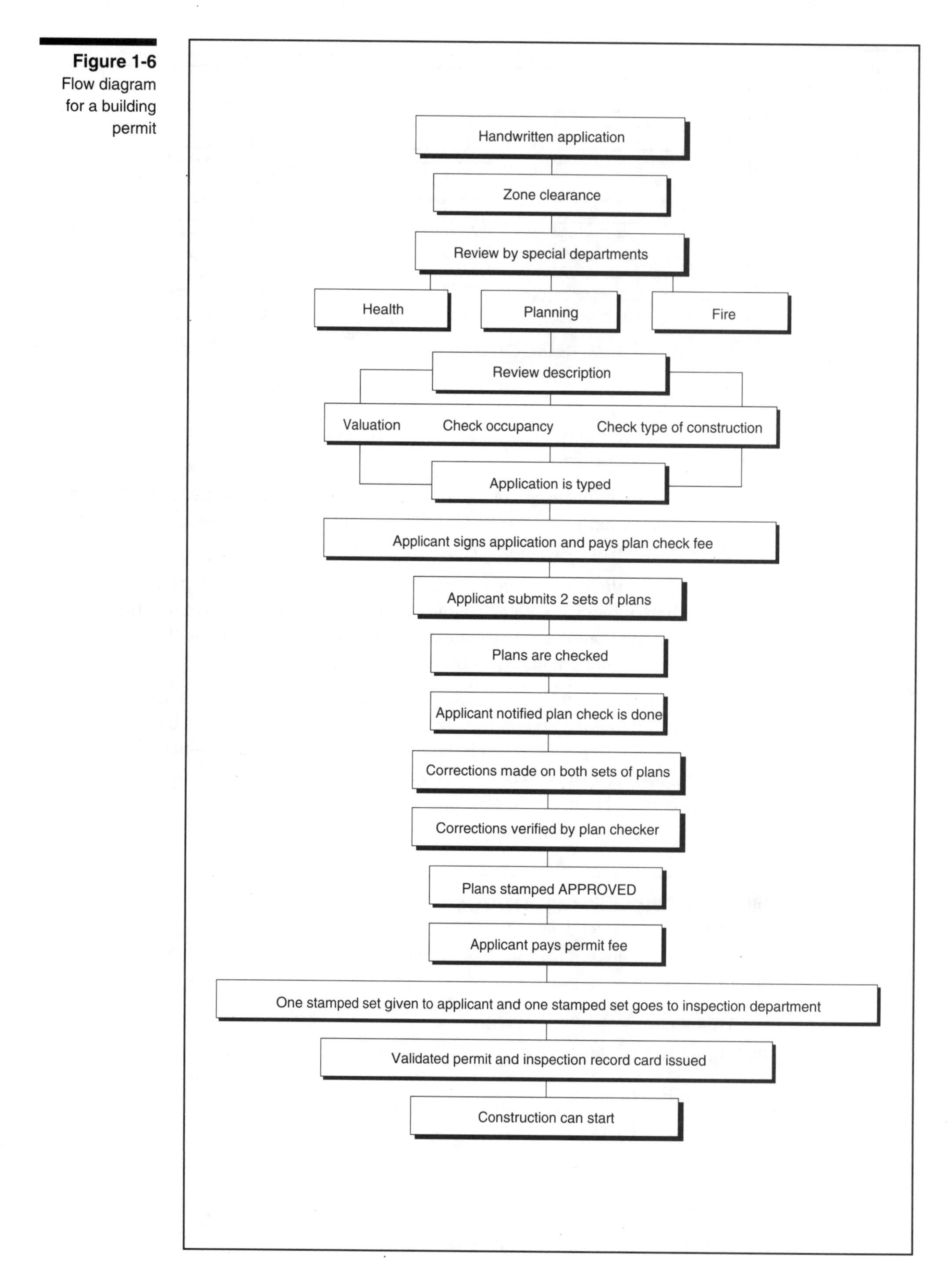

Figure 1-6 Flow diagram for a building permit

Group R: Residential, including hotels, apartments, and dwellings

Group S: Storage

Group U: Utilities

These classifications may vary somewhat in different codes. Each of these groups is further divided into different levels of hazard. This book considers only Group B and Group R. A B-1 occupancy includes repair garages while B-2 includes almost all types of stores, restaurants, and workshops. Group R-1 covers apartment houses and hotels while R-3 includes dwellings.

Types of Construction

Types of construction are classified by the fire-resistive properties of the major elements of a building. These include the structural frame, walls, ceilings, floors, and roofs. Type I construction is the most fire-resistive while Type V is the least.

The terms that describe how a building will react to a fire are *combustible*, *non-combustible*, *fire-resistive*, and *not fire-resistive*. The words *protected* and *non-protected* may be used in place of the last two terms. For example, a wood frame structure is combustible while a steel frame one is non-combustible. A coat of plaster over a wood or steel beam provides some fire-resistive protection. The degree of protection is based on the time required for a fire to weaken the member. Typically, fire-resistance may be one, two, three, or four-hour protection. Various thicknesses of coatings produce different degrees of fire-resistance. Fireproof coatings may be of gypsum or portland cement, or similar mineral materials.

Buildings are classified into five general types of construction. These are based on whether the structural frame is combustible or non-combustible and how the frame is protected against fire:

Type I fire-resistive construction

This is the most fire resistant construction. Frames are non-combustible with three-hour fire-resistive protection. In some areas, buildings made using this type of construction are called Class A buildings.

Type II fire-resistive 1-hour, or non-protected, construction

These are the next levels of fire resistant construction. Frames are made of a non-combustible material and have two-hour, one-hour, or no fire-resistive protection.

Type III 1-hour, or non-protected, construction

Frames are combustible with one-hour or no fire resistive protection.

Type IV heavy-timber construction

Frames are combustible, but made of heavy timbers which are slow to burn.

Type V 1-hour, or non-protected, construction

These are the most vulnerable to fire. Frames are made of combustible light wood framing, with one-hour or no fire-resistive protection.

Figure 1-7A shows the basic allowable area, in square feet, for one-story buildings with Group B-2 and R-1 occupancies for different types of construction. A building area is considered to be the area within the exterior walls of a building exclusive of vent shafts and cores. Figure 1-7B shows the height of buildings, in feet, for Group B-2 and R-1 occupancies for different types of construction.

It's obvious that the allowable area and height of a building increases with the more fire-resistant types of construction. For example, a six-story apartment building like the one in Figure 1-5 must have Type I or II fire resistive construction.

A one-story apartment building with Type V 1-hour construction can have a basic area of 10,500 square feet. For multiple stories, the total allowable area may be increased to 200 percent of the basic allowable area, or 21,000 square feet. You can

Figure 1-7A
Basic allowable floor area for buildings one story in height[1] (in square feet)

	TYPES OF CONSTRUCTION								
	I	II			III		IV	V	
OCCUPANCY	F.R.	F.R.	ONE-HOUR	N	ONE-HOUR	N	H.T.	ONE-HOUR	N
A-1	Unlimited	29,900	Not Permitted						
A-2-2.1[2]	Unlimited	29,900	13,500	Not Permitted	13,500	Not Permitted	13,500	10,500	Not Permitted
A-3-4[2]	Unlimited	29,900	13,500	9,100	13,500	9,100	13,500	10,500	6,000
B-1-2-3[3]	Unlimited	39,900	18,000	12,000	18,000	12,000	18,000	14,000	8,000
B-4	Unlimited	59,900	27,000	18,000	27,000	18,000	27,000	21,000	12,000
E-1-2-3	Unlimited	45,200	20,200	13,500	20,200	13,500	20,200	15,700	9,100
H-1	15,000	12,400	5,600	3,700	Not Permitted				
H-2[4]	15,000	12,400	5,600	3,700	5,600	3,700	5,600	4,400	2,500
H-3-4-5[4]	Unlimited	24,800	11,200	7,500	11,200	7,500	11,200	8,800	5,100
H-6-7	Unlimited	39,900	18,000	12,000	18,000	12,000	18,000	14,000	8,000
I-1.1-1.2-2	Unlimited	15,100	6,800	Not Permitted[8]	6,800	Not Permitted	6,800	5,200	Not Permitted
I-3	Unlimited	15,100	Not Permitted[5]						
M[6]	See Chapter 11								
R-1	Unlimited	29,900	13,500	9,100[7]	13,500	9,100[7]	13,500	10,500	6,0007
R-3	Unlimited								

[1]For multistory buildings, see Section 505 (b).
[2]For limitations and exceptions, see Section 602.
[3]For open parking garages, see Section 709.
[4]See Section 903.
[5]See Section 1002 (b).
[6]For agricultural buildings, see also Appendix Chapter 11.
[7]For limitations and exceptions, see Section 1202 (b).
[8]In hospitals and nursing homes, see Section 1002 (a) for exception.

N—No requirement for fire resistance **F.R.**—Fire resistive **H.T.**—Heavy timber

Figure 1-7B
Maximum height of buildings

OCCUPANCY	TYPES OF CONSTRUCTION								
	I	II			III		IV	V	
	F.R.	F.R.	ONE-HOUR	N	ONE-HOUR	N	H.T.	ONE-HOUR	N
	MAXIMUM HEIGHT IN FEET								
	Unlimited	160	65	55	65	55	65	50	40
	MAXIMUM HEIGHT IN STORIES								
A-1	Unlimited	4	Not Permitted						
A-2-2.1	Unlimited	4	2	Not Permitted	2	Not Permitted	2	2	Not Permitted
A-3-4[1]	Unlimited	12	2	1	2	1	2	2	1
B-1-2-3[2]	Unlimited	12	4	2	4	2	4	3	2
B-4	Unlimited	12	4	2	4	2	4	3	2
E[3]	Unlimited	4	2	1	2	1	2	2	1
H-1[4]	1	1	1	1	Not Permitted				
H-2[4]	Unlimited	2	1	1	1	1	1	1	1
H-3-4-5[4]	Unlimited	5	2	1	2	1	2	2	1
H-6-7	3	3	3	2	3	2	3	3	1
I-1.1[5]-1.2	Unlimited	3	1	Not Permitted	1	Not Permitted	1	1	Not Permitted
I-2	Unlimited	3	2						
I-3	Unlimited	2	Not Permitted						
M[7]	See Chapter 11								
R-1	Unlimited	12	4	2[8]	4	2[8]	4	3	28
R-3	Unlimited	3	3	3	3	3	3	3	3

[1]For limitations and exceptions, see Section 602 (a).
[2]For open parking garages, see Section 709.
[3]See Section 802 (c).
[4]See Section 902.
[5]See Section 1002 (a) for exception to number of stories in hospitals and nursing homes.
[6]See Section 1002 (b).
[7]For agricultural buildings, see also Appendix Chapter 11.
[8]For limitations and exceptions, see Section 1202 (b).

N—No requirement for fire resistance **F.R.**—Fire resistive **H.T.**—Heavy timber

increase this even more by planning wider yards around the building or by adding automatic fire sprinklers. For property located in a relatively low-density area, the code may permit a larger building area.

In summary, the conditions that set the maximum allowable area for a building are:

- type of construction
- group occupancy
- yards
- sprinklers
- fire zone
- number of stories

Figures 1-8 to 1-13 show permit forms similar to those used in many building department offices. Notice that forms include applications for buildings, demolition, grading, plumbing, and electrical work.

Figure 1-8
Sample building permit

CITY OF LOS ANGELES DEPT. OF BUILDING AND SAFETY

1 APPLICATION FOR INSPECTION — WORKSHEET — OF NEW BUILDING AND FOR CERTIFICATE OF OCCUPANCY

INSTRUCTIONS: 1. Applicant to Complete Numbered Items Only. 2. Plot Plan Required on Back of Original.

1. LEGAL DESCR.	LOT	BLOCK	TRACT	COUNCIL DIST. NO.	DIST. MAP / CENSUS TRACT

2. PURPOSE OF BUILDING ( ) — ZONE

3. JOB ADDRESS — FIRE DIST.

4. BETWEEN CROSS STREETS AND — LOT (TYPE)

5. OWNER'S NAME PHONE — LOT SIZE

6. OWNER'S ADDRESS CITY ZIP

7. ENGINEER BUS. LIC. NO. ACTIVE STATE LIC. NO PHO — ALLEY

8. ARCHITECT OR DESIGNER BUS. LIC. NO. ACTIVE STATE LIC. NO. PHONE — BLDG. LINE

9. ARCHITECT OR ENGINEER ADDRESS CITY ZIP — AFFIDAVITS

10. CONTRACTOR BUS. LIC. NO. IVE S TE LI NO. PHONE

11. SIZE OF NEW BLDG. WIDTH LENGTH — STORIES — HEI — NO. STING BUILDINGS ON LOT AND USE

12. MATERIAL OF CONSTRUCTION — EXT. WALLS — OOF — FLOOR — P. C. REQ'D

1 **13.** JOB ADDRESS — DIST. OFFICE

14. VALUATION TO INCLUDE FIXED EQUIPMENT REQUIRED T OPE TE AND USE PROPOSED BUI ING $ — SEISMIC STUDY ZONE

GRADING — FLOOD

HWY. DED. — CONS.

SAMPLE

PURPOSE OF BUILDING		STORIES	HEIGHT	ZONED BY
TYPE	GROUP OCC.	FLOOR AREA	PLANS CHECKED	FILE WITH
DWELL. UNITS	MAX. OCC.	TOTAL	APPLICATION APPROVED	TYPIST
GUEST ROOMS	PARKING REQ'D	PARKING PROVIDED STD. COMP.	INSPECTION ACTIVITY COMB GEN MAJ. S. CONS	INSPECTOR

P.C.	G.P.I.	CONT. INSP.
S.P.C.	P.M.	SPRINKLERS REQ'D SPECIFIED
B.P.	E.I.	
I.F.	F.H.	
O/S	O.S.S.	
DIST. OFFICE	S.O.S.S.	
P.C. NO.	C/O	
	ENERGY	

B & S B-1A (R 3.86)

PLAN CHECK EXPIRES ONE YEAR AFTER FEE IS PAID. PERMIT EXPIRES TWO YEARS AFTER FEE IS PAID OR 180 DAYS AFTER FEE IS PAID IF CONSTRUCTION IS NOT COMMENCED.

Bureau of Engineering	ADDRESS APPROVED	
	DRIVEWAY	
	SEWERS	SEWERS AVAILABLE
		NOT AVAILABLE
		SFC PAID
	SFC NOT APPLICABLE	SFC DUE
Conservation	APPROVED FOR ISSUE ☐ NO FILE ☐ CLOSED FILE ☐	

Figure 1-9
Sample building permit

CITY OF LOS ANGELES DEPT. OF BUILDING AND SAFETY

3 APPLICATION FOR INSPECTION

— WORKSHEET —

TO ADD-ALTER-REPAIR-DEMOLISH AND FOR CERTIFICATE OF OCCUPANCY

INSTRUCTIONS: 1. Applicant to Complete Numbered Items Only.

1. LEGAL DESCR.	LOT	BLK.	TRACT	COUNTY REF. NO.

2. PRESENT USE OF BUILDING ( ) | NEW USE OF BUILDING ( )

3. JOB ADDRESS

4. BETWEEN CROSS STREETS | AND

5. OWNER'S NAME | PHONE

6. OWNER'S ADDRESS | CITY | ZIP

7. ENGINEER | BUS. LIC. NO. | ACTIVE STATE LIC. NO. | PHONE

8. ARCHITECT OR DESIGNER | BUS. LIC. NO. | ACTIVE STATE LIC. NO. | PHONE

9. ARCHITECT OR ENGINEER'S ADDRESS | CITY | ZIP

10. CONTRACTOR | BUS. LIC. NO. | ACTIVE STATE LI NO. | PHO

11. SIZE OF EXISTING. BLDG. WIDTH LENGTH | STORIES | HEIGHT | NO. EXISTIN BUILD S ON T AND USE

12. FRAMING MATERIAL OF EXISTING BLDG. → | EXT. WALLS | OOF | FLOOR

3 13. JOB ADDRESS | STREET GUIDE

14. VALUATION TO INCLUDE A FIXED EQUIPMENT REQUIRED TO RATE AND USE PROPOSED BUIL NG

15. NEW WORK (Describe)

DIST. MAP	
CENSUS TRACT	
ZONE	
FIRE DIST.	COUN. DIST.
LOT (TYPE)	
LOT SIZE	
ALLEY	
BLDG. LINE	
AFFIDAVITS	
DIST. OFF.	P.C. REQ'D
GRADING	SEISMIC
HWY. DED.	FLOOD
FILE WITH	
ZONED BY	
TYPIST	
INSPECTOR	

NEW USE OF BUILDING | SIZE OF ADDITION | STORIES | HEIGHT

TYPE	GRO OCC.	FLOOR AREA	PLANS CHECKED
DWELL. UNITS	MAX OCC.	TOTAL	APPLICATION APPROVED
GUEST ROOMS	PARKING REQ'D.	PARKING PROVIDED STD. COMP.	INSPECTION ACTIVITY CS / GEN. / MAJ. S. / EQ.
P.C.	G.P.I.	CONT. INSP.	
S.P.C.	P.M.	SPRINKLERS REQ'D SPECIFIED	
B.P.	E.I.		
I.F.	F. H.		
O/S	O.S.S.		
ISSUING OFFICE	S.O.S.S.		
P.C. NO.	C/O		
	ENERGY		

B & S B-3A (R.9/88)

PLAN CHECK EXPIRES ONE YEAR AFTER FEE IS PAID. PERMIT EXPIRES TWO YEARS AFTER FEE IS PAID OR 180 DAYS AFTER FEE IS PAID IF CONSTRUCTION IS NOT COMMENCED.

Bureau of Engineering	ADDRESS APPROVED		
	DRIVEWAY		
	SEWERS	SEWERS AVAILABLE	
	RES. NO.	NOT AVAILABLE	
	CERT. NO.	SFC PAID	
	SFC NOT APPLICABLE	SFC DUE	
Comm. Safety	APPROVED FOR ISSUE ☐ NO FILE ☐ FILE CLOSED ☐		

Figure 1-10
Sample grading permit

G APPLICATION FOR INSPECTION — CITY OF LOS ANGELES — DEPT. OF BUILDING AND SAFETY — OF GRADING AND FOR GRADING CERTIFICATE

— WORKSHEET —

INSTRUCTIONS: 1. Applicant to Complete Numbered Items Only. 2. Plot Plan Required on Back of Original.

1. LEGAL DESCR. — LOT — BLK. — TRACT — COUNTY REF. NO. — DIST. MAP — CENSUS TRACT
2. PURPOSE OF GRADING ( ) — ZONE
3. JOB ADDRESS — FIRE DIST. — COUN. DIST.
4. BETWEEN CROSS STREETS AND — LOT (TYPE)
5. OWNER'S NAME — PHONE — LOT SIZE
6. OWNER'S ADDRESS — CITY — ZIP
7. PLANS BY CIVIL ENGR. — BUS. LIC. NO. — ACTIVE STATE LIC. NO. — PHONE — ALLEY
8. CIVIL ENGR. ADDRESS — CITY — ZIP — BLDG. LINE
9. ENGR. GEOLOGIST — BUS. LIC. NO. — ACTIVE STATE LIC. NO./CERT. NO. — PHONE — AFFIDAVITS
10. SOIL ENGR.—TESTING AGENCY — BUS. LIC. NO. — ACTIVE STATE LIC. NO. — PHONE
11. CONTRACTOR — BUS. LIC. NO. — ACTIVE STATE LIC. NO. — PHONE
12. CONTRACTOR'S ADDRESS — CITY — ZIP

G
13. JOB ADDRESS — STREET GUIDE — DIST. OFF. — P.C. REQ'D
14. NUMBER OF CUBIC YARDS — CUT — FILL — GRADING — SEISMIC
15. MAXIMUM SLOPE CUT FILL — RETAINING WALL REQUIRED YES NO — BOARD FILE NO. — HWY. DED. — FLOOD

FILL DENSITY TESTS & CERTIFICATION ☐ 90% REQUIRED ☐ NOT REQUIRED — FILE WITH

IMPORT/EXPORT REQ. — ZONED BY

CALIF. ENVIRONMENTAL QUALITY ACT REQUIREMENTS EXEMPT COMPLETED — YARDAGE APPROVED — TYPIST

BOND AMOUNT____ — PLANS CHECKED — G.P.I. INSPECTOR

☐ CASH DATE POSTED ____ — APPLICATION APPROVED — INSPECTOR

☐ SURETY CA # ____ — 08—B-100A (R.8/89)

P.C.	G.P.I. + NP
S.P.C.	I.F.
G.P.	O.S.S.
ISSUING OFFICE	S.O.S.S.
P.C. NO.	
SYS	SSYS

PLAN CHECK EXPIRES ONE YEAR AFTER FEE IS PAID. PERMIT EXPIRES TWO YEARS AFTER FEE IS PAID OR 180 DAYS AFTER FEE IS PAID IF CONSTRUCTION IS NOT COMMENCED.

Bureau of Engineering		
	ADDRESS APPROVED	
	DRIVEWAY	
	SEWERS AVAILABLE YES NO	
	FLOOD CLEARANCE	
	DRAINAGE TO WATERCOURSE APPROVED	
	GRADING IN WATERCOURSE APPROVED	
GRADING	PRIVATE SEWER SYSTEM APPROVED	

SAMPLE

Figure 1-11
Sample plumbing permit

P 4 CITY OF LOS ANGELES DEPARTMENT OF BUILDING AND SAFETY

APPLICATION FOR PLUMBING INSPECTION AND PLAN CHECK

TOILETS	BATHS	SHOWERS	LAVATORIES	SINKS	GARB. DISP.	DISH-WASH.
CLOTHES WASHERS	TRAYS	FLOOR DRAINS	URINALS	PRESSURE REGULATORS	SOFTENERS	WATER PIPING
LAWN SPRINKLERS NO. OF VALVES:		SWIMMING POOLS	PUB.: PRIV.:	FILTER	ROOF DRAINS	
FLOOR SINKS	BACKFLOW DEVICES		TRAP PRIMERS		WATER USING DEVICES	
CRITICAL SOIL SURVEY	CAP SEWER	FILL SEPTIC TANK AND CESSPOOL		SEWER TO:	PRIVATE SEWER DISPOSAL SYSTEM	
WATER HEATERS		W.H. VENTS	GAS SYSTEMS NO.	☐ LP ☐ MP ☐ HP	GAS OUTLETS NO.	
FIRE SPRINKLERS HEADS:		UNDERGROUND SERVICE		STAND PIPE OUTLETS, CLASS:	I	II III
PLAN CHECK	TOTAL DRAINAGE FIXTURE UNITS		WATER PIPE SIZE		GAS PIPE SIZE	

MISCELLANEOUS

SUPPLEMENTARY TO APPLIC. FEE NO.: Date:

CASHIER'S USE ONLY

INVESTIGATION FEE	APPLICATION APPROVED
$ ISSUING FEE	BLDG. INSPECTION APPLICATION NO.
PLAN CHECK FEE	ENG. PERMIT NO.
SUBTOTAL	IND. WASTE PERMIT NO. / DATE
ONE STOP SURCHARGE	PLAN CHECK NO.
TOTAL FEE DUE	

SUPPLEMENTARY APPL. NO,S

NO. OF STORIES

BUSINESS TAX REGISTRATION CERT. NO.

DISTRICT INSPECTOR | DIST. NO.

JOB ADDRESS (Suite or tenant) ZIP CROSS STREETS &

SAMPLE

GAS RISERS ☐ LP ☐ MP ☐ HP | HOUSE DRAINS | FIRST GAS | FINAL GAS

G. W. HEATER VENT | GAS WATER HEATER

FIRE SPRINKLERS

UNDERGROUND | SEPTIC TANK | SEWER

FIRST PLUMBING | FINAL PLUMBING | CESSPOOL-SEEPAGE PIT | MISCELLANEOUS

PLUMBING OUTLETS INSTALLED			
TOILETS			
BATHS			
LAVATORIES			
SHOWERS			
SINKS			
TRAYS			
CLO. WASHERS			
DISHWASHERS			
FLOOR DRAINS			
TOTAL			

WATER METER | H.S. | RELIEF VALVE

FIXTURE UNITS OLD | NEW | PRESSURE REG.

QUALIFIED INSTALLER | LICENSE NO. | TYPE | PHONE NO.

NAME (Qualified Installer/Qualified Plan Submitter) (Please Print — Use ink)

ADDRESS

CITY STATE ZIP

OWNER OR LESSEE PHONE NO.

ADDRESS CITY

(CHECK) NEW ☐ OR EXISTING ☐

ONE OR TWO FAMILY DWELLING ☐

APT. OR COMMERCIAL ☐

NO. OF DWELL UNITS

Applicant certifies that the information given is correct. Application expires if work is not commenced within 180 days after fee is paid or if work is suspended for a period of more than 180 days.

B & S P-4 (R.1/88)

Figure 1-11 (cont.)
Sample plumbing permit

DECLARATIONS AND CERTIFICATIONS

LICENSED CONTRACTORS DECLARATION

I hereby affirm that I am licensed under the provisions of Chapter 9 (commencing with Section 7000) of Division 3 of the Business and Professions Code, and my license is in full force and effect.

Date ________ Lic. Class ________ Lic. No. ________ Contractor's/Agents Signature ________

OWNER-BUILDER DECLARATION

I hereby affirm that I am exempt from the Contractor's License Law for the following reason (Sec. 7031.5, Business and Professions Code: Any city or county which requires a permit to construct, alter, improve, demolish, or repair any structure, prior to its issuance, also requires the applicant for such permit to file a signed statement that he is licensed pursuant to the provisions of the Contractor's License Law (Chapter 9 (commencing with Section 7000) of Division 3 of the Business and Professions Code) or that he is exempt therefrom and the basis for the alleged exemption. Any violation of Section 7031.5 by any applicant for a permit subjects the applicant to a civil penalty of not more than five hundred dollars ($500).):

☐ I, as owner of the property, or my employees with wages as their sole compensation, will do the work, and the structure is not intended or offered for sale (Sec. 7044, Business and Professions Code: The Contractor's License Law does not apply to an owner of property who builds or improves thereon, and who does such work himself or through his own employees, provided that such improvements are not intended or offered for sale. If, however, the building or improvement is sold within one year of completion, the owner-builder will have the burden of proving that he did not build or improve for the purpose of sale.).

☐ I, as owner of the property, am exclusively contracting with licensed contractors to construct the project (Sec. 7044, Business and Professions Code: The Contractor's License Law does not apply to an owner of property who builds or improves thereon, and who contracts for such projects with a contractor(s) licensed pursuant to the Contractor's License Law.).

☐ I am exempt under Sec. ________, B. & P. C. for this reason ________

Date ________ Owner's Signature ________

WORKERS' COMPENSATION DECLARATION

I hereby affirm that I have a certificate of consent to self-insure, or a certificate of Worker's Compensation Insurance, or a certified copy thereof (Sec. 3800, Lab. C.).

Policy No. ________ Insurance Company ________

☐ Certified copy is hereby furnished.

☐ Certified copy is filed with the Los Angeles City Dept. of Bldg. & Safety.

Date ________ Applicant's Signature ________

Applicant's Mailing Address ________

CERTIFICATE OF EXEMPTION FROM WORKERS' COMPENSATION INSURANCE

I certify that in the performance of the work for which this permit is issued, I shall not employ any person in any manner so as to become subject to the Workers' Compensation Laws of California.

Date ________ Applicant's Signature ________

NOTICE TO APPLICANT: If, after making this Certificate of Exemption, you should become subject to the Workers' Compensation provisions of the Labor Code, you must forthwith comply with such provisions or this permit shall be deemed revoked.

CONSTRUCTION LENDING AGENCY

I hereby affirm that there is a construction lending agency for the performance of the work for which this permit is issued (Sec. 3097, Civ. C.).

Lender's Name ________

Lender's Address ________

I certify that I have read this application and state that the above information is correct. I agree to comply with all city and county ordinances and state laws relating to building construction, and hereby authorize representatives of this city to enter upon the above-mentioned property for inspection purposes.

I realize that this permit is an application for inspection, that it does not approve or authorize the work specified herein, that is does not authorize or permit any violation or failure to comply with any applicable law, that neither the city of Los Angeles nor any board, department, officer or employee there of make any warranty or shall be responsible for the performance or results of any work described herein or the condition of the property or soil upon which such work is performed. (See Sec. 94.0203 LAMC)

Signed ________ (Owner or agent having property owner's consent) ________ Position ________ Date

Figure 1-12
Sample HVAC permit

5 HEATING VENTILATING AIR CONDITIONING REFRIGERATION

CITY OF LOS ANGELES — DEPT. OF BUILDING AND SAFETY

APPLICATION FOR INSPECTION AND PLAN CHECK

FEE	NO.	TYPE APPLIANCE OR EQUIPMENT			BTU
		COMFORT COOLING COMP.	H.P.		INCIDENTAL GAS
		REFRIGERATION COMP.	H.P.		SMOKE DAMPERS
		AIR INLETS - OUTLETS			APPL. VENTS
		COMM. COOKING VENT. SYSTEMS			FIRE DAMPERS
		HOODS - VENT. SYSTEMS			SMOKE DET.
		OTHER VENT. SYSTEMS			EVAP COOLERS
		AIR HANDLING UNITS			BOILER VENT (LISTED)
	ISSUING FEE				BOILER VENT (UNLISTED)
	INVESTIGATION				
	SUPPLEMENTAL				
	PLAN CHECK				
	MISC. PERMIT				
	SUB TOTAL				
	SURCHARGE				
	TOTAL FEE				

HEATING - VENTILATION - AIR CONDITIONING - REFRIGERATION
☐ ALTERED ☐ REPAIRED ☐ ADDED

CASHIER'S USE ONLY

SAMPLE

JOB ADDRESS | SUITE OR ROOM NO. | DIST. NO.

OWNER | OWNER'S PHONE NO.

OWNER'S ADDRESS

CITY | STATE | ZIP

CITY USE ONLY
CHECKED AND APPROVED
INSPECTION CALLED

IS BUILDING
☐ NEW
☐ EXISTING
☐ ADDITION
☐ ONE OR TWO FAM. DWELL.
☐ APARTMENT
☐ COMMERCIAL

NO. OF DWELL. UNITS

QUALIFIED INSTALLER | LICENSE NO. | TYPE | PHONE NO.

PLANCHECK APPLICANT (Please Print Use Ink)

NAME ____________

ADDRESS ____________

CITY ____________ STATE ________ ZIP ________

Bldg. Insp. Appl. No.

ROUGH O.K.

BUSINESS TAX REG. CERT. NO.

FINAL O.K.

DIST. OFFICE | USE OF BUILDING | LOCATION OF EQUIPMENT IN BUILDING | PLAN CHECK NO.

Permit expires if work is not commenced within 180 days after fee is paid or if work is suspended for a period of more than 180 days.

Plan check expires after one year unless a permit has been issued.

B & S H-5 (R.2/88)

CHANGE OF ADDRESS

From ____________

To ____________

Old Application # ____________

Figure 1-12 (cont.)
Sample HVAC permit

DECLARATIONS AND CERTIFICATIONS

LICENSED CONTRACTORS DECLARATION

I hereby affirm that I am licensed under the provisions of Chapter 9 (commencing with Section 7000) of Division 3 of the Business and Professions Code, and my license is in full force and effect

Date ________ Lic. Class ________ Lic. No ________ Contractor's Agents' Signature ________

OWNER-BUILDER DECLARATION

I hereby affirm that I am exempt from the Contractor's License Law for the following reason (Sec. 7031.5, Business and Professions Code: Any city or county which requires a permit to construct, alter, improve, demolish, or repair any structure, prior to its issuance, also requires the applicant for such permit to file a signed statement that he is licensed pursuant to the provisions of the Contractor's License Law (Chapter 9 (commencing with Section 7000) of Division 3 of the Business and Professions Code) or that he is exempt therefrom and the basis for the alleged exemption. Any violation of Section 7031.5 by any applicant for a permit subjects the applicant to a civil penalty of not more than five hundred dollars ($500).):

☐ I, as owner of the property, or my employees with wages as their sole compensation, will do the work, and the structure is not intended or offered for sale (Sec. 7044, Business and Professions Code: The Contractor's License Law does not apply to an owner of property who builds or improves thereon, and who does such work himself or through his own employees, provided that such improvements are not intended or offered for sale. If, however, the building or improvement is sold within one year of completion, the owner-builder will have the burden of proving that he did not build or improve for the purpose of sale.).

☐ I, as owner of the property, am exclusively contracting with licensed contractors to construct the project (Sec. 7044, Business and Professions Code: The Contractor's License Law does not apply to an owner of property who builds or improves thereon, and who contracts for such projects with a contractor(s) licensed pursuant to the Contractor's License Law.).

☐ I am exempt under Sec. ________, B. & P. C. for this reason ________

Date ________ Owner's Signature ________

WORKERS' COMPENSATION DECLARATION

I hereby affirm that I have a certificate of consent to self-insure, or a certificate of Worker's Compensation Insurance, or a certified copy thereof (Sec. 3800, Lab. C.).

Policy No. ________ Insurance Company ________

☐ Certified copy is hereby furnished.

☐ Certified copy is filed with the Los Angeles City Dept. of Bldg. & Safety.

Date ________ Applicant's Signature ________

Applicant's Mailing Address ________

CERTIFICATE OF EXEMPTION FROM WO RS' MP NSATION SU NCE

I certify that in the performance of the work for which th per is issu , I sh l not employ person in any manner so as to become subject to the Workers' Compensation s of lifor

Date ________ Applicant's S ture ________

NOTICE TO APPLICANT: If, after making this tificat o Exem on, you should become subject to the Workers' Compensation provisions of the Labor e, you m t rthw h mply ith such provisions or this permit shall be deemed revoked.

ONSTR CT L DING AGENCY

I hereby affirm th is a co truc lend ag for the performance of the work for which this permit is issued (Sec. 3097, Civ. C

Lender's Name ________

Lender's Address ________

I certify that I have read this ppl ation and state that the above information is correct. I agree to comply with all city and county ordinanc and s laws relating to building construction, and hereby authorize representatives of this city to enter upon the above- d property for inspection purposes.

I realize that this permit is an application for inspection, that it does not approve or authorize the work specified herein, that it does not authorize or permit any violation or failure to comply with any applicable law, that neither the city of Los Angeles nor any board, department, officer or employee thereof make any warranty or shall be responsible for the performance or results of any work described herein or the condition of the property or soil upon which such work is performed. (See Sec. 95.0220 LAMC)

Signed ________ (Owner or agent having property owner's consent) ________ Position ________ Date

Figure 1-13
Sample electrical permit

6 EQUIPMENT TO BE INSPECTED — City of Los Angeles — Department of Building and Safety — **APPLICATION FOR ELECTRICAL INSPECTION**

FEE		
$		15-20A 120 V LT. OR REC. BR. CIR. AND DWELL APPL. BR. CIR. (15 TO 50A) AND NONDWELL MTRS. OR APPL. NOT EX. 3-HP-KVA
		UTILIZATION EQUIP. EXISTING CIR. (0-3 KW)
		15-20 A 208 V TO 277 V LT. BR. CIR.
		NONDWELL POW. EQUIP. HP OR KVA 3.1-5
		5.1-20 ... 20.1-50
		50.1-100 ... OVER 100
		SERVICES 0-200A .... 201-400 .... 401-600 .... 601-1200 ..... OVER 1200
		SERVICES OVER 600V ... MISC.
		SWITCHBOARDS ... PANEL BOARDS
		F.A./EMER./COMM. DEVICES ... F.A./CONTROL PANELS
		SMOKE DETECTORS—NO. OF RESIDENTIAL UNITS
$		NO. OF UNITS
$		INVESTIGATION FEE
$		SUPPLEMENTAL FEE
$		ISSUING FEE
$		SUBTOTAL
$		ONE STOP SURCHARGE
$		TOTAL FEE DUE

DESCRIPTION OF WORK

CASHIER'S USE ONLY

SAMPLE

JOB ADDRESS	SUITE OR ROOM NO.	OFFICE	DIST. NO.

			ACTION	DATE:	BY:
OWNER					
OWNER'S ADDRESS			ENERGY CONS. O.K.		
CITY	STATE	ZIP	APPL. O.K.		
USE AND AREA OF BUILDING	☐ NEW ☐ EXIST.	NO. OF DWELL UNITS	ROUGH O.K.		
QUALIFIED INSTALLER	LICENSE NO. / TYPE	PHONE NO.	TEMP O.K.		
NAME (Qualified Installer)	(Please Print — Use ink)		OTHER		
ADDRESS			FINAL O.K.		
CITY	STATE	ZIP	BLDG. PER. NO.		
DIST. OFFICE	BUSINESS TAX REGISTRATION CERT. NO.		ADDED METER INFO.	F	C

O. H.	U. G.	RES.	COML.	LITE	POW.	1 Ø	3 Ø	3 W	4 W	120/208
120/240	240	277/480	480	NO. MTRS.	CTS.	NEW	CHANGE	RESET	RESEAL	REROUTE

DISTRIBUTION: Original—Inspector White—Cashier Pink—File Yellow—Applicant B & S E-6 (R.12/87)

Figure 1-13 (cont.)
Sample electrical permit

DECLARATIONS AND CERTIFICATIONS

LICENSED CONTRACTORS DECLARATION

I hereby affirm that I am licensed under the provisions of Chapter 9 (commencing with Section 7000) of Division 3 of the Business and Professions Code, and my license is in full force and effect.

Lic. Class ________ Lic. Number ____________ Contractor's/Agents Signature ______________________ Date ________

OWNER-BUILDER DECLARATION

I hereby affirm that I am exempt from the Contractor's License Law for the following reason (Sec. 7031.5, Business and Professions Code: Any city or county which requires a permit to construct, alter, improve, demolish, or repair any structure, prior to its issuance, also requires the applicant for such permit to file a signed statement that he is licensed pursuant to the provisions of the Contractor's License Law (Chapter 9 (commencing with Section 7000) of Division 3 of the Business and Professions Code) or that he is exempt therefrom and the basis for the alleged exemption. Any violation of Section 7031.5 by any applicant for a permit subjects the applicant to a civil penalty of not more than five hundred dollars ($500).):

☐ I, as owner of the property, or my employees with wages as their sole compensation, will do the work, and the structure is not intended or offered for sale (Sec. 7044, Business and Professions Code: The Contractor's License Law does not apply to an owner of property who builds or improves thereon, and who does such work himself or through his own employees, provided that such improvements are not intended or offered for sale. If, however, the building or improvement is sold within one year of completion, the owner-builder will have the burden of proving that he did not build or improve for the purpose of sale.).

☐ I, as owner of the property, am exclusively contracting with licensed contractors to construct the project (Sec. 7044, Business and Professions Code: The Contractor's License Law does not apply to an owner of property who builds or improves thereon, and who contracts for such projects with a contractor(s) licensed pursuant to the Contractor's License Law.).

☐ I am exempt under Sec. __________, B. & P. C. for this reason ______________________

Date ____________________ Owner's Signature ______________________

WORKERS' COMPENSATION DECLARATION

I hereby affirm that I have a certificate of consent to self-insure, or a certificate of Worker's Compensation Insurance, or a certified copy thereof (Sec. 3800, Lab. C.).

Date ________________ Policy No. ________________ Insurance Company ________________

☐ Certified copy is hereby furnished.

☐ Certified copy is filed with the Los Angeles City Dept. of Bldg. & Safety.

Applicant's Signature ______________________

Applicant's Mailing Address ______________________

CERTIFICATE OF EXEMPTION FROM WORKERS' COMPENSATION INSURANCE

I certify that in the performance of the work for which this permit is issued, I shall not employ any person in any manner so as to become subject to the Workers' Compensation Laws of California.

Date ________________ Applicant's Signature ______________________

NOTICE TO APPLICANT: If, after making this Certificate of Exemption, you should become subject to the Workers' Compensation provisions of the Labor Code, you must forthwith comply with such provisions or this permit shall be deemed revoked.

CONSTRUCTION LENDING AGENCY

I hereby affirm that there is a construction lending agency for the performance of the work for which this permit is issued (Sec. 3097, Civ. C.).

Lender's Name ______________________

Lender's Address ______________________

I certify that I have read this application and state that the above information is correct. I agree to comply with all city and county ordinances and state laws relating to building construction, and hereby authorize representatives of this city to enter upon the above-mentioned property for inspection purposes.

I realize that this permit is an application for inspection, that it does not approve or authorize the work specified herein, that is does not authorize or permit any violation or failure to comply with any applicable law, that neither the city of Los Angeles nor any board, department, officer or employee there of make any warranty or shall be responsible for the performance or results of any work described herein or the condition of the property or soil upon which such work is performed. (See Sec. 93.0203 LAMC)

Signed ______________________ ______________ ______________
(Owner or agent having property owner's consent) Position Date

Permit expires if work is not commenced within 180 days after fee is paid or if work is suspended for a period of more than 180 days.

USE OPPOSITE SIDE OF SHEET FOR ELECTRICAL PLAN

Figure 1-14 Departments that review building plans

Department	Jurisdiction
Planning	Parking, zoning, building lines and areas, and occupancy
Grading	Fill and excavation, cut and fill slopes
Public works	Streets, underground structures, curbs, storm drains
Sanitation	Sanitary sewers, manholes, sewage lift stations
Fire	Fire hydrants, water mains
Flood control	Catch basins, manholes, culverts, channels
Health	Septic tanks, sewage disposal
Water	Water main, pressure regulator, meters, backflow preventers
Public utilities	Gas main, pressure reducers, meters, overhead and underground cables

No permit will be issued until several departments at city hall have had a chance to review your plans. Figure 1-14 is a list of departments that may review your plans, and what they'll be looking for.

Additional permits may be required for building security, accessibility for the handicapped, and energy conservation. Permits may also be required by state safety agencies for hazardous work such as asbestos removal and high rise scaffolding.

Unfortunately, the foregoing list may not be complete. Other public agencies may have the right to review your plans and hold up issuance of the permit until you've met their conditions. No wonder getting a permit issued can take a year!

What Is an Engineer?

The term *engineer* is sometimes used for salespeople, service technicians, and maintenance workers to add dignity to their job titles or imply education and experience. For example, people driving trash collection trucks may refer to themselves as "sanitary engineers." That's fine with me. But I'm going to use the term in a narrower sense. People become engineers when they're licensed by a government agency to provide engineering services to the public. Of course, there are many types of engineers. Figure 1-15 lists some of the common engineering professions and their areas of expertise in building construction.

The National Council of Engineering Examiners (NCEE) defines an engineer as a person with special knowledge, education, and experience in mathematics, physics, engineering sciences, and methods of engineering analysis and design.

The NCEE also says that a professional engineer is an engineering specialist who is registered and licensed as a professional engineer. Some professional engineering specialists working in construction are: civil engineers, soils engineers, engineering geologists, structural engineers, mechanical engineers, electrical engineers, and land surveyors. Here are descriptions of some engineering professions and their work.

Figure 1-15 Consulting engineers on specific building components

Area	Civil Engineer	Structural Engineer	Mechanical Engineer	Electrical Engineer	Soils Engineer	Architect
Survey	✔					
Grading	✔				✔	
Demolition	✔	✔				
Concrete	✔	✔				
Steel	✔	✔				
Wood	✔	✔				✔
Masonry	✔	✔				✔
Roofing						✔
Electrical				✔		
HVAC			✔			
Elevator			✔			✔

Civil engineer

This is probably the oldest engineering profession. It includes design and supervision for constructing foundations, framed structures, and buildings. It covers all studies and activities connected with fixed water works, water supply, waterways, and harbors.

Soils engineer

This field developed from civil engineering. A soils engineer investigates proposed building foundations and site grading for housing, industry, and municipal improvements. In some states, a soils engineer must be a registered civil engineer.

Structural engineer

A structural engineer is usually a registered civil engineer who is specially qualified in structural engineering. He or she investigates and designs force-resistant and load-supporting structural members. These may be foundations, walls, beams, columns, trusses, and other framing portions of buildings and structures.

Mechanical engineer

Mechanical engineers belong to a broad profession which takes in many specialties. In building construction, mechanical engineers plan the heating, ventilating, refrigeration, air conditioning, and plumbing systems.

Electrical engineer

An electrical engineer designs and supervises the electrical services and wiring in construction projects. This includes illumination, protective devices, instrumentation, control, electrical machinery, and communication.

Architect

The architect is the prime professional on many large building projects who plans and designs buildings, identifies construction problems, and gathers background information. He or she evaluates this information and creates a design that satisfies the client's requirements. The architect prepares plans, specifications, and other contract documents for the builder and craft subcontractors. As the prime professional, the architect coordinates the engineering consultants on the project.

When Do You Need an Engineer?

Although the building official (head of the building department) may let an architectural firm accept responsibility for all engineering work on a project, in practice each engineering specialist is responsible for his or her own specialty. The UBC, BOCA, and SBCCI don't identify exactly when an engineer is, or isn't, required. Under the code, that's largely up to the chief building official at your local building department. Of course, most building departments prefer to be cautious, sometimes especially cautious. Suits against cities for negligence by their building departments have become very common.

State laws may govern when an engineer is necessary. For example, the California Professional Engineers Act requires:

All civil engineering plans, specifications, reports, or documents shall be prepared by a registered civil engineer or by a subordinate employee under his direction, and shall be signed by him to indicate his responsibility for them.

Many states have similar requirements for preparing electrical and mechanical engineering plans, specifications, reports, and documents.

Exceptions to these state requirements usually include:

- single or multiple dwellings not more than two stories and basement in height
- garages and other structures used with these buildings
- farm and ranch buildings
- any one-story building where the span between bearing walls does not exceed 25 feet and the building isn't steel frame or concrete

An engineer may be required by inference. For example, the building code may state:

Any method or system of construction to be used shall be based on a rational analysis in accordance with well established principles of mechanics.

The code may say that sufficient strength must be demonstrated by structural calculations, or that energy calculations prepared by a professional engineer are required for habitable buildings.

When you need the opinion of an engineer to get a permit issued, it's a good idea to file a written record of the computations which justify the design. And each drawing sheet and written record of computation should be signed by, or bear the approved stamp of, an engineer or architect licensed in the state for the type of service performed. Many larger cities require a licensed structural engineer to sign plans for buildings over 160 feet in height.

Some cities take another approach. They may not require the signature of a registered engineer or architect on structures that require some structural design. But the person responsible for the design may be required to make engineering calculations and sign those calculations and the sheets of the plans which show those details.

Building departments provide a great deal of *free engineering*. This includes the tables and charts in the codes. You can use these tables to size rafters, ceiling joists, floor joists, floor girders, roof and floor sheathing, and foundations for one and two-story buildings. Also, many building departments provide the public with simple detail drawings which show how to build typical structures, such as:

- one- and two-story wood frame buildings
- attached carports or patio covers
- masonry and concrete retaining walls under 4 feet in height.

When Do You Need a Land Surveyor?

Before building on any lot, it's good practice to have a licensed land surveyor mark the lot corners. The surveyor can also make a topographic survey which shows the contours of the land and precise location of any existing buildings or other improvements.

If the land has to be split or subdivided, get a land surveyor or civil engineer to survey, document, and record the subdivision. *Don't build before you have the survey.* After the property boundaries are established, it's O.K. to have construction layout done by someone who's unlicensed but trained to do that type of surveying. Chapter 2 discusses this type of field engineering in depth.

When Do You Need a Civil or Structural Engineer?

Building department regulations may identify when a civil or structural engineer is required. The more risk, the more complicated the construction, the more likely an engineer will be required. In some offices, any building with a beam over a certain length or retaining wall over a certain height will require an engineer's approval.

In structures where high-strength concrete, masonry, or welding are called for, a registered engineer may be required to issue a certificate of compliance when work is completed. This document certifies that the work was done according to the plans. This normally happens when plans call for concrete having a compressive strength over 2000 psi. Standard strength concrete has a minimum compressive strength of 2000 psi. All concrete stronger that this is considered to be high strength concrete.

You may also need an engineer's certificate when the design calls for full-strength masonry or welding. Without this certificate, the design may have to be based on one-half of the maximum allowable strength of the masonry or welding. This may mean that the size of weld or the amount of steel reinforcing must be increased.

On jobs involving prestressed concrete construction, some codes say that the completed work must be certified by a licensed civil or structural engineer.

Normally a civil engineer must design and supervise all construction work within public property (public roads, driveways, water supply, and sewage systems). So any portion of a project that's at least partly on public property, or is to be dedicated to the public, needs a civil engineer.

Since the recent California earthquakes, certain construction jobs within Seismic Zones 3 and 4 need a civil or structural engineer to check the structure during construction. These jobs include constructing essential or hazardous facilities, and buildings with habitable rooms that are over 75 feet above grade.

The engineer must visit the site to make sure the approved structural drawings are followed. At job completion, he or she has to submit a statement in writing to the building department certifying compliance.

If a design was made by a civil or structural engineer, special inspections will be made by a deputy, or special, inspector during construction. The owner provides for special inspection. The types of work requiring a special inspector are:

- work using concrete with strength over 2000 psi
- foundations with concrete having strength over 2500 psi
- pneumatically-placed concrete and Shotcrete
- moment-resisting frames, or building structures in which members and connections can resist forces by flexure
- welding of structural steel or reinforcing bars
- any use of high-strength bolts, or bolts that develop the strength of a connection by means of friction between the parts being connected rather than by shearing strength in the bolt
- piling, drilled piles, and caissons

If there is any doubt about the safety of an existing structure, the building official may order a structural investigation by analysis, load test, or both.

When Do You Need Soils Engineers and Engineering Geologists?

Grading specialists in the building department will require approval by soils engineers or engineering geologists on critical grading jobs. They will also require periodic inspections to make sure the grading work is being done according to plan. These inspections usually include the following services:

- Initial inspection: before any grading or brush clearing is started.
- Toe inspection: after natural ground is exposed and prepared to receive fill, but before any fill is placed.
- Excavation inspection: after excavation is started, but before the vertical height of lifts or layers exceeds 10 feet.
- Drainage device inspection: after forms for drainage devices and pipe are in place, but before any concrete is placed.
- Rough grading inspection: when all rough grading has been completed.
- Final inspection: when all work has been completed.

Always have a soils engineer test and certify compacted fills. You'll probably have to submit grading plans when:

- an excavation exceeds 5 feet in vertical depth at its deepest point
- a fill is over 3 feet in vertical depth at its deepest point
- the total volume of earth you need to cut or fill is more than 50 cubic yards.

A soils engineer or engineering geologist should certify the safety of cut or fill slopes that are steeper than 2:1 (horizontal to vertical).

In summary, construction and engineering are dynamic fields. The building codes change all the time. Codes are revised and upgraded as new experience and knowledge are acquired. Catastrophic earthquakes, floods, and tornadoes result in new subjects being added to the UBC, BOCA, and SBCCI. Some of these revisions include seismic and seismic-isolated structures, flood-resistant construction, construction in high-wind areas, and freezing and thawing conditions.

The following chapters contain tables and charts used for rule-of-thumb design and as checklists for building construction. There are simple methods for sizing foundations, concrete, and wood and steel structural members. There are also basic rules for sizing plumbing and heating, ventilating, and air-conditioning systems.

CHAPTER 2

Surveying for Construction

It's *very* important that you know *exactly* where the property lines are before beginning construction. Make sure that any excavating, stockpiling, or construction you do is within these property lines. Building a wall on the wrong lot may sound funny. But it's never funny if it happens to you.

Another common error is excavating or stockpiling building materials on public property without permission. The result will usually be a citation and a hefty fine.

Unfortunately, it's easy to make these mistakes. For example, you might take a surveyor's offset marker as the corner marker. Be sure you understand the surveyor's monuments. Compare the survey stakes with the plot plan. A surveyor will often set a marker a short distance away from the true position because the true point is inaccessible. He may also put a reference or witness stake near the true corner monument in case the main marker gets destroyed during construction. If you use one of the reference stakes or offset markers as the property corner, you'll be building in the wrong place.

Once you've found the true property corners, set the foundation lines. The foundation will be either parallel to, or at a specified angle to, one of the property lines. Use bench marks set within the construction site to control the elevation of foundations, pavements, sewers, and other work. These bench marks will be set at a specific elevation relative to a known elevation on a public street or sewer line. Follow these elevations carefully so rainwater and sewage drain off the site as planned.

Figure 2-1 shows typical offset reference stakes or hubs set around a main corner marker. In this way, if the main hub gets disturbed you can relocate it by drawing strings across from points on the opposing reference hubs.

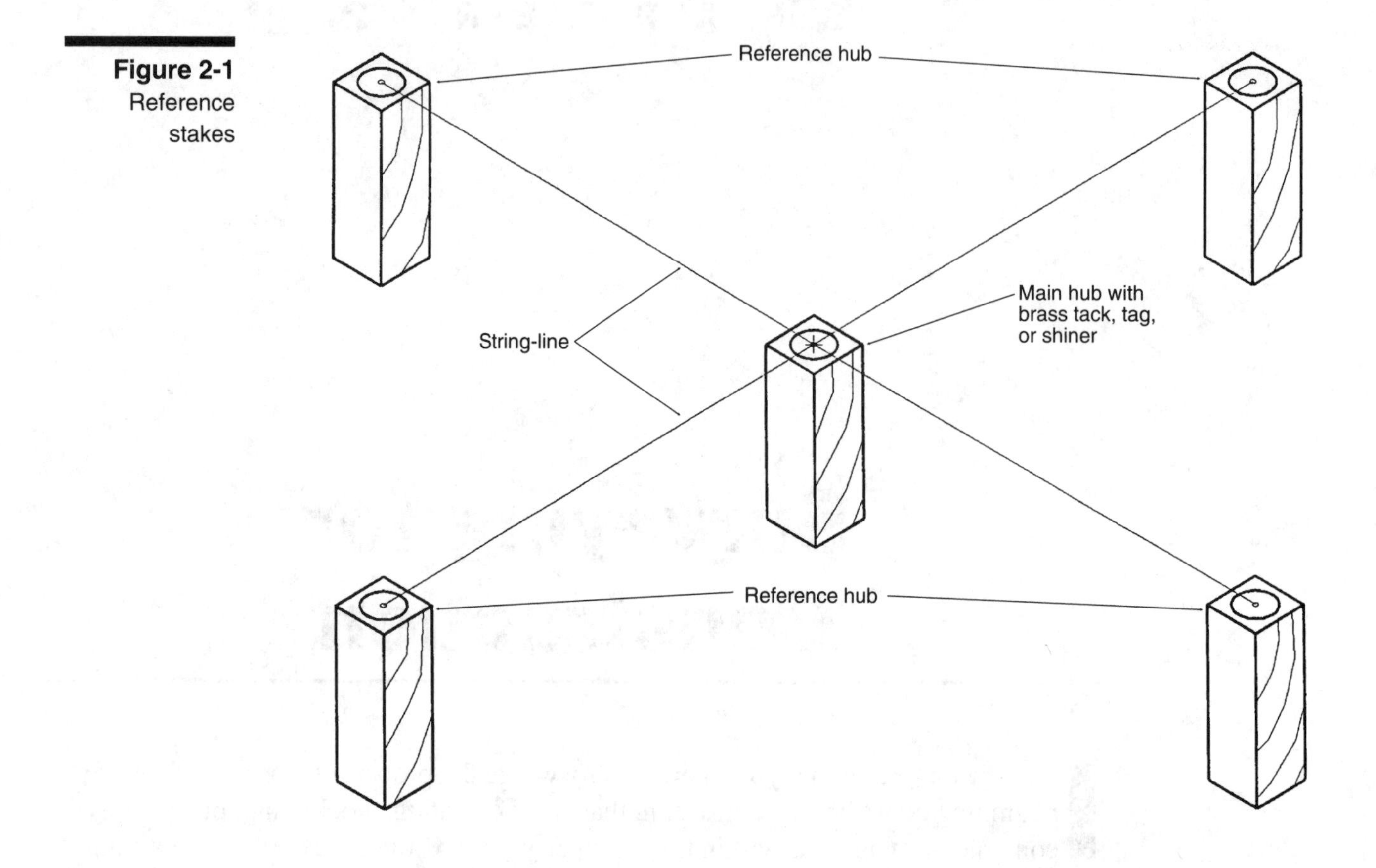

Figure 2-1 Reference stakes

Survey stakes are usually made from dressed 1 x 2 softwood boards pointed at one end. These are driven into the ground near monuments or hubs and usually extend about 24 inches above grade. Survey information or notes, which describe the adjacent marker, are often marked on the side of the stake.

Survey hubs are normally made of 2 x 2 hardwood, pointed at one end and painted white at the other end. Hubs are usually driven close to the ground surface so they won't be accidentally disturbed. Survey points are marked on the top of the hub with a stake tack, flat shiner and tack, or imprinted survey tag and nail. The terms hubs and stakes are sometimes used interchangeably.

Most land surveyors set a brass tag engraved with their license number on top of corner monuments. This may be indicated as L.S. 1234, or R.C.E. 1234 if a civil engineer performed the survey. You can find a surveyor through the Roster of Professional Engineers and Land Surveyors. This roster is published and available through the State Board of Registration of Professional Engineers. All licensed surveyors and engineers are listed in both alphabetical order by name and in numerical order by their license number. The numerical listing also shows which certificates have expired or whether the certificate holder is deceased.

The distance and direction of an offset point are always shown on a surveyor's plat. In Figure 2-2, only the northwest property corner is actually marked at the true position. Unfortunately, some plot plans don't give you this information, so you

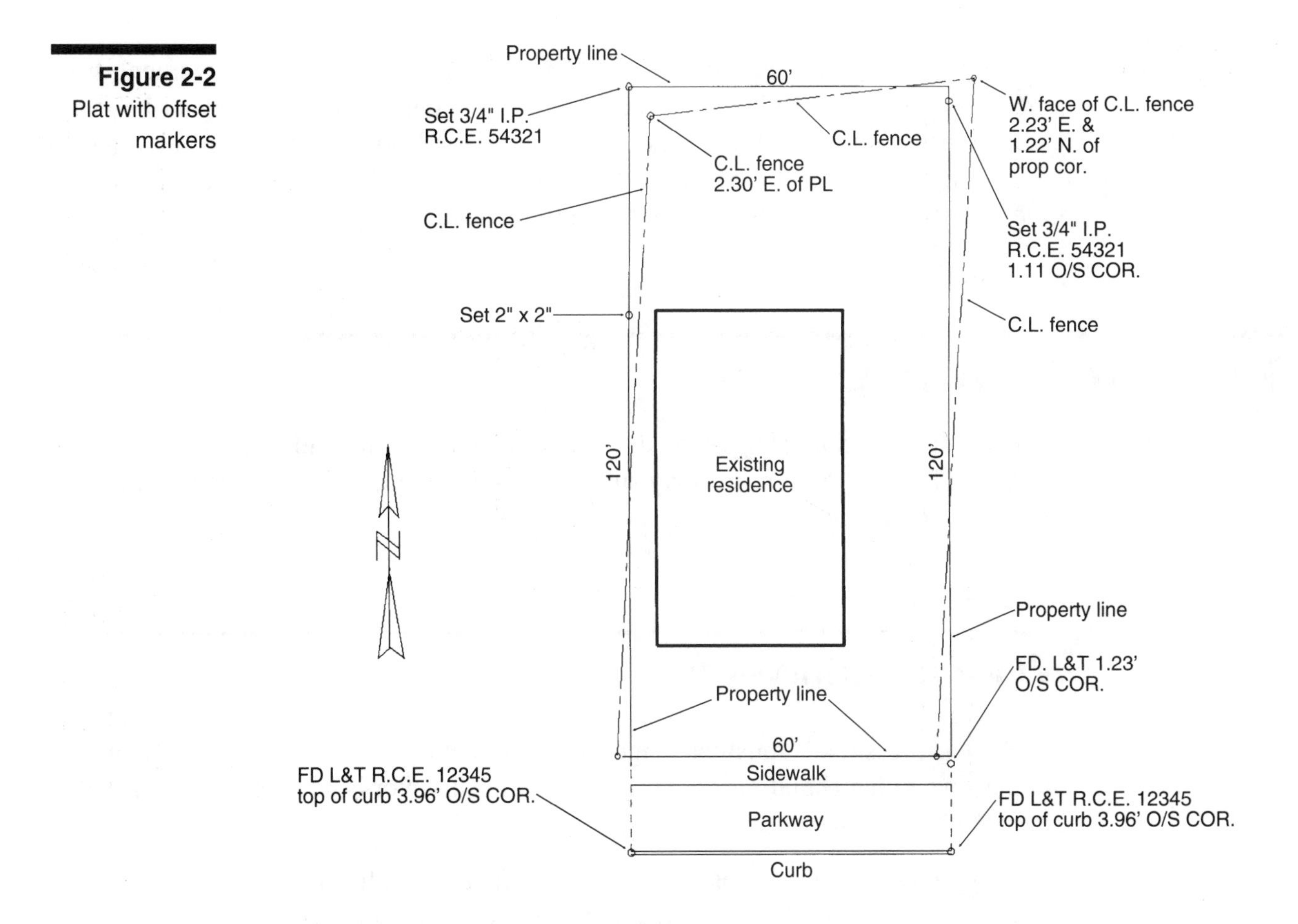

Figure 2-2
Plat with offset markers

could still get into trouble using the wrong marker. In Figure 2-2, the survey showed that the fence was put in the wrong place. The surveyor measured to the left, or west, side of the fence.

Here's the key for the symbols shown on Figure 2-2:

- ¾" I.P. — a ¾-inch diameter iron pipe driven into the ground
- R.C.E. 54321 — a brass tag with the Registered Civil Engineer's registration number
- O/S — the marker is offset from the true point
- L&T — lead and tack set in a concrete surface
- FD — found, meaning that the surveyor found an existing monument
- SET — set, meaning that the surveyor placed a new monument
- COR. — corner, or intersection of property lines
- C.L.F. — chain link fence
- PL — property line

Lead and tack is a means of marking a point in a concrete curb or pavement. It's done by chipping a small hole in the concrete with a chisel and pounding a ribbon of lead into the hole. Then a brass tack is driven into the lead. When a surveyor wants to identify himself, he takes a brass tag that's engraved with his license number, and drives the tack through a hole in the tag, into the lead.

Types of Surveying

There are two general types of surveying: land surveying and engineering surveying. By law, land or property surveying is done by a licensed land surveyor or a registered civil engineer.

Licensed Surveyor

These professionals must pass rigorous examinations that cover both mathematics and the laws of land subdivision. A licensed land surveyor is accountable to the general public for his work.

Property boundaries are usually located by measuring from the nearest recorded land subdivision monuments. These points are normally at the corners of a tract or at the intersections of street centerlines.

A land surveyor prepares maps to divide land into smaller tracts. He lays out streets and rights-of-way and prepares and interprets land descriptions for deeds, title policies, and other legal documents. The most important thing a land surveyor will do for a builder is to set corner monuments around the property boundary.

Engineering Surveyor

Engineering surveying (sometimes called *field surveying*) is part of the construction process and usually begins after land surveying is completed. You'll usually subcontract field surveying to a licensed land surveyor. An engineering surveyor doesn't have to be a licensed surveyor or a registered civil engineer. But he (or she) should know how to use surveying instruments and how to use mathematics to calculate distances on the ground.

The most important rules of trigonometry used by an engineering surveyor are shown in Figure 2-3. Using these basic formulas and an inexpensive pocket calculator with trigonometric functions, surveyors can calculate the length of a side of a right triangle when they can measure one side and an angle.

Figure 2-3 Basic trigonometric formulas

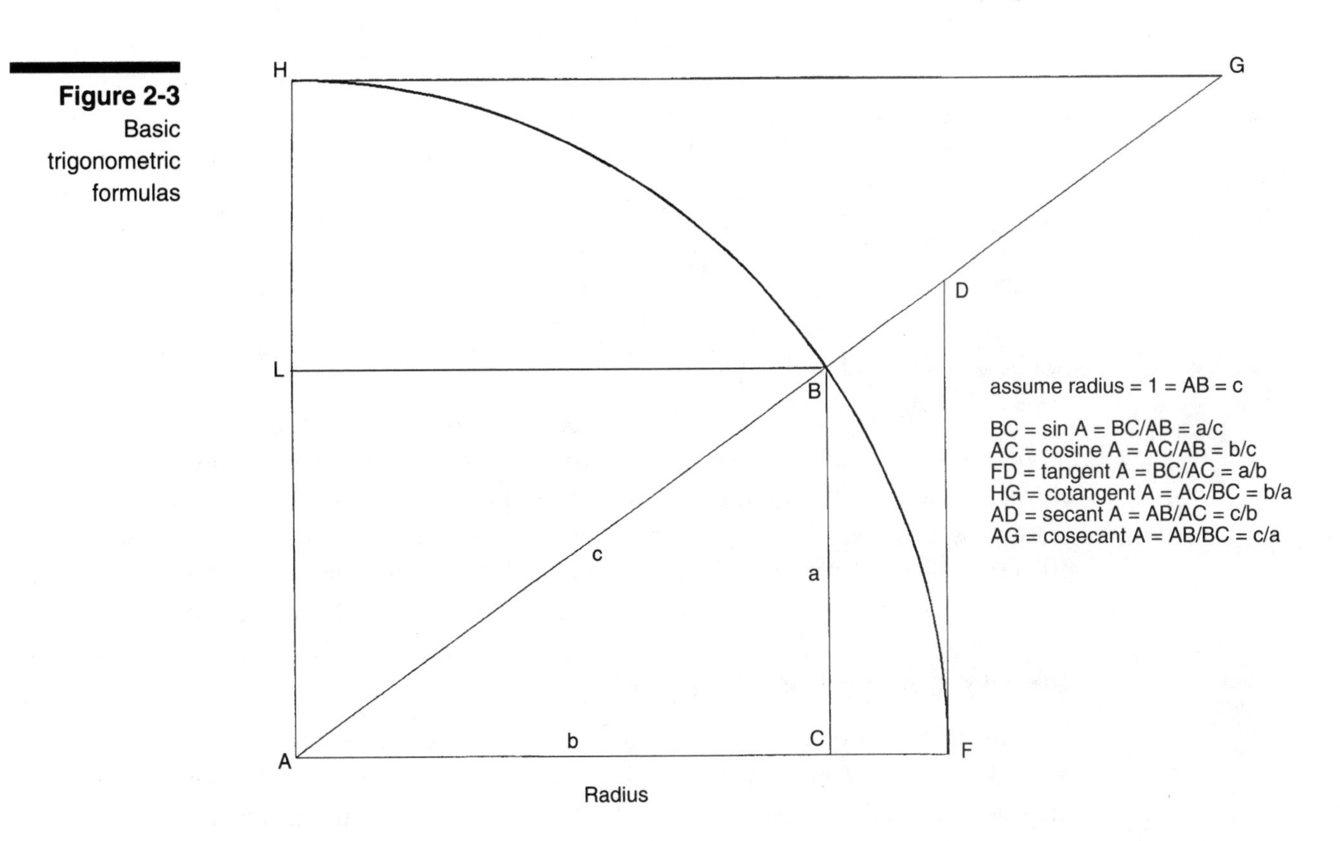

You can use these easy rules to remember the trigonometric relationships of a right triangle:

- The sine (sin) of an interior angle is equal to the length of the side opposite the angle divided by the length of the hypotenuse of the triangle.
- The cosine (cos) of an interior angle is equal to the length of the side adjacent to the angle divided by the length of the hypotenuse of the triangle.
- The tangent (tan) of an interior angle is equal to the length of the side opposite the angle divided by the length of the side adjacent to it.
- The cotangent (cot) of an interior angle is the reciprocal of the tangent of the angle, or the length of the side adjacent to the angle divided by the length of the side opposite to it.
- The secant (sec) of an interior angle is the reciprocal of the cosine of the angle, or the length of the hypotenuse of the triangle divided by the length of the side adjacent to the angle.
- The cosecant (csc) of an interior angle is the reciprocal of the sine of the angle, or the length of the hypotenuse of the triangle divided by the length of the side opposite the angle.

Engineering surveying includes:

- control of horizontal and vertical distances in construction
- topographic surveys for grading

- profiles and cross sections for roads
- quantity and measurement surveys of earthwork
- building layouts
- as-built surveys

The engineering surveyor sets up all transit lines, coordinate lines, and bench marks necessary for construction. These lines are used to lay out the position of roads, utilities, and buildings.

The field surveyor should study the plot plan to see how the land surveyor marked the property corners, and then locate the datum elevation shown on the plot plan. A datum elevation is a point marked on a permanent feature, such as a concrete curb or sidewalk. It's used as an arbitrary elevation for a project. It's usually assigned a basic elevation of 100.00 feet. The plans should indicate the true elevation in reference to the public street and sewer elevations shown on the plot plan. This may be on the top of a street curb or a manhole cover frame. A properly-drawn plot plan should show the location of all buildings, including their setbacks, and side and rear yard lines.

The field surveyor should set extra reference stakes so a second survey won't be needed if the main control marks are disturbed during construction. He should highlight the stakes with brightly-colored strips of cloth or plastic (called flagging), to make them more visible. He can make hubs and stakes stand out by using shiners, or shiny metal disks. He should record the location of all stakes placed on the job in a field logbook.

Survey Documents

The survey information shown on a plot plan includes data from the record of survey, title policy, and the plat prepared by the land surveyor.

Record of Survey

The record of survey is a black ink map, usually 18 by 26 inches in size, made of linen or polyester film. It shows the following information:

- all monuments that were found, set, reset, replaced, or removed by the surveyor
- description of the kind and location of the monuments
- basis of bearings, usually the adjoining street
- bearings and lengths of lines
- scale of the drawing
- certificate showing the signature of the licensed surveyor or registered civil engineer
- date, page, and book number recorded, name of county recorder's office
- signature of the county recorder

- any other data necessary for interpretation of the items and location of points, lines, and areas shown, such as street lines, block corners, section corners, and midpoints.

Zoning laws

Zoning laws control the type of building you can erect on a site, and where the building has to be on that site. Zoning regulations specify the minimum front setback, size of side yards, and use of a property. Tract restrictions may impose additional limits on how the property is used.

Tract restrictions and CC&R's

Tract restrictions and CC&R's (Conditions, Covenants, and Restrictions) are usually created by the developer of a subdivision. These restrictions can't permit what zoning ordinances prohibit. Instead, they impose additional requirements such as minimum building setback and the size of side yards.

Zoning ordinances are enforced by governments. Tract restrictions are enforced by other property owners. For example, I know of a builder who thought the plans were wrong when they called for a 25-foot building setback. He knew the zoning ordinance required only a 20-foot setback. What he didn't know was that tract restrictions required a 25-foot setback. So he built a house 20 feet from the front property line, because he knew the owner would appreciate a larger rear yard. The building inspector had no objection to the 20-foot setback and construction proceeded normally until the architectural committee of the subdivision discovered the problem. They filed a lawsuit against the owner to move his house back 5 feet. Needless to say, the owner didn't appreciate the favor the builder had done for him.

The CC&R's may include other restrictions, such as:

- type of fencing and walls
- location of mechanical equipment, such as air conditioning compressors and condensers
- site drainage
- materials that may be used in construction. Exterior walls may be restricted to brick, wood siding, or plaster. Roofs may only be made of concrete or clay tile.

As a contractor, you have to be aware of all tract restrictions that apply as well as the building and zoning laws.

Plans and Plats

A *plat* is a map that describes a parcel of land. The word plat is derived from "platting" or plotting a property. Most states have laws which govern platting or subdividing a tract of land into smaller parcels. The plat, for instance, should show the boundary line of the tract. The corners should be marked with permanent monuments. These monuments may be iron pipes set in concrete or brass caps embedded in concrete blocks.

Before a plat can be recorded, it must be reviewed and approved by representatives of the departments of subdivision, zoning, grading, streets, flood control, health, and others. The plat must also be signed by the landowner, civil engineer, or licensed surveyor. The plat becomes a legal document only when it's recorded in the county recorder's office.

Tract map

A tract map is a recorded plan showing the subdivision of land into blocks, lots, streets, and easements. Permanent markers are set at the intersection of street centerlines and where the centerlines intersect the tract boundary line. The blocks are divided into lots and marked with semipermanent monuments. These may be iron pipes or wood hubs driven into the ground.

The tract map must be certified by the land surveyor or civil engineer. Figure 2-4 is an example of a tract map certification. The wording of a tract certification may vary with different governmental agencies.

Figure 2-4 Typical certification of a tract map

I, Max Schwartz, hereby certify that I am a Registered Civil Engineer of the State of Califonia; that this map consisting of one sheet correctly represents a survey made under my supervision during August 1990; that such survey is true and complete; and that the monuments shown thereon actually exist, are of the character and do occupy the positions indicated and are sufficient to enable the survey to be retraced, except that the monuments of lot corners are not yet set; that they will be set in the positions indicated thereon within one year from the date of recordation of said map.

____

Registered Civil Engineer No. 8820

Accepted for record and recorded in Book ______ of Maps, Page ______ in the Office of the County Recorder of the County of Kern, this day of _________, 1990, at ____________.

____

County Recorder of the County of Kern

Figure 2-5
A typical tract map

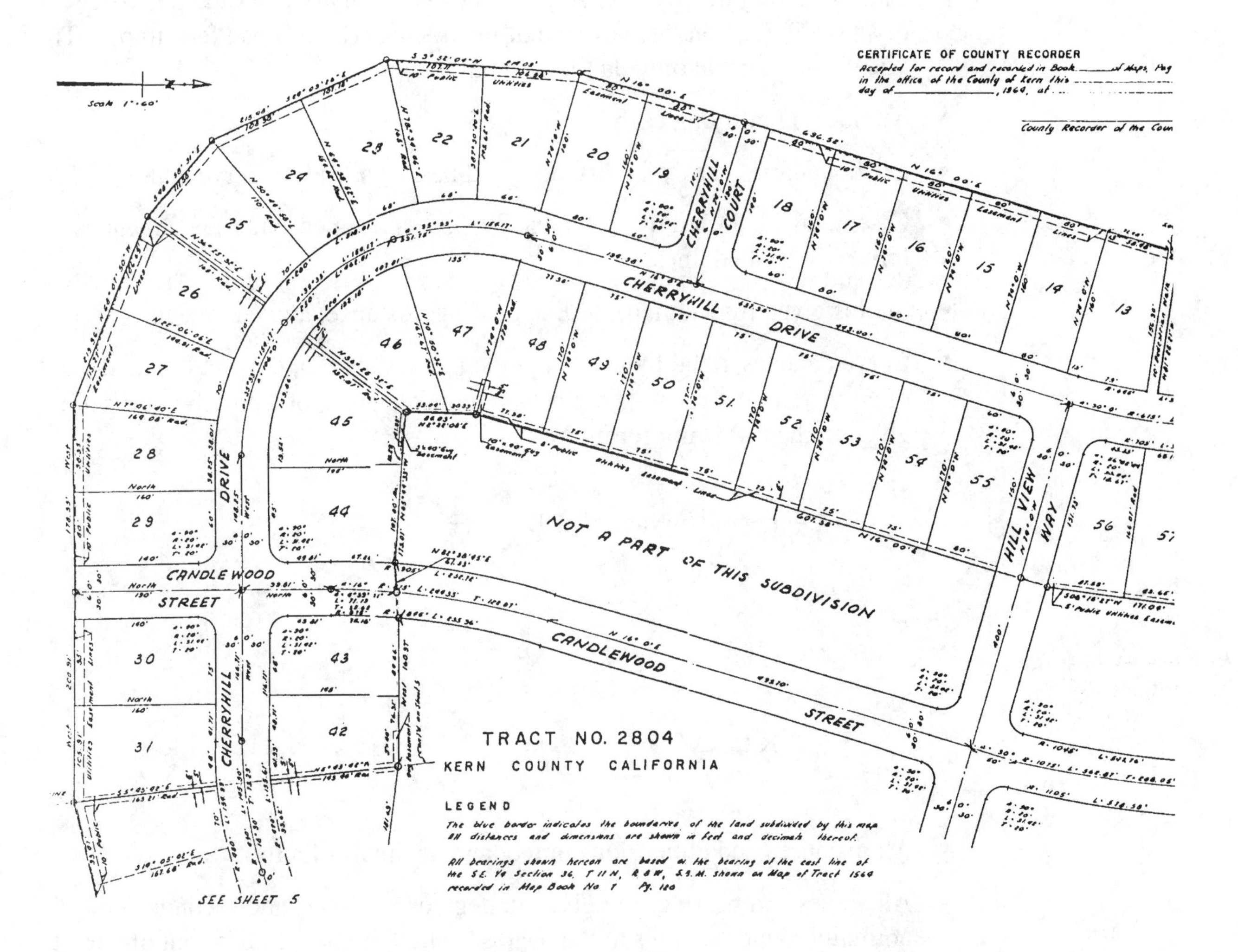

Figure 2-5 shows a typical tract map. It defines the boundary of the tract and the boundary of every individual lot and easement by distances and bearings. When a property is adjacent to a curved street, the property line is described by the properties of the curve. This includes the angle of deflection (also called interior angle, delta, or Δ), and the length, tangent, and radius of the curve.

To describe a property line around a curve, you start the course with the point at the beginning of curve, then you proceed along a radial line to the center of that curve, then follow the next radial line to the end of the curve. The angle between the two radial lines is called the interior angle, and is also equal to the angle between the projection of the two lines on each side of the curve, which is called the angle of deflection. The interior angle and the angle of deflection are also called delta.

Note the following data in Figure 2-5:

- Tract boundary line shown in distances and bearings. Boundary lines are usually highlighted by a broad blue line. You usually mark the boundary line of a tract in blue on the backside of the tracing with a broad felt-tip pen. This line may be visible only on the original tracing.
- Numbers identifying each lot
- Lot boundaries shown in distances and bearings, and by curve data
- Easements identified by width and purpose. Easement lines are shown as broken or dashed lines.
- Streets given by width, centerline bearing, distance, and curve data
- Curve data described by interior angle (Δ), radius (R), tangent (T), and length of curve (L). The relationship between the elements of a circular curve are given in the following formulas:

$$\text{Degree of Curvature, or } D = \frac{5729.58}{R}$$

$$L = \frac{100\,\Delta}{D}$$

$$T = \frac{R\tan\Delta}{2}$$

- All distances and dimensions in feet and decimals of a foot
- All angles and bearings are given in degrees, minutes, and seconds. You normally round bearings to the nearest minute in the initial layout of a tract or parcel, or when they're adjusted for closure. When the outline of a tract is closed by mathematical means, these bearings are reduced to seconds. Staking out property lines to the nearest second is usually not practical. A 1-degree bearing error in 100 feet equals 1.57 foot offset, which is a significant error. A 1-minute error in 100 feet equals 0.026 foot offset, and a 10-second error in 100 feet is equal to 0.00433 feet or 0.052 inch offset. A 1-second error in 100 feet, or 0.0052 inch, would be insignificant in the field.
- Monuments found by the surveyor, marked with a cross and solid circle (solid circles are monuments set prior to construction and are outside the area shown in Figure 2-5)
- Concrete monuments set by the surveyor, marked with a solid circle
- Concrete monuments the surveyor will set, marked with an open circle
- U.S. government monuments found by the surveyor, marked with an open circle and triangle

Plat map

This is another form of a record of survey for a parcel of land. It's prepared by a licensed land surveyor or registered civil engineer. It describes the boundary by metes and bounds, the oldest type of land surveying. The term *metes* means measurements, and *bounds* means direction. Metes and bounds are based on the principle that the various sides of a parcel of land close to form a complete self-contained area. The method consists of measuring the length and direction of each side of a parcel and then placing permanent monuments at the corners of the parcel.

Plot plan or site plan

This drawing is normally prepared by the architect or civil engineer for the construction project. It may have all the information shown on the plat map, and all existing and proposed structures. Fences, underground piping, and other items to be constructed are also shown.

The plot plan should show the location and depth of the public sewer and storm drain. You'll need this to set the right grade so the house sewer or storm drain will go into the public mains. If possible, you should get copies of the sanitary sewer plan, storm drain plans, and street plans from the public agencies responsible for them.

Here's a suggested checklist of utilities that exist or will be installed:

- water supply, including the location of the water main, water meter, size, and water pressure
- electrical service, including location of the power pole, underground main, and nearest vault
- gas service, including location of gas main, gas meter, pipe size, and gas pressure
- sanitary sewer, including location, size, slope, station, nearest manhole, and depth of main sewer
- private sewage system, including size and location of septic tank and leach field or pit
- storm drain, including location, size, slope, station, nearest manhole, and depth
- telephone and cable television lines, including location of pole, underground main, and nearest vault

Zone map

Most building departments have zoning maps available for inspection at their public counter. These maps show areas designated for residential, commercial, and industrial uses. Use the map in the building department to verify the zone on the site where you plan to build.

Topographic plan

This plan shows the contours of the land, and existing and finished grades. It may also show all the existing features on the site, such as fences, driveways, drainage channels, and buildings.

Street plans

These plans are the property of the government agency responsible for constructing and maintaining public streets. You can get copies for a nominal fee. Use these plans to find the location and elevation of the curb, gutter, catch basins, and other features found on public streets.

Sewer and storm drain plans

These plans belong to the government agencies that build and maintain sanitary sewers and storm drains. You can get copies for a nominal fee. Use these plans to find the location and elevation of the sanitary sewer and storm drain.

Survey Equipment

Transits

A transit is used to locate horizontal and vertical angles. You can also use it to set elevations, but it's not as effective as a builder's level because its telescope has less magnifying power. Also, the transit has more moving parts than the level, so it's more difficult to set up and keep steady.

Figure 2-6 shows the main parts of a transit; the telescope, standards, plate, magnetic compass, vertical circle, horizontal circle, and verniers. The horizontal and vertical circles are divided into degrees and 30-minute increments (one-half of a degree). The surveyors' transit is sometimes identified by the smallest angle that can be directly read on the vernier. You can read a 1-minute transit to 1 minute of angle in either the vertical or horizontal circle.

The major components of a transit are the telescope, horizontal circle, vertical circle, and the leveling base. A telescope is classified by its magnification power, minimum focus distance, field of view, and stadia ratio. The telescope is suspended between standards and rotates 360 degrees. A leveling vial is attached to the underside of the telescope.

The horizontal circle contains two movable plates which the surveyor rotates or locks. The angles between the two circles are graduated and read with a vernier. Movement of the horizontal circle is done by adjusting screws. Two leveling vials are attached to the top of the horizontal circle.

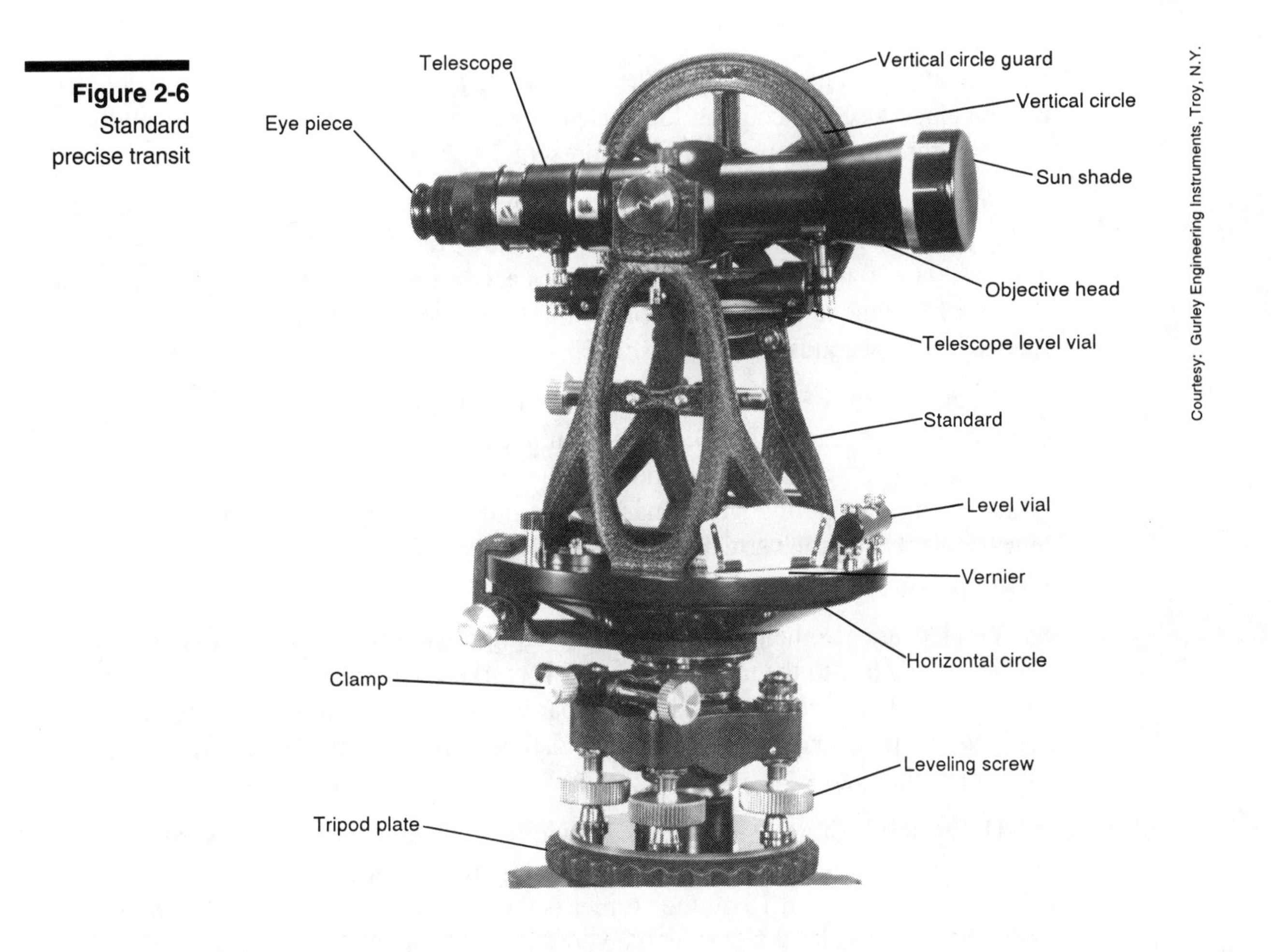

Figure 2-6
Standard precise transit

The vertical circle contains two movable plates which are attached to the side of the standards. The angles between the two circles are graduated and read with a vernier. Movement of the vertical circle is done by an adjusting screw.

The leveling base contains leveling screws for adjusting the horizontal circle relative to the base and tripod.

This instrument comes with a high visibility case and a carrying strap. It also has a sunshade you put on the end of the telescope, a dust cap to protect the objective lens when it's not in use, and a rain cover. There's a plumb bob to center the instrument over a selected point, a magnifying glass to help you read the vernier, adjusting pins for tightening and correcting the leveling vials, a screw driver for adjusting the other components of the instrument, and an instruction book.

Reading verniers

A built-in vernier is a graduated scale used to measure parts of a circular scale. It can be used to read angles as small as 5 minutes. Verniers are used for measuring horizontal and vertical angles on a transit. An ordinary transit is usually graduated to read in single minutes, but instrument readings to 30, 20, 10, or even 5 seconds are also made. This is done by use of the vernier marked on the inner circle. The outer ring of the horizontal circle can be locked so it's stationary with the tripod. The inner circle rotates with the telescope. The outer circle is graduated from 0 to 360 degrees in both clockwise and counterclockwise directions. Each degree is further subdivided into two, three, or four parts, which represent 30, 20, or 15 minutes.

The inner ring is marked with two indicating arrows (A and B) on opposite sides of the circle. On both sides of the arrow there are graduations marked from 0 to 30 minutes, 0 to 20 minutes, or 0 to 15 minutes according to the subdivision of the outer ring. Each minute in turn is subdivided into two or three parts, representing 30 seconds or 20 seconds.

When the arrow is directly over one of the graduations on the outer circle, you can take a direct reading of 30, 20 or 15 minutes. When the arrow falls between two graduations, use the vernier to subdivide the space between the two graduations. Read the angle by noting which mark on the outer circle lines up with a mark in the inner circle. Using the vernier, you can read to the accuracy of 1 minute and 30 or 20 seconds, depending upon the instrument.

Verniers are also helpful in reading leveling rods. A brass plate engraved with a vernier is attached to the target. If the rod is marked with bars every one-hundredth of a foot in width, the vernier divides each bar width by 10. Therefore, readings can be made to one-thousandth of a foot. It can also be used on a horizontal scale such as a caliper or other gauge.

The principle of a vernier is that the widths of the divisions on the vernier are slightly less than the divisions on the scale. Figure 2-7 shows this principle. The distance between 0 and 10 on the vernier is the same as the distance between 3.00 and 3.90 on the scale. When 0 on the vernier coincides with 3 on the scale, the reading is 3.00 as shown in Figure 2-7A. When 1 on the vernier coincides with 3.1 on the scale, the reading is 3.11 as shown in Figure 2-7B, and when 8 on the vernier lines up with 3.8 on the scale the reading is 3.88 as shown in Figure 2-7C.

Using the transit

The telescope on a transit has an eyepiece, cross hairs, and focusing screws, plus a level for leveling the telescope.

A transit sits on a wood tripod. A plumb bob, or plummet, hangs from underneath the transit to a set point on the ground. To position a transit, set the tripod firmly on the ground with the telescope approximately at eye level and the bob directly over the intended point. Adjust the leveling screws so the compass box is horizontal and the plate stays level no matter what direction you point the telescope. See Figure 2-8.

You'll need to know about angles, bearings, and azimuths to use a transit. And you should be able to add and subtract bearings to find angular differences, and to interpret the verniers on the horizontal and vertical circles for close readings of angles.

A bearing is determined by the angle between a line and the north-south line. A bearing is never more than 90 degrees. For example, a bearing of 30 degrees 20 minutes west of north is written N 30°20'W. A line 45 degrees 30 minutes east of south is written S 45°30'E. A bearing used in a property description is based on a previously recorded bearing, such as a street or section line.

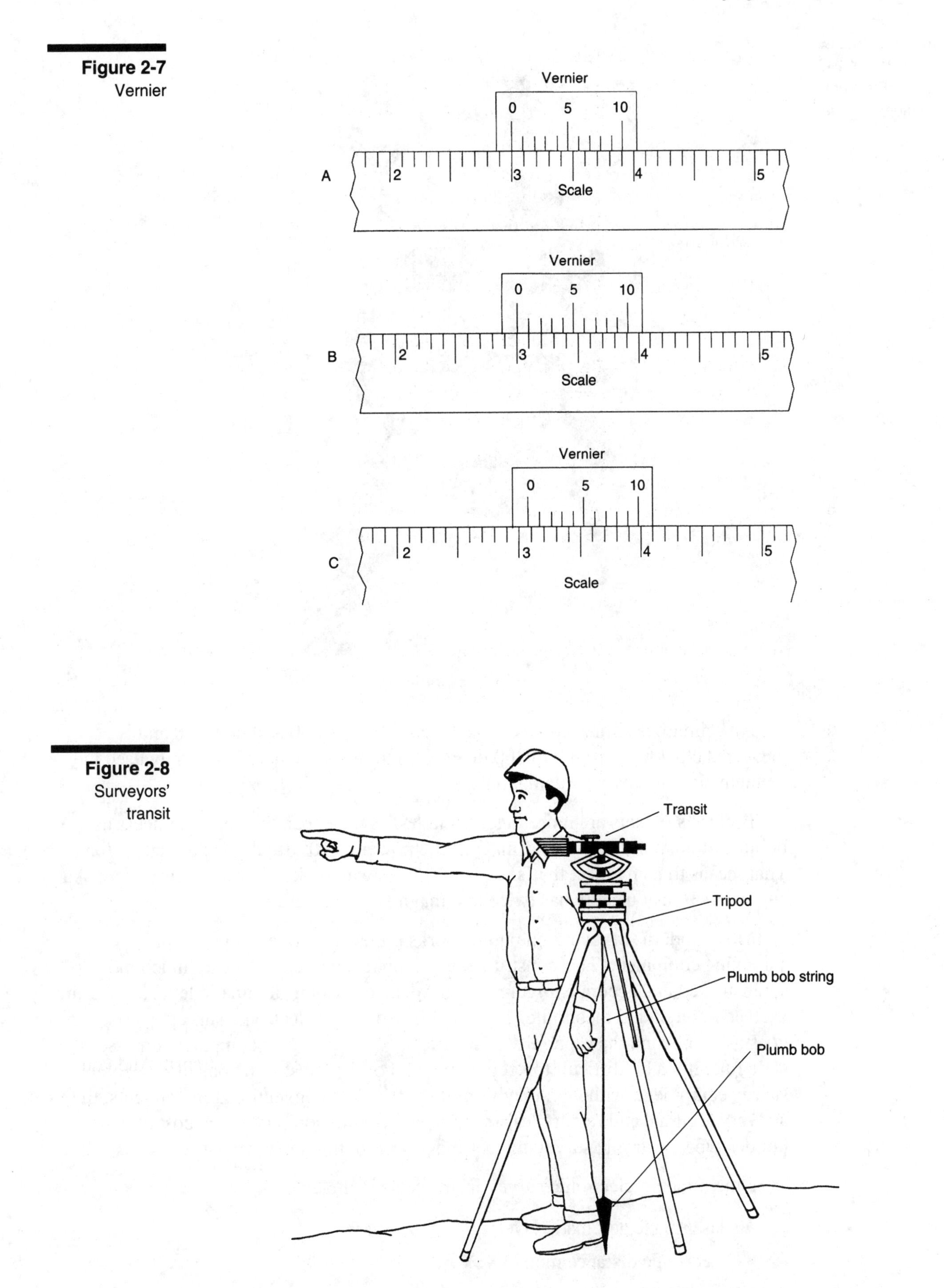

Figure 2-7
Vernier

Figure 2-8
Surveyors' transit

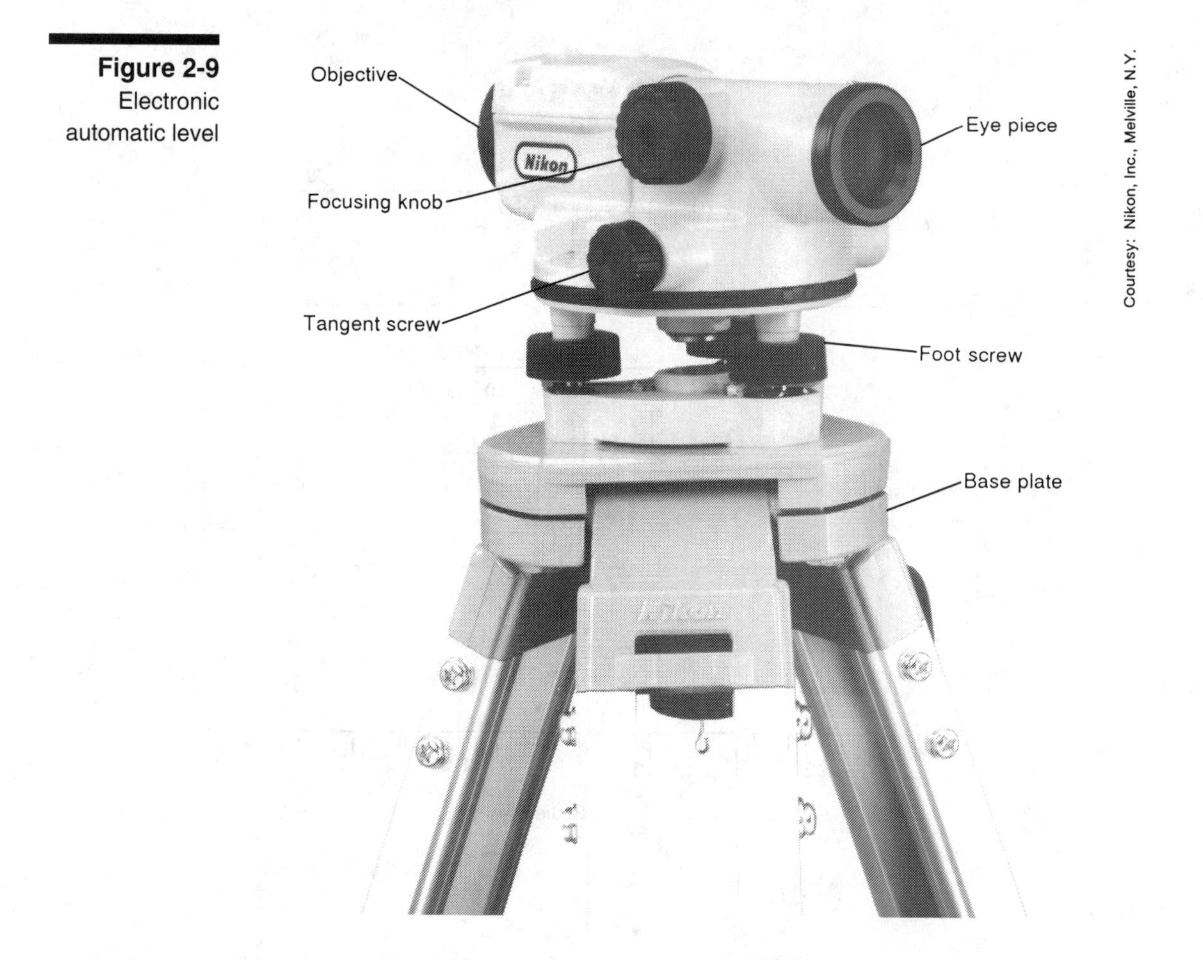

Figure 2-9 Electronic automatic level

An azimuth is similar to a bearing except it's always based on north and is measured clockwise from 0 to 360 degrees. Azimuths are more commonly used in mathematical surveying calculations.

Builder's transits are more rugged and less precise than the ones engineers use. A builder's transit normally has a magnifying power of 10 (usually expressed as *10X*). That means that an object that's really 120 feet away looks like it's 12 feet away. An engineer's transit usually has twice that magnifying power.

Most modern surveying on public works projects is done with electronic surveying equipment. These instruments require a small crew and are much more sophisticated and accurate. Figure 2-9 shows an electronic automatic level and Figure 2-10 shows a digital theodolite. The capability of these electronic survey instruments goes as far beyond the capability of the transit and tape as the computer surpasses the slide rule. It's a bit difficult to keep up with all of the latest functions of electronic survey equipment. Although they're more costly than conventional instruments, they are very accurate and require less manpower. As with computers, the cost of builder-type electronic surveying instruments is continuously decreasing.

Other modern electronic survey instruments include:

- laser plane automatic level
- electronic distance meter

Figure 2-10 Digital theodolite

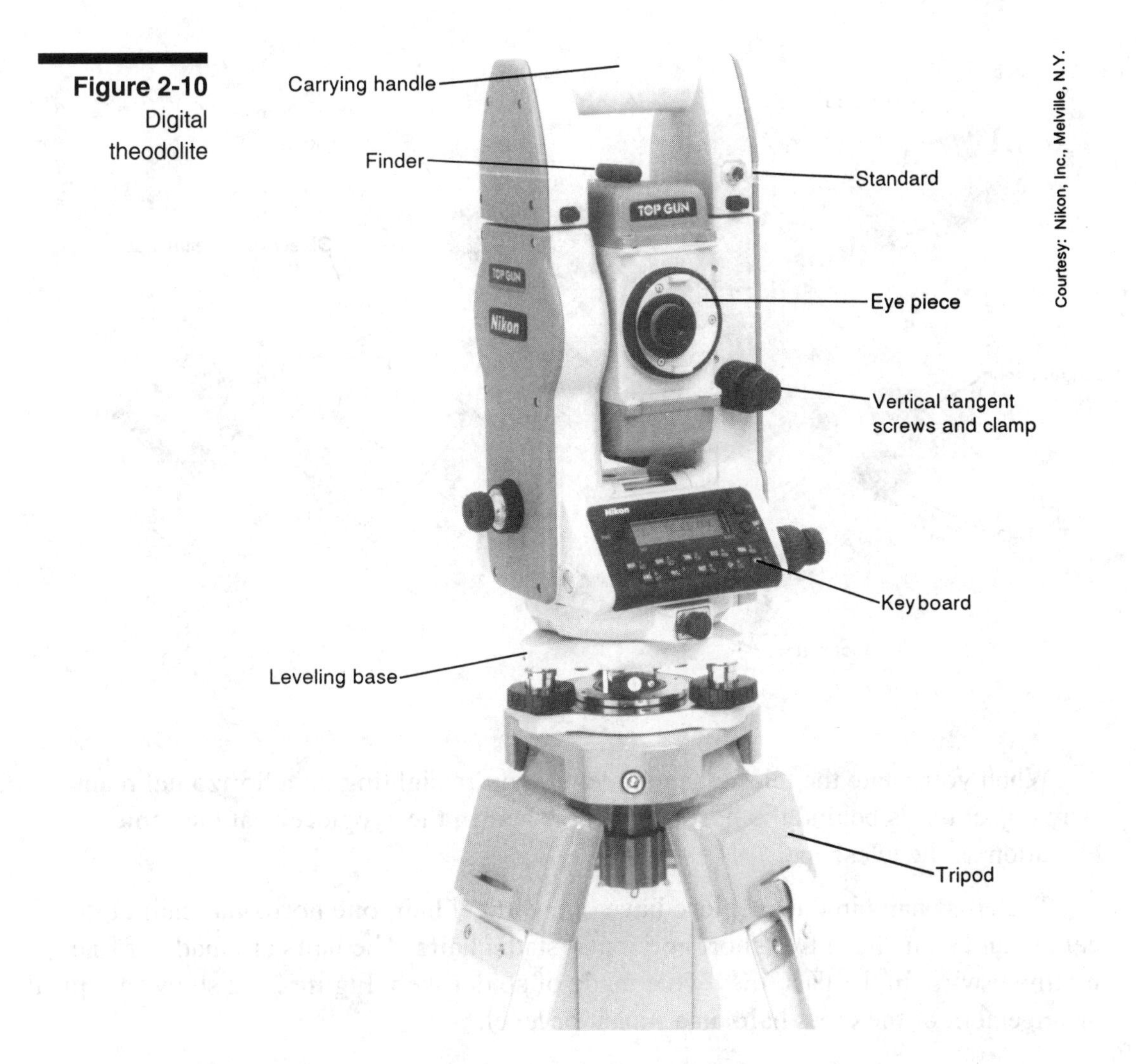

- electronic reduction tachometer
- infrared distancer
- laser theodolite

Levels

The level is similar to the transit, except that its telescope is longer and has greater magnifying power. Figure 2-11 is a photograph of the type of level used in construction. It's used to set the height of concrete forms and floor slabs, and to set the depth of sewer pipes. A level is more practical than a transit in setting elevations because it has fewer moving parts and is easier to keep steady. Most levels don't measure vertical or horizontal angles.

A basic level has a telescope with an attached level vial. The telescope is mounted on the ends of a straight bar. The bar rotates around its center on a vertical axis. Use the adjusting screws on the tripod to make the plate exactly horizontal.

Figure 2-11 Type of level used in construction

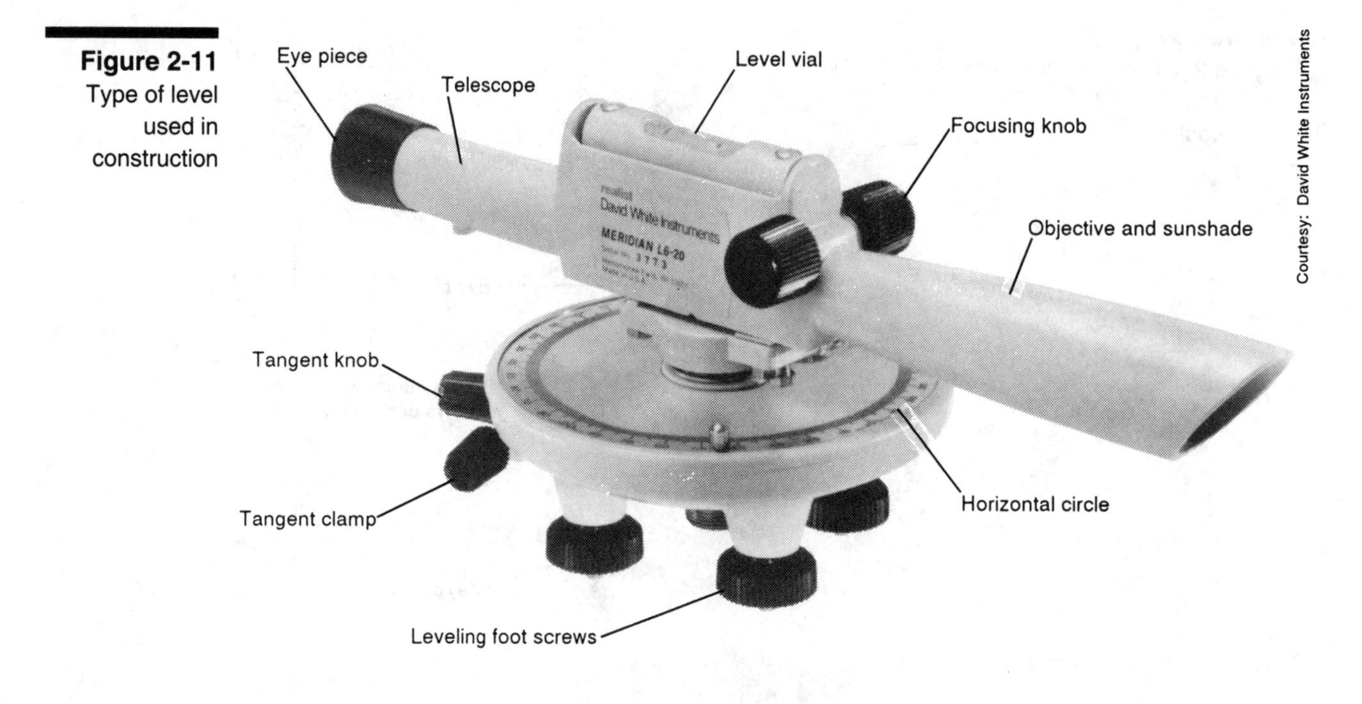

When you rotate the telescope on a level, you're sighting on a horizontal plane. Any object that's behind the horizontal cross hair in the eyepiece is at the same elevation as the telescope.

The cross hairs in the eyepiece have one vertical hair, one horizontal hair at the center, and some have two short horizontal stadia hairs. The hairs are made of fine platinum wire. In the past, they were made of spider web. Figure 2-12 shows a typical arrangement of the cross hairs in a transit or level.

Figure 2-12 Telescope cross hairs

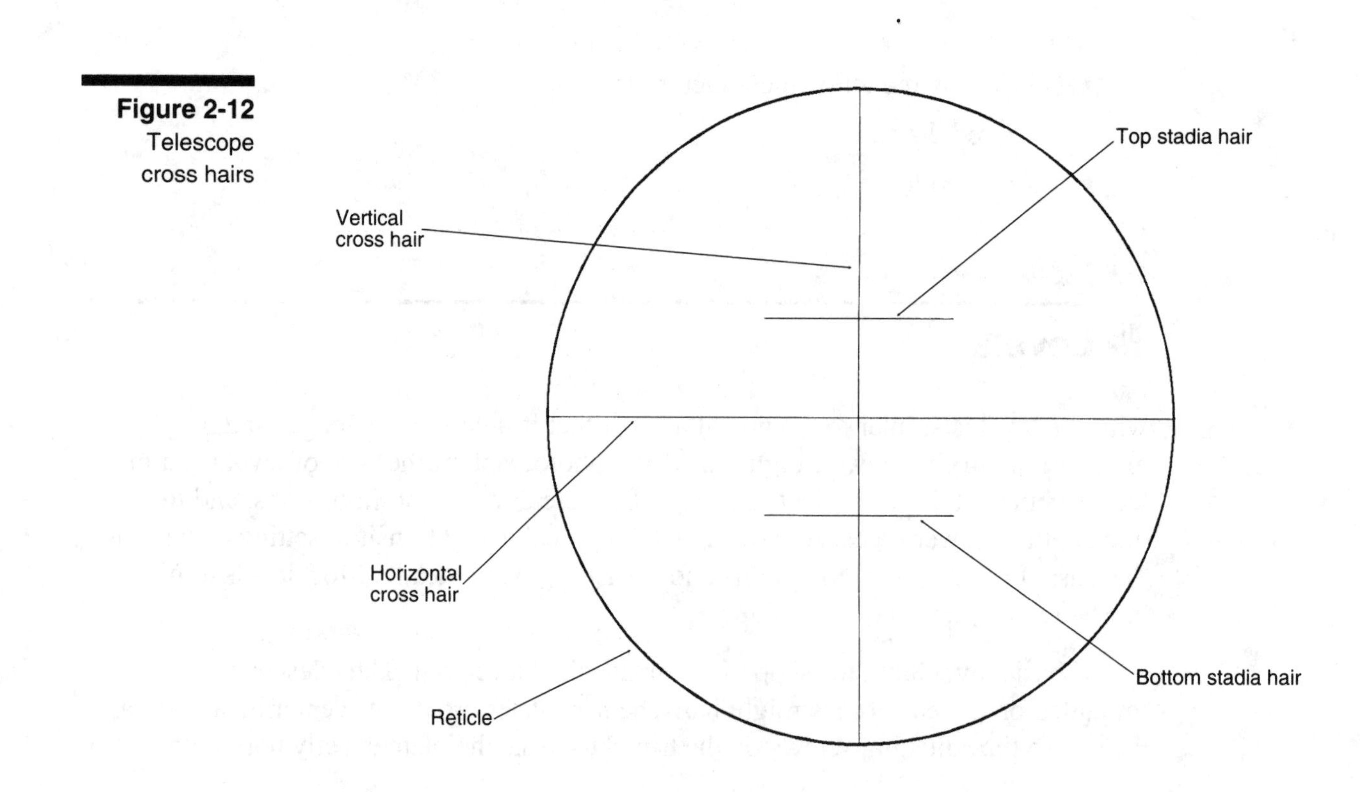

You can use the stadia hairs to find the approximate distance from the instrument to the stadia rod. When the stadia cross hairs are preset to 1:100, a 1-foot interval read on the stadia rod means that the rod is 100 feet from the instrument.

A typical builder's level can be read to an accuracy of 3⁄16 inch at a 150-foot distance. Use it for any of the following types of work:

- bench mark leveling
- transfer of elevations from bench marks to work above or below the ground (i.e., street and drainage works)
- profile leveling and cross sectioning
- grading work

The builder's transit-level serves as both instruments. It's more rugged than either the surveyors' transit or level. The smallest angle that can be read with it may be from 5 to 15 minutes.

The automatic laser level is an instrument that lets you do leveling work by yourself. It's an automatic level with electronics and infrared technology, mounted on a tripod. It has a battery with a charger. It generates a reference plane around the instrument using a rotating emitter. Some laser levels have a hinged top so they can also generate a vertical reference plane.

You use a bull's eye bubble to initially level an instrument. It automatically keeps itself level after that. If you accidentally knock this type of instrument out of level, it stops rotating and a light flashes.

Measure the height of this type of instrument with a yardstick you set between the ground point and a mark on the housing. Then use the detector or sensor mounted on the staff to take measurements. The detector has a battery-powered display and tone generator. You can read the sensor from front or back. Several people can use the same automatic laser level at one time.

Tripods

Most modern electronic survey equipment is very accurate. It should be mounted on heavy hardwood frames. The legs of a tripod are extendable and have metal shoes with spurs and hardened steel points. Use the spurs to help drive the shoes into the ground. To keep a tripod steady, spread its legs apart and press them into the ground. On slopes, you can place two legs of the tripod on the lower part of the slope and extend the third leg upward, as shown in Figure 2-13.

Builders usually use lightweight tripods which come in all degrees of sturdiness and in several materials. The wide-frame hardwood tripod is the most sturdy. It has heavy extendable legs which can be locked at any length. It's painted bright yellow for easy visibility. The straight leg hardwood tripod is lighter and you can't adjust the length of its legs. Some tripods are made of a combination of hardwood and aluminum.

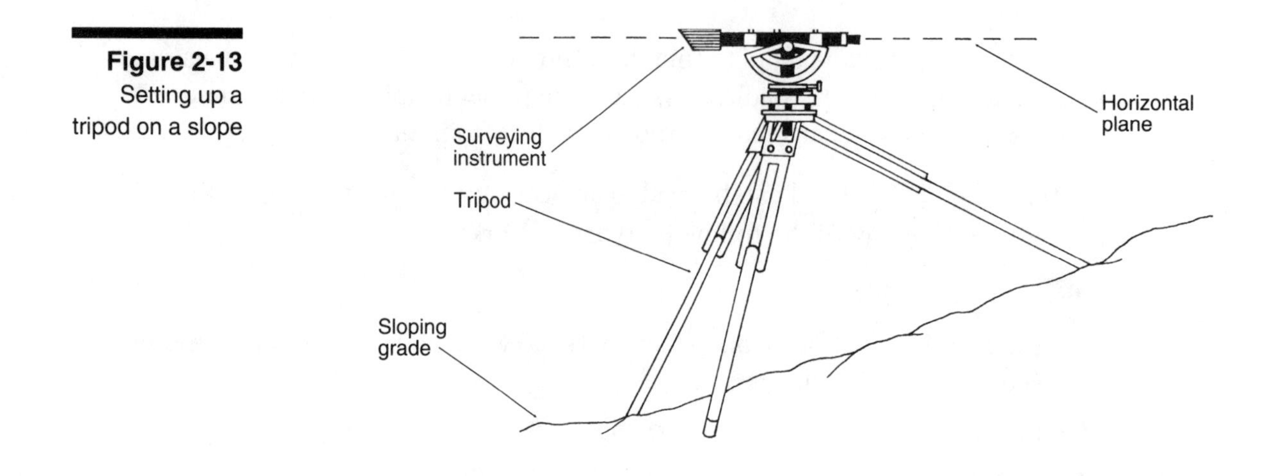

Figure 2-13 Setting up a tripod on a slope

Metallic tripods are made of steel or aluminum. The builder's tripod is probably the simplest. It's made of enamel-coated steel. The legs are either fixed or adjustable, and are painted to look like hardwood.

Hand Level

This is a small hand-held metal tube with lenses that don't magnify. Use it to measure approximate differences in elevation over short distances. Hand levels have a bubble that you see at the same time as the object you're leveling. Cross hairs split the image of the bubble when you hold the instrument level.

Plumb Bob or Plummet

This is an ancient tool with a solid brass or steel cone weight which hangs from a cord. The force of gravity pulls the cord in a vertical line. The cord is usually made of linen or nylon, in white or bright colors so it's easy to see.

Plumb bobs are used to extend a line that's absolutely vertical. You can also use a plumb bob to drop a point on a steel tape to the ground, and to position the center of a transit directly over a monument.

Leveling and Stadia Rods

There are many types of rods used for vertical measurement. We'll describe Philadelphia rods, builder's rods, and stadia rods.

The Philadelphia rod

The most common type of leveling rod is called the Philadelphia rod. It's made of hardwood to resist shrinking or swelling. The facing is made of Mylar with markings and numbers that won't rub off. Every number indicating a foot is in red, tenths of a foot are marked in black. Hundredths of a foot are marked in black bars along one edge of the rod. In the standard leveling rod, the numbers increase from the bottom up. In the direct elevation rod, the numbers increase from top to bottom.

Philadelphia rods come equipped with a target with a brass micrometer that's graduated into hundredths of a foot. The target can slide along the rod until you set it in the selected position with a lock screw.

A device called a rod level can also be attached to any type of rod to indicate when the rod is in a true vertical position. This consists of a bull's eye circular leveling vial attached to a clamp which slides on the rod.

The builder's rod

Builder's rods are smaller but more ruggedly built than Philadelphia rods. The builder's rod is usually 12 feet long, and can be folded into two sections. It's marked in feet and inches to the nearest eighth of an inch. Engineer's rods are marked in feet and decimal parts to the nearest hundredth of a foot. Rods are painted white, with feet indicated in red, and fractions in black. A round target painted black and red over a white background is attached to the rod.

The stadia rod

The stadia rod is a highly-visible calibrated board. You aim at the rod with the cross hairs in the telescope of a level. A stadia rod can be used over a greater distance than a leveling rod. You find the distance from the eyepiece to the rod by reading the interval between the horizontal cross hairs. Stadia surveying is often used for rough topographic mapping.

The Measuring Pole and the Range Pole

Another version of the rod is the measuring pole. Figure 2-14 shows a digital measuring pole. This device extends up to 26 feet in height and is self-reading so you can measure height and depth using only a measuring pole.

Digital measuring poles are aluminum poles which can be extended to as much as 26 feet. You read distances on a digital dial attached to the lower section of the pole. These poles are useful in measuring the height of telephone poles, high ceilings, or widths between walls. Using this device, one person can make measurements in difficult locations where normally two people and a ladder would be needed.

Figure 2-14
Using a digital measuring pole

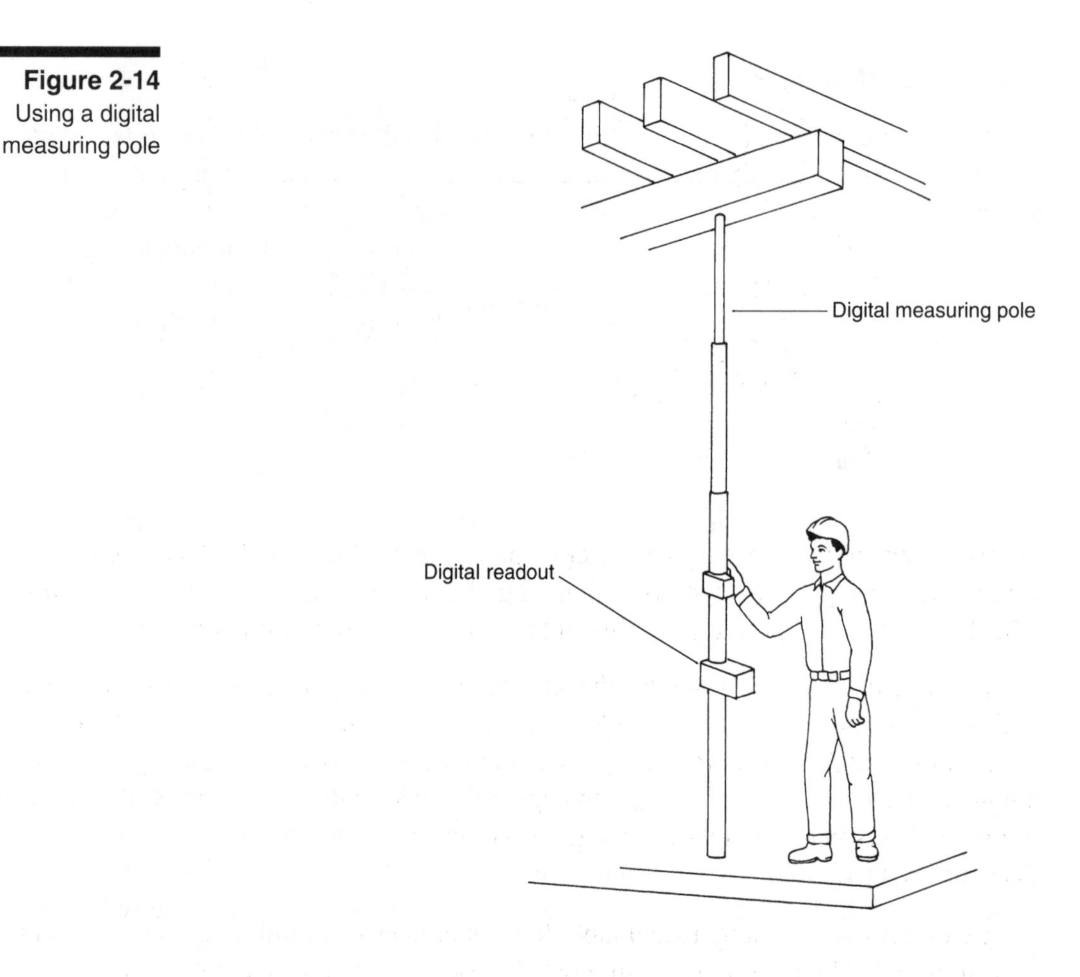

Range poles are round poles 4 or 8 feet in length and about $1\frac{1}{8}$ inches in diameter. They're painted in alternate red and white bands so they're easy to see. Use the metal point on the bottom of a range pole to set it in the ground. You can add extensions to range poles to make them longer. Use range poles to make it easier to see distant points. You can also use them in a series of poles to extend a straight line for tape measurement.

Steel Tapes

Surveyors usually use steel tapes to measure distances. Tapes may be band chains, flat steel wire tapes, ordinary steel tapes, or metallic coated tapes. Many old-time surveyors still call the steel tape a chain. This is because the early American surveyors used a chain made of metal links to measure distances. The term *chaining* came to mean measuring distances. Tapemen were called chainmen. Even today, some manufacturers of steel tapes call their product chain tapes.

Most steel tapes are either 100, 200, or 300 feet long. They're marked at 1-foot intervals. One end is marked in tenths and hundredths of a foot, reading from right to left. The 100-foot tapes are kept on reels when not in use. Leather thongs and holding clamps are used to keep the tape tight. Tension handles help reduce sag in the tape.

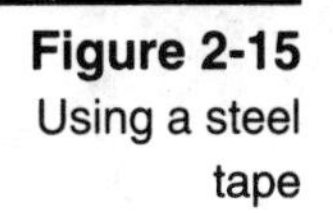

Figure 2-15
Using a steel tape

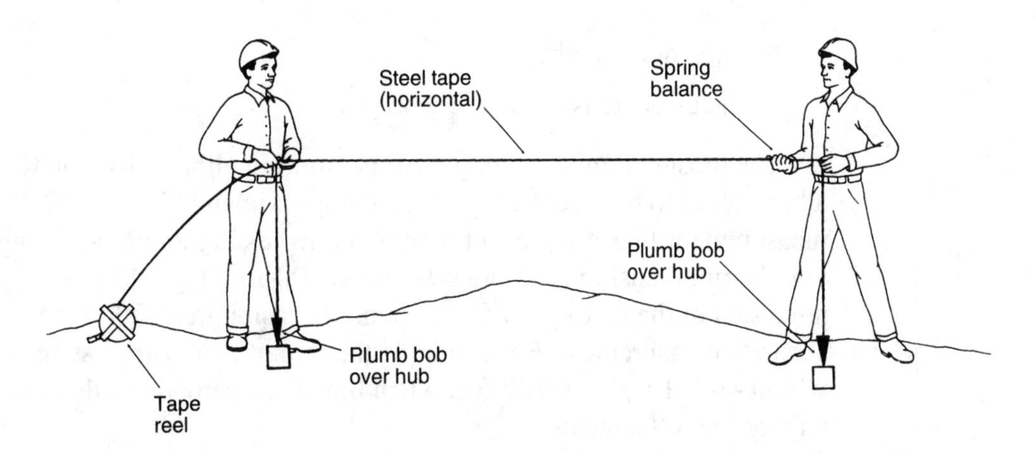

Another important accessory for a steel tape is a repair kit. This kit has a punch pliers, eye-setting pliers, eyelets, Allen wrench, pierced plain tape, dies, and rivets. This lets you repair a broken or kinked tape in the field without losing a lot of time.

For measuring level terrain, lay the tape directly on the ground and mark the points with steel pins or scratches on the sidewalk or pavement. In rough terrain or where there's vegetation, hold the tape horizontally above the ground as shown in Figure 2-15. Attach a spring balance to one end of the tape to apply tension. Carry the points to the ground with plumb bobs suspended from a cord. Figure 2-16 shows how to measure horizontal distances along a slope.

Be careful when using a steel tape. It's easier than you think to get an incorrect measurement. The tape must be supported throughout its entire length, at a temperature of 68 degrees F., and at a tension of 10 to 12 pounds. If you just suspend the tape from both ends, it'll sag, adding a few inches to the true measurement. A light 100-foot tape, weighing 1 pound, will be shortened about 0.042 feet unless you put 18 pounds of tension on it. A heavy tape, weighing 3 pounds, would be shortened about 0.0375 feet. This would be difficult to correct with a hand-held spring, so you'd need to apply a mathematical correction to the measured distance.

Most errors in measuring horizontal distances are the result of:

- incorrect reading of tape
- tape not held horizontal

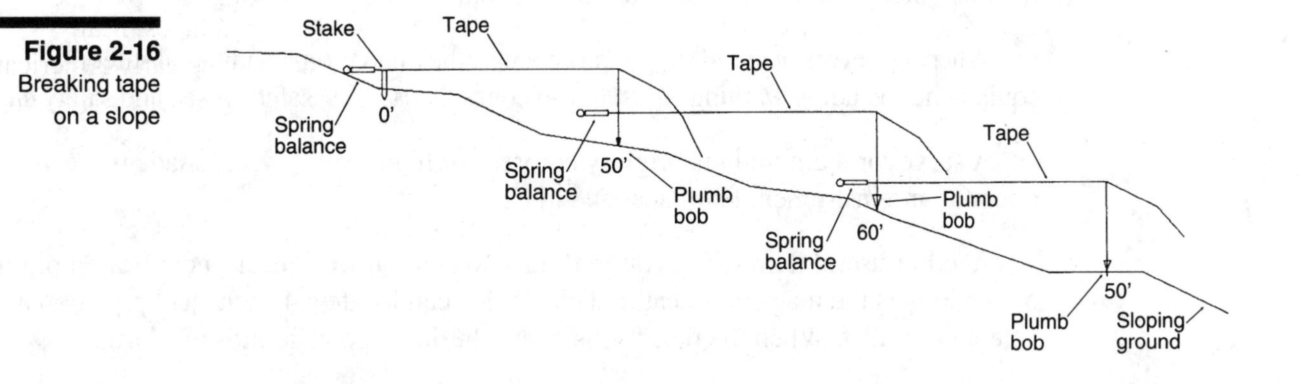

Figure 2-16
Breaking tape on a slope

- tape not straight
- incorrect tension on the tape

The most common reading error is omitting either 1 foot or 100 feet. You can use a hand level to be sure the tape is held approximately horizontal. To help keep a measurement line straight, set temporary stakes in the ground along the measuring line. Normal tension is 15 pounds with a 100-foot tape. Most steel measuring tapes are tested in the factory for the amount of tension in a 100-foot length required for accurate measurement. For example, the manufacturer may state that the accuracy is within +/- 0.1 inch per 100 feet when the tape is supported throughout at 15 pounds tension and 68 degrees F.

Miscellaneous Survey Equipment

Besides the surveying equipment we've described, there are field accessories that all surveyors should have. First is a plumb bob and leather carrying scabbard. Besides using it to carry a point from a steel tape to the ground and placing the transit, you can check that forms or studs are vertical.

Along with the plumb bob, a spool of plumb bob cord should be available. The cord is made of linen or nylon. Plumb bob targets are brightly colored paper or plastic cards. The cards are attached to the plumb bob cord to make it more visible to the instrument man.

When stakes are set in brush, they're hard to locate. Plastic or cloth flagging attached to the stakes make them more visible.

There are various tools available for marking reference or cut stakes, including large felt tip pens, marking paint, lumber crayons, and carpenters pencils.

Carry spray paint cans to mark on pavement. Use a concrete or masonry nail for temporary marking in concrete. A more permanent marker is the lead and tack described earlier. It requires a concrete chisel.

Other recommended hand tools are a machete, bush hook, hand axe, pry bar, pick hammer, sledge hammer, a rock or concrete chisel, and safety goggles.

When surveyors are working in traffic areas, they need some additional safety equipment, including warning signs, traffic cones, barricades, safety vests, and safety caps.

A surveyor's umbrella is brightly colored for high visibility. Its shade protects the operator and instrument from heat and glare.

Another useful item when you're doing excavation work near underground piping or conduits is the magnetic locator. This device can locate a 4-inch steel pipe down to a depth of 8 feet. When the detector is over a buried pipe, it sounds an alarm.

Taking Care of the Equipment

Transit and level

Here are some suggested rules for caring for the transit and level:

- If it rains while you're surveying, cover the instrument with a plastic hood or a plastic bag.
- Don't store a wet instrument in its metal carrying can or case. Let it dry indoors before you store it.
- After surveying in cold weather, don't bring the instrument directly into a warm room.
- Don't expose the instrument to sudden changes in temperature.
- If the instrument fogs up, warm it gradually with a heat lamp.
- When you carry an instrument and tripod on your shoulder, tighten the clamps slightly to keep it from moving. Set it down gently so you don't break the cross hairs.
- Stow the needle on the magnetic compass when you carry the instrument.
- Keep spirit levels dry to prevent warping from moisture penetration.
- Never tap the level with a hammer or other tool. This may damage the plaster of paris in which the vials are set.

Tripod

- Check that the shoes, hinges, or places where the side pieces enter the metal sockets are tight. If they're loose, tighten them with a tripod socket wrench.
- Don't set a tripod on asphalt pavement during hot weather. The sharp points may sink into the pavement, making the tripod uneven.
- Make sure the shoes don't slip when you set the tripod on concrete.
- Rub a wood tripod regularly with linseed oil.

Steel tape

It's easy to damage a steel tape or make it unreadable. Here are some rules for taking care of a steel tape:

- Don't let vehicular traffic cross a steel tape.
- Don't jerk or step on a tape.
- Try not to kink a steel tape.
- Band tapes should be done up in 5-foot loops, figure-of-eight form.
- Wipe etched tapes clean and dry at the end of each day's work, and oil them.
- Use a tape repair kit to repair a tape if it's kinked or broken by a vehicle.

Plumb bobs or plummets

To take care of a plummet:

- Protect the point of the bob.
- Try to keep knots out of the string.
- When not in use, carry the plumb bob in its leather holster.

Control stakes

Protect the permanent survey marks set by the land surveyor, and the control stakes set by the engineering surveyor. Preserve property line and corner survey markers when you can. Sometimes it's not possible. Notify the surveyors before you start any work that's likely to disturb a marker. Most surveyors charge for another visit when they have to replace markers damaged in construction.

Layout Tools

You'll need these tools when you lay out buildings and do other construction work:

Chalk line

A chalk line is made of a twisted cord and reel. The cord is coated with white chalk. The line is stretched tight just above the surface between two points to be connected by a straight line. When you snap the cord, it leaves a trace of white chalk directly between the two points. You'll hear this referred to as "snapping a line."

Spacing rule

This has been a favorite tool of carpenters and surveyors for decades. It's a folding wood rule, usually 6 feet long when extended and 6 inches long when folded. It's made of hardwood, with steel brass-plated spring joints that lock the rule in position. The surface has an enamel finish with black and red numbers printed on a white background. A carpenter's spacing rule is marked in feet and inches to the nearest eighth of an inch. A surveyor's spacing rule is marked in feet and tenths of a foot to the nearest hundredth of a foot.

Tracing tape

This tape is a ribbon of 1-inch-wide cotton, usually 200 feet long. You'll use it to lay out excavation or foundation lines.

Spirit level

This device is also called a carpenter's or mason's level. It's a straightedge made of wood, aluminum, magnesium, or plastic, in lengths of 2, 4, and 8 feet. You use it to make sure a surface is either level (exactly horizontal) or plumb (exactly vertical). The level is horizontal or vertical when a bubble is centered exactly in a glass tube or vial.

Straightedge

This tool is used by masons and cement finishers. It's made of tapered wood or metal, with hand holes. It's trapezoidal in shape, 30 inches long at the bottom edge and 10 inches long at the top edge. You can use this tool as a long level by resting a shorter spirit level on the top edge. Usually you'll use it to lay out straight lines between points.

Line level

This is a small spirit level that can be suspended from a string line. When the string line is level, the bubble will be centered in the vial.

Land Surveying

Permanent survey marks are usually set at the boundary corners on tracts and at intersections of the centerlines of public streets. Permanent marks are usually brass caps or steel pipes set in concrete or attached to anything likely to be stationary for many years.

When a subdivision is first laid out and recorded with the county recorder, lot corners will be marked in the field. These marks may be pipes, wood hubs, or stakes driven into the ground. Unfortunately, many of these markers are lost during rough grading or other work. Any missing corner markers should be reset by a surveyor before the building contractor starts work. To resurvey a specific lot, the land surveyor must extend measurements from two of the nearest permanent monuments. The distances and directions will be based on data shown on the recorded tract map.

Elevations on construction projects are controlled from permanent monuments, or bench marks. These may be lead and tack in a concrete curb, the edge of a manhole cover ring, or government monuments. On large jobs, the engineering surveyor will probably set bench marks at several locations on the site.

Field Surveying

You've probably paced off distances many times — 30 paces down the first base line on a baseball diamond, 180 paces to the green on a golf course. That's a simple type of measuring. It isn't accurate enough for most construction work, but it's fast and requires no special tools. If you do much field measuring, you should know the length of your natural pace. Here's one way to find out what it is:

1) Mark out a 400-foot distance with a tape.
2) Pace this distance four times in each direction, marking down the number of paces each time.
3) Calculate the average number of paces. The length of your pace is 400 divided by the average number of paces.

Simple survey tools on a construction site usually include:

- a range finder
- a hand level
- a magnetic compass
- range or prism poles
- a steel tape
- a garden hose
- a measuring wheel

A range finder is an optical instrument that finds distance by triangulation. It has an accuracy of about plus or minus 1 percent at 300 feet, plus or minus 2 percent at 300 to 500 feet, and plus or minus 15 percent at 500 to 1000 feet.

The hand level provides a simple way to measure differences in elevation. You can also use it to keep a tape horizontal while chaining distances.

The magnetic compass lets you find the approximate location of bearings and angles between lines.

Use range poles to extend a straight line over long distances without using a transit. Or use a prism pole, which is similar to a range pole but it's used to mount a prism for electronic distance meters (EDM). The prism is used as a sighting target. The prism pole tripod is a lightweight aluminum tripod that holds a prism pole, unattended, in a vertical position. A staff is a simple hardwood or aluminum pole with a metal shoe used to mount a prism or pocket transit.

A translucent garden hose makes a good level when you have to compare the elevation of two nearby points that can't be seen from a single point. Fill the hose with water. Hold both ends of the hose up. Water at both ends of the hose will stabilize at exactly the same level. Use the hose to level any two points anywhere you can crawl or climb, even around corners or in basements.

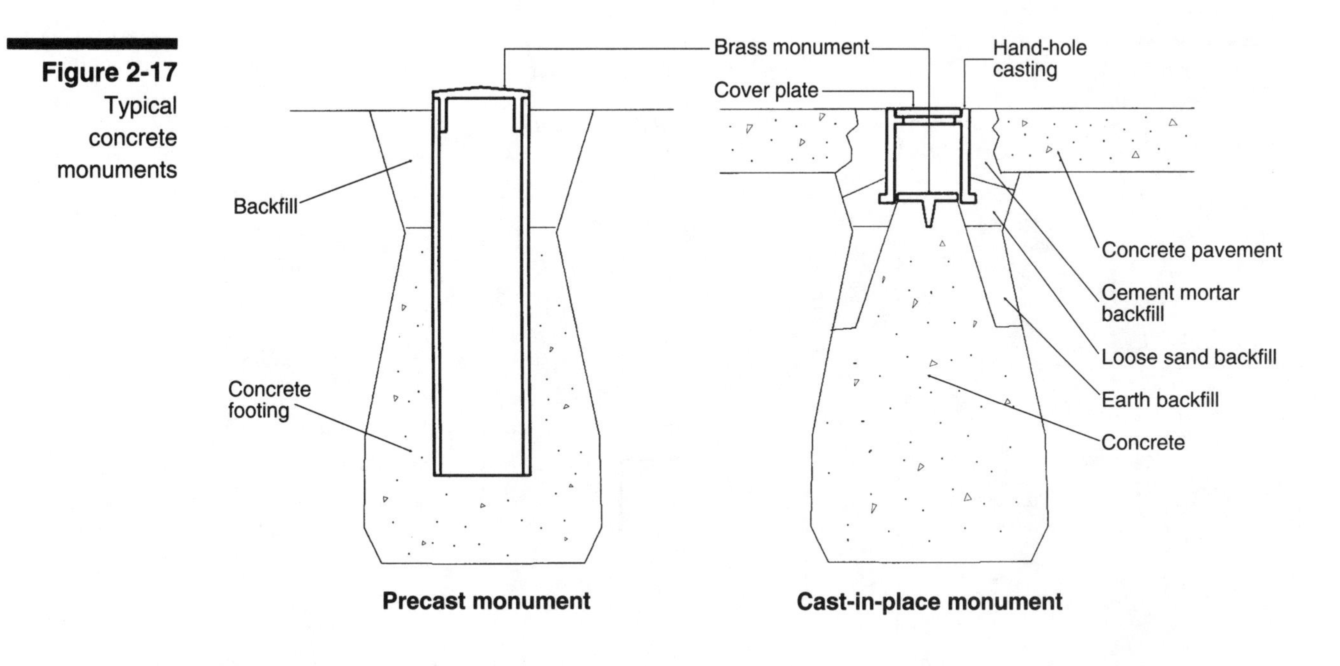

Figure 2-17 Typical concrete monuments

Horizontal Control

Measurement of horizontal distances on a construction project begins at some permanent survey monument. These monuments usually have a concrete base with a bronze cap cast in the top. A monument should be unaffected by settling and frost action. When a monument is set in a roadway or paved surface, a cast iron hand-hole is often placed over the cap for protection. Figure 2-17 shows two types of monuments.

Governmental monuments used as control points on triangulation or traverse systems are usually made using first order survey accuracy. See the section on Construction Tolerances for more information on the accuracy of surveys. The monuments are normally 3½ inches in diameter and have a red brass cap. The cap is engraved with letters and symbols describing the name, number, and type of monument. Some monuments are designed to be cast into concrete, and others fit into cast iron sleeves which are cast in concrete.

Coordinate system

When a construction project is large and has many buildings and other structures, a coordinate system is very useful. You can make a coordinate system by setting up a grid over the site, usually in 100-foot increments. If possible, make the grid parallel to two property lines which are at right angles to each other. See Figure 2-18.

Base or transit line

Plans for any project that's long and narrow, such as a road, will usually include a transit line. The line runs through the center of the project and is marked with station numbers every 100 feet. The first station will be 0+00. The next stations will be 1+00, 2+00, and so on.

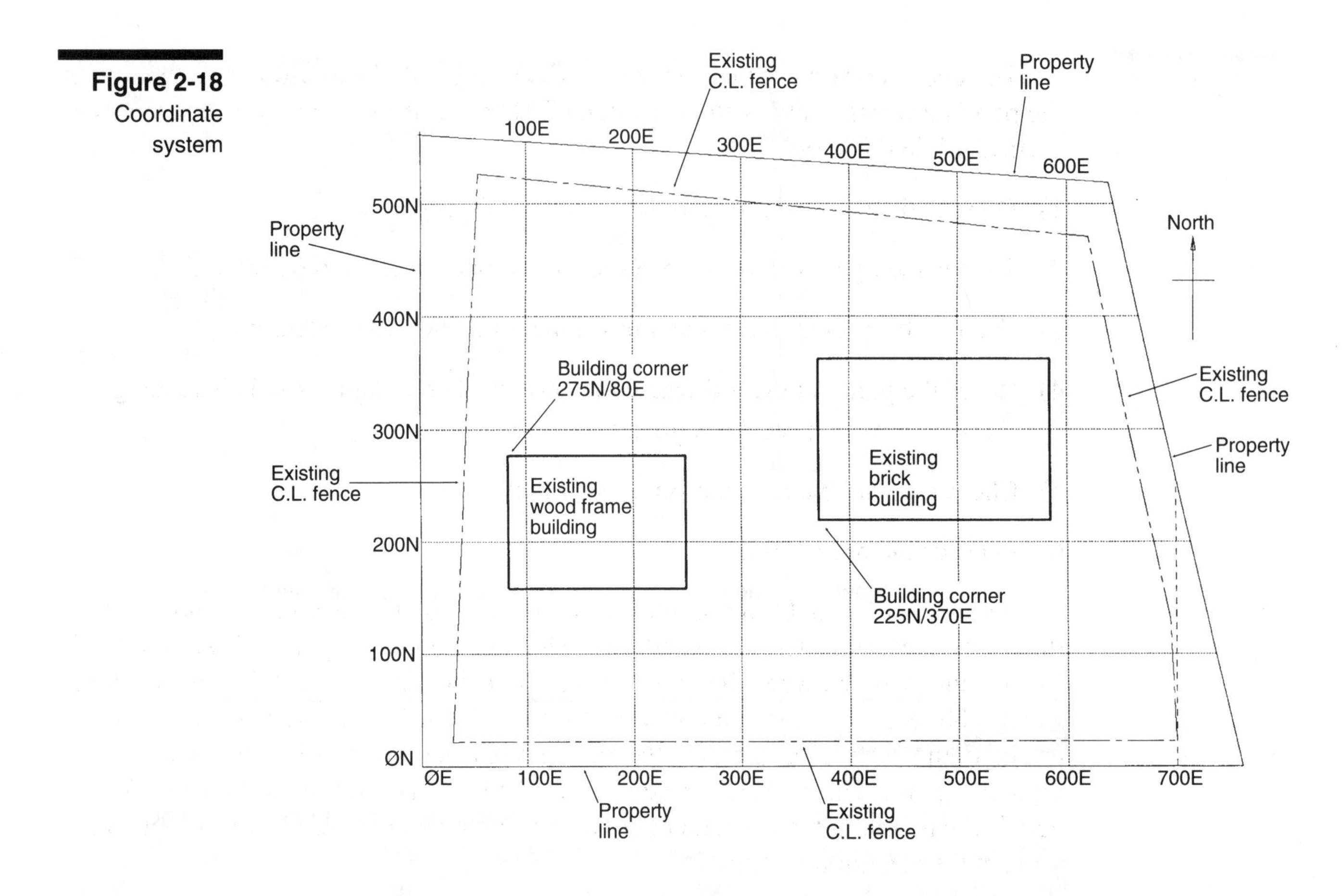

Figure 2-18 Coordinate system

The location of points between stations can be identified with additional numbers, such as Station 1+32.21, which is 132.21 feet from the beginning point, station 0+00. Every point on the site can be identified by a distance 90 degrees to the right or left of some station on the transit line. This system is often used in the rough plumbing layout in large buildings. In some cases, two transit lines are set at right angles to each other and marked on the site. The lines are called “Line A” and “Line B.” You measure key points in the rough plumbing system as offsets from the respective transit lines.

Vertical Control

Measurement of vertical distances on larger construction projects is usually based on some government bench mark. This may be a U.S. Geological Survey monument which shows the elevation above sea level. Elevation is very important when planning and building public streets, sanitary sewers, and storm drains.

On smaller projects, some point may be assumed to be Elevation 100.00. Every other point on the project can be assigned an elevation either above or below 100. Where you see this system used, you’ll probably find a reference to the actual elevation of Elevation 100 somewhere on the plans. For example, the plans might say, *Datum Elevation 100.00 = U.S.G.S. Elevation 396.27.*

To check elevations using the level and leveling rod, the assistant sets the rod on the point he wants to check the elevation of. He holds the rod steady and perfectly vertical while the surveyor:

1) Focuses the telescope to read the figures on the rod.
2) Swings the telescope until the vertical cross hair is centered on the rod.
3) Centers the bubble in the leveling vial between the two graduations.
4) Reads the point on the rod where the horizontal cross hair falls. This reading should be to the nearest hundredth of a foot.
5) Checks the bubble to make sure it's still level.
6) Records the rod reading.

Figures 2-19A, B, C, and D show four possible conditions you can find when doing hand leveling. The first condition, which is shown in Figure 2-19A, has the instrument set up between the known point, A, and unknown point, B. The leveling rod can be read from a single instrument setup. The arithmetic involved is included on the figure. The backsight is to the known point, and the foresight is to the unknown point. The backsight reading, plus the known elevation of the point A, equals the height of the instrument or H.I. By subtracting the foresight from the H.I., you get the elevation of point B.

In the second condition, shown on Figure 2-19B, the instrument is set up directly over a known point. The height of instrument is determined by measuring vertically from the known point, or benchmark, to the center of the telescope, and adding this value to the elevation of the benchmark. Then you subtract the rod reading over the unknown point, B, from the H.I. to get the elevation at point B.

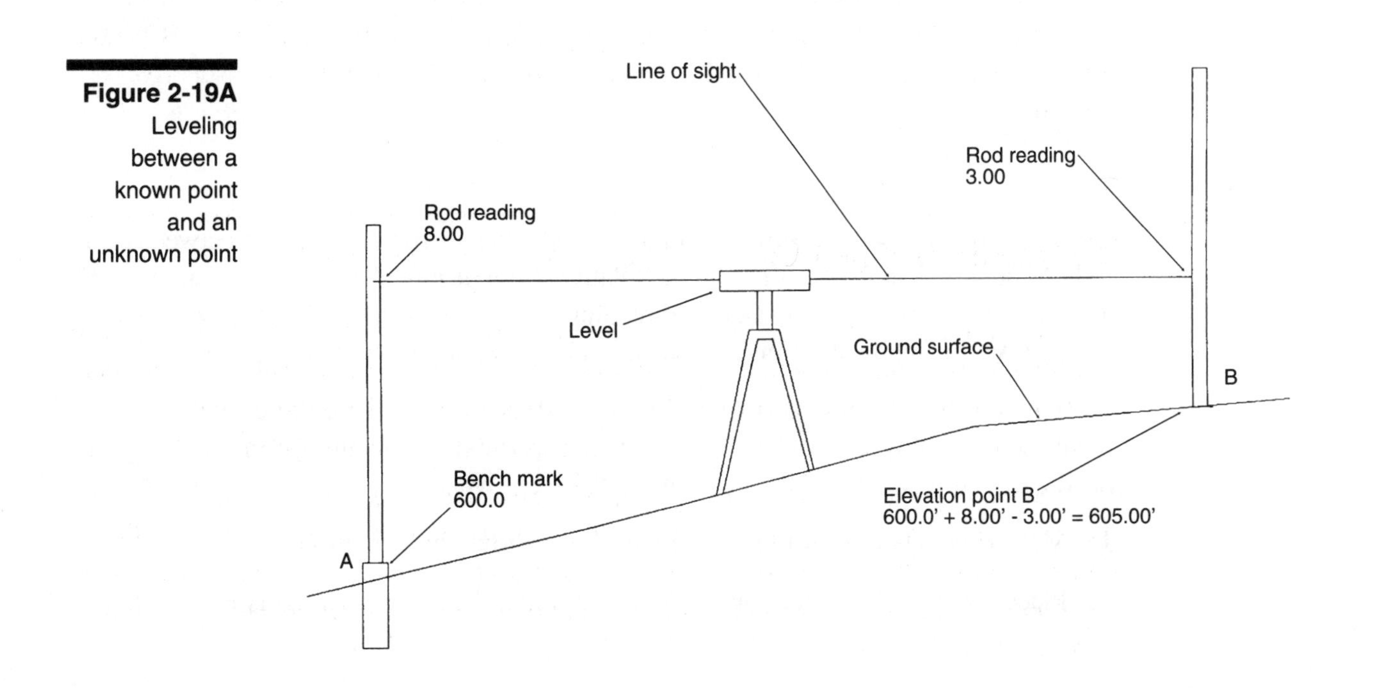

Figure 2-19A Leveling between a known point and an unknown point

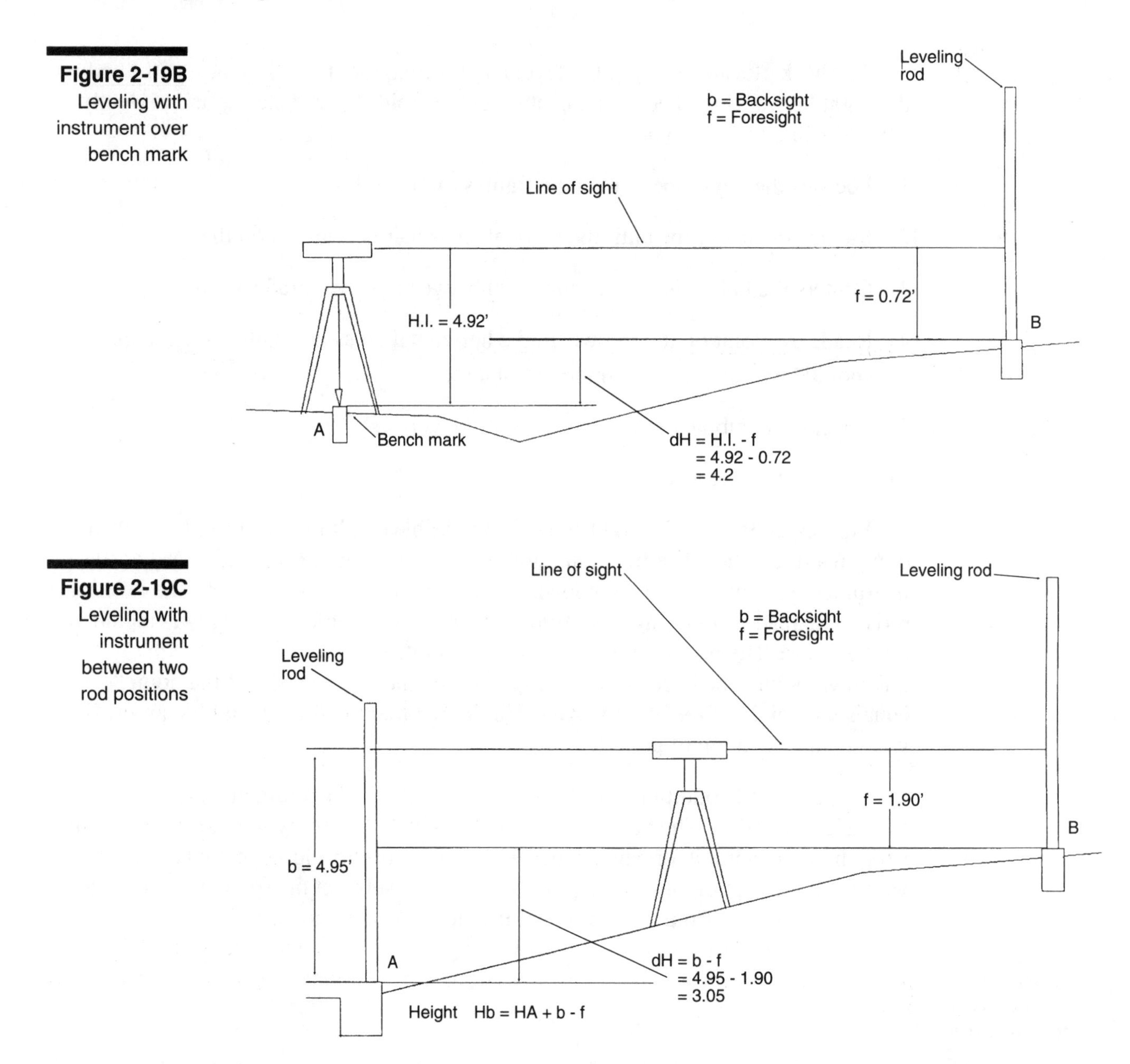

Figure 2-19B Leveling with instrument over bench mark

Figure 2-19C Leveling with instrument between two rod positions

In the third condition, shown on Figure 2-19C, the instrument is set up between two rod positions. The purpose is to find the relative difference in elevation between points A and B. By subtracting the foresight rod reading of point B from the backsight rod reading of point A, you get the difference in elevation between points A and B.

The fourth condition, shown on Figure 2-19D, is more difficult because the instrument is below one of the points. Since a vertical angle must be measured, you need a transit instead of a level. Here are the steps to take:

1) Measure the height of the instrument, H.I., above the benchmark.
2) Place the movable target on the leveling rod at the same reading as H.I., or H.I. equals Z.

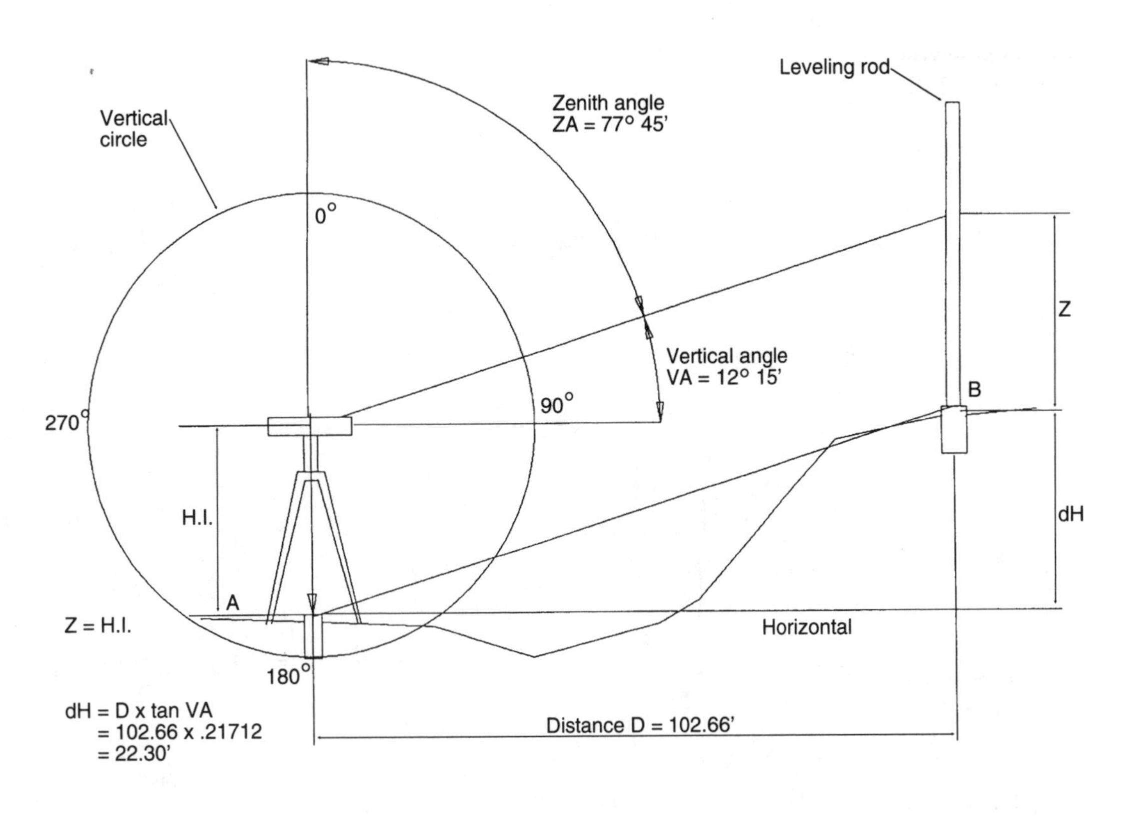

Figure 2-19D Leveling with instrument below rod bottom

3) Measure the horizontal distance from the bench mark to the unknown point, B.

4) Sight the telescope on the target and read the vertical angle, VA.

5) The difference in elevation between the benchmark and the unknown point, B, is equal to distance times the tangent of the vertical angle, VA. Measurements and calculations are noted on the figure.

Figure 2-20 shows a simple setup for hand leveling. A hand level is used for rough leveling. A typical method for recording a hand leveling procedure is shown in Figure 2-21 The terms used in these notes are defined as follows:

- *B.M. #1* is bench mark #1, a concrete monument, used as the datum elevation.
- *+S* is a backsight reading of the rod when set over B.M. #1.
- *H.I.* is the height of the instrument, obtained by adding the back-sight reading, +S, to the elevation of B.M. #1.
- *T.P. 1* is turning point #1, which is a newly-created point replacing B.M. #1.

The same procedure is repeated until T.P. 13 is located. The final -S reading is to the original bench mark, B.M. #1. This provides a quick check of how accurate the survey is. The error is 0.3 feet, (690.7 - 690.40) which is within the allowable margin of error of 0.4 feet.

These notes are similar to those used in more precise leveling done with a level mounted on a tripod. The procedure and note-keeping are the same. The only difference is that elevations are given in hundredths of a foot.

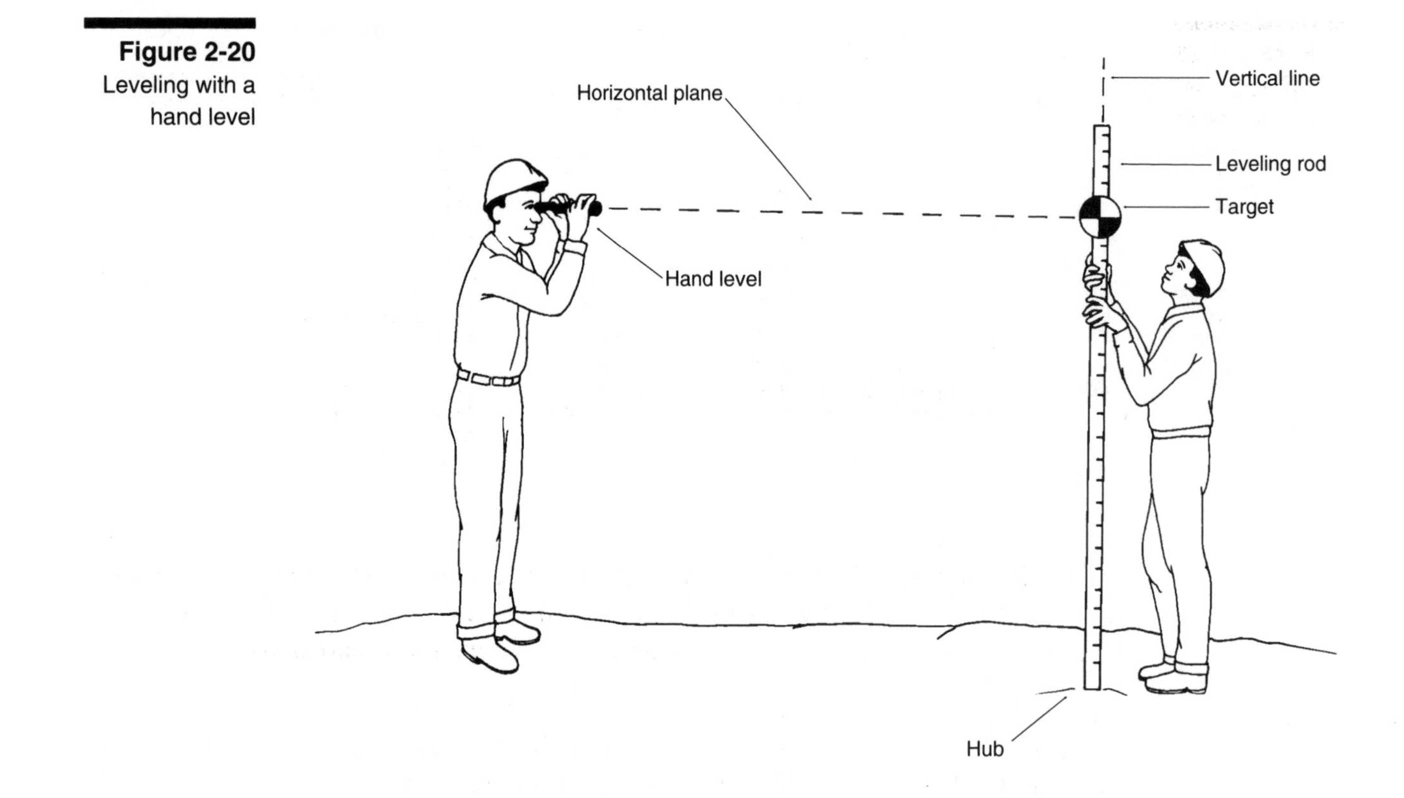

Figure 2-20 Leveling with a hand level

Figure 2-21 Field notes on leveling with a hand level

Point	+S	H.I.	-S	Elevation	Notes
B.M. #1	8.5	698.9		690.4	conc monument
T.P. 1	6.6	697.5	8.0	690.9	
T.P. 2	6.0	697.3	6.2	691.3	
T.P. 3	5.2	698.5	4.0	693.3	
T.P. 4	4.8	700.3	3.0	695.5	fence
T.P. 5	4.4	702.9	1.8	698.5	
T.P. 6	4.5	705.7	1.7	701.2	
T.P. 7	6.0	709.8	1.9	703.8	
T.P. 8	3.7	710.9	2.6	707.2	ridge
T.P. 9	3.9	710.7	4.1	706.8	
T.P. 10	5.0	709.9	5.8	704.9	
T.P. 11	4.2	707.2	6.9	703.0	wall
T.P. 12	3.3	701.9	8.6	698.6	
T.P. 13	3.0	694.5	10.4	691.5	
B.M. #1			3.8	690.7	conc monument

Error of closure = 690.7 - 690.4 = 0.3'
Allowable error = $.01 \times \sqrt{N} = 0.4'$
N = number of points

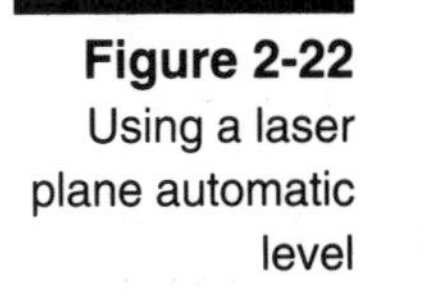

Figure 2-22 Using a laser plane automatic level

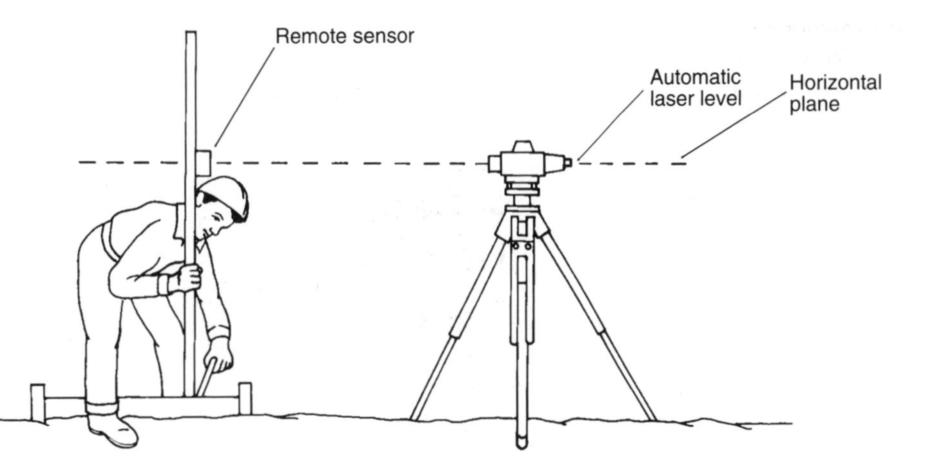

Modern high-production leveling is done with a laser plane automatic level, as shown in Figure 2-22. One person equipped with an automatic level, transmitter, receiver, and rod, can set grades very quickly for foundations and slabs.

The automatic level has an accuracy of ¼-inch error in 100 feet. Once you set it up, it automatically maintains its steady position by an internal magnetically-dampened compensator. You shouldn't have to adjust this instrument often.

Pipeline Staking

Grade sheets, also called cut sheets, are prepared to locate the alignment and depth of a sloping underground pipeline. The typical procedure is as follows:

1) With a transit, lay out a line parallel to the intended centerline of the pipe, but offset a few feet from the line. The offset should be enough to provide space for stockpiling the excavated soil and room for the trenching equipment to maneuver.
2) Drive wood hubs along this transit line flush with the ground at convenient intervals, usually 10 to 20 feet apart.
3) Take level readings to the top of the hubs to the nearest hundredth of a foot and record this data on the grade sheet.
4) Calculate the vertical distance between the top of the hub and the pipeline adjacent to the hub. Record this distance on the grade sheet.
5) Set reference stakes next to each hub. Mark the stake with the vertical distance from the top of the hub to the bottom of the pipe or trench, or use a cut sheet as shown in Figure 2-23.

The information shown on the cut sheet in Figure 2-23 is:

1) Stakes are set 6 feet to the right of the centerline of a 12-inch-diameter pipeline.
2) Stakes are set 25 feet apart (See the *Sta.* column in Figure 2-23).

Figure 2-23 Typical cut sheet

Main Sewer
Stakes 6 ft right

Sheet 1 of 1 sheets
Notes Book ____________ Page ________

Sta.	Size	Grade	Invert	Stake	Feet	1/100 ft	Notes
0+00	12"	0.0025	105.50	110.75	5	25	M/H
0+25	12"	0.0025	105.56	112.30	6	74	
0+50	12"	0.0025	105.63	109.70	4	07	stub
0+75	12"	0.0025	105.69	110.35	4	66	
1+00	12"	0.0025	105.75	111.99	6	24	

3) The slope of the pipe is 0.0025, or 0.25 feet per 100 feet. (See the *Grade* column in Figure 2-23.)

4) The calculated elevation of the invert, or bottom, of the pipe is given at 25-foot intervals. (See the *Invert* column in Figure 2-23.)

5) The elevation of each stake is noted at 25-foot intervals. (See the *Stake* column in Figure 2-23.)

6) The vertical distance between the stake and the invert of the pipe is given in feet and hundredths of a foot. (See the *Feet* and *1/100 ft* columns in Figure 2-23.)

7) The beginning elevation is the invert at the manhole (M/H).

Figure 2-24 shows another way to install sewer pipe using grade boards that straddle the pipe trench. Figure 2-25 is the field book page showing the computations for this technique.

1) Set a controlling string line 8 feet above the invert and centerline of the pipe (shown in the *Invert Elevation* column). Then set a transit line parallel and offset to the centerline of the sewer.

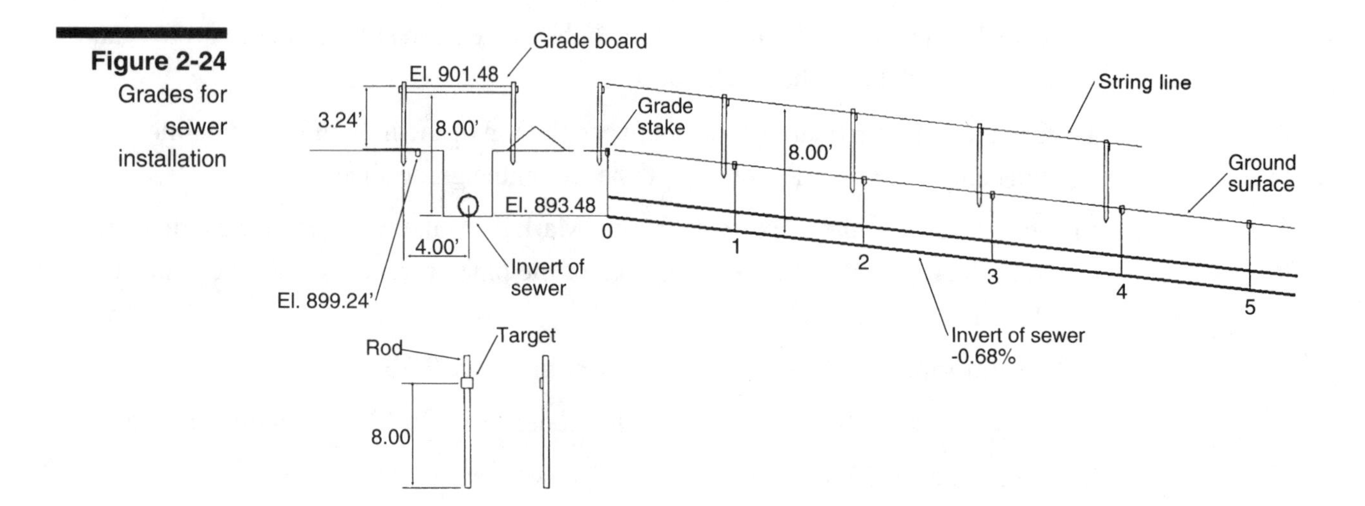

Figure 2-24 Grades for sewer installation

Figure 2-25 Computations for sewer grades

Station	Invert elevation	Invert elevation plus 8 feet	Elevation grade stake	Stake to grade board
0	893.48	901.48	898.24	3.24 ft
1	892.80	900.80	897.42	3.38 ft
2	892.12	900.12	897.04	3.08 ft
3	891.44	899.44	896.73	2.71 ft
4	890.76	898.76	896.25	2.51 ft
5	890.08	898.08	895.49	2.59 ft

Note: All grade boards are 8.00 feet above sewer invert.

2) Drive hubs at intervals along the transit line and take profile levels to the top of these hubs. Look in the *Elev. Grade Stake* column in Figure 2-25.

3) Then add the vertical height of 8 feet to the invert elevation (*Invert Elev. Plus 8 Feet*).

4) The *Stake to Grade Board* column shows the difference in elevation between the grade boards and hubs. You can add another column to convert these differences from decimal parts of a foot to feet, inches, and fractions of an inch.

5) After the trench is excavated, drive pairs of new stakes, 4 to 5 feet long, into the ground. Set them on opposite sides of the trench at the same stations as the hubs on the transit line. Then clamp straight boards to the stakes. Level them with a carpenter's spirit level, then measure from the top of the hub to the top of the board.

6) Finally, stretch a string across the top of the boards over the centerline of the pipe.

7) Then it's a simple job to measure down from the string to any section of the pipe. Use a narrow board with a metal shoe at the bottom and a small target set 8 feet above the bottom. When the target is on the string line, the metal shoe is at the correct invert elevation of the pipe.

Contour Lines

Contour lines show the elevation of earth surfaces. All points on a contour line have the same elevation. Existing and finish contour lines are usually shown with different symbols on a grading plan. Existing elevations are normally shown in parentheses (322'), while finish elevations are noted without parentheses. Existing contour lines are generally shown as dashed lines, while finish contour lines are shown as solid lines.

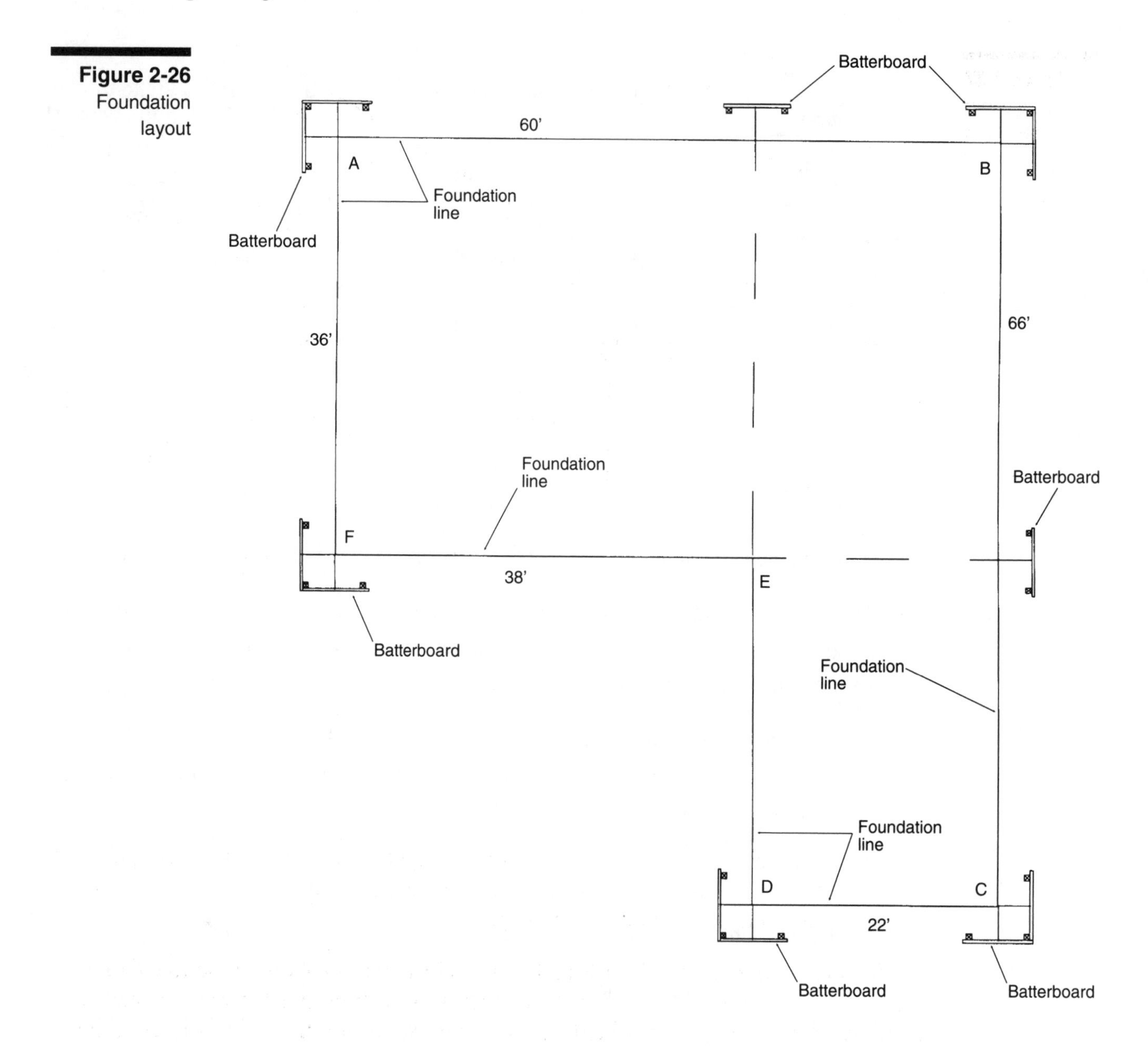

Figure 2-26 Foundation layout

Laying Out Buildings

Batterboards

There are several good ways to lay out the location of a foundation for a building. One simple method is to stake out a rectangle around the major outer dimensions of the structure. Then stake out the irregularities of the building shape using smaller rectangles. This is shown in Figure 2-26.

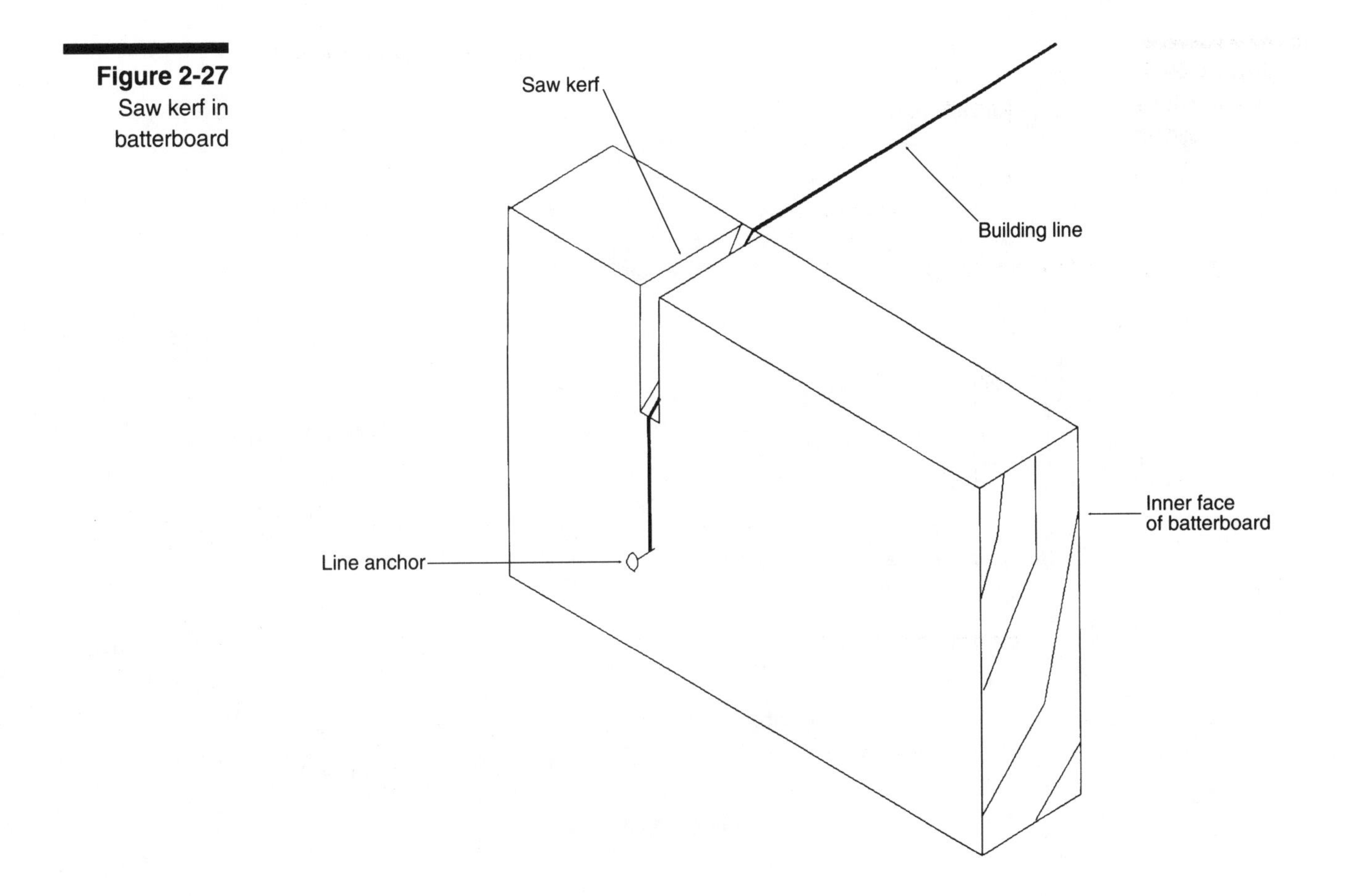

Figure 2-27
Saw kerf in batterboard

Mark the building corners with corner hubs, right-angle batterboards, and posts. Place batterboards 3 to 4 feet outside of a corner stake. Use posts made of 2 x 4s and batterboards made of 2 x 4 or 2 x 6 lumber.

Set the top of the batterboards at the same elevation as the top of the foundation wall. Drive nails into the top of the batterboards. Run string line between the nails to show the perimeter of the foundation. The point where two strings cross is a building corner. Drive a 2 x 2 hardwood hub into the ground where the strings intersect. Hammer a nail into the top of the hub to mark the location of the true corner. One way of attaching cord to a batterboard is shown on Figure 2-27.

Figure 2-28 shows another way to stake out a building. Here's how to do it:

1) Locate or set property corner stakes. These corner stakes should be set by a licensed surveyor.

2) Run a line across the front of the property and along the two sides. Measure from these two lines.

3) Drive stakes at the approximate location of building corners.

4) Measure from the front property stake along a side property stake a distance equal to the front setback. Drive a stake at this point.

Figure 2-28 Staking out a building

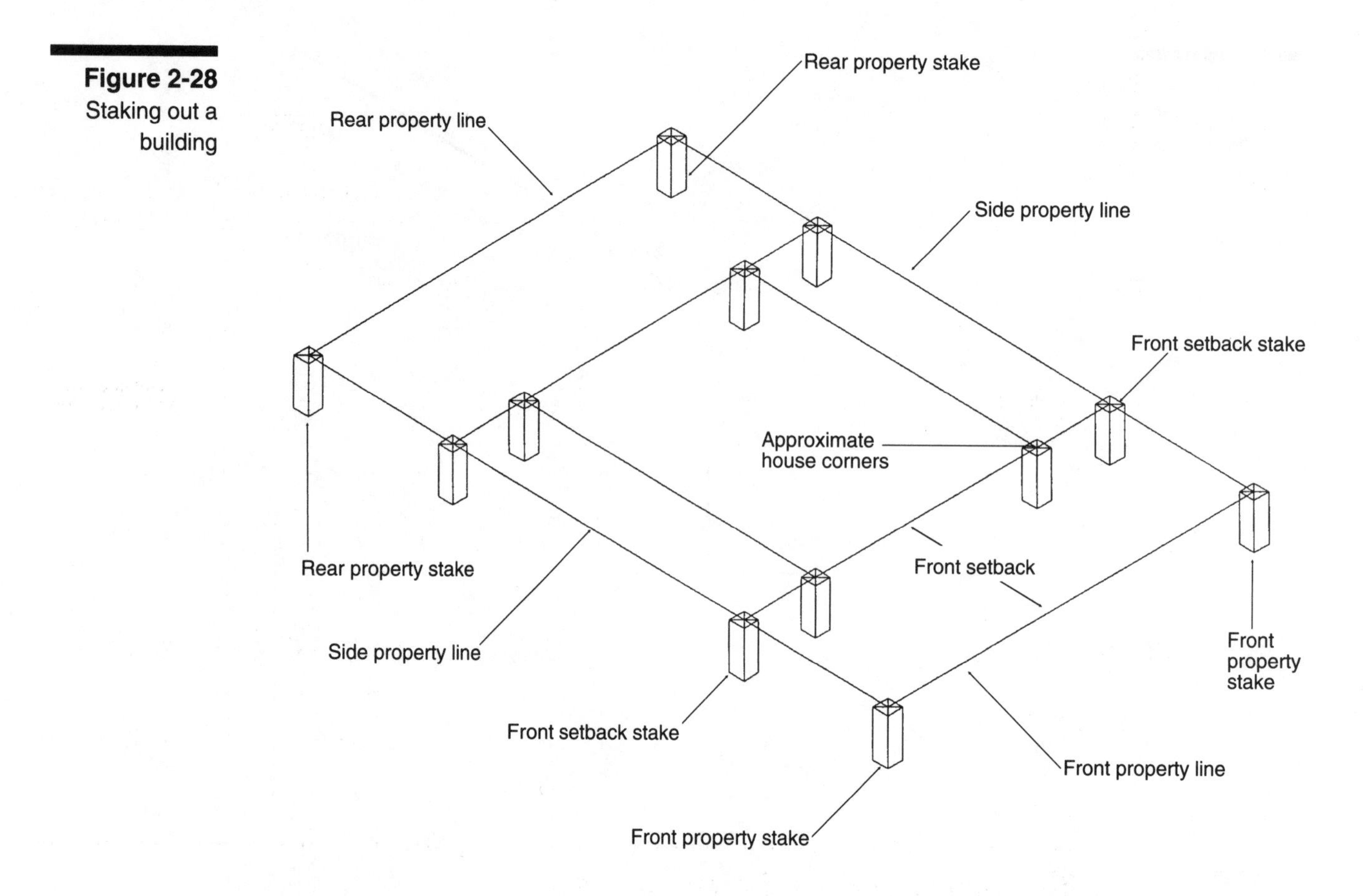

5) Measure from this stake along the side property line a distance equal to the depth of the building from the front to the back. Drive another stake.
6) Do this again along the other side property line.
7) Measure in from the front and rear setback stakes on the side property line a distance equal to the side yard. Drive stakes at these two points. This sets up two building corners.
8) Do this again along the other side property line. This sets up the other two building corners.
9) Remove all building lines.
10) Install batterboards around each corner stake using three stakes and two horizontal boards set at right angles. Set the top of the batterboards at the same level as the top of the foundation, if possible. You can set the boards above or below the top of foundation if it's more convenient. Set the batterboards far enough from the building corners so they won't be in the way of foundation excavation.
11) Mark the exact corner locations on the batterboards. You can check the accuracy of your batterboard layout by measuring triangular distances, as shown in Figure 2-29. If the layout is a true rectangle, the diagonal measurements will be identical.

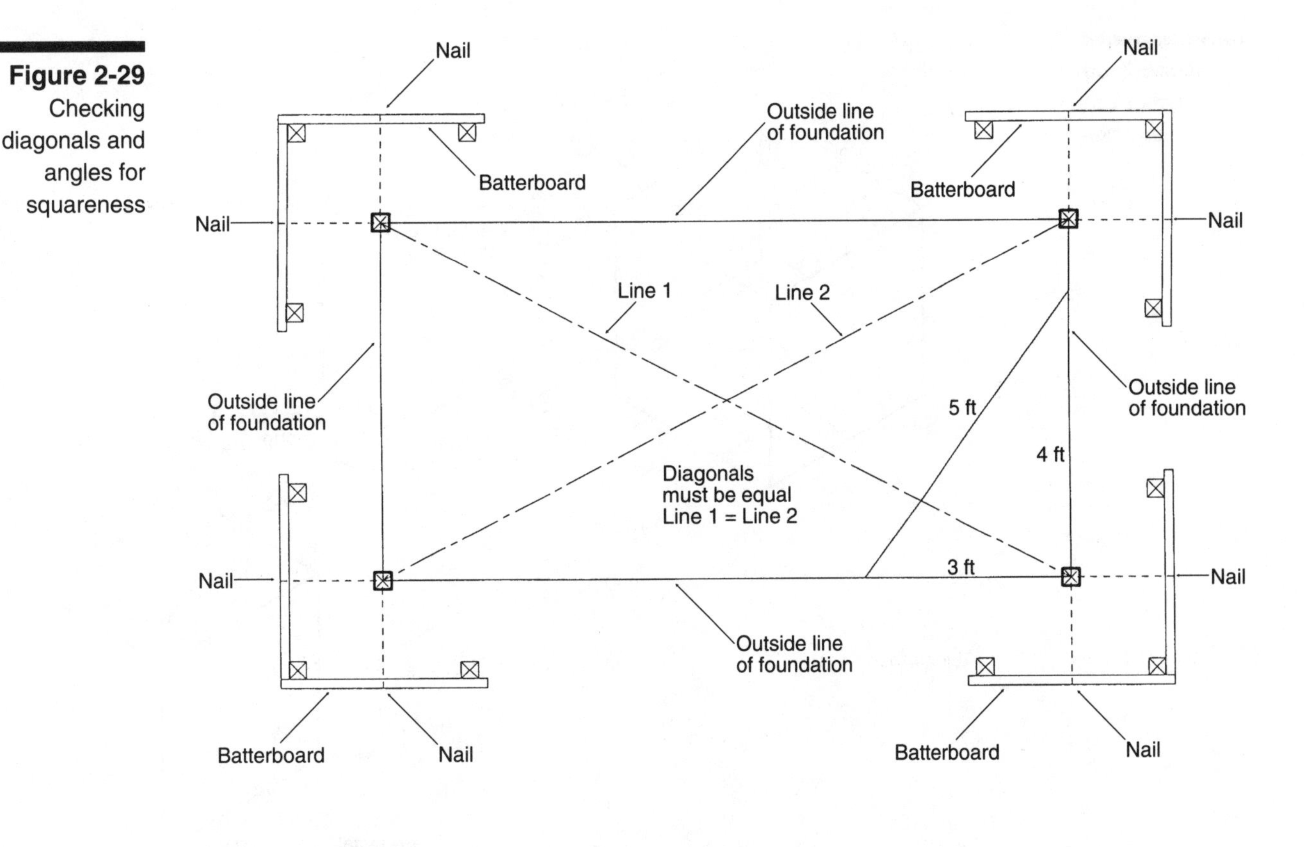

Figure 2-29 Checking diagonals and angles for squareness

Construction Tolerances

How accurate should your layout be? Is a 1-inch error in 200 feet too much? That's a very important question. Of course, distances can be calculated mathematically to a much greater precision than they can physically be measured in the field. Remember too that each building trade has its own tolerance for dimension error. Subcontractors may demand extra pay when errors in field measurements are more than their usual tolerance.

Surveyors generally use three levels of accuracy. Only the second applies in construction work. First order survey accuracy is for geodetic-type governmental surveys. Second order survey accuracy covers land subdivision surveys. It's used for property boundary surveys and allows a tolerance of one part in 10,000, or 1:10,000. A tract boundary that's 1000 feet in length must be accurate to within a tenth of a foot. A distance of 100 feet must be accurate to within a hundredth of a foot. The third order may be too approximate to use in construction.

Standard structural steel fabrication and erection require a relatively close accuracy to place anchor bolts. For example, the tolerance between any two anchor bolts in a group of bolts is ⅛ inch. The tolerance between the center of adjacent groups of anchor bolts is ¼ inch. The tolerance in a 100-foot column line is ¼ inch and the tolerance between the center of an anchor bolt group from the column line is ¼ inch. Figure 2-30 shows the allowable errors in setting anchor bolts for steel construction.

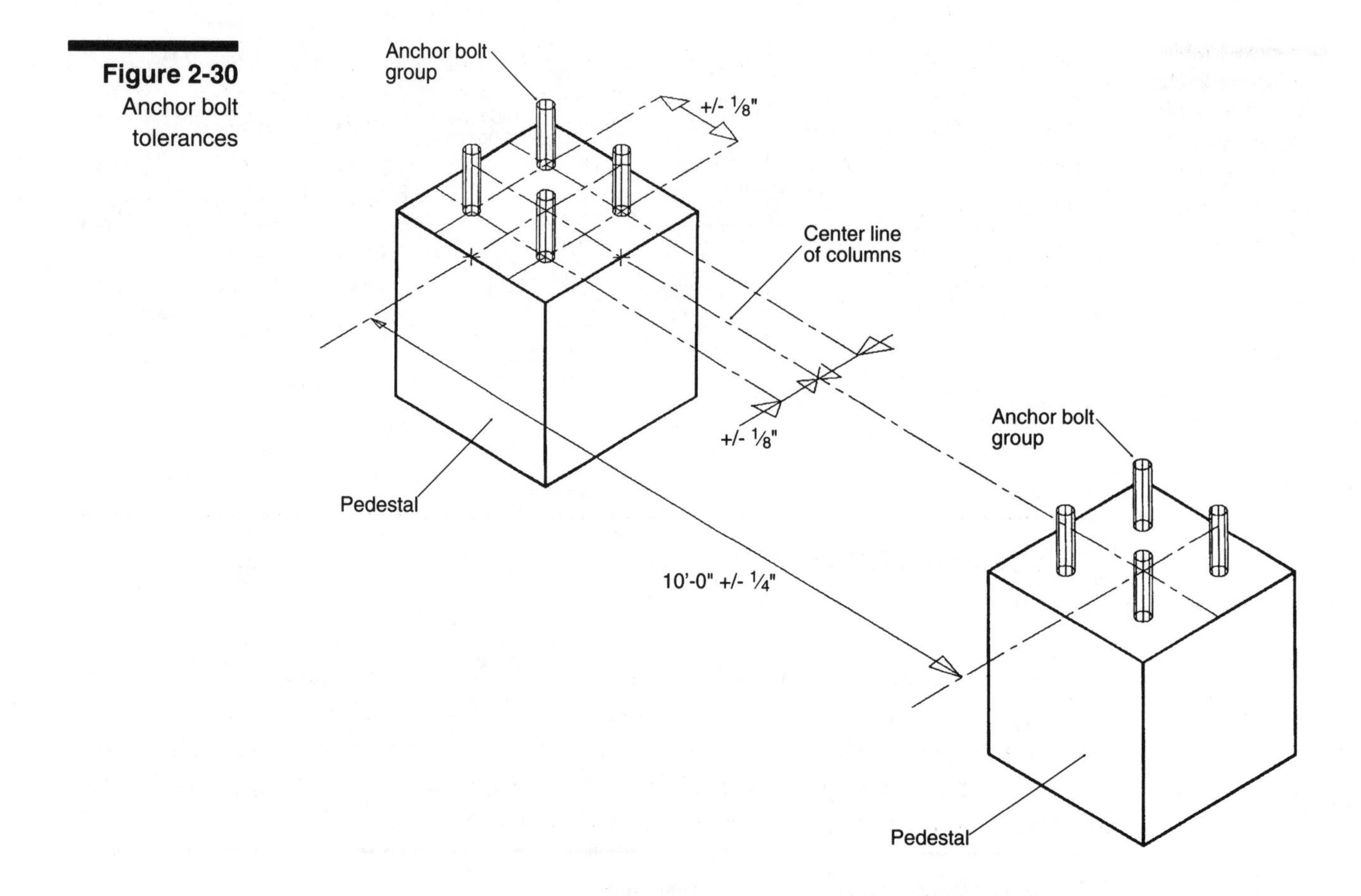

Figure 2-30
Anchor bolt tolerances

Some masons consider 1 inch as "close enough" for the foundation of a masonry wall. Manufactured concrete unit masonry allows a tolerance of 3⁄8 inch in the thickness of the block. Therefore, if a building line is the outside surface of a block wall, the interior surface may vary by 3⁄8 inch. This would affect dimensions taken to the interior face of a concrete block wall.

Grading work has its own standards. Rough grading is done to one-tenth foot accuracy, while elevations for pavements are set at one-hundredth foot accuracy.

Concrete slabs on grade are considered straight if there is less than 1⁄8-inch variation in a 10-foot distance. In addition, three consecutive points should be set on the same slope so that any variation from a straight grade can be detected. Reinforced concrete beams and elevated slabs are usually constructed with an upward camber. This is to allow for future deflection from dead and live loads. Therefore, these structural members may have an upward curve when there is no live load. Keep that in mind when taking measurements from the top surface of a cambered beam or slab.

Steel fabrications are usually made to a 1⁄16-inch tolerance at the factory. Masonry is laid with a much greater tolerance for error. That can create problems when steel beams are supported by masonry walls. You'll need some way to make field adjustments. Use slotted holes in the masonry or weld shims to the steel to make the adjustments you need. If the plans call for steel to be embedded in masonry walls without room for any adjustment, warn the designer that there may be a problem.

Figure 2-31 Conversion of fractions of an inch to decimals of an inch and a foot

Fraction of an inch	Decimal of an inch	Decimal of a foot	Fraction of an inch	Decimal of an inch	Decimal of a foot	Fraction of an inch	Decimal of an inch	Decimal of a foot
1/16	.0625	.0052	**3/8**	.3750	.0313	**11/16**	.6875	.0573
1/8	.1250	.0104	**7/16**	.4375	.0365	**3/4**	.7500	.0625
3/16	.1875	.0156	**1/2**	.5000	.0417	**13/16**	.8125	.0677
1/4	.2500	.0208	**9/16**	.5625	.0469	**7/8**	.8725	.0729
5/16	.3125	.0260	**5/8**	.6250	.0521	**15/16**	.9375	.0781

Units of Measurement

Some construction trades and professions use different measuring units. For example, architects use feet and inches to the nearest sixteenth of an inch. Civil engineers use feet and decimal parts of a foot to the nearest hundredth of a foot. Road and sewer designers use stations that represent 100 feet subdivided into hundredths of a foot. Figure 2-31 lists conversions of fractions of an inch to decimals of an inch or foot.

Slopes are also described differently. Carpenters and plumbers use inches and fractions of an inch of rise per foot of horizontal run. For example, a roof might have a slope of 3 inches per foot (called 3 in 12). Sewer contractors and civil engineers use decimal parts of a foot or percent to indicate slope (such as slope = 0.002, or 0.2 percent). Steel detailers measure in feet and inches to the nearest $\frac{1}{32}$ of an inch. Slopes of lines in structural steel details are given as the rise in a 1-foot run. A rise is measured in inches to the nearest $\frac{1}{32}$ of an inch. Slopes are described by rise and run using *Smoley's Handbook* rather than conventional trigonometry. Figure 2-32 shows conversion of slopes to percent slopes.

Figure 2-32 Conversion of slopes

Inch rise per foot	Slope	Percent slope	Inch rise per foot	Slope	Percent slope
1/16	.0052	0.52	**9/16**	.0469	4.69
1/8	.0104	1.04	**5/8**	.0521	5.21
3/16	.0156	1.56	**11/16**	.0573	5.73
1/4	.0208	2.08	**3/4**	.0625	6.25
5/16	.0260	2.60	**13/16**	.0677	6.77
3/8	.0313	3.13	**7/8**	.0729	7.29
7/16	.0365	3.65	**15/16**	.0781	7.81
1/2	.0417	4.17	**1**	.0833	8.33

Angular measurements may be in degrees-minutes-seconds, bearings, or azimuths. A line may be described as 15°30'15" to the right of some other line. A line also may be described as S 32°23'10" W. The same line may be described as having an azimuth of 212°23'10". Some angular measurements are given in degrees and decimal parts of a degree.

The most common unit of measurement in U.S. construction surveying is feet and decimal parts of a foot. The conventional unit for angular measurements is degrees, minutes, and seconds.

All the units of measurement you're likely to encounter are explained in the glossary in the back of the book.

Sample Checklist for Surveying

Figure 2-33 is a checklist you can use for all of your surveying projects. I've filled it in so you can see how it's used. There's a blank copy at the end of the chapter.

The Language of Land Description

Here's a brief background on the language of land description. Originally all government land was divided into townships and sections, as shown in Figure 2-34. This was called the *United States Rectangular System*. The origin of each system was based on a specific baseline and meridian within each state. Each principal meridian is a true north-south line and the baseline is a true east-west line.

Ranges consist of 6-mile-wide columns running parallel to the principal meridian. Townships consist of 6-mile-wide rows running parallel to the baseline. Each 6-mile square is identified by its township and range, such as Township 1 North and Range 1 East, written T1N, R1E.

Each township is further divided into 36 mile-square sections, as shown in Figure 2-35. The sections are consecutively numbered from the northeast corner to the southeast corner. The full identification of a section is by its number, township, range, principal meridian, and baseline. As an example, a section may be called: Section 36 of Township 1 North, Range 2 West of the San Bernardino Baseline and Meridian. This may be abbreviated as Sec 36, T1N, R2W, SBBM.

Sections may be further divided into half sections or quarter sections and so forth. These may be described as west-half or east-half, northeast quarter or southwest quarter of section 36. The divisions can continue to very small parcels.

When parcels are not a fraction of a section, they are usually described by metes and bounds. Figure 2-36 shows the outline of a parcel of land that's described by distances and bearings.

Figure 2-33
Checklist for all of your surveying projects

Surveying Checklist

Property Survey

- [ ] Lot number: 65
- [ ] Block number: None
- [ ] Tract number or name: 2804
- [ ] Corners: 1" Iron pipe
- [ ] Bearings: N 76° 00' E & N 14° 00' W
- [ ] Basis of bearings: E line of SE 1/4 Sec 36, T 11 N, R 8 W, SBM
- [ ] Distances between points: 170' and 74'
- [ ] Easements: 5' Public Utilities Easement east side of lot
- [ ] Utilities: gas, elec., sanitary, water
- [ ] Gas: 3" pipe 10' East of center of street
- [ ] Electric: Easement - power poles
- [ ] Sanitary sewer: 8" V.C.P. at center of street
- [ ] Water main: 8" A.C.P. at 20' of center of street

Construction Survey

- [ ] Building setback: 25'
- [ ] Side yard: 5'
- [ ] Rear yard: 25'
- [ ] Topographic survey: yes
- [ ] Orientation of buildings: NS/EW
- [ ] Exterior dimensions of buildings: 45' x 60'
- [ ] Bench mark or reference of elevations: U.S.G.S. Mon NE Corner Sect. 36
- [ ] Elevation of property corners: 2516, 2521, 2532, 2518
- [ ] Curb and gutter elevations: See plans
- [ ] Finish elevations of paved areas: See plans
- [ ] Finish floor elevations: 2533
- [ ] Centerline and width of roads: 60'
- [ ] Coordinate system: None
- [ ] Utilities: See above
- [ ] Flow line elevation of sewer: 2508
- [ ] Flow line of storm drain: None
- [ ] Electric service: P.P. in easement
- [ ] Water service: See above
- [ ] Gas service: See above

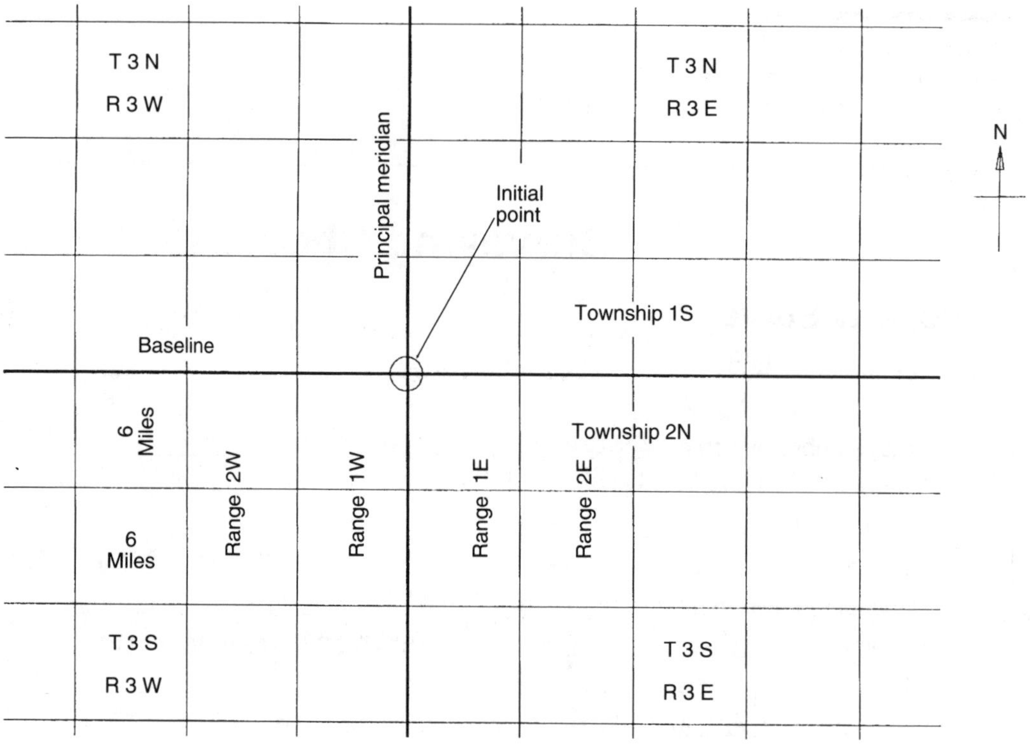

T 3 N
R 3 W
T 3 N
R 3 E
Principal meridian
Initial point
Township 1S
Baseline
Township 2N
6 Miles
Range 2W
Range 1W
Range 1E
Range 2E
6 Miles
T 3 S
R 3 W
T 3 S
R 3 E
N

Figure 2-34
U.S. Rectangular System

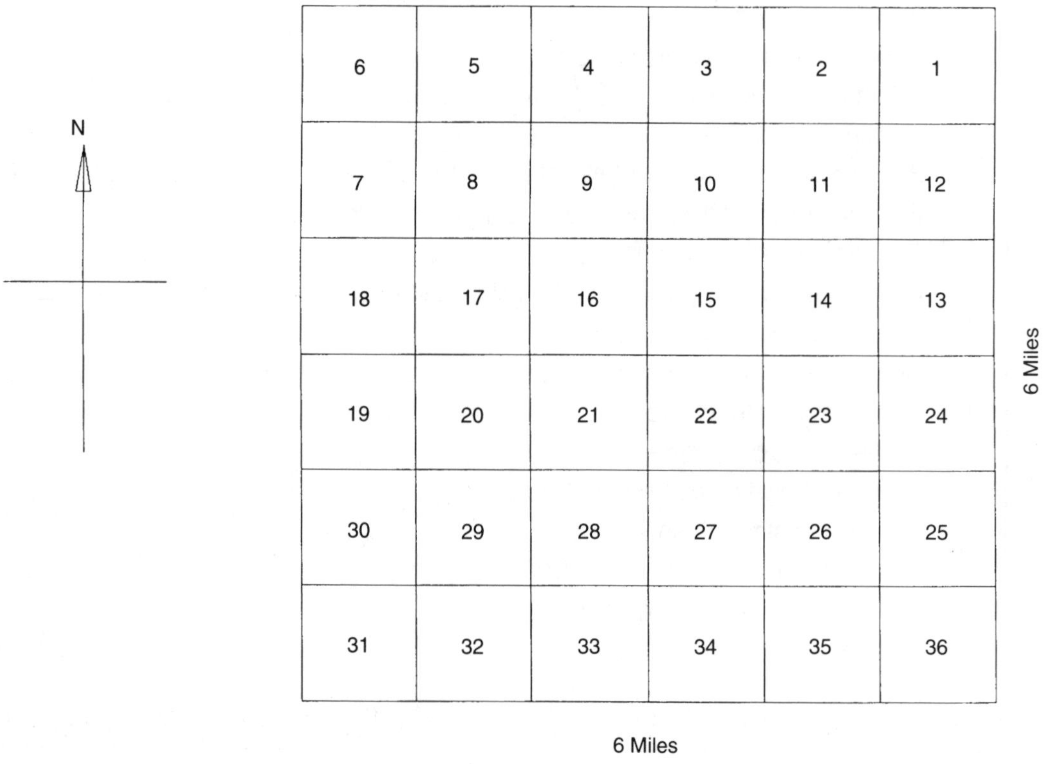

N

6	5	4	3	2	1
7	8	9	10	11	12
18	17	16	15	14	13
19	20	21	22	23	24
30	29	28	27	26	25
31	32	33	34	35	36

6 Miles

6 Miles

Figure 2-35
Sections

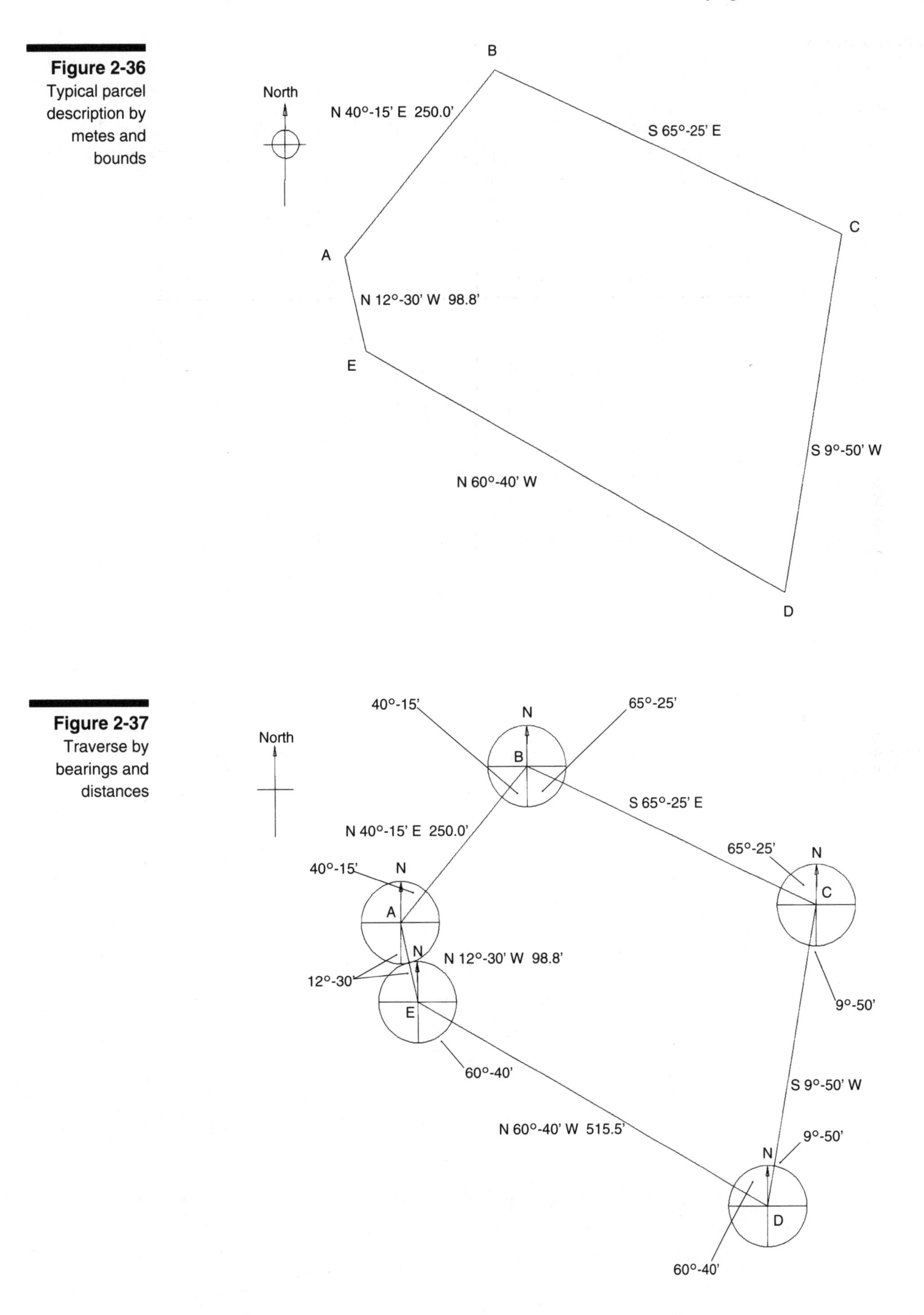

Figure 2-36 Typical parcel description by metes and bounds

Figure 2-37 Traverse by bearings and distances

Figure 2-37 shows how the bearing of each course was developed. The direction of a course is defined by the angle between that line and the north or south line.

The traverse shown in this figure was described in a clockwise direction. If the description were in a counterclockwise direction, the bearing would be reversed. The first course N40-15E would be read as S40-15W.

Surveying Checklist

Property Survey

- ☐ Lot number: ____________
- ☐ Block number: ____________
- ☐ Tract number or name: ____________
- ☐ Corners: ____________
- ☐ Bearings: ____________
- ☐ Basis of bearings: ____________
- ☐ Distances between points: ____________
- ☐ Easements: ____________
- ☐ Utilities: ____________
- ☐ Gas: ____________
- ☐ Electric: ____________
- ☐ Sanitary sewer: ____________
- ☐ Water main: ____________

Construction Survey

- ☐ Building setback: ____________
- ☐ Side yard: ____________
- ☐ Rear yard: ____________
- ☐ Topographic survey: ____________
- ☐ Orientation of buildings: ____________
- ☐ Exterior dimensions of buildings: ____________
- ☐ Bench mark or reference of elevations: ____________
- ☐ Elevation of property corners: ____________
- ☐ Curb and gutter elevations: ____________
- ☐ Finish elevations of paved areas: ____________
- ☐ Finish floor elevations: ____________
- ☐ Centerline and width of roads: ____________
- ☐ Coordinate system: ____________
- ☐ Utilities: ____________
- ☐ Flow line elevation of sewer: ____________
- ☐ Flow line of storm drain: ____________
- ☐ Electric service: ____________
- ☐ Water service: ____________
- ☐ Gas service: ____________

CHAPTER 3

Concrete

Concrete is one of the most common construction materials. It's permanent, it resists decay, moisture, and corrosion, and it's easy to install. But it's also the most unforgiving material. If you make a mistake with concrete, corrections can be very expensive. For example, the wrong mixture of concrete can develop holes. If you don't put the reinforcing bars in the right place, the concrete may spall or crack. If the formwork isn't strong enough, the concrete may be misshapen. And you're sure to have delays and back charges if you set the anchor bolts in the wrong position.

Concrete work requires precision. If one part of a foundation is wrong, the whole building may be affected. With wood or steel framing, you can remove and replace anything that isn't right. With concrete you'll probably have to completely demolish what's wrong and start over. Once concrete has cured or set up, it's hard to correct mistakes. This chapter will help you avoid the more common mistakes and get the job done right — the first time.

Some parts of a building should be made with stronger concrete than others. Columns and beams use higher strength concrete than pavements and foundations. Stronger concrete costs more because it uses more cement, and cement costs more than sand and crushed stone.

For many types of work, the strength of the concrete is very important and has to be measured accurately. To test its strength, pour a sample of the concrete into a cylinder 6 inches in diameter and 12 inches long. After a set curing time, test the sample in a machine that applies weight to the sample until it breaks. Concrete test cylinders should have a minimum three-day curing period. Usual practice is to provide enough concrete cylinders so that the concrete can be tested after three, seven, and 28 days of curing. Record the force (in pounds) it took to break the sample. Divide the force by the cross-sectional area of the cylinder (in square inches) to find the strength of the concrete in pounds per square inch — such as 2000 psi or 3000 psi. Engineers often use the term kip in place of 1000 pounds to simplify calculations by reducing the number of zeros.

Concrete Materials

Reinforced concrete work has three major parts:

- concrete mix — cement, sand, gravel, water, and admixtures
- reinforcement — usually deformed steel bars and welded wire mesh (also called electric welded wire mesh, EWWM, or wire fabric)
- formwork and shoring — sheathing, framing, bracing, and supports

Concrete is a solid mixture of cement, sand, and gravel. The gravel, or crushed rock, and sand mixture is called an aggregate. Cement, also known as portland cement, holds the aggregate together. Figure 3-1 shows how the coarse aggregate should be uniformly dispersed.

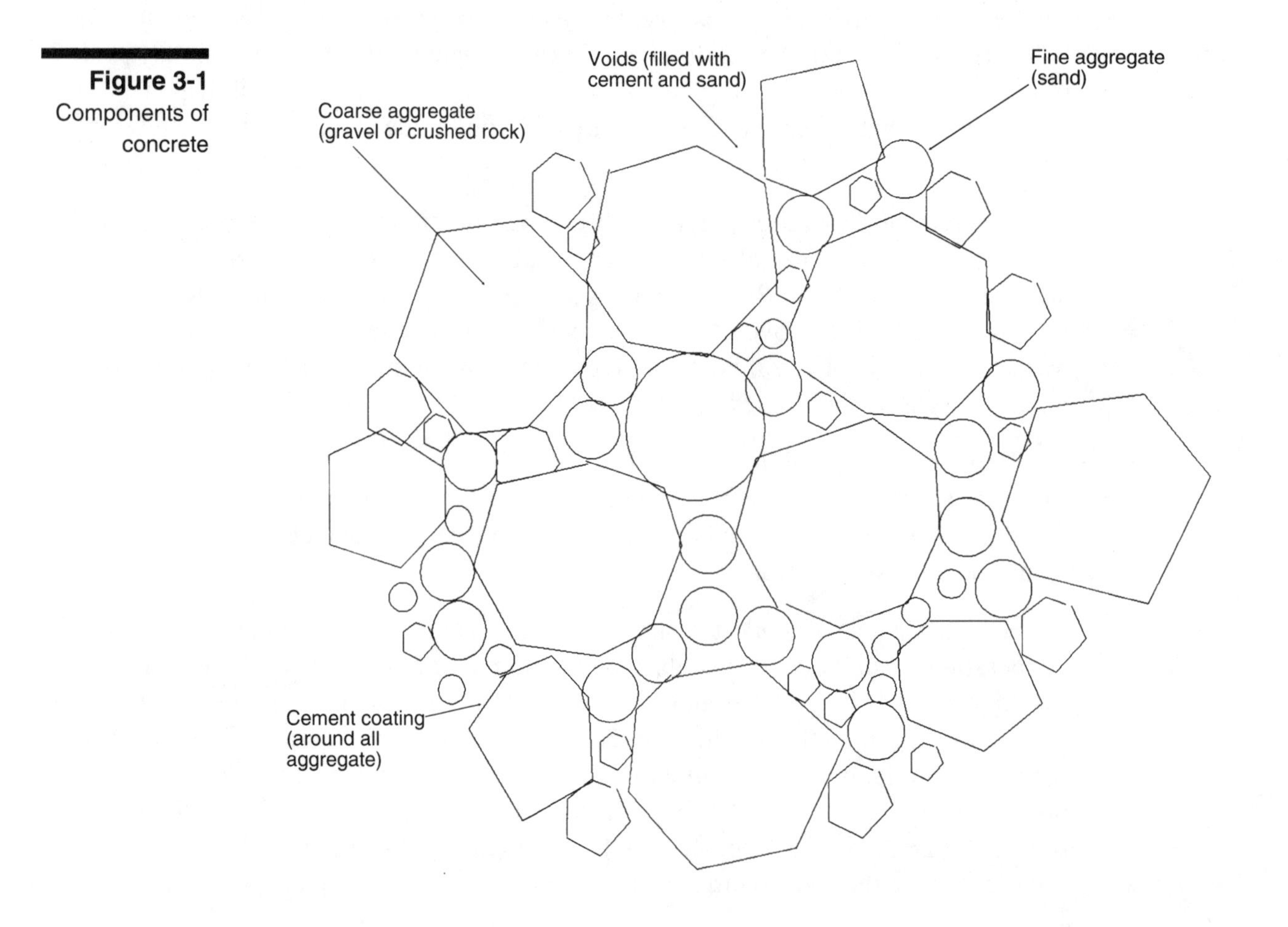

Figure 3-1 Components of concrete

Cement

Cement is made from limestone, which is mainly calcium carbonate. Limestone was formed millions of years ago when the shells of microscopic sea animals were compressed together in layers of sediment. Limestone was also formed by the precipitation of calcium carbonate from ground water. Limestone is mined and processed by crushing and roasting, or calcining, it with clay in rotating kilns. This leaves a dehydrated porous material called clinkers. The clinkers are ground into a fine powder called cement, which is used to make concrete.

Portland cement was patented by Joseph Aspdin, an English stone mason, in 1824. He named it after the gray stone found on the Isle of Portland. There are five common types of portland cement:

- Type I-Normal
- Type II-Modified
- Type III-High Early
- Type IV-Low Heat
- Type V-Sulfate Resistant

Use Type I-Normal portland cement for general construction where you don't need the special properties of the other types. Use it in pavements and buildings that aren't exposed to acids or salts. This type of concrete reaches nearly its full strength in 28 days. Figure 3-2 shows the relationship between curing time and strength in Type I-Normal cement.

If the structure you're building will be exposed to acids or salts, use Type II-Modified cement. You'd use this on a beach-front building where concrete must resist salty mist. This cement cures more slowly than Type I-Normal. In 28 days it's about 85 percent of the strength of Type I-Normal cement. It may take three months to reach its full strength.

If you need concrete that cures quickly, use Type III-High Early cement. You'll need this when traffic or access to a building must wait until the concrete is cured. Type III-High Early cement is 190 percent stronger than Type I-Normal cement in three days, 130 percent stronger in 28 days, and 115 percent stronger in three months.

Most building contractors don't need Type IV-Low Heat cement. It's used mainly for massive concrete pours, such as a dam, where the concrete will generate a lot of heat as it cures.

Use Type V-High Sulfate Resistant cement if the structure will be directly exposed to sea water or other severe conditions. This cement cures slowly, reaching only 65 percent of the strength of Type I-Normal cement in 28 days, and 85 percent in three months. Figure 3-3 lists the relative strength of different types of cement as a percentage of the strength Type I-Normal cement would have in a particular number of days. For example, concrete made with Type I cement may have a compressive strength of 2000 psi in 28 days, 1340 psi (67 percent of 2000) in seven days, and 920 psi (46 percent of 2000) in three days.

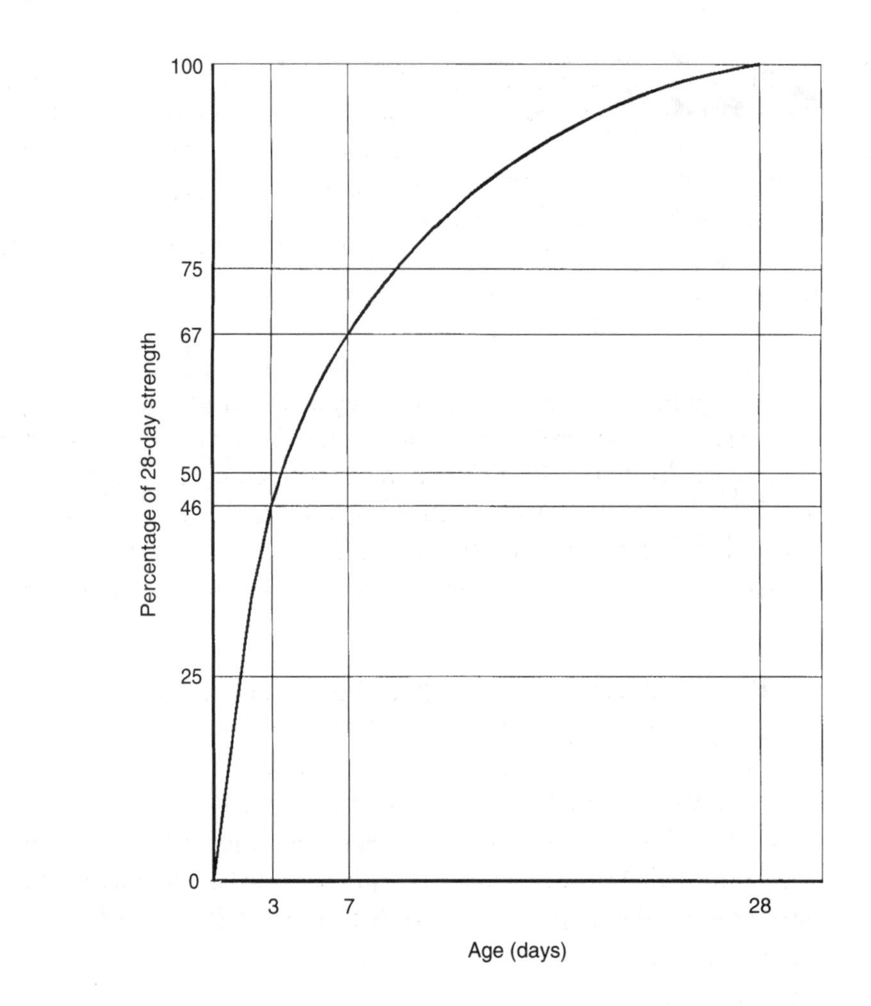

Figure 3-2 Normal concrete strength vs. time

Figure 3-3 Relative curing rates of portland cement

Type	3 days	28 days	3 months
I	100%	100%	100%
II	80%	85%	100%
III	190%	130%	115%
IV	50%	65%	90%
V	65%	65%	85%

Concrete made with the same amount of Type II cement would have a compressive strength of 1700 psi (85 percent of 2000) in 28 days and 736 psi (80 percent × 920) in three days. If concrete is made with Type II, it'll take three months before it reaches a strength of 2000 psi.

On the other hand, the same concrete mix made with Type III-High Early cement has 190 percent more strength than Type I cement has in three days, or 1748 psi (190 percent of 920). It'll be 130 percent stronger than Type I cement in 28 days (130 percent of 2000), or 2600 psi.

Figure 3-4 Typical concrete mixes

Grade of work	Sacks of cement	Wet sand (lbs)	Gravel (lbs)	Wet sand (cubic ft)	Gravel (cubic ft)
1	5	1245	1935	2¾	4¼
2	6	1180	1915	2¼	3½
3	7	1125	1895	1¾	3

The ratio of portland cement to aggregate in concrete directly affects the strength and cost of the concrete. The more cement in a mix, the stronger and more expensive it is. Normally, the ratio is given in sacks of cement per cubic yard of concrete. A six-sack mix will be stronger than a similar five-sack mix. The ratio of cement to concrete is also given by weight. A sack of cement contains 1 cubic foot and weighs about 94 pounds. A cubic yard of concrete contains 27 cubic feet and weighs about 2 tons.

Use a five-sack mix for 2500 to 3000 psi concrete in foundations, walls, and footings. Use a six-sack mix of 3500 to 4000 psi concrete for driveways, slabs, and columns. Use a seven-sack mix for highly detailed architectural items, such as concrete ledges, sills, and posts where you need 4500 and 5000 psi concrete. The word pound is also used instead of psi to tell the compressive strength of concrete, for example 2000-pound concrete. The amount of cement, sand, and gravel for different classes of concrete work are shown in Figure 3-4.

Water

The key to properly mixed concrete is the water. It changes the loose mixture into a rock-like solid material. When you mix water with cement, it makes a paste. The paste goes around the particles and fills in the empty spaces. Some of the water joins chemically with the cement and starts a chemical reaction. When the chemical conversion is completed, or cured, the mixture becomes rock-hard concrete. Use clean drinkable (potable) water if you can when you mix concrete.

The amount of water in a mix also determines the strength of the concrete. Measure the ratio between water and cement by the number of gallons of water per sack of cement. Water/cement ratios may go from 4 to 1 for high-strength concrete, to 8.5 to 1 for a low-strength mix. The total amount of water in a mix includes the free water in the aggregate and admixture as well as the added water. Strength of concrete and the water/cement ratio relationships by weight are shown in Figure 3-5. Water weighs 62.4 pounds per cubic foot or 8.6 pounds per gallon.

Only a little water is needed to start the chemical reaction in cement. But you have to add more water to be able to work with it. If you don't use enough water, it'll be too stiff to work with. Too much water will make it weak and the cement may

Figure 3-5 Compressive strength vs. water/cement ratio

Compressive strength (psi)	Water/cement ratio by weight
2000	0.80
3000	0.69
4000	0.57
5000	0.47
6000	0.40

wash out of the mixture. If too much water evaporates out of a mixture, you can get air pockets in the concrete. Try to use as little water as possible and still have a workable mixture.

A slump test measures how easy a concrete mix will be to work with. It shows how fluid, and how soft or wet, a batch of concrete is. Figure 3-6 shows how to make a slump test using a 12-inch-high sheet metal cone that's open at both ends. The bottom of the cone is 8 inches in diameter, and the top is 4 inches in diameter.

Start the slump test by moistening the inside of the metal cone. Then place the cone on a flat smooth surface with the large end down and fill it with a concrete sample in three layers. Tamp each layer before you add the next. After the cone is completely full, lift the cone so the wet concrete slumps down. Measure the height of the concrete cone. Subtract this height from the original 12-inch height to give the result of the slump test.

If the number is high, the concrete is a wet mix. If it's low, the concrete is a dry mix. Recommended slumps for different types of concrete work are shown in Figure 3-7. If the concrete isn't thoroughly vibrated (tamped), increase the values in the table by 50 percent. The slump should never be more than 6 inches.

Figure 3-6 Slump test

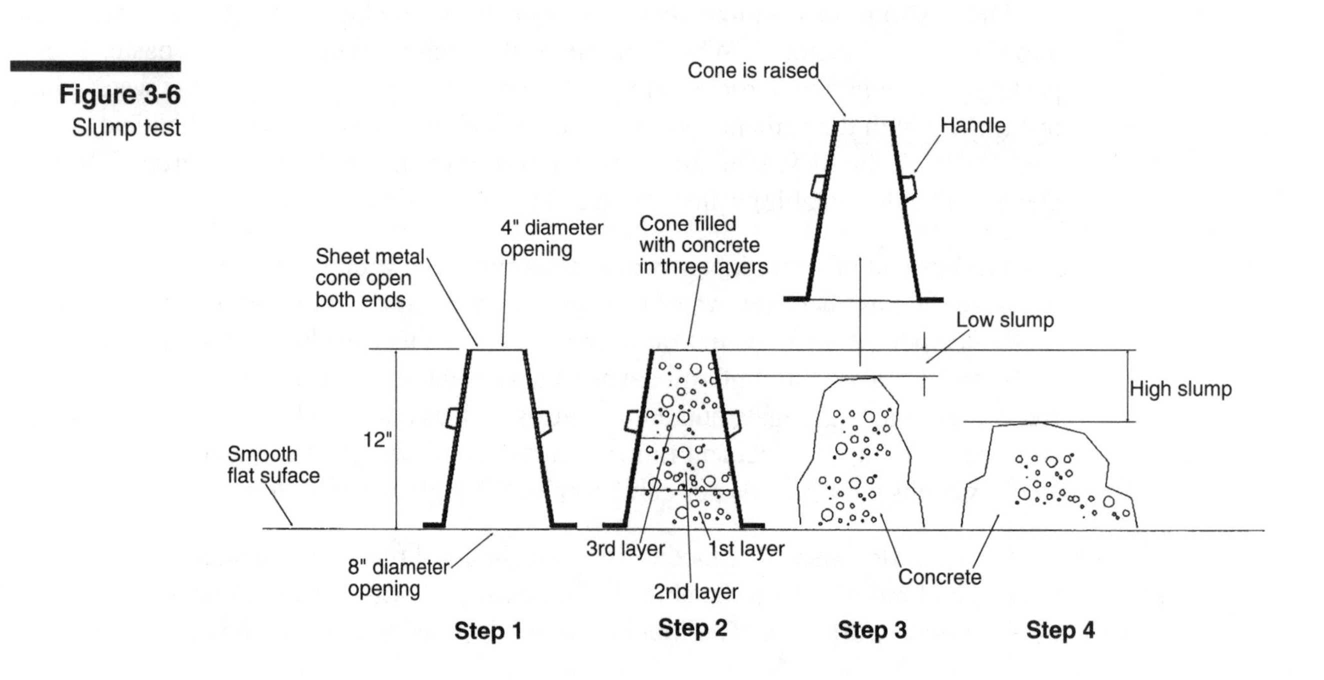

Figure 3-7
Maximum and minimum concrete slumps

Type of concrete work	Maximum slump (inches)	Minimum slump (inches)
Reinforced foundations	4	2
Plain foundations and caissons	3	1
Concrete slabs, beams, walls and columns	5	2
Pavements	2	1
Mass concrete	2	1

Aggregate

The best concrete mixture is the densest one. All the spaces between the coarse aggregate should be filled. Aggregate, gravel, or crushed stone larger than ¼ inch is considered coarse. Fine aggregate, which includes sand, is smaller than ¼ inch across. Coarse aggregate may be classified as No. 1, 2, 3, and 4 gravel, depending on the size of the particles. Sand may be natural, or manufactured by crushing rock. Sands are classified as coarse, medium, and fine.

All aggregate should be clean hard rock with no attached silt or clay. The surface of the particles should be angular and sharp. Don't use round smooth particles, such as beach sand. It doesn't stick together well.

There are two weights of aggregate — regular and lightweight. Regular weight aggregate is made from crushed granitic rock. Lightweight aggregate has air pockets. It's made from volcanic material, or by roasting blast furnace slag, shale, perlite, or slate. Regular concrete weighs from 145 to 150 pounds per cubic foot. Lightweight concrete weighs between 95 and 115 pounds per cubic foot. Lightweight concrete is usually used in high-rise buildings to reduce the weight of the structure.

Admixtures

An admixture is anything in concrete that isn't water, aggregate, or portland cement. Admixtures change the properties of concrete by:

- improving its workability
- speeding up the curing period
- slowing down the curing period
- resisting the effect of cold weather
- reducing the amount of water in the mix
- increasing the air entrainment
- improving hardening, bonding, and sealing the concrete

You can harden the surface of a concrete slab by sprinkling an admixture of finely-ground iron particles on it. One common surface hardener for concrete is called Masterplate. It's an admixture, manufactured by Master Builders, which contains iron powder. You trowel the iron particles into the surface of the concrete to make the slab more resistant to abrasion and wear. Do this just before you float the surface, and again before you do the final troweling. Masterplate is also often used in warehouses where there's a lot of traffic with steel wheel equipment.

Be careful with admixtures that have a lot of chlorides or salts. They can corrode the steel reinforcing bars, which in turn can cause spalling. Always check on the chloride content of an admixture.

To get more strength by reducing the amount of water and still maintaining workability, use a Pozzalith admixture. This material increases the slump of a concrete mix without adding more water.

Mixing, Placing, and Curing Concrete

To get the best and strongest concrete mixture, use the proportions of cement, sand, and gravel recommended by the experts who have tested the admixtures you plan to use. A typical laboratory data sheet will provide the following information:

- maximum aggregate size in inches
- air content, as a percentage
- recommended maximum and minimum slump range in inches
- exposure conditions (severe or mild) in air, fresh water, sea water, or sulfate (acidic)
- maximum water/cement for exposure as a percentage
- specified design strength in psi
- water/cement ratio for strength

When you mix concrete at the job site, use this rule of thumb proportion for 2000 psi concrete (parts are given by volume or weight):

- one part cement
- three parts sand
- four parts 1-inch rock
- maximum of 8.5 gallons of water per sack of cement

Concrete is usually delivered to the job site by transit mix truck. A mixer truck has a rotating drum, a water tank, and measuring devices mounted on a truck chassis. Agitator trucks are similar to mixer trucks but they don't have water tanks. The dry cement, aggregate, and admixtures are loaded into the truck at the plant. Water is added and it's all mixed as the truck travels to the job site.

Another procedure is called shrink mixing. The material is agitated in a stationary mixer at the plant just enough to lightly mix the ingredients. Then the mixing is completed while the mixer truck travels to the site. The shrink method reduces the amount of time the concrete must be mixed in transit. This gives you better control over the mixture.

Usually, you get a load ticket with each batch of premixed concrete delivered to you at the job site. Here's an example of what you'll usually find on a load ticket:

- Amount of cement: 4230 pounds
- Amount of sand: 9825 pounds
- Amount of No. 2 gravel: 5850 pounds
- Amount of No. 3 gravel: 7560 pounds
- Amount of No. 4 gravel: 1710 pounds
- Amount of water added at plant: 180 gallons
- Total allowed water: 270 gallons
- Free moisture: 66.8 gallons
- Concrete produced: 7.5 cubic yards

Placement

Here's a checklist to review before you pour any concrete:

- Make sure any excavation you did for the foundation is dry and free of water.
- Make sure the reinforcing steel and other embedded items are securely in place.
- Set the openings and sleeves for ducts, pipes, and conduits and have them checked by the job superintendent.
- Have the forms and reinforcing bars inspected and approved by the building inspector.

Here are some rules to follow as you pour concrete:

- Mix the concrete thoroughly as you pour it. You can use a rod, but a mechanical vibrator works better. High frequency vibration will keep the coarse and fine aggregate from separating.
- Don't dump concrete directly from the mixer into the bucket. This throws the heavier and larger rock to one side.
- When the concrete comes out of the buggy (a cart used to carry fresh concrete) into a formed wall, unload it in a concentric fashion using a vertical chute. Don't use a sloping chute. Figure 3-8 shows using the metal chute and down pipe when pouring concrete into wall forms. Pour the concrete in even horizontal layers, 6 to 24 inches in depth. Be sure to place the next layer before the initial set takes place. If you don't use drop chutes, don't let the

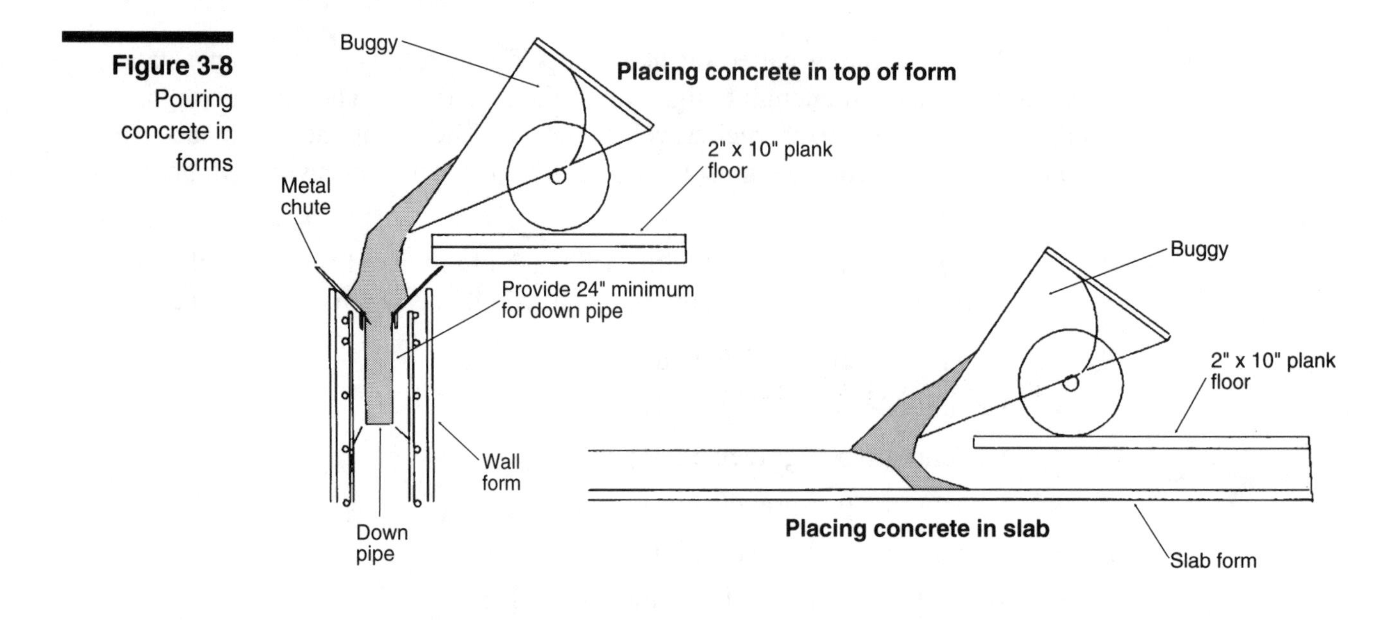

Figure 3-8 Pouring concrete in forms

concrete free fall more than 3 to 5 feet. Pour the concrete as close as possible to its final position. Don't move the concrete horizontally. Pour the concrete at the far end of slabs against the previously placed concrete.

- Avoid bouncing the mixture off one side of the forms. This makes the heavier aggregate bounce farther than the fine particles so you'll get some separation in the mixture.

- Pump the concrete from the mixer truck to the forms if access to the site is limited. Dump the concrete from the transit mix truck, by chute, into a hopper that feeds a pump that forces the mix through a hose to the forms. You can mount the pump on a trailer or truck chassis. It's better if the hopper, pump, and hose are mounted on a specially-designed truck with an articulated boom.

Other methods for transporting and pouring concrete are:

- Crane and bucket. Capacity of the bucket may be from ½ to 4 cubic yards.

- Belt conveyors, where a mobile unit can pour as much as 100 cubic yards of concrete per hour, up to three stories above or below grade.

- Pneumatic, also called Gunite and shotcrete, often used to build swimming pools and curved and sloping surfaces. The pneumatic system uses high pressure to send out a very dry mix. This makes a dense and strong concrete.

- Telescopic boom truck, when you need to pour concrete many floors above the street in a high-rise building. Some mobile concrete pumps can put out as much as 80 cubic yards of concrete per hour, 320 feet vertically or 1250 feet horizontally.

Curing the Concrete

After you've poured the concrete, the next important phase is curing. Curing concrete takes time. It continues as long as there's water present, and stops when the concrete is dry. Normally, concrete gains most of its full strength in 28 days. It will continue to get stronger, but not a lot.

Concrete that dries out too soon won't develop its full potential strength. For maximum strength, keep concrete moist during the entire curing period. In hot, dry weather, cover it with wet burlap or straw. You can also spray a fresh concrete surface with a curing compound made of a blend of oils, resins, waxes, and solvents. When the solvent evaporates, it leaves a membrane on the surface that seals in the original mixing water.

If concrete has dried too soon, spray the surface with water. The chemical reaction will continue, but the concrete won't be as strong as if you had kept it moist all the time. The longer a mixture is kept dry, the weaker it gets.

The outside temperature will also affect how concrete cures and how strong it'll be. It cures faster when the weather is warm than when it's cold. Below 40 degrees F, it'll cure slowly. The best temperature range for pouring concrete is between 50 and 70 degrees F.

Forming and Shoring

Poor formwork can ruin the best concrete mix. Concrete details on the plans usually specify what the interior surfaces of the forms should be. But it's the contractor's responsibility to design and build the forms correctly. On major construction jobs, engineers may have to design and approve the forms and shoring. But most formwork is designed by carpenters.

After you install all of the reinforcing steel bars and they're inspected, place the concrete. When the concrete has partially cured or set, strip, clean, and stack the forms so you can use them again.

Each part of a forming system has a function. The entire system must be stable, watertight, and resistant to the pressure of wet concrete. Formwork for concrete includes several distinct operations:

1) form design
2) fabrication
3) erection
4) alignment
5) bracing

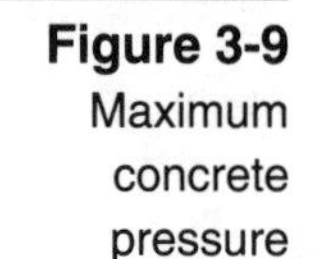

Figure 3-9
Maximum concrete pressure

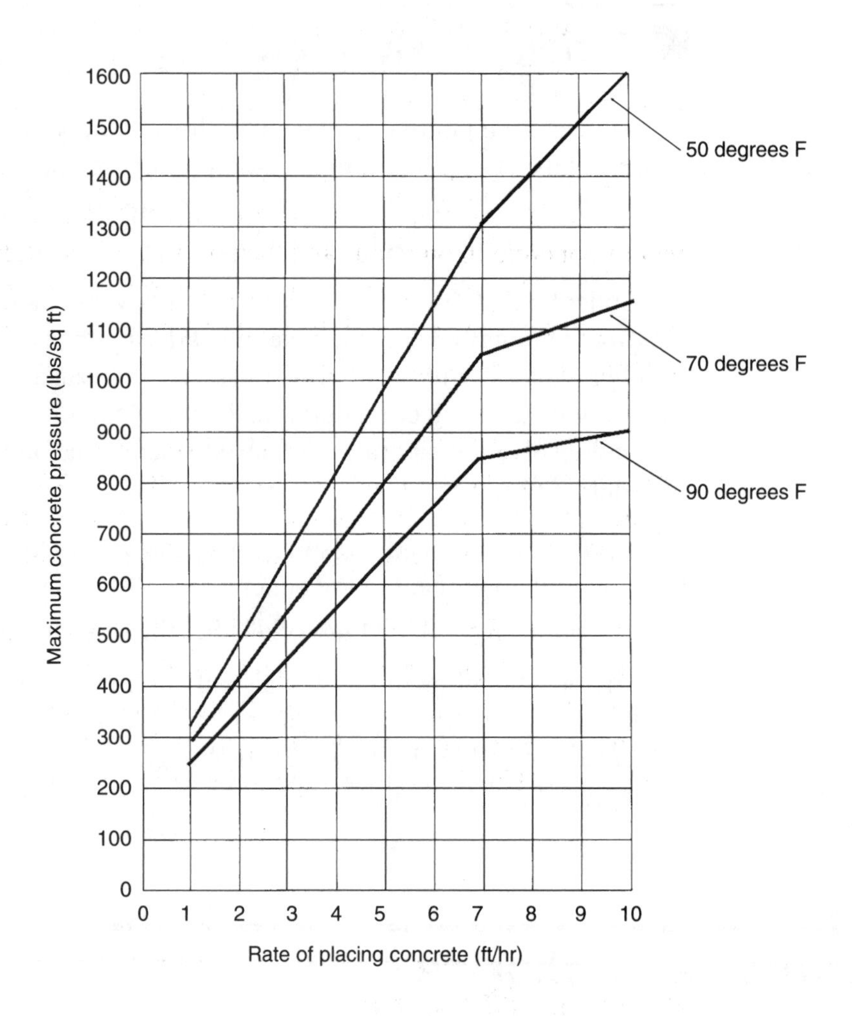

When you place concrete in a form, it's in a semiliquid state, and it exerts an outward pressure on the forms. As with water, this pressure is called hydrostatic pressure. One of the main objectives in good form design is to resist this pressure. When you pour walls or columns, the maximum pressure occurs at the bottom of the forms as shown in Figure 3-9. The rate you pour the concrete at, in ft/hr, (called the rate of lift or height of pour) is the capacity of the mixer (in cu. ft/hr) divided by the area of the form (in sq. ft). The lateral pressure on wall forms is affected by the:

- height of pour
- pour rate
- weight of concrete
- temperature
- type of cement
- vibration
- concrete slump (water/cement ratio)
- chemical additives in the concrete

Before concrete hardens, it pushes out against the forms holding it. The amount of pressure at any point on the forms is determined by the height and weight of concrete above the point. Pressure isn't affected by the thickness of the wall.

Figure 3-9 shows how the outward, or lateral, pressure of concrete on forms increases with the height of concrete placed in one hour. For example, if you place 5 feet of concrete in one hour and the temperature is 70 degrees F, the maximum pressure on the lower portion of the lift is 800 psf (pounds per square foot). If you place 7 feet of concrete and the temperature is 70 degrees F, the maximum pressure at the bottom of the pour is 1050 psf.

At higher temperatures, concrete cures faster and the pressure is reduced. So, pressure depends on the air temperature and how fast you pour the concrete. Hydrostatic pressure increases with the height of pour unless the concrete has taken its initial set. At moderate temperatures, this takes about 90 minutes.

Fresh concrete weighs between 150 and 160 pounds per cubic foot. A 1-foot thick concrete wall 10 feet high would weigh about 1550 pounds per linear foot. If fresh concrete were placed in the forms for this wall within one hour, the maximum hydrostatic pressure on the lowest part of the forms could reach 1550 pounds per square foot. This is a rather large force to resist with form framing.

Form Materials

The basic construction materials used in building forms usually include wood, steel, and fiberboard. Each of these materials should have certain properties to function properly. You should use straight, sound, smooth, kiln-dried lumber to make wood forms. Very dry wood may swell when it comes in contact with wet concrete. Studs and wales should be surfaced on four sides.

Rented forms are usually made of steel sheets or a combination of steel and plywood. Steel forms should be thick enough to hold their shape under repeated use and high pressure from concrete. Don't use metal forms that have rough or bent surfaces. Manufacturers of prefabricated metal forms usually tell you how to use their forms. Round columns are often formed with curved steel sheets that are flanged at their edges so they can be bolted together. These provide a smooth interior surface that's easy to strip.

Fiber reinforced plastic (FRP) is also used in rented forms. These have a smooth surface and are easy to strip. Custom-designed column forms, such as hexagonal-shaped columns, are also made with FRP. Fiber forms, made of asphalt impregnated cardboard, are used for casting round columns.

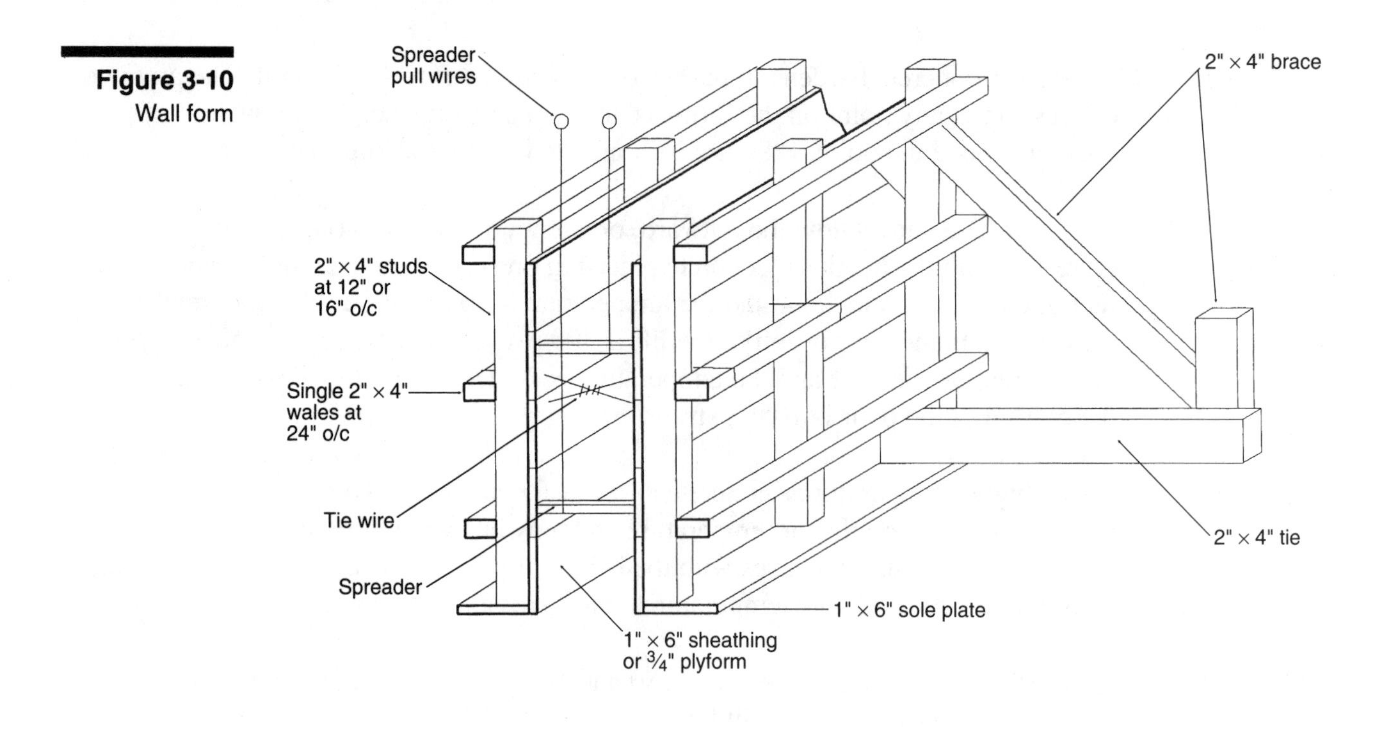

Figure 3-10
Wall form

Sheathing

Plywood used for form sheathing is also called plyform. It should be made with exterior type glue. Its surfaces should be sanded. Install plywood with the face grain perpendicular to the supporting members.

One type of plywood is overlaid with a layer of plastic that has a very smooth surface so you can use it again and again. This is called High Density Overlaid plywood, or HDO. It's made with a phenolic resin-impregnated fiber surface that is applied to both sides of an exterior-type plywood core.

Sheathing for field fabricated forms is usually made of wood boards or plywood. Boards may be 1 to 2 inches thick. Plywood is usually ½ to 1 inch thick, depending on the span between supporting studs.

To get a patterned architectural finish on concrete, use a sheathing that has a texture on its interior surface. If an aggregate finish is specified, you can bond special gravel to the sheathing with an adhesive.

Wall Forms

A form for casting concrete building walls is shown in Figure 3-10. A form for casting concrete foundations is shown in Figure 3-11.

Gang forms are job-fabricated modular wall forms that you fasten together to cover large areas. Each prefabricated module includes the sheathing, studs, wales, and top and bottom plates. The maximum recommended size of a gang form is 24

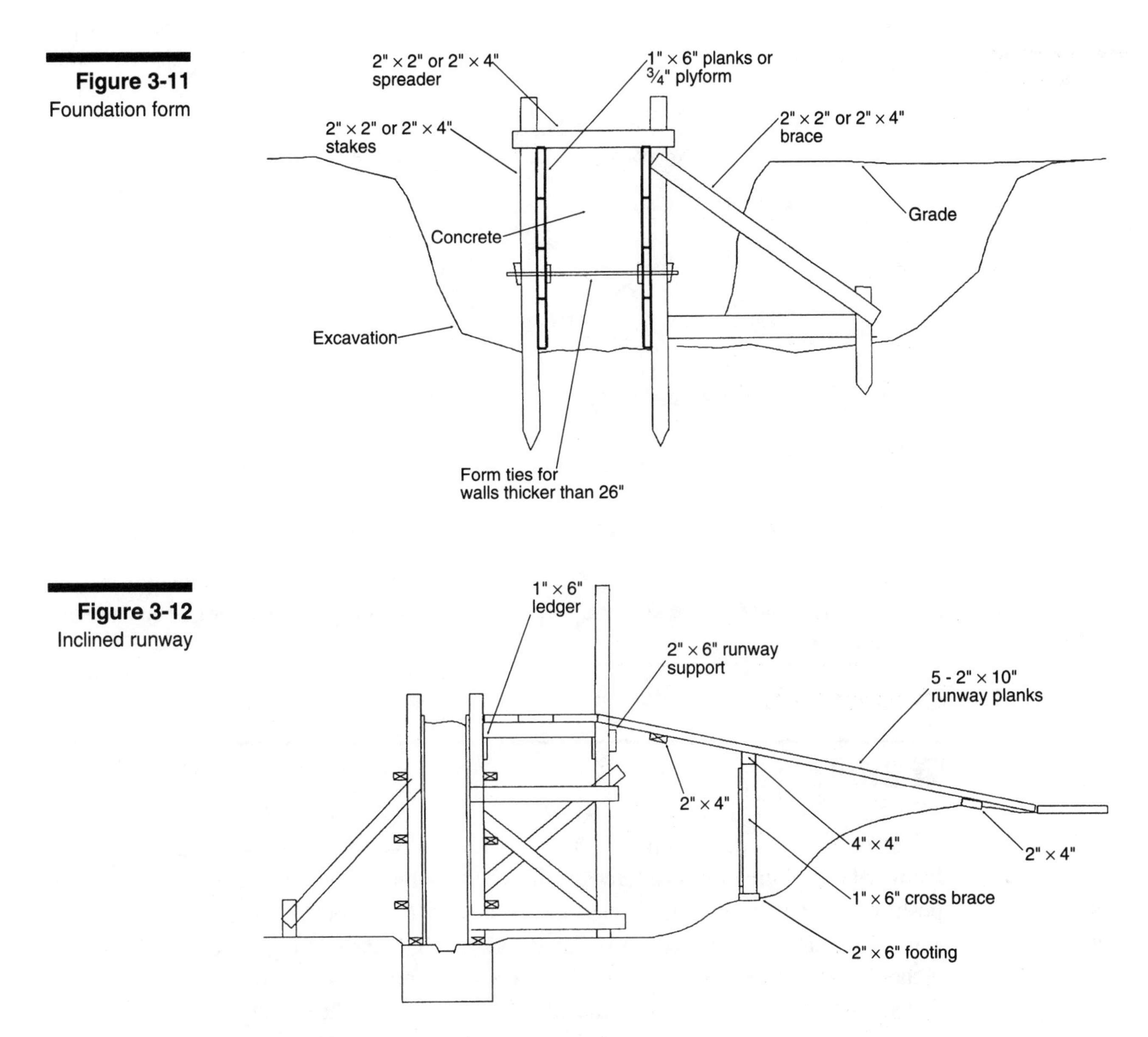

Figure 3-11
Foundation form

Figure 3-12
Inclined runway

feet square. Each gang-form module, or panel, usually comes in 8 or 12 foot lengths. To make the panels stiffer, nail the studs to the sheathing. Attach wall panels to each other by nailing the end studs together with 16d double-headed nails.

You can use radius forms to cast curved concrete walls. Normally you rent these prefabricated steel forms. Some types have a fixed radius, and others are adjustable to form walls with different radii.

Use battered faced forms to construct concrete walls that vary in thickness. Retaining walls are a common example of battered walls.

When forms are over 6 feet in height, build a work platform on one side of the form to make it easier for the buggy to pour the concrete. The platform is usually constructed of wood joists, plank decking, and a handrail built of 2 × 4 lumber. Figures 3-12 and 3-13 show two typical types of runways for carts and wheelbarrows

Figure 3-13
Runway along a wall

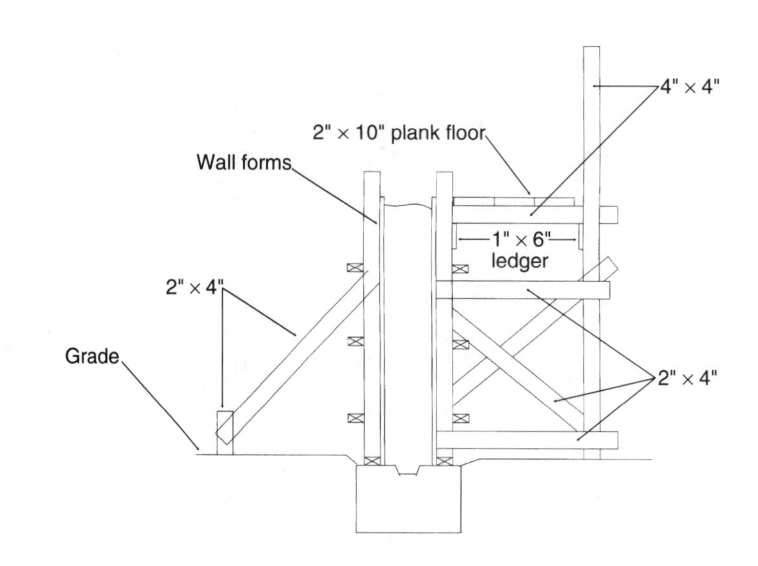

that make it easier to place concrete in forms. The first is an inclined runway for access to the runway alongside the form. The second is a section through the runway alongside a wall.

Studs

Studs are the vertical members that reinforce and support the sheathing. The hydrostatic pressure of wet concrete pushes the sheathing outward and the studs resist that push. Wood studs are normally made of 2 × 4 or 2 × 6 lumber. Maximum spacing of studs is 32 inches, but when you use plywood sheathing the spacing is usually 16 or 24 inches. Stud spacings for wall forms are shown in Figure 3-14. The maximum spacing depends on the type of sheathing and lateral pressures from the concrete.

The lateral pressure at any point on a form depends on the height of the concrete above that part of the form. For example, using Figure 3-14, if the maximum pressure is 800 psf and you're using ¾-inch plywood sheathing, you should space the studs 12 inches or less apart.

Steel studs are often used in prefabricated steel forms. They are usually cold-formed into channels which are then welded to the steel sheet. Each panel is framed with a steel channel so you can bolt it to the adjoining panel.

Wales

Wales, or walers, reinforce studs that are over 4 feet in height. The word wales comes from an old English shipbuilding term for the reinforcement wood planking on a ship's hull.

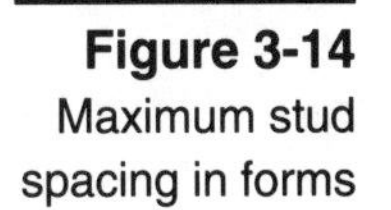

Figure 3-14
Maximum stud spacing in forms

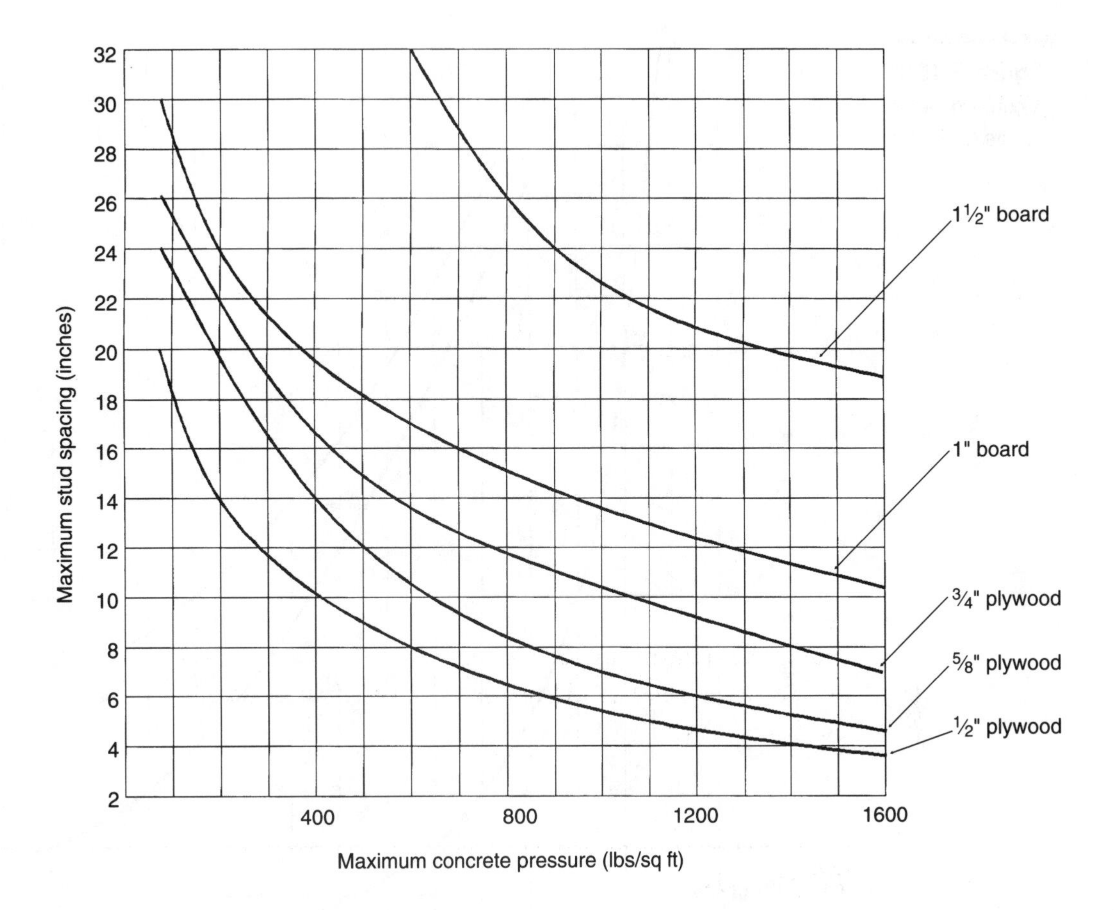

Wales are usually made of single or double 2 × 4 lumber. Some wales are made of laminated lumber for better strength. When you put forms on site, as with gang forms, fasten the prefabricated modules together with 16d double-headed nails through the wales. This helps keep the forms in a straight line and plane. You can also put wood wales, horizontally, directly on the sheathing. Usually you should set horizontal 2 × 4 wales 16 inches apart. Spacing of wales on wall forms is shown in Figure 3-15A. Wale spacing ranges from 10 to 48 inches.

Metal wales may be box-type, channel, or U-shaped. They are made of aluminum or steel. Wales are supported by strongbacks. Maximum spacing between 2 × 4 strongbacks is usually 6 feet.

Bracing

Bracing resists the outward push of wet concrete against the sheathing, studs, and wales. A typical brace is a diagonal 2 × 4 nailed to the wale at the upper end, and to a stake driven into the ground. You can use turnbuckles or screwed jack ends to adjust the length of the braces. A turnbuckle is a metal assembly with a threaded frame and a threaded rod at each end. The ends of the threaded rods are connected to opposite

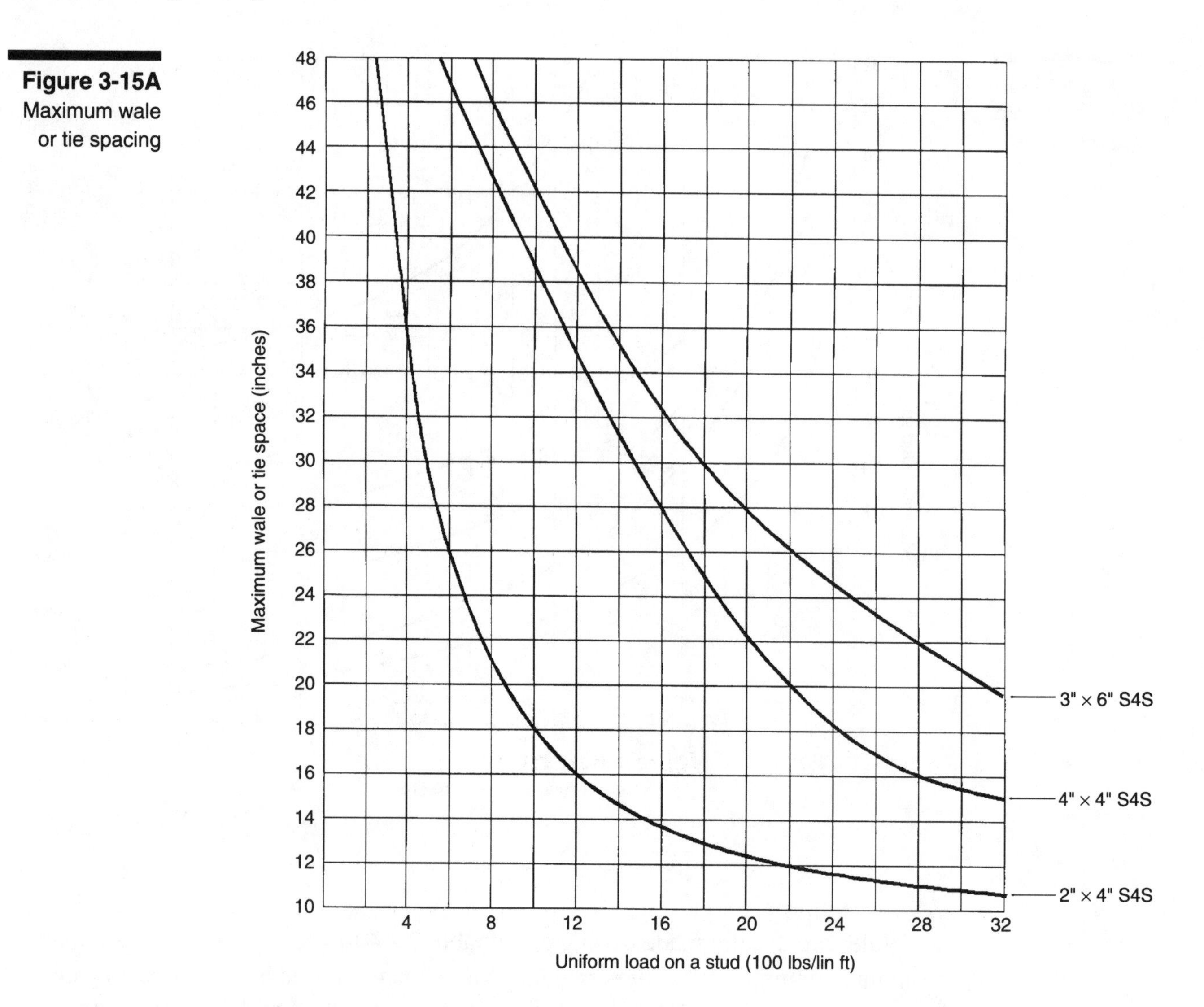

Figure 3-15A Maximum wale or tie spacing

sections of the shore post. Rotating the metal frame in one direction, moves both rods outward extending the length of the post. Rotating the frame in the other direction, pulls the rods in, shortening the length of the post. This assembly is also used in adjustable braces.

Stakes used to anchor braces to the ground may be made of pointed 2 × 4 lumber or steel bars with punched holes. It's a good idea to use stake removers to get steel stakes from the ground easily.

Miscellaneous Form Parts

Use shoe plates, or bottom plates, to make a level foundation for the form studs and sheathing. Nail the studs to the shoe plate. Nail top plates on the top of the studs to hold them in a straight line.

Spreaders are small removable pieces of wood you set between the sheathing to hold the sheathing apart. Spreaders are held in place by friction, but you can pull them out through the wet concrete by an attached wire.

Attach the tie wires to each stud at the wales on both sides of the forms. Tighten the wires by twisting them with a wedge to keep the forms from spreading. The wires are made from No. 8 to 11 gauge black, annealed iron. Spacing of tie wires for wall forms is shown in Figure 3-15A.

Double-headed 16d nails are commonly used to hold adjacent wall panels together. These nails are easy to drive in, and they hold tightly. They're also easy to take out.

In addition to the above items, form accessories include:

- waler brackets for connecting walers to studs
- form clamps for rectangular column forms
- snap, coil, and taper ties for holding wall forms in position
- inserts for embedding pickups, reglets, and bracing connections

Building a Wood Form for a Wall

Here's a suggested list of things you should figure out before you build wood forms for concrete walls:

1) Determine what materials are available.
2) Determine the rate of delivery of concrete to the job site in cubic yards.
3) Determine the area to be enclosed by the concrete in square feet.
4) Determine the rate of pour in the forms in vertical feet per hour.
5) Estimate the air temperature at the time of placement.
6) Determine the maximum concrete pressure in the forms in pounds per square foot (see Figure 3-9).
7) Determine the maximum stud spacing in feet (see Figure 3-14).
8) Determine the unit load on the stud in pounds per linear foot.
9) Determine the maximum wale spacing in feet.
10) Determine the uniform load on the wales in pounds per linear foot.
11) Determine the tie wire spacing based on the wale size.
12) Determine tie wire spacing based on wire strength.
13) Determine the maximum tie spacing.
14) Compare the maximum ties spacing with the maximum stud spacing.
15) Determine the number of studs and wales for one side of the form.

Column Forms

The major components of column forms include:

- Templates, which are set on the foundation for positioning the column form.
- Forms for rectangular columns, which are usually made of wood, prefabricated steel, a combination of steel and plywood, or fiber reinforced plastic (FRP).
- Forms for round columns, which are usually made of fiberboard, paper, or steel sheet. For round columns that flare out at the top, add conical forms.
- Bracing, usually of 2 × 4 lumber, which stabilizes and holds the form vertically.

The accessories required for column forms include chamfer strips to smooth out the sharp corners of rectangular columns, and yokes, or frames, to hold the sheathing in place and resist the pressure of the concrete. Yokes may be built of wood or steel, or a combination of both. Yoke locks hold the yokes together. Space these frames according to the largest dimension of a rectangular column cross section and the height of the column. Spacing may vary from 12 to 30 inches for columns 16 to 36 inches in size and up to 20 feet in height. Figure 3-15B shows the suggested spacing of column yokes. Here's how to use the table:

1) Select height of column in feet (left column of diagram).
2) Select the largest dimension of a rectangular column (top row).
3) Find the intersection of that column and row.
4) Read the maximum spacing of the column yokes in the block at that intersection.

   For example, if the column height is 12 feet and the largest dimension of a 16 × 18-inch column is 18 inches, the maximum yoke spacing is 18 inches.

Column yokes serve the same purpose as column scissor clamps. You can also use Figure 3-15B for spacing various types of metal column clamps.

Here's the way to design and install a column form:

- Determine what materials are available.
- Determine the height of the column.
- Determine the cross-sectional dimensions of the column.
- Determine the yoke spacing.
- Install template at base of the column.
- Install column form for a rectangular or round column.
- Install bracing frame for a rectangular column.

Figure 3-15B Maximum spacing of column yokes

Largest dimension of column (inches)								
Height (ft)	16	18	20	24	28	30	32	36
1			27					
2		29			21	20	19	17
3	31			23				
4						19	18	15
5		28	26		20			12
6					18	18	17	11
7	30			22	15		13	10
8			24		13	12	12	
9	29	26		16	12	10	10	8
10			19					
11		20	16	14	10	9	8	7
12	21	18		13	9	8		
13	20		15	12			7	6
14		16	14			7		
15	18	15	12	10	8		6	
16	15	13	11		7	6		
17	14			9	6			
18		12		8				
19	13		10					
20		11						
	12		9					

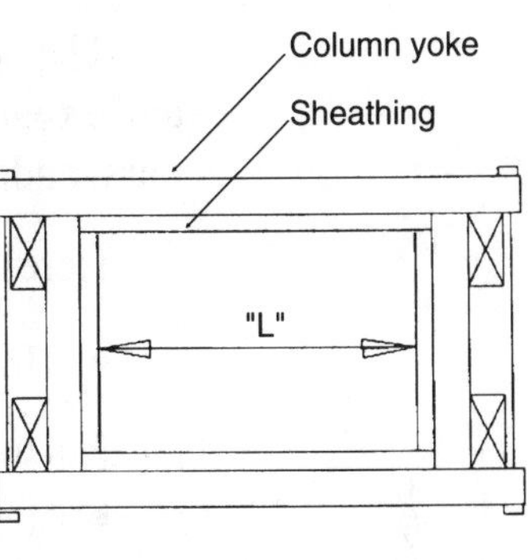

- Install bracing.
- Install reinforcing steel bars and ties.

Forms for Elevated Slabs

The major parts used to form an elevated concrete slab are:

- Wood or steel shores
- Wood shores that are adjustable using a turnbuckle assembly or Ellis clamps. Ellis clamps are metal devices used to adjust the length of wood shores by a clamping action. There are many types of steel tubular shores which adjust from 6 to 14 feet. The manufacturers of various tubular shores provide tables which list the capacity of each type of shore at various heights. For example, a shore that's 6 feet high may hold 7500 pounds, but the same type of shore that's 14 feet high can hold only 2000 pounds. The more you extend a shore, the less carrying capacity it'll have.
- Steel shores that are made of steel tubes with holes that you put pins into to vary the height of the shore.
- An adjusting screw at one end for final height adjustment.

- Scaffolding that's made of prefabricated braced supports.
- Stringers, beams, and trusses that support the sheathing. Wood stringers are also called joists. Some beams are made of lightweight aluminum, and adjust from 4 to 20 feet in length.
- Wood sheathing for elevated slabs that's made of planks or plywood sheets. Forms for concrete slabs are also made of ribbed or corrugated steel decking. Forms for concrete joist or waffle slabs are made of steel, FRP, or cardboard forms made of multiple layers of heavy paper bonded with waxes and resins.

The accessories used with elevated slab forming include screeds for construction and expansion joints, sleeves for pipes and conduits, chairs to support reinforcing bars, and metal embedments.

Here's the way to erect an elevated slab form:

- Install 4 × 4 posts, steel shores, or scaffolds.
- Install mud sills if shoring is supported on earth.
- Install stringers.
- Install bracing between posts.
- Install joists.
- Frame forms for beams and girders using 2 inch S4S sill, 1 inch sheathing and chamfer strips, 2 × 4 wales, and temporary spreaders.
- Install decking.
- Install 1 × 6 tongue and groove (T&G) boards or plywood.
- Oil interior surfaces of sheathing.
- Install wood or metal screeds for construction and expansion joints.
- Install chairs to support reinforcing bars.
- Install embedded metal items.

Recommended Safety Rules for Scaffolding and Shoring

Here are the safety rules you should follow when shoring and scaffolding:

- Follow local ordinances, codes, and regulations on shoring.
- Use manufacturer's recommended safe working loads for the type of shoring frame and height from supporting sill to formwork.
- Keep a shoring layout at the job site. A shoring layout is a drawing that shows the location and size of shores, bracing and joists, or stringers. It's usually prepared by the job superintendent or the job engineer. The designer of a shoring must consider the load of wet concrete, the strength of the joists and

sheathing, and the capacity of the shores to prevent accidents. When you use rented shoring equipment, the shoring layout is usually prepared by the company that rents out the shoring equipment.

- Provide and maintain solid footing to distribute loads properly.
- Use adjustment screws to adjust for uneven grade conditions.
- Don't exceed the shore frame spacing or tower heights shown on the shoring layout.
- Plumb and level all shoring frames as you build them.
- Fasten all braces securely.
- Inspect erected shoring and forming before and during pouring, and after concrete has set.
- Have the reshoring procedure approved by a qualified engineer.
- Don't remove braces or back off on adjustment screws until the job superintendent or job engineer authorizes you to do so.

Common Defects in Concrete Forming

Some of the common defects in concrete forming are:

- inadequate diagonal shore bracing
- placing concrete too quickly without regard to temperature
- unstable soil under mudsills supporting shoring
- insufficient nailing
- shoring not plumb
- unlocked locking devices on metal shoring
- vibration from adjacent moving loads
- supports removed too soon
- rough contact surface on forming
- concrete with knotholes, cracks, or other blemishes from the sheathing surface

Tests

Since so many things can go wrong with concrete that are difficult to fix, you should test it often. Local building codes often require that all concrete having a strength higher than 2000 psi should be continuously inspected during pouring. Also, representative samples from each concrete batch must be tested by an approved laboratory.

To do this, make at least three sample cylinders from each concrete batch you pour. Take these samples to a laboratory to cure in a wet room. The specimens will be tested for their compressive strength after three, seven and 28 days in accordance with the American Society of Testing and Materials (ASTM) Specifications C39 and C42. The results of these tests are sent to the building department, engineer, and contractor.

Normally, concrete made with Type I Normal cement reaches about 46 percent of its full strength in three days, 67 percent in seven days, and full strength in 28 days. If the early tests show low results, there's still time to remove the defective concrete. One accepted formula for figuring out the strength of concrete is:

$$S28 = S7 + 30 \times \sqrt{S7}$$

where

S28 = 28 day strength

S7 = 7 day strength

For example, if the seven-day test results in 900 psi, the 28-day strength will be 900 + (30 × 30) or 1,800 psi.

When a job requires high-strength reinforcing steel, the bars may have to be tested for bending and tensile strength. Any wire and cables you use in pretensioned or post-tensioned concrete work will also need to be tested. The testing procedures are listed in American Society of Testing and Materials ASTM A416 and ASTM A421, respectively.

Reinforcement

Although concrete is strong in compression, it's weak in tension. Without steel reinforcement, concrete can crack and break. To develop its tensile strength, you need to put steel bars or mesh in it.

The most common types of concrete reinforcement are deformed and plain steel bars and welded wire fabric. Other types are fibers of steel, glass, and plastic. Welded wire fabric, also called wire mesh, is identified by the gauge and spacing of the wires. For example, 4 × 12³⁄₆ stands for longitudinal wires 4 inches on center, transverse wires 12 inches on center, longitudinal wires No. 3 gauge, transverse wires No. 6 gauge. Welded wire fabric comes in lengths of 150 to 200 feet. The width is usually 84 inches. The gauge of a wire indicates its diameter.

You can identify a plain undeformed steel bar by its diameter, such as a ½-inch bar, a ⅝-inch bar, and a ¾-inch bar. A deformed bar has irregular surfaces to keep it from slipping in the concrete. It's identified by a number which stands for its average diameter in eighths of an inch. For example, a No. 5 deformed bar has the same cross-sectional area as a ⅝-inch diameter plain bar. Figure 3-16 lists the properties of

Figure 3-16 Standard reinforcing bars

Bar size, designation no.	Nominal diameter (inches)	Cross-sectional area (sq inches)	Weight (lbs/ft)
2	1⁄4	0.05	0.167
3	3⁄8	0.11	0.376
4	1⁄2	0.20	0.668
5	5⁄8	0.31	1.043
6	3⁄4	0.44	1.502
7	7⁄8	0.60	2.044
8	1	0.79	2.670
9	1 1⁄8	1.00	3.400
10	1 1⁄4	1.27	4.303
11	1 3⁄8	1.56	5.313

most standard deformed reinforcing bars. Standard mill lengths of reinforcing bars are 20, 40, and 60 feet. You can get nonstandard lengths by special arrangement with the supplier.

Reinforcing bars come in several grades. These used to be called structural grade, intermediate grade, and hard grade. Now they're called:

- Grade 60, standard grade, which has a yield strength of 60,000 psi
- Grade 40, lower strength, which has a yield strength of 40,000 psi
- Grade 75, premium grade (usually not stocked by suppliers), which has a yield strength of 75,000 psi. When you put more tension on a reinforcing bar than its tensile yield strength, the bar will be permanently deformed.

Usually you put reinforcing steel in a concrete member where the tensile force is the greatest, for example at the bottom of a simple beam, and at the top of a cantilever beam near the support. In continuous beams, put the bars at the top, over the support, and at the bottom at the middle of the span.

When the bars are at the bottom of a beam or slab, they are called positive steel. Bars at the top are called negative steel. The terms positive and negative come from the way the beam is bent under a load. When the beam is bent concave upward, it's said to have a positive bending moment. When it's bent concave downward, it has a negative bending moment. Bending moments cause one part of a beam to compress, and the other part to stretch with a tensile stress. The strength of a structural member depends on its ability to resist the compressive stress and the tensile stress.

The compression area of a structural member is the part of its cross-sectional area that's subjected to compressive stresses. Reinforcing bars in the compression area of a member are called compression bars. They increase the compressive strength of the member. To cut down on the depth of a concrete beam, you can put bars in the compression area.

The tension area of a structural member is the part of its cross-sectional area subjected to tensile stresses. Reinforcing bars in the tensile area of a member are called tensile bars. They provide the only tensile strength to a concrete member. Positive reinforcing bars are the tensile bars at the bottom of a beam and negative reinforcing bars are the tensile bars at the top of a beam. A continuous beam has both positive and negative tensile bars.

Another type of reinforcement that helps keep slabs and walls from cracking due to shrinkage as they cure is temperature reinforcement. This reinforcement also resists stresses from thermal expansion and contraction. See the section on Slab Reinforcement for more information on temperature reinforcement.

Concrete cover for reinforcement

When a reinforcing bar gets wet, it rusts and gets larger in diameter. If the bar is close to the surface of concrete, this expansion can make cracks so the concrete spalls off in large chunks. To keep this from happening, you should protect reinforcing bars from outside moisture with an adequate cover of concrete. In a severely-corrosive environment, such as salt water exposure, use galvanized or epoxy-coated reinforcing steel bars. Galvanized wire mesh is also available for severe conditions. There are a couple of rules for putting a concrete cover over reinforcement.

First, support horizontal reinforcing bars, which are in a slab placed on the earth, with concrete pads or mortar blocks. Concrete pads are thin precast concrete blocks that keep reinforcing bars off the ground. Mortar blocks are similar to concrete pads but they're made only with cement and sand and no coarse aggregate.

Second, when a concrete slab is cast in a form, use special metal chairs or bolsters that are galvanized or plastic-coated to support the reinforcement. Make these just high enough to give the proper height above the form surface. See Figure 3-17.

Here are four conditions that generally determine how much cover is required:

1) concrete cast against soil

2) concrete cast in forms but exposed to the ground or weather

3) concrete exposed to a hostile environment such as sea water or acids

4) concrete exposed to an interior noncorrosive environment

Here's the minimum protective covering of concrete you should put on all reinforcing steel (except stirrups and ties):

- Put at least a 3-inch cover on concrete that's placed directly against soil.
- Put at least a 2-inch cover on concrete that's placed against forms and exposed to weather or soil.
- Put at least a 2-inch cover on column reinforcement that's exposed to the weather.
- Put at least a 2-inch cover on walls that are in contact with the ground or exposed to the weather.

Figure 3-17
Wire bar supports

Courtesy: Concrete Reinforcing Steel Institute

TABLE I — TYPICAL TYPES AND SIZES OF WIRE BAR SUPPORTS

SYMBOL	BAR SUPPORT ILLUSTRATION	BAR SUPPORT ILLUSTRATION PLASTIC CAPPED OR DIPPED	TYPE OF SUPPORT	TYPICAL SIZES
SB	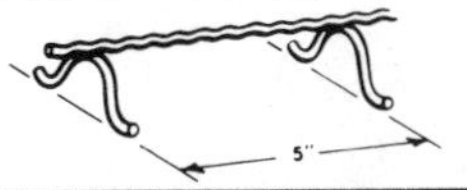 5"	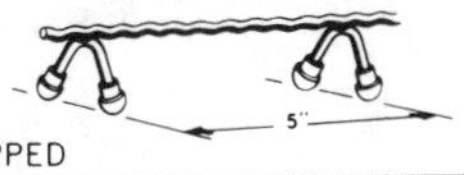 CAPPED 5"	Slab Bolster	¾, 1, 1½, and 2 inch heights in 5 ft. and 10 ft. lengths
SBU*	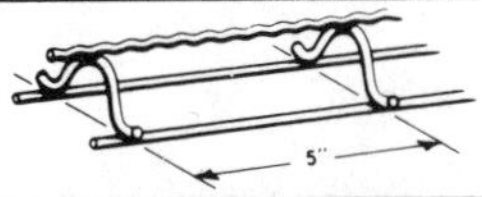 5"		Slab Bolster Upper	Same as SB
BB	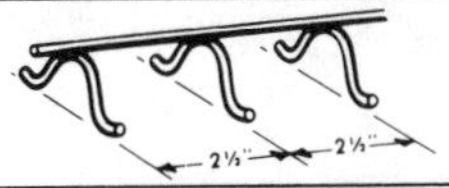 2½" 2½"	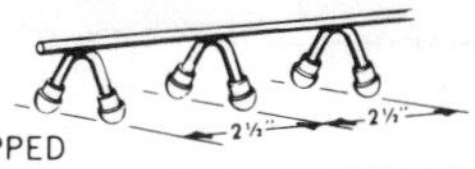 CAPPED 2½" 2½"	Beam Bolster	1, 1½, 2, over 2" to 5" heights in increments of ¼" in lengths of 5 ft.
BBU*	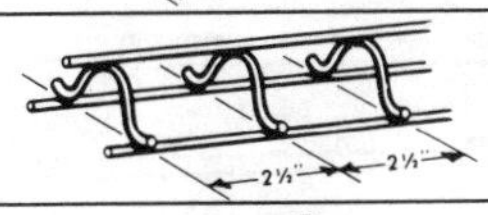 2½" 2½"		Beam Bolster Upper	Same as BB
BC	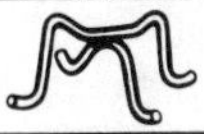	 DIPPED	Individual Bar Chair	¾, 1, 1½, and 1¾" heights
JC	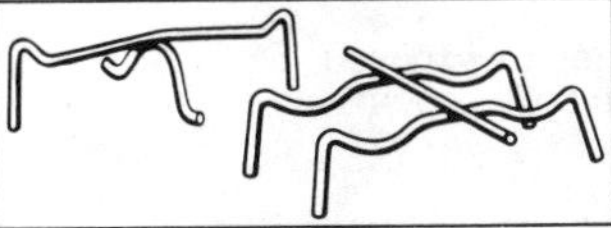	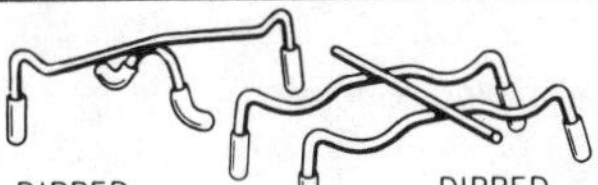 DIPPED DIPPED	Joist Chair	4, 5, and 6 inch widths and ¾, 1 and 1½ inch heights
HC	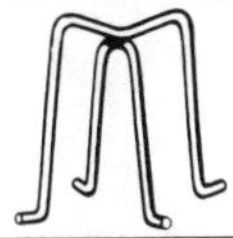	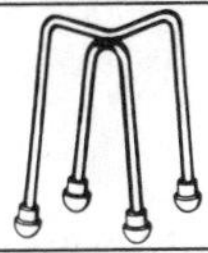 CAPPED	Individual High Chair	2 to 15 inch heights in increments of ¼ inch
HCM*	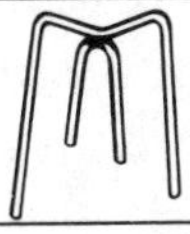		High Chair for Metal Deck	2 to 15 inch heights in increments of ¼ in.
CHC	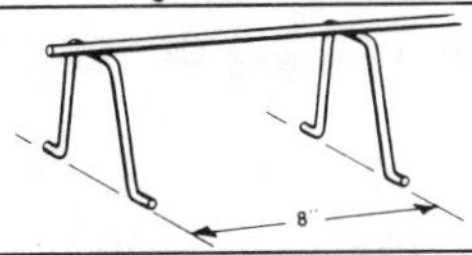 8"	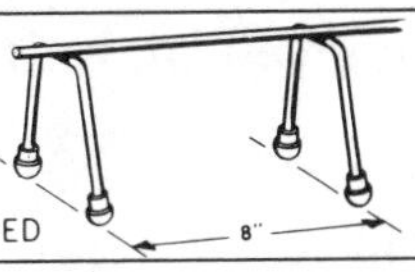 CAPPED 8"	Continuous High Chair	Same as HC in 5 foot and 10 foot lengths
CHCU*	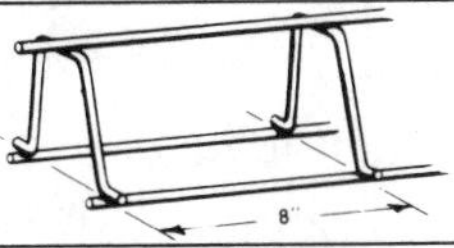 8"		Continuous High Chair Upper	Same as CHC
CHCM*			Continuous High Chair for Metal Deck	Up to 5 inch heights in increments of ¼ in.
JCU**	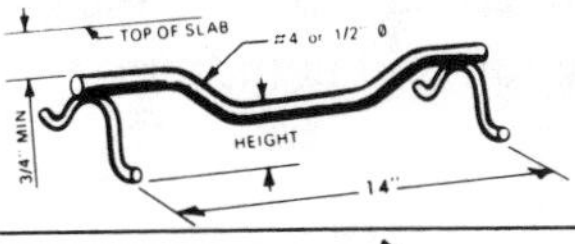 TOP OF SLAB, #4 or 1/2" Ø, HEIGHT, 3/4" MIN, 14"	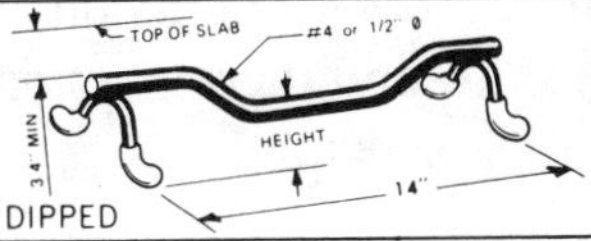 TOP OF SLAB, #4 or 1/2" Ø, HEIGHT, 3/4" MIN, 14", DIPPED	Joist Chair Upper	14" Span Heights −1" thru +3½" vary in ¼" increments
CS	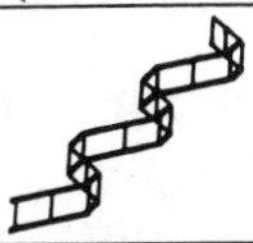		Continuous Support	1½" to 12" in increments of ¼" in lengths of 6'-8"

*Usually available in Class 3 only, except on special order.
**Usually available in Class 3 only, with upturned or end bearing legs.

Figure 3-18 Defects in concrete balcony

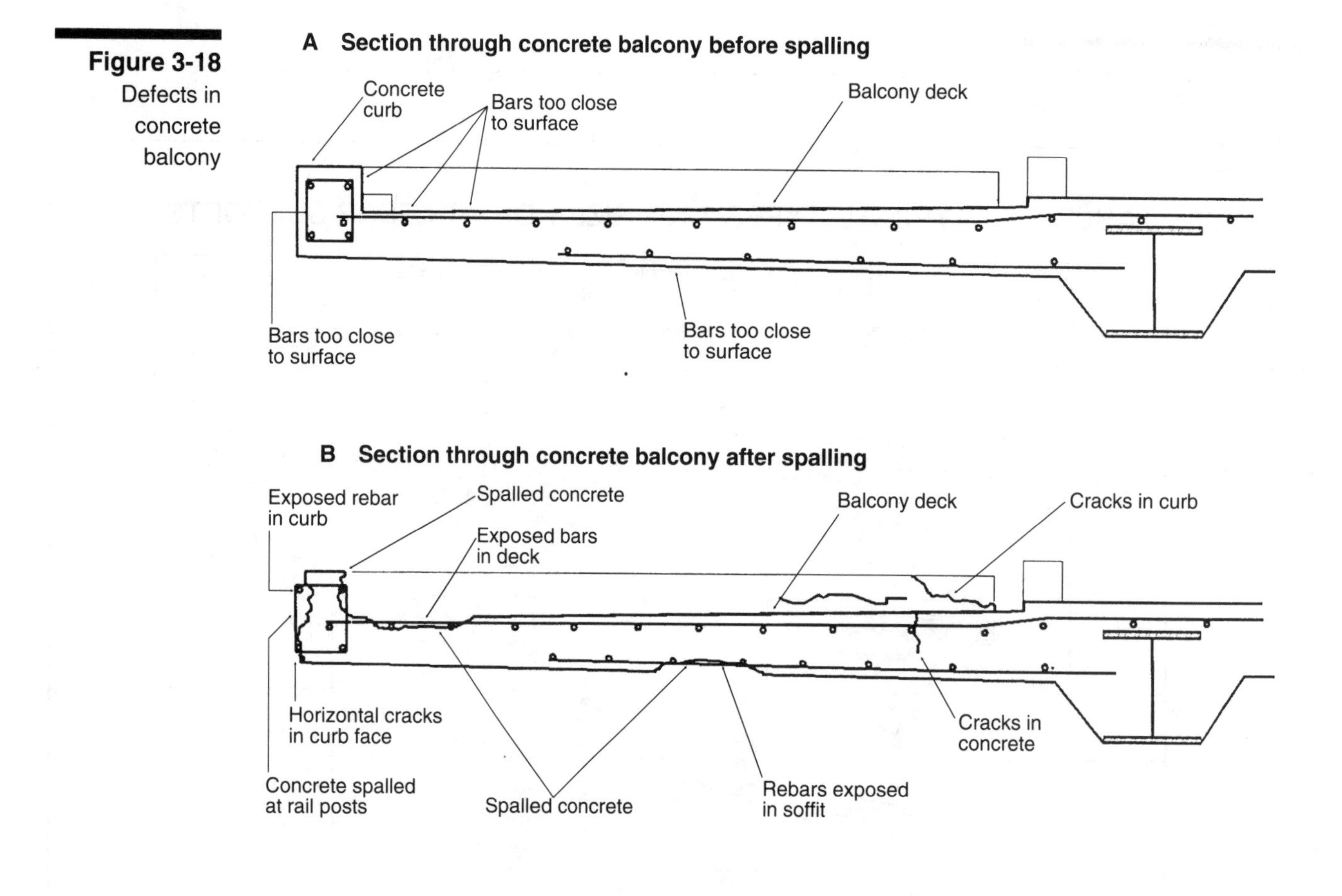

- Use at least as much concrete cover over steel reinforcement as that required for proper fire-resistant construction. (Fire-resistant construction relates to the time the effect of a fire can be retarded before the steel is weakened.)
- Make sure reinforcement isn't less than one bar diameter from the surface of the concrete.
- Use enough concrete cover to provide fire protection to reinforcing bars.

Figures 3-18 shows a section through a concrete balcony which didn't have enough concrete to cover the reinforcing bars. Figure 3-18A shows the balcony before spalling occurred. Figure 3-18B shows the same balcony after spalling due to corrosion of the reinforcing bars. Note that the spalling and cracking occurred at the locations where there's not enough concrete cover over the bars. A combination of water penetration, salts, and thermal expansion and contraction made the bars rust and the concrete crack and spall.

Steel reinforcing bars must also be protected against fire hazard. Steel loses a large part of its strength at temperatures above 800 degrees F. Therefore, depending on the type of construction, Type I, II, III, IV, or V, the building code requires a minimum amount of concrete cover over reinforcing bars. The amount of concrete cover required should be given on the construction plans.

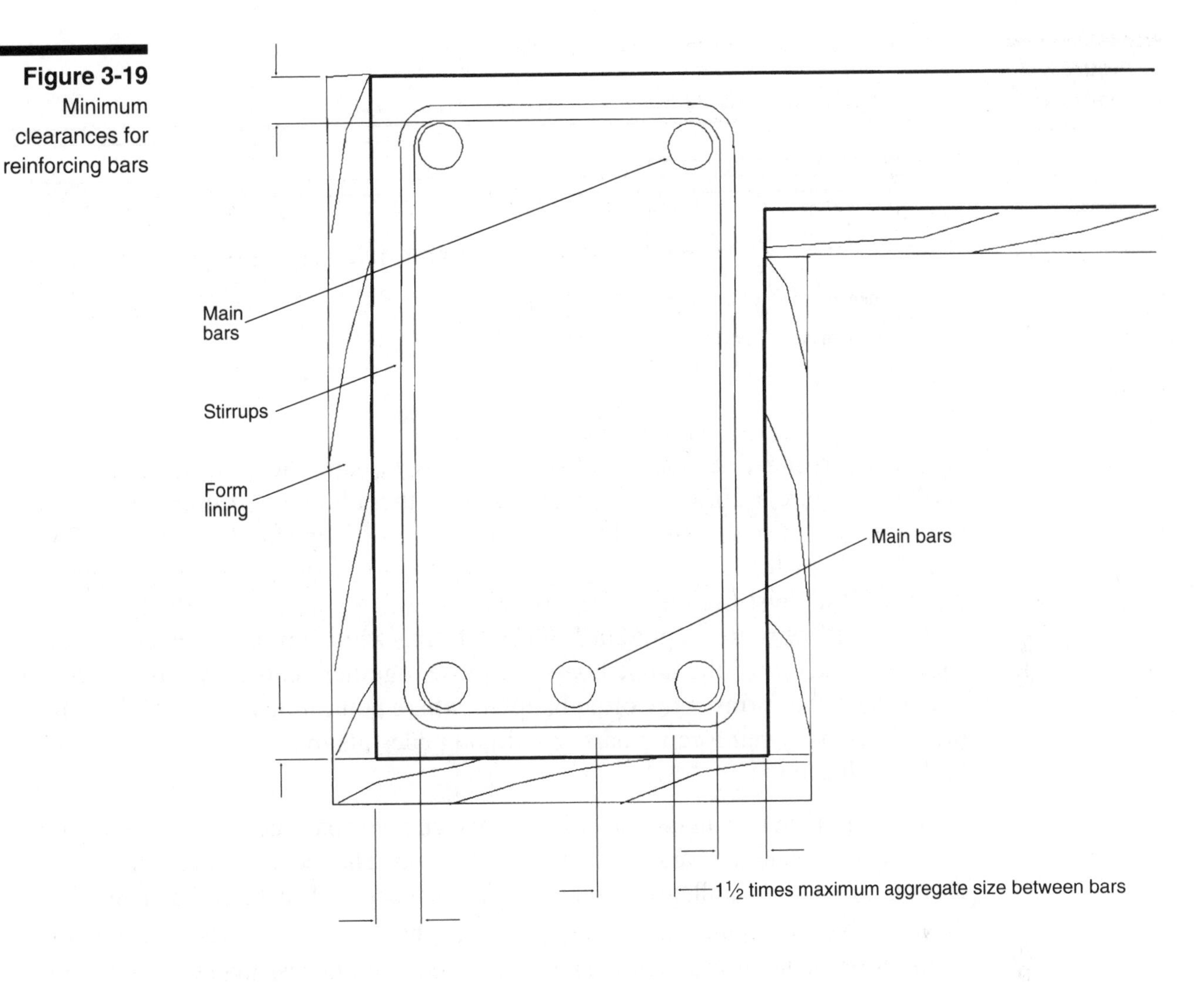

Figure 3-19
Minimum clearances for reinforcing bars

Where chloride salts may get into the concrete from deicing or exposure to salt spray, use more concrete cover than the list recommends. Figure 3-19 shows concrete cover over reinforcing bars in columns and beams. The dimensions of the cover will vary according to:

- required amount of fire resistance (from Table 43A of the *Uniform Building Code*)
- type of exposure (from Section 2607 (3h) of the UBC)
- size of aggregate you use

If the building code specifies different amounts of concrete cover for an area, use the largest requirement. For example, if the minimum cover for concrete exposed to the weather is 2 inches, and the minimum fire protection cover is ¾ inch and the bar is 1 inch in diameter, use the larger cover — 2 inches.

Placing reinforcement

If you put reinforcement in the wrong place in a member, it can cause severe cracking or structural failure. In most structural members, concrete with no steel reinforcement won't resist high tensile stresses.

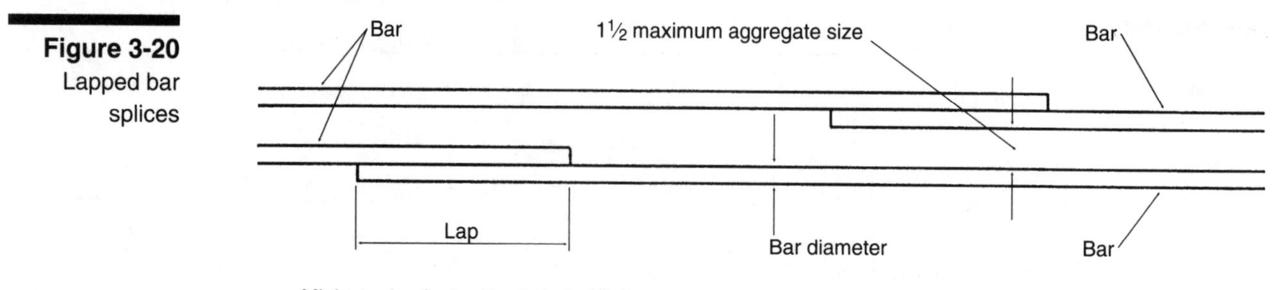

Figure 3-20 Lapped bar splices

Steel bars must be continuous or spliced by lapping the ends of the bars, as shown in Figure 3-20. If you lap bars to make them longer or continue them through joints, use this rule of thumb: the amount of lap should be at least 25 times the diameter of the bars, or a minimum of 12 inches, whichever is the greatest. You must use the greater dimension. Also, you must tie the lapped bars together with wire to keep them from separating while you pour the concrete. Be careful not to crowd the lapped bars too close together. This may push some bars too close to the forms. This often happens where continuous beams and columns meet. Too many reinforcing bars in column pedestals may make it difficult to put in the anchor bolts.

If you put the bars in parallel layers, leave enough space between them to let the concrete go around the bars and make a good bond. The recommended minimum clearance between parallel bars in horizontal members is 1 inch, or the diameter of the larger bar. In columns, minimum clearance should be 1½ inches, or 1.5 times the diameter of the larger bar. The clear distance shouldn't be less than 1.33 times the maximum size of the coarse aggregate. Use the largest required distance.

Foundations

It's said that a building is no better than its foundation. A foundation is the lowest part of a structure. It carries the loads from the roof and floors to the soil. A footing is the lowest part of a foundation. But the two terms are often used interchangeably. When the entire foundation is in contact with the soil, it may be called a footing. When there are two distinct parts to a foundation, such as a wall and a footing or a pedestal and a footing, the entire combination is called a foundation and the lower part, a footing. The most common classes of foundations are:

- isolated foundations
- continuous foundations (grade beams)
- structural mat
- cantilever foundations
- caissons

Figure 3-21
Allowable foundation bearing pressures

TABLE NO. 29-B—ALLOWABLE FOUNDATION AND LATERAL PRESSURE				
Class of materials[2]	Allowable foundation pressure lbs./sq. ft.[3]	Lateral bearing lbs./sq./ft./ft. of depth below natural grade[4]	Lateral sliding[1] Coefficient[5]	Lateral sliding[1] Resistance lbs./sq. ft.[6]
1. Massive crystalline bedrock	4000	1200	.70	
2. Sedimentary and foliated rock	2000	400	.35	
3. Sandy gravel and/or gravel (GW and GP)	2000	200	.35	
4. Sand, silty sand, clayey sand, silty gravel and clayey gravel (SW, SP, SM, SC, GM and GC)	1500	150	.25	
5. Clay, sandy clay, silty clay and clayey silt (CL, ML, MH and CH)	1000[7]	100		130

[1]Lateral bearing and lateral sliding resistance may be combined.
[2]For soil classifications OL, OH and PT (i.e., organic clays and peat), a foundation investigation shall be required.
[3]All values of allowable foundation pressure are for footings having a minimum width of 12 inches and a minimum depth of 12 inches into natural grade. Except as in Footnote No. 7 below, increase of 20 percent allowed for each additional foot of width or depth to a maximum value of three times the designated value.
[4]May be increased the amount of the designated value for each additional foot of depth to a maximum of 15 times the designated value. Isolated poles for uses such as flagpoles or signs and poles used to support buildings which are not adversely affected by a ½-inch motion at ground surface due to short-term lateral loads may be designed using lateral bearing values equal to two times the tabulated values.
[5]Coefficient to be multiplied by the dead load.
[6]Lateral sliding resistance value to be multiplied by the contact area. In no case shall the lateral sliding resistance exceed one half the dead load.
[7]No increase for width is allowed.

- piles
- machinery foundations

A foundation is sized to minimize subsidence and to prevent differential settlement, i.e. parts of the building settling at different rates. The size of footing you need depends on the bearing capacity of the soil and the amount of load. The bearing capacity is the amount of weight per square foot the soil can support without settling too much.

Most building damage happens when parts of a building settle unevenly. Footings are sized to be equally loaded to reduce this differential settlement. So, large and small footings should have the same unit bearing pressure per square foot on the soil. To find out the unit bearing pressure of a member, divide the total load by the area of the footing. For example, a column that carries a 90 kip (90,000 lbs) load to a 5-foot square footing will have a unit bearing pressure of $^{90}/_{25} = 3.6$ ksf (kips per sq ft), or 3600 psf (pounds per sq ft) unit bearing pressure.

A 4-foot square footing should carry four times the load of a 2-foot square footing.

Use the soil classification table published by your local building department as a preliminary guide when you pick what size foundation to build. Figure 3-21 shows the allowable foundation bearing pressure and lateral bearing pressure of five major classes of soil (Table 29-B UBC). Allowable foundation bearing pressure is also called allowable soil pressure, or allowable soil bearing value. Allowable lateral bearing pressure is the allowable soil bearing in the horizontal direction at the sides of a footing, as in a retaining wall footing. The allowable foundation bearing pressure is usually based on a footing that's at least 12 inches deep, but you can get higher soil bearing values by making the footing deeper or wider.

To get a really accurate classification of any soil, you'll need to hire a soils engineer. He'll get samples of the soil tested in a soils laboratory for classification, allowable bearing pressures, expansion from moisture, and other important properties.

In very cold areas, you have to make sure the bottom of a footing goes below the frost line. Wet soil expands when it freezes and it can push up a building. The frost line varies throughout the United States from 1 foot deep in the southwest, to over 7 feet deep in the northeast. Your building department can tell you about the local frost line.

Selecting a Foundation Type

Use Figure 3-22 as a checklist to help you select what type of foundation you should use for a building. The checklist also gives some good advice on how to handle drainage and slopes.

The cheapest foundation you can use on a building site will depend on three things:

- loading conditions
- area available
- soil conditions

Loading conditions are the amount of dead, live, and earthquake loads the foundation must carry. The dead load is permanent, but it's applied slowly as the building is constructed. The live load is variable and temporary, and it's applied over a shorter period of time. Soils usually rebound after live loads are gone. Earthquake loads are a very short period load and the soil will absorb part of the pressure. Therefore, the allowable foundation pressure is usually increased by one-third for earthquake and wind loads.

The shape of a foundation is often controlled by adjacent obstructions. For example, columns within a building will require a different shape foundation than columns near the property line or adjacent structures.

Soil condition has a great influence on the design of a foundation. For example, the level of the natural grade determines the depth of the foundation. The moisture condition of the soil is important in cold climates where upheaval of frozen soil may occur. Also, some soils, like adobe, expand when wet. Additional reinforcement is usually required.

Figure 3-23 shows various soil conditions and types of foundations for each condition. This is a simplified version of the many soil classifications. When a soil with moderate allowable bearing pressure, 1000 psf to 2500 psf, is near the surface, you can use a structural slab grade beam continuous foundation. If the frost line is over 4 feet deep, you'll have to put the footing below the frost line. When there's a relatively stable fill over 10 feet in depth, you may be able to use a belled caisson. If

Figure 3-22
Foundation checklist

Foundation Checklist

- ☐ Is there a soils report? If not, use the soil classification and bearing values in the building codes when you make a preliminary design.
- ☐ Check the type of soil classification used for nearby building permits.
- ☐ Is there a contour map of the original grade? Compare the original contours with the final contours. This will tell you whether the site has been filled or cut.
- ☐ If there is a fill, what's it like, and how far does it go? If the fill is over 3 feet deep, check with the grading department on whether the fill was certified or not. If it was, find out from the grading department what the allowable soil pressure is. If it wasn't, don't assume that it's good enough to support the building. Extend the foundations to natural grade. If the uncertified fill is deeper than 3 feet, call a soils engineer to make an investigation and recommend what type of foundation you should use.
- ☐ Find out if there are underground or above ground obstructions. Adjacent buildings can get in the way of any excavating you'll need to do.
- ☐ Find out from the building department how deep the frost line is.
- ☐ How deep is the water table? A high water level can flood footing excavations and caisson holes. Find out if the people who built nearby found ground water when they were constructing in the area. If there's a possibility of high water table, hire a soils engineer to bore some test holes.
- ☐ What are the elevations of finish grade, column foundations, and building floor? You'll need at least a 6- to 8-inch high foundation above finish grade.
- ☐ What is the soil bearing value at the depth of the footing? If you didn't get a soils engineer's report on the soil, dig a trench or pit to find out if there's any fill.
- ☐ If there's a slope on part of the property, or adjacent to the property, check the angle of slope. A cut or fill slope shouldn't be steeper than 2:1 (horizontal to vertical) unless the slope was approved by the building department or a soils engineer.
- ☐ Check with the grading department on how close you can build to the edge of a fill or cut slope.
- ☐ Check the building code to find out how far below the natural or finish grade the footing has to be. In expansive soil, such as adobe, you have to put in a deeper footing to keep moisture from reaching the bottom of the footing.
- ☐ If there are cut or fill slopes, check with the neighbors and grading department if there have been landslides, soil-slumps, or mudflows in the nearby lots.
- ☐ Inspect for visible cracks in adjacent buildings and surrounding ground. These signs are an indication of expansive-soil, deep-fill settlement, or soil-bedrock down slope creep.
- ☐ Check for evidence of burrowing rodents. This may be an indication of loose fill. The combination of loose fill and a network of rodent holes makes the slope vulnerable to saturation and slump-type failure.
- ☐ In landslide, questionable, or hazardous areas, no building or grading permit is normally issued until the grading department get reports from a soils engineer which certify the safety of the area.
- ☐ Don't allow drainage over a slope unless it's controlled by drainage devices such as pipes or channels. Drain rainwater toward the street and away from the slope if possible.
- ☐ Plant cut and fill slopes to prevent erosion. Check with the grading department for the recommended types of plants for slope erosion control. Install a sprinkler system.
- ☐ If a building site is on a hillside, check the location of natural water courses. Install debris basins, debris fences, or other devices in the water courses to protect the site from mud slides.
- ☐ Don't do any grading on hillside lots during the rainy season.

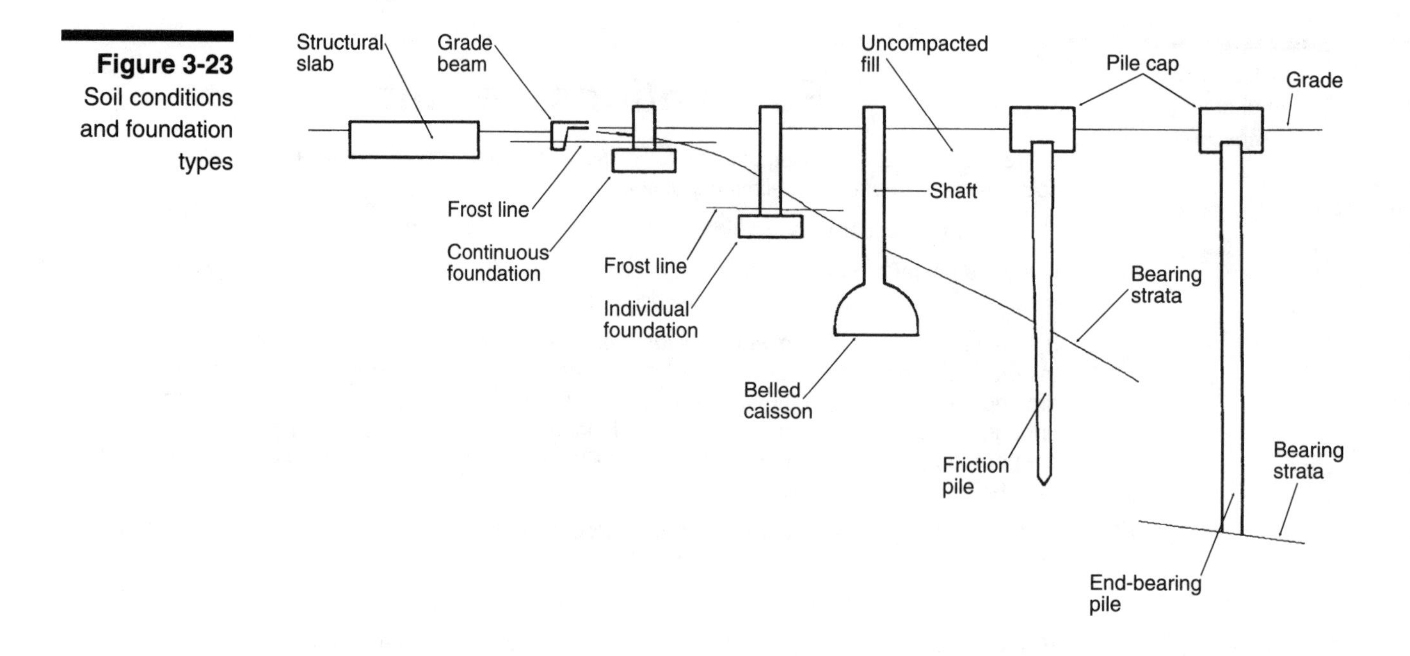

Figure 3-23 Soil conditions and foundation types

the fill is too unstable, or there's high ground water level, you'll need friction piles. When bedrock is within 40 or 50 feet of the surface, end bearing piles may be the best type of foundation.

Isolated and Continuous Foundations

An isolated foundation is simply a rectangular footing and a pedestal. The pedestal carries the column load to the footing. The footing spreads the load onto the soil. The terms isolated foundation or individual foundation are used interchangeably. They both mean that the foundation stands alone. Both types are considered shallow foundations. When the soil bearing value is moderate, such as it is with sandy loam and clay-sand soils, you can usually use an isolated foundation.

Figures 3-24A and 3-24B show how you would reinforce a typical isolated foundation using three groups of bars — the horizontal mat near the bottom of the footing, the vertical bars or dowels in the pedestal, and the ties around the pedestal bars. Mat bars are a group of reinforcing bars placed in a horizontal grid pattern and wired together. On large slabs, mats may be prefabricated by welding the bars together, so you can place the entire mat as a unit.

Dowels are bars that project from one structural component into another structural member. The dowels are spliced to matching bars in the adjacent structural element. Examples of dowels are the bars that project upward from a footing that are lapped and spliced to pedestal or column bars.

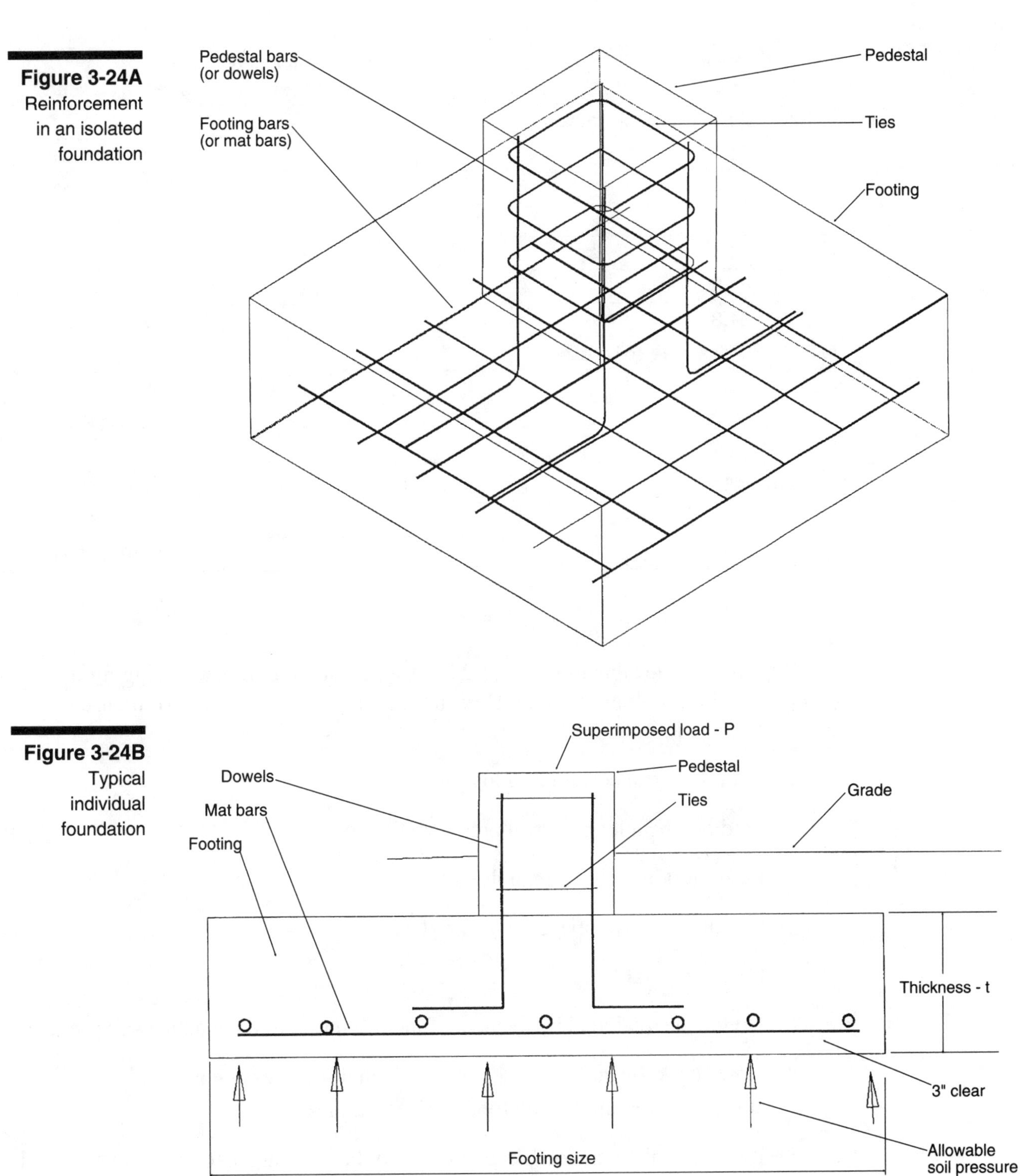

Figure 3-24A Reinforcement in an isolated foundation

Figure 3-24B Typical individual foundation

The horizontal mat reinforcing bars are the ones that are placed near the bottom of the footing. These bars resist the tensile stresses created in the footing as it bends upward from the soil pressure. The pedestal bars, or dowels, tie the pedestal to the footing, and also resist bending stresses in the pedestal from horizontal forces in the column. The ties in the pedestal hold the vertical bars in place and prevent them from buckling from compressive forces.

Figure 3-25 Safe carrying capacity of square individual column footings

Capacity (kips)	Footing size (feet)	Footing thickness (inches)	Mat bars each way
1000 psf allowable soil pressure			
21.7	5	10	#4 @ 13" o.c.
31.3	6	10	#5 @ 13" o.c.
42.3	7	11	#6 @ 13" o.c.
54.4	8	12	#6 @ 11" o.c.
67.9	9	13	#6 @ 10" o.c.
81.3	10	15	#7 @ 12" o.c.
2000 psf allowable soil pressure			
46.6	5	11	#5 @ 9" o.c.
66.2	6	13	#6 @ 11" o.c.
88.8	7	15	#6 @ 9" o.c.
115.2	8	16	#7 @ 11" o.c.
160.2	9	18	#8 @ 14" o.c.
176.3	10	19	#8 @ 11" o.c.

You can use the table in Figure 3-25 to figure out a preliminary sizing for a square individual column footing. Here are the suggested steps you would use to design a square footing:

1) Find out the superimposed load (P).
2) Find out the allowable soil pressure (SP) for the soil.
3) Pick a footing size from table.
4) Calculate the volume and weight of foundation.
5) Calculate total load.
6) Divide the total load by the area of the footing to find the actual soil pressure.
7) Now check the table to find out if the actual soil pressure is less than allowable pressure for the footing size you picked.
8) If not, calculate the total load again with the next larger footing. Do this until you find a soil pressure that is less than the allowable pressure for the footing.

Here's a sample calculation for a 30,000-pound load on soil with an allowable soil pressure of 1000 psf. Will a 6-foot × 6-foot × 10-inch footing and 10-inch × 10-inch × 24-inch pedestal work? Converting all the dimensions of the footing and pedestal to feet, the weight of the footing and pedestal for concrete weighing about 150 pcf (pounds per cu ft) is:

$$(6' \times 6' \times 0.83' + 0.83' \times 0.83' \times 2') \times 150 \text{ pcf} = 4689 \text{ pounds}$$

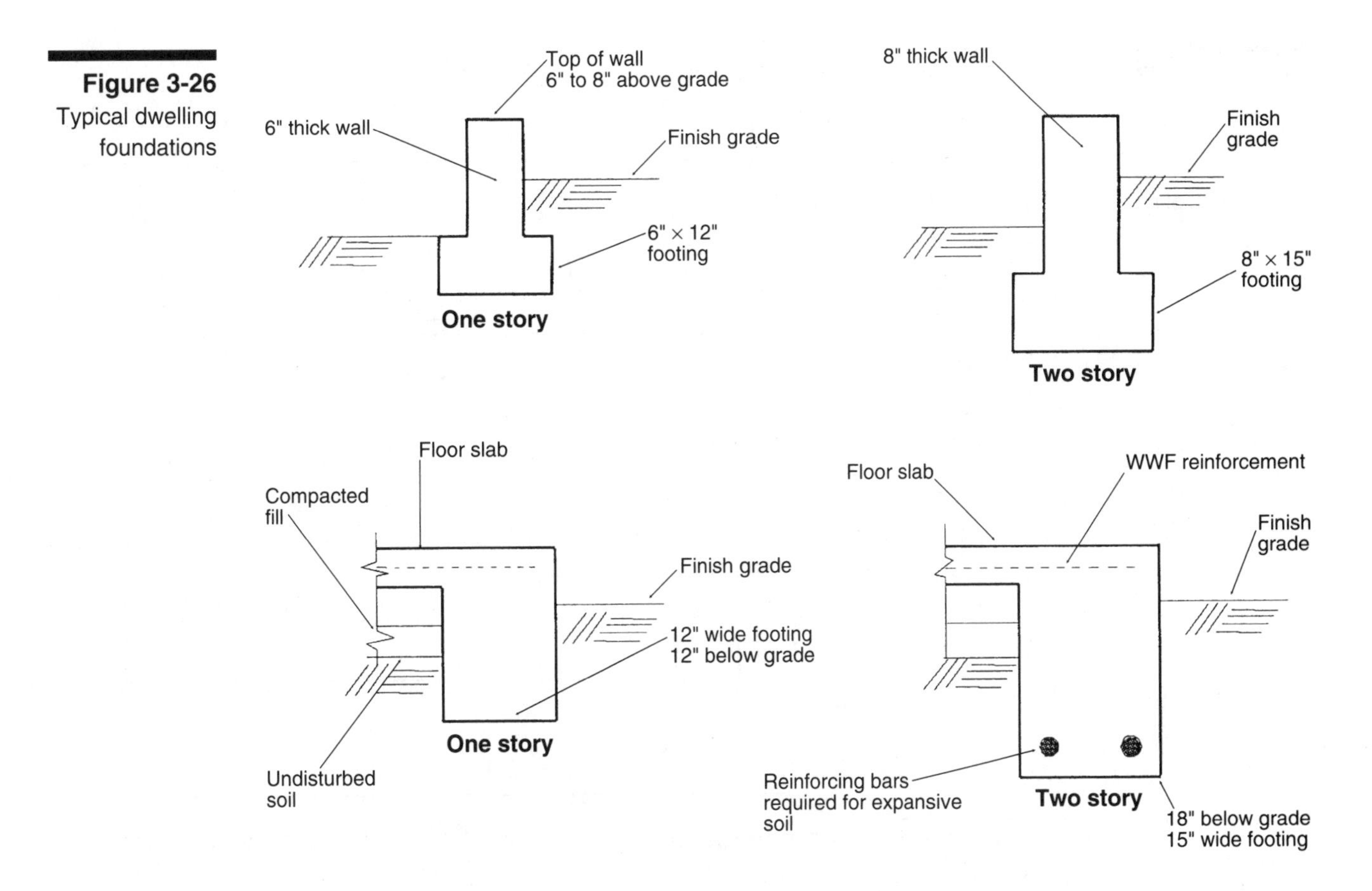

Figure 3-26
Typical dwelling foundations

The total load is 30,000 + 4,689 = 34,689 pounds. The soil pressure per square foot is 34,689 / 36, or 964. The problem was to find out whether a 6-foot square footing was adequate. Since 964 psf is less than 1000 psf, the 6 × 6 × 10 footing is adequate.

The usual sizes and depths of dwelling foundations are shown in Figure 3-26. If the soil is expansive, you'll need deeper footings and steel reinforcement in the foundation. For example, the Los Angeles City Building Code says that foundations in adobe soil must be at least 24 inches below the finish grade for exterior walls, and 18 inches below for interior walls. Also, there should be at least two #4 bars, placed 4 inches from the bottom and 4 inches from the top of the foundation. See Figure 3-26 for information on individual foundations.

A continuous foundation is a concrete wall and footing under a building wall. The wall carries the weight of the building to the footing, which transfers the load to the soil. A continuous foundation or footing, rather than a column, is normally used to support a bearing wall. A typical example is a one- or two-story dwelling with interior and exterior bearing walls. Buildings with slab floors often have a footing cast integrally with the slab. These are also called grade beams.

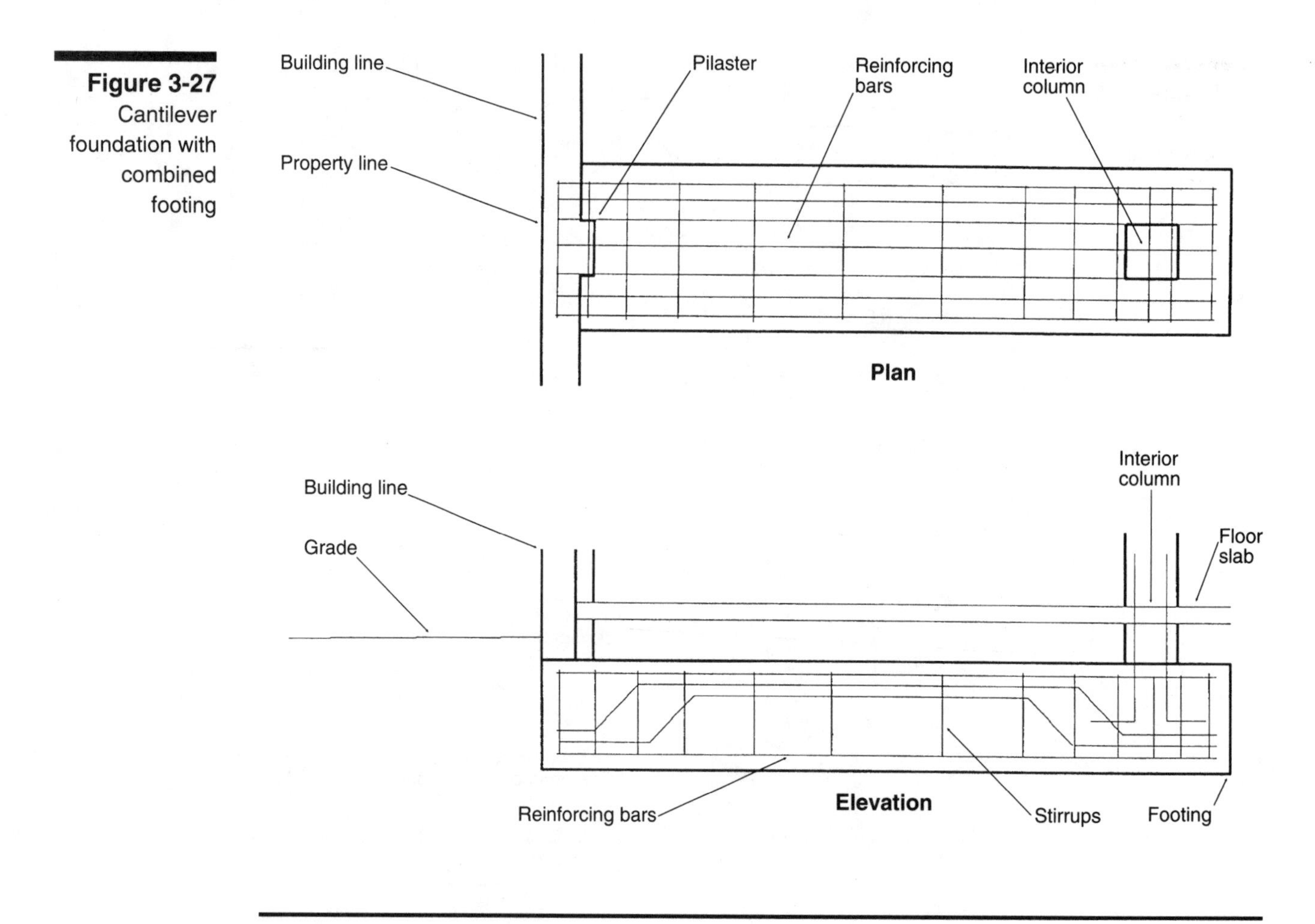

Figure 3-27 Cantilever foundation with combined footing

Structural Mat

Another type of foundation is the structural mat or slab foundation. It's a thick reinforced concrete slab which rests under the entire building. The building walls and columns are supported by the concrete mat. The mat is very rigid. It transfers all of the loads from the building equally to the soil. You can say that the structural mat floats on the soil.

Cantilever Foundations

Cantilever foundations are special isolated foundations, often used at a property line. When there's not enough room for a symmetrical foundation, it's a good idea to use a cantilever foundation. When a new building is close to the property line or an existing structure, a cantilever foundation may be the only type of foundation you can use. Figures 3-27 and 3-28 show two types of cantilever foundations.

If there's not enough room for a symmetric continuous or isolated footing, the eccentricity of the load must be counter balanced by an interior column or by the weight of the cantilevered footing. The foundation shown in Figure 3-27 uses the weight of the interior column to stabilize the eccentrically-loaded wall foundation.

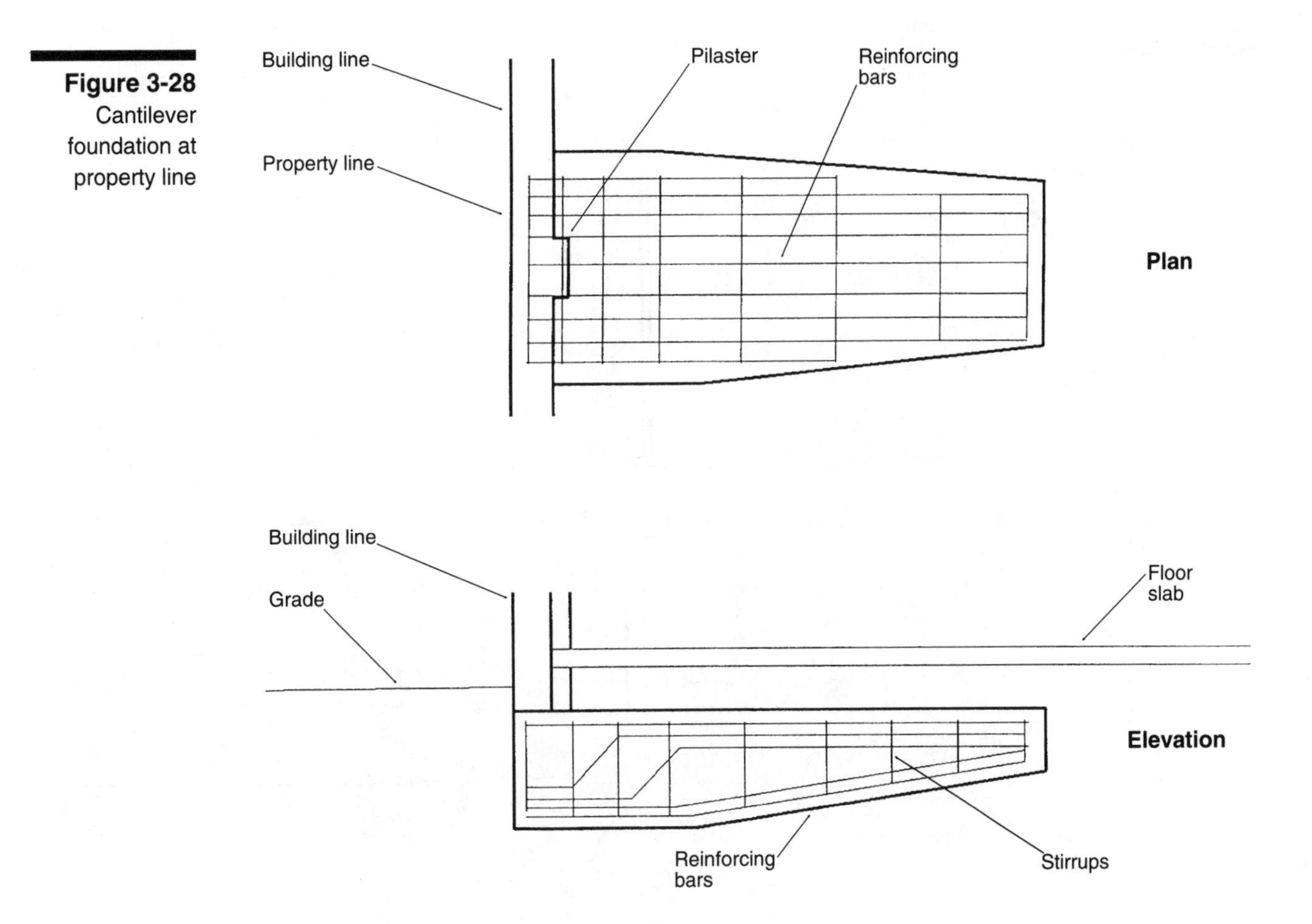

Figure 3-28 Cantilever foundation at property line

This prevents the foundation from rotating counterclockwise. The foundation shown in Figure 3-28 uses its own dead weight to counter the eccentric load from the wall which would rotate the foundation counterclockwise.

Caissons

A caisson acts as a long column that carries the weight of a building deep into the soil. Belled caissons are commonly used where soil conditions require deep foundations. Use a belled caisson foundation where there's a stiff layer of earth below soft soil or uncompacted fill.

The area of a belled caisson depends on the total of the dead and live loads, and the allowable soil bearing pressure at the bottom of the bell. The diameter of the shaft of a belled caisson depends on the size of the bucket used to form the bell. The diameter of the hole will also depend on the allowable bearing pressure of the lower stiff layer and the amount of load.

To make a belled caisson, drill a cylindrical hole to the depth you need. There are cutting blades at the bottom of the bucket which cut the soil and fill the bucket. There are also hinged cutting blades attached to the bottom of the bucket that rotate

Figure 3-29 Caisson drilling rig

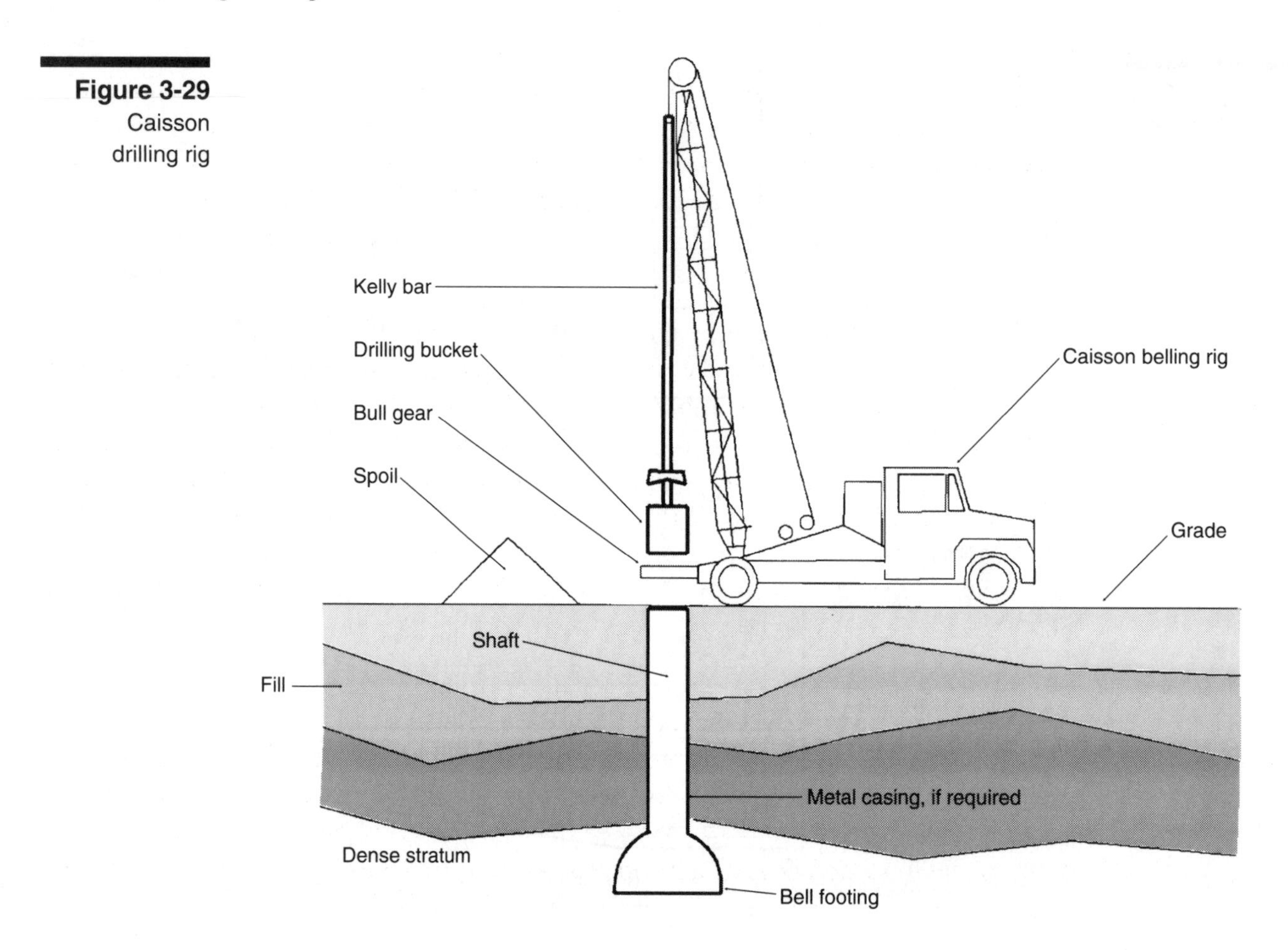

downward to cut the bell out of the soil. When the bucket is full, lift it up and dump the spoil material near the hole. Then, expand the bottom of the hole to about twice the size of the shaft.

Because of the shape of the hole, you rarely need to put in any reinforcement. If the top layers of soil are sandy, put a thin metal or fiber casing in the hole to stop cave-ins. After the completed hole has been inspected, fill the shaft and footing with concrete. Figure 3-29 shows a caisson being drilled through the layers of earth under an unstable soil.

Piles

A pile is a load-carrying column that's driven into the soil to support the weight of a structure. A pile can be made of treated or untreated wood, cast-in-place or precast concrete, steel, or a combination of materials. There are two types of piles — end-bearing and friction. An end-bearing pile is a long column that extends down to solid bedrock, and works like a leg on a table. A friction-type pile is more like a nail driven into wood. Friction between the pile's surface and the soil supports the load on the pile.

The load-carrying capacity of a pile is usually given in tons. The load-carrying capacity of a friction type pile is generally determined in two ways. The first method, which is the most common, counts the number of hammer blows needed to drive the pile 1 foot into the ground. This number is converted into tons by an approved driving formula based on a gravity-drop or power-actuated hammer.

The other method is called a static load test. With this method, a known load is put on an in-place pile. Then the pile is measured to see how much it settled under the load after 24 hours. The allowable load on the pile shouldn't be more than 50 percent of the load that will make the pile settle disproportionately. This is called the yield point of the pile. For example, suppose you have a pile that settles 0.05 inch for each additional ton of test load up to a total of 40 tons. Then, when you add more weight on it, the rate of settlement goes up to 0.08 inch for each ton of additional load. That means you've have reached the yield point of the pile, and the allowable load on it is 50 percent of 40 tons, or 20 tons.

The allowable load also shouldn't cause a net settlement of more than 0.01 inches per ton of the test load. In addition, the amount that the pile settles should remain constant over a 24-hour period. You need to use the static test method on the piles you drive into the ground with a vibratory pile driver. Don't use the formulas that apply to the force and drop of a hammer when you're not using a hammer to drive the piles.

If there are no ground heave conditions in the area where you're driving piles, you can fill the shells that are close to the new piles you're driving right away. If there are ground heaves in the area, don't fill the shells until you're driving the new piles in an area that is beyond the range of the ground heaves and you've redriven any pile shells that needed redriving. Driving new piles too close to previously driven piles may cause the previously driven piles to move out of position. If they move out of position, you'll have to start over and drive them in again. A soils engineer can determine which soils are more susceptible to ground heaves.

Wood piles

Wood piles deteriorate due to decay, insect attack, and marine-borer attack. If you bury piles in dry ground, only decay and insect attack can damage them. If you put the piles partially in water, all three can damage them.

Decay in wood piles is caused by fungi which need moisture and air to live. If a pile is completely under water, there's no air and fungi can't grow. If a wood pile is alternately wet and dry, it can deteriorate. Treat any piles that will be alternately submerged and exposed to air to keep the wood from rotting.

Metal piles

A popular type of metal-cased pile is a combination of steel shell and concrete fill, usually called a step-taper pile. It has a closed-end corrugated steel shell which is driven in with a full-length steel mandrel. A mandrel is a heavy steel tip inserted into a pile casing which is driven into the soil. Once the pile is far enough in, the shell is filled with concrete.

Shell sections in step-piles come in basic lengths of 4, 8, 12, and 16 feet. Join shell sections to make up the pile length you need. The diameter of the pile increases from the tip (bottom) to the butt (top) at the rate of 1 inch per section length. The rate of taper, or pile shape, may vary with the sections of length used. Within practical limits, many section lengths can be combined into a single pile. Tip diameters are usually 8 to 11 inches, but you can get larger ones.

Step-taper piles are a good choice in most types of soil. They're used both as friction and end-bearing piles. Usually, these piles are supported by a combination of friction and direct (or end) bearing. The pile's tapered shape lets it carry a higher load than similar piles without a taper. Usually you won't need any internal steel reinforcement in step-taper piles or caissons. The exception is piles that are subjected to uplift, high lateral loads, or those extending through air, water, or very fluid soils.

Concrete piles

Concrete piles can be cast under water. Concrete is poured through a tremie into a drilled hole without forms. Special techniques and design mixes are used for placing concrete into piles or caissons to make sure the pile shell is filled completely and uniformly. The most common is the tremie, which is a pipe with a funnel-shaped upper end into which the concrete is fed. The tremie should be lifted slowly to let the concrete flow out. A hose is used to carry concrete to the bottom of the casing. The pipe must be long enough to reach from the working platform or ground to the lowest point the concrete will be poured. You should keep the bottom of the discharge end buried in newly-placed concrete the entire time you are pouring the concrete. Pumping from a mixer is the best way of supplying concrete. Tremies may be suspended by a crane boom and can be raised and lowered by the boom.

The concrete should have a slump of about 6 inches. About 60 percent of the total aggregate should be sand and the maximum coarse aggregate should be 1½ to 2 inches. A plasticizing agent in the mix helps prevent arching by providing a uniform consistency to the concrete.

Before selecting any type of pile foundation, consider the properties of the soil you'll be working in. Here's some of the questions you should ask yourself:

- What is the depth of the bearing stratum? The bearing stratum is the top of the natural soil which is capable of supporting the weight of the building, such as the bottom of an uncompacted fill.

- What is the friction factor of the soil? The friction factor of a soil is the amount of resistance it has to the surface of a pile. The total load-carrying capacity of a friction pile is equal to the friction factor of the soil times the total surface area of the pile.

- Is there a fill that may make the piles sink? Piles that are driven through a sinking or subsiding fill can be dragged down by the fill.

- Are batter, or sloping, piles necessary for lateral loads?

- What are the expected loads on the piles? Expected loads on piles are the same as for other types of foundations — dead load, live load, and lateral loads from wind or earthquake.

- How many piles are required in each group? When the total column load is much greater than the carrying capacity of an individual pile, additional piles must be used in a cluster. For example, if the total building column load is 60 tons, and the carrying capacity of a pile is 20 tons, a three-pile cluster is required.

- What is the minimum spacing between piles? Normally piles are not closer than 3 feet or 3 pile diameters, whichever is greater. Resistance piles are usually tapered outward from the tip to the butt. Once you know what length pile you need to carry a particular load, you'll also know the diameter at the top of the pile. Using the diameter of the tip and the amount of taper for a specific type of pile (wood, precast concrete or steel case), you can find the butt diameter by using the pile's length. For example, if the pile diameter increases by 1 inch per 10 feet from the tip to the butt and the tip is 8 inches in diameter, a 30-foot pile would have an 11-inch diameter butt ($8 + {}^{30}\!/_{10} = 11$). The number of piles and their diameters govern the shape and size of the pile cap. A pile cap is similar to a footing except that it's supported by the top of the pile cluster and not the soil. Figure 3-30 shows a typical pile cap.

- What is the minimum depth of the pile cap above top of pile? The top of piles are usually embedded into the bottom of the pile cap a minimum of 4 inches. The thickness of the pile cap above the top of the pile cap is determined by a 45-degree plane between the pile and the pedestal on the pile cap. Figure 3-30 shows the location of the 45-degree plane between the pile and the pedestal. The steel reinforcing mat at the bottom of the pile cap is a minimum of 3 inches above the top of the pile. When you have all this information, you'll know the number, length, and diameter of piles and the size of the pile cap.

Figure 3-30 shows three kinds of pile caps — clusters of two, three, and four piles. The top of a pile goes at least 4 inches into the cap, and the reinforcing mat is not less than 3 inches above the top of the pile. This figure doesn't show the grade beams that tie the pile caps together in both directions.

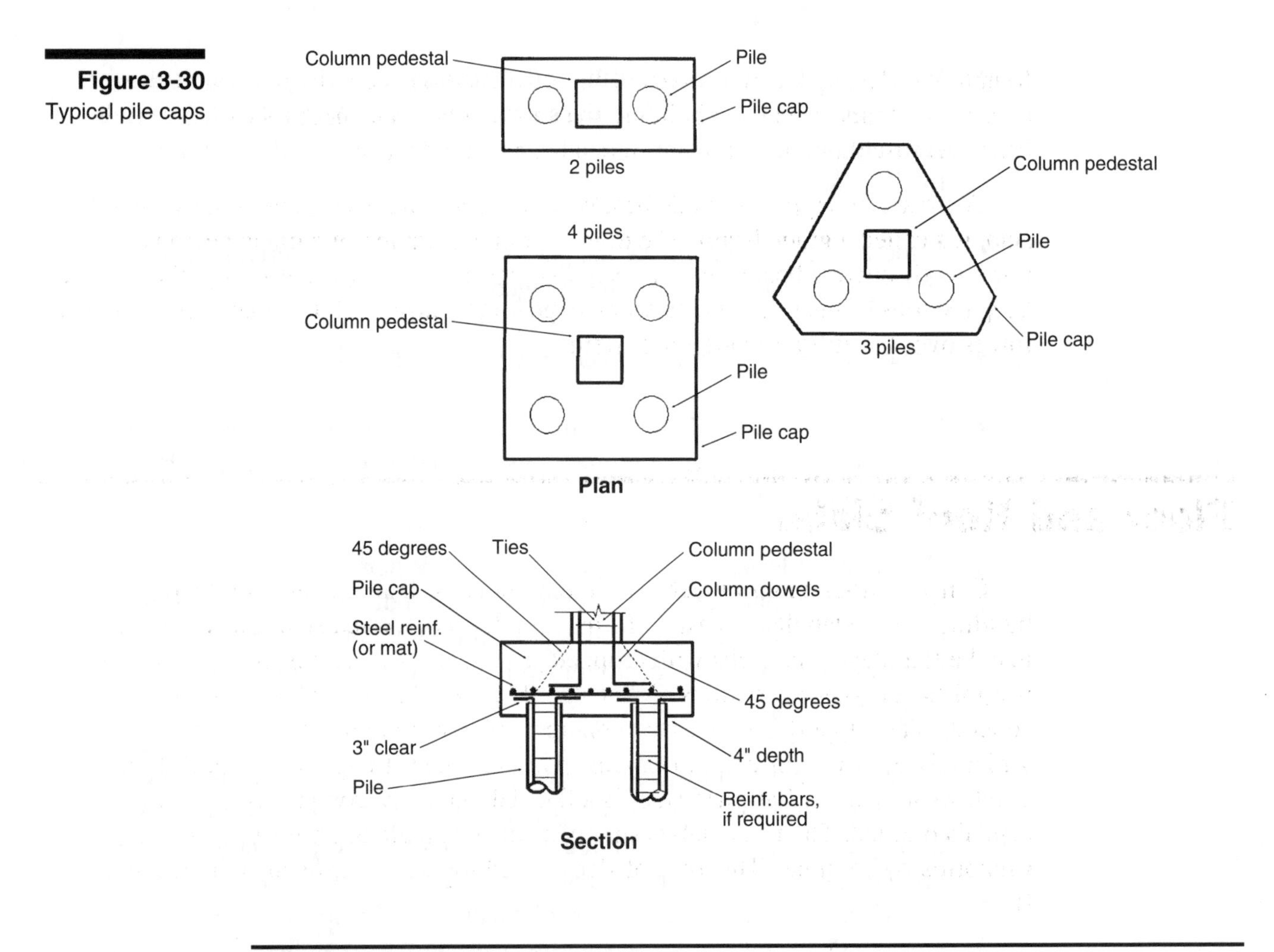

Figure 3-30
Typical pile caps

Machinery Foundations

Unlike other foundations, supports for heavy dynamic machinery are often sized by the total weight of the concrete. As a rule of thumb, the weight of a block foundation for a reciprocating machine should be at least five times the weight of the machine. A reciprocating machine has a greater dynamic motion than a rotary machine. Typical examples of reciprocating machines are piston-driven compressors and gasoline engines. Rotary machines include centrifugal pumps and electric motors. Therefore, more mass in the concrete foundation is required to absorb the vibrations of a reciprocating machine. The concrete block should weigh at least three times the weight of a rotating machine. The block has to be massive enough to absorb and dampen the vibration of the machine. Usually vibration isolators will be installed between the machine and the foundation. You can use neoprene pads, ductile iron pads, and air cushion or spring-type isolators.

Grade Beams

Grade beams are the rectangular concrete members placed around the outside of a building. They support the exterior walls. The two lower footings shown on Figure 3-26 show a grade beam and floor slab cast together. This is called a monolithic foundation. The upper two details of Figure 3-26 show a continuous foundation. A continuous foundation for a building with a slab floor is sometimes called a grade beam because the foundation acts like a beam at grade level.

There are two types of grade beams. When a footing is cast integrally with a floor slab, it's called a grade beam. The more common meaning of a grade beam is a reinforced concrete beam cast on grade that ties pile caps or caissons together. These keep the tops of the piles and caissons in position and spread the wind and earthquake forces over the entire foundation system.

Floor and Roof Slabs

Concrete floor and roof slabs are important parts of single- and multistory buildings. Any slab that is cast on forms is called an elevated slab. Elevated slabs may be flat plates, flat slabs with dropped panels, one-way and two-way solid slabs, pan joists, and other variations. A flat slab floor is a solid or ribbed reinforced concrete slab supported directly by concrete columns without the aid of beams or girders. Drop panels or column capitals provide added strength to the slab at the points of support. A flat-plate floor is a flat slab with the drop panels or column capitals omitted. The floor slab is simply a concrete plate of uniform thickness supported by columns. This type of slab is used for lighter loads than the flat slab floor.

A pan joist slab is a type of slab with ribs, or joists, formed by metal pans. When the ribs run in both directions, the slab is called a waffle slab. One- and two-way slabs are discussed in the next section.

Slab Reinforcement

All of these slabs are reinforced with either steel bars or wire fabric. Tension steel reinforcing is used to carry loads, and temperature steel reinforcing is used to keep concrete from cracking due to thermal expansion and contraction. It's always a good idea to use temperature reinforcement. It helps prevent cracking as the concrete cures as well as contraction and expansion cracking caused by temperature changes.

There are two types of temperature reinforcement — deformed steel bars and welded wire fabric. The minimum amount of temperature steel is determined by the thickness of the slab. Figure 3-31 lists the average temperature steel reinforcement required for one-way slabs that are 2 to 8 inches thick. Figure 3-32 shows preliminary sizing of thickness and reinforcement for one-way floor slabs in residential buildings. The final sizing of a slab and reinforcement should be done by a qualified structural designer or registered engineer.

Figure 3-31 Typical temperature reinforcement for one-way slabs

Slab thickness (inches)	Roof slabs	Floor slabs
2	6X6/6-6 WWM	6X6/6-6 WWM
2.5	6X6/4-4 WWM	6X6/6-6 WWM
3	#3 @ 16" o.c.	6X6/4-4
3.5	#3 @ 12" o.c.	#3 @ 12" o.c.
4	#3 @ 12" o.c.	#3 @ 15" o.c.
4.5	#3 @ 12" o.c.	#3 @ 16" o.c.
5	#4 @ 18" o.c.	#3 @ 13" o.c.
5.5	#4 @ 18" o.c.	#3 @ 12" o.c.
6	#4 @ 16" o.c.	#3 @ 12" o.c.
6.5	#4 @ 14" o.c.	#4 @ 18" o.c.
7	#4 @ 13" o.c.	#4 @ 16" o.c.
7.5	#4 @ 12" o.c.	#4 @ 14" o.c.
8	#4 @ 12" o.c.	#4 @ 14" o.c.

Figure 3-32 Typical one-way slab reinforcement

Span (feet)	Thickness (inches)	Reinforcing steel
4	3	#3 @ 7" o.c.
5	3	#3 @ 7" o.c.
6	3	#3 @ 7" o.c.
7	3½	#3 @ 5½" o.c.
8	3½	#3 @ 5½" o.c.
9	4	#3 @ 4½" o.c.
10	4½	#4 @ 7½" o.c.
11	5	#4 @ 6½" o.c.
12	5½	#4 @ 5½" o.c.
13	6	#4 @ 5" o.c.

A typical example of temperature reinforcing occurs in a slab cast on grade. If the slab is 4 inches thick, every foot width of the slab contains 48 square inches of concrete. The minimum amount of temperature reinforcement is 0.002 × 48 or 0.096 square inches per foot width of slab. As a #3 bar has a cross-sectional area of 0.11 square inches, a #3 bar at 12 inches on center would satisfy this requirement. A 4 × 4-W3 × W3 welded wire mesh would also be enough reinforcement. You can use a mat of welded reinforcing bars instead of wire mesh.

You'll need to take special care with slabs cast on grade. You have to keep moisture out, and prevent temperature changes from cracking them. Use a minimum thickness of 3½ inches, although this will usually be specified as a 4-inch nominal thickness. Put welded wire fabric in the concrete to keep it from cracking as it cures. Use an area of temperature steel that is at least 0.2 percent of the area of the concrete slab.

If there's moisture under a slab, the slab will act like a blotter, drawing the moisture to the surface. Use a bed of coarse aggregate that's 3 to 6 inches thick under a slab to keep water from soaking into the concrete. The coarse aggregate used under a slab on grade is usually between ¼ inch to ½ inch in diameter. On top of the aggregate, put a waterproof membrane under a slab as an added precaution.

Sizing a Slab

To figure the preliminary size of an elevated floor slab such as a multistory residential building, use a live load of 40 pounds per square foot. A live load means any temporary load such as occupants, appliances, or furniture. The minimum live load for roof slabs is 20 pounds per square foot in most areas. In places where snow can build up, use a live load of 30 pounds per square foot. The live load for parking garage floors is usually 50 pounds per square foot. Each slab must also carry a dead load which is the weight of all permanent structural and nonstructural components such as the slab itself plus the ceilings, partitions, piping, and flooring.

Figures 3-33A, 3-33B, and 3-33C show Table No. 23-A (Uniform and Concentrated Loads), Table No. 23-B (Special Loads), and Table No. 23-C (Minimum Roof Live Loads) from the 1991 *Uniform Building Code*.

One-Way Slabs

The most common concrete slab is the one-way solid slab, which spans between two parallel beams. The beams may be steel or concrete. Load-carrying reinforcement is placed near the bottom of the slab, perpendicular to the direction of the beams. The temperature reinforcement, which doesn't carry the load, goes near the top of the slab, parallel to the beams.

Two-Way Flat Slabs

A two-way slab is a concrete slab which spans between two sets of parallel beams that are 90 degrees from each other. Tension bars run in both directions. Buildings with columns, in which the bays are nearly square, may use a two-way flat slab system. A bay is the horizontal space between adjacent columns and walls. This kind of slab has load-carrying reinforcement in both directions. Some slabs have dropped panels, or thickened slabs, over the columns.

Figure 3-33A
Uniform and Concentrated Loads

TABLE NO. 23-A—UNIFORM AND CONCENTRATED LOADS			
USE OR OCCUPANCY		UNIFORM LOAD[1]	CONCENTRATED LOAD
Category	Description		
1. Access floor systems	Office use	50	2,000[2]
	Computer use	100	2,000[2]
2. Armories		150	0
3. Assembly areas[3] and auditoriums and balconies therewith	Fixed seating areas	50	0
	Movable seating and other areas	100	0
	Stage areas and enclosed platforms	125	0
4. Cornices, marquees and residential balconies		60	0
5. Exit facilities[4]		100	0[5]
6. Garages	General storage and/or repair	100	[6]
	Private or pleasure-type motor vehicle storage	50	[6]
7. Hospitals	Wards and rooms	40	1,000[2]
8. Libraries	Reading rooms	60	1,000[2]
	Stack rooms	125	1,500[2]
9. Manufacturing	Light	75	2,000[2]
	Heavy	125	3,000[2]
10. Offices		50	2,000[2]
11. Printing plants	Press rooms	150	2,500[2]
	Composing and linotype rooms	100	2,000[2]
12. Residential[7]		40	0[5]
13. Restrooms[8]			
14. Reviewing stands, grandstands, bleachers, and folding and telescoping seating		100	0
15. Roof decks	Same as area served or for the type of occupancy accommodated		
16. Schools	Classrooms	40	1,000[2]
17. Sidewalks and driveways	Public access	250	[6]
18. Storage	Light	125	
	Heavy	250	
19. Stores	Retail	75	2,000[2]
	Wholesale	100	3,000[2]

[1]See Section 2306 for live load reductions.

[2]See Section 2304 (c), first paragraph, for area of load applications.

[3]Assembly areas include such occupancies as dance halls, drill rooms, gymnasiums, playgrounds, plazas, terraces and similar occupanies which are generally accessible to the public.

[4]Exit facilities shall include such uses as corridors serving an occupant load of 10 or more persons, exterior exit balconies, stairways, fire escapes and similar uses.

[5]Individual stair treads shall be designed to support a 300-pound concentrated load placed in a position which would cause maximum stress. Stair stringers may be designed for the uniform load set forth in the table.

[6]See Section 2304 (c), second paragraph, for concentrated loads.

[7]Residential occupancies include private dwellings, apartments and hotel guest rooms.

[8]Restroom loads shall not be less than the load for the occupancy with which they are associated, but need not exceed 50 pounds per square foot.

Figure 3-33B
Special Loads

TABLE NO. 23-B—SPECIAL LOADS[1]			
USE		VERTICAL LOAD	LATERAL LOAD
Category	Description	(pounds per square foot unless otherwise noted)	
1. Construction, public access at site (live load)	Walkway, see Sec. 4406	150	
	Canopy, see Sec. 4407	150	
2. Grandstands, reviewing stands, bleachers, and folding and telescoping seating (live load)	Seats and footboards	120[2]	See Footnote No. 3
3. Stage accessories (live load)	Gridirons and fly galleries	75	
	Loft block wells[4]	250	250
	Head block wells and sheave beams[4]	250	250
4. Ceiling framing (live load)	Over stages	20	
	All uses except over stages	10[5]	
5. Partitions and interior walls, see Sec. 2309 (live load)			5
6. Elevators and dumbwaiters (dead and live load)		2 × Total loads[6]	
7. Mechanical and electrical equipment (dead load)		Total loads	
8. Cranes (dead and live load)	Total load including impact increase	1.25 × Total load[7]	0.10 × Total load[8]
9. Balcony railings and guardrails	Exit facilities serving an occupant load greater than 50		50[9]
	Other		20[9]
10. Handrails		See Footnote No. 10	See Footnote No. 10
11. Storage racks	Over 8 feet high	Total loads[11]	See Table No. 23-P
12. Fire sprinkler structural support		250 pounds plus weight of water-filled pipe[12]	See Table No. 23-P
13. Explosion exposure	Hazardous occupancies, see Sec. 910		

[1]The tabulated loads are minimum loads. Where other vertical loads required by this code or required by the design would cause greater stresses, they shall be used.

[2]Pounds per lineal foot.

[3]Lateral sway bracing loads of 24 pounds per foot parallel and 10 pounds per foot perpendicular to seat and footboards.

[4]All loads are in pounds per lineal foot. Head block wells and sheave beams shall be designed for all loft block well loads tributary thereto. Sheave blocks shall be designed with a factor of safety of five.

[5]Does not apply to ceilings which have sufficient total access from below, such that access is not required within the space above the ceiling. Does not apply to ceilings if the attic areas above the ceiling are not provided with access. This live load need not be considered as acting simultaneously with other live loads imposed upon the ceiling framing or its supporting structure.

[6]Where Appendix Chapter 51 has been adopted, see reference standard cited therein for additional design requirements.

[7]The impact factors included are for cranes with steel wheels riding on steel rails. They may be modified if substantiating technical data acceptable to the building official is submitted. Live loads on crane support girders and their connections shall be taken as the maximum crane wheel loads. For pendant-operated traveling crane support girders and their connections, the impact factors shall be 1.10.

[8]This applies in the direction parallel to the runway rails (longitudinal). The factor for forces perpendicular to the rail is 0.20 × the transverse traveling loads (trolley, cab, hooks and lifted loads). Forces shall be applied at top of rail and may be distributed among rails of multiple rail cranes and shall be distributed with due regard for lateral stiffness of the structures supporting these rails.

[9]A load per lineal foot to be applied horizontally at right angles to the top rail.

[10]The mounting of handrails shall be such that the completed handrail and supporting structure are capable of withstanding a load of at least 200 pounds applied in any direction at any point on the rail. These loads shall not be assumed to act cumulatively with Item 9.

[11]Vertical members of storage racks shall be protected from impact forces of operating equipment, or racks shall be designed so that failure of one vertical member will not cause collapse of more than the bay or bays directly supported by that member.

[12]The 250-pound load is to be applied to any single fire sprinkler support point but not simultaneously to all support joints.

Figure 3-33C
Minimum Roof Live Loads

TABLE NO. 23-C—MINIMUM ROOF LIVE LOADS[1]

Roof slope	METHOD 1			METHOD 2		
	TRIBUTARY LOADED AREA IN SQUARE FEET FOR ANY STRUCTURAL MEMBER			UNIFORM LOAD[2]	RATE OF REDUCTION r (Percent)	MAXIMUM REDUCTION R (Percent)
	0 to 200	201 to 600	Over 600			
1. Flat or rise less than 4 inches per foot. Arch or dome with rise less than one eighth of span	20	16	12	20	.08	40
2. Rise 4 inches per foot to less than 12 inches per foot. Arch or dome with rise one eighth of span to less than three eighths of span	16	14	12	16	.06	25
3. Rise 12 inches per foot and greater. Arch or dome with rise three eighths of span or greater	12	12	12	12	No Reductions Permitted	
4. Awnings except cloth covered[3]	5	5	5	5		
5. Greenhouses, lath houses and agricultural buildings[4]	10	10	10	10		

[1]Where snow loads occur, the roof structure shall be designed for such loads as determined by the building official. See Section 2305 (d). For special-purpose roofs, see Section 2305 (e).
[2]See Section 2306 for live load reductions. The rate of reduction r in Section 2306 Formula (6-1) shall be as indicated in the table. The maximum reduction R shall not exceed the value indicated in the table.
[3]As defined in Section 4506.
[4]See Section 2305 (e) for concentrated load requirements for greenhouse roof members.

Figure 3-34 shows a typical two-way slab floor plan. You can think of a two-way flat slab as strips of concrete running perpendicular to each other. The slab is divided into panels, and the panels into strips. The panels are called either exterior or interior panels. The strip over the centerline of the columns is called the column strip. The width of the column strip is one-half of the span in the opposite direction. The strip between the column strips is called the middle strip. The width of the middle strip is also equal to one-half the span in opposite direction. The strip along side the exterior wall or columns is called the exterior strip and its width is one-quarter of the span in opposite direction.

The abbreviations shown on Figure 3-34 are:

L = span of slab between supports or columns.

L/2 = width of column strip in feet and equal to one-half of L.

C/L = centerline of columns

Col = column

Because each area has different stress conditions, the thickness and reinforcement are based on the loads and locations of the panels and strips. The steel reinforcement must resist the different tensile stresses in each strip. The reinforcement of each column, middle, or exterior strip is based on assuming that the strip is a separate beam carrying its proportional amount of load. The reinforcement in each strip is

Figure 3-34
Plan of a typical two-way flat slab

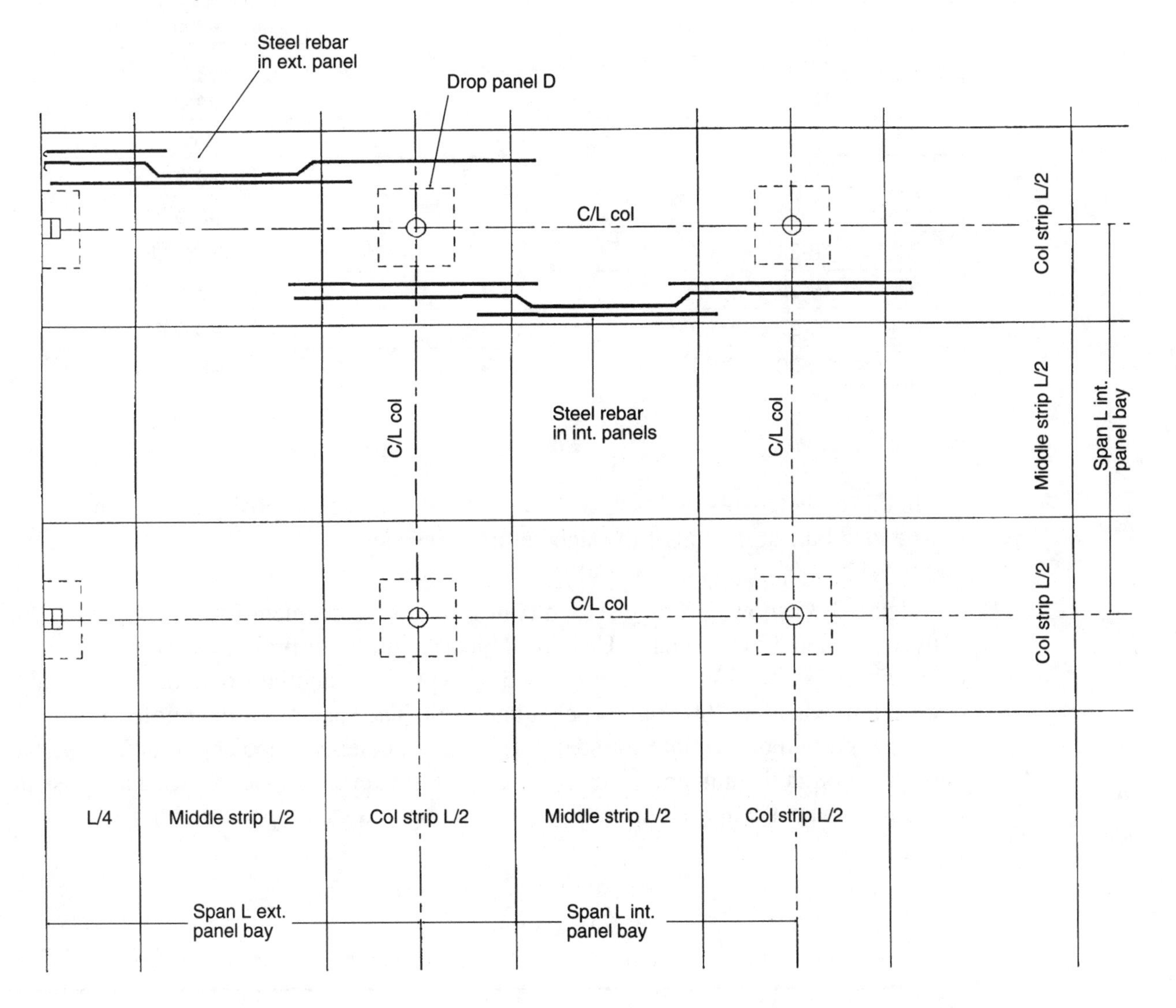

dependent upon the strength of concrete, the thickness of slab, the width of the strip, the span of the strip and loading condition on the strip. The selection of reinforcement should be done by a qualified structural designer or registered engineer.

Reinforcing steel in two-way slabs is identified as:

- positive steel in the column strip
- negative steel in the column strip
- positive steel in the middle strip
- negative steel in the middle strip
- temperature steel

Figure 3-35
Typical two-way concrete slab reinforcement

Spans, range (feet)	Thickness (inches)	Reinforcement, top and bottom
10-13	4	#3 & 6" o.c.
10-14	4.5	#4 & 8" o.c.
11-15	5	#4 & 7" o.c.
12-16	5.5	#4 & 6" o.c.
13-17	6	#4 & 5.5" o.c.
15-20	6.5	#4 & 5" o.c.
16-23	7	#5 & 7" o.c.
18-24	7.5	#5 & 6.5" o.c.
18-25	8	#5 & 6" o.c.
19-27	8.5	#6 & 8" o.c.
20-30	9	#6 & 7" o.c.

In this case, tensile load bars at the bottom of the slab are positive steel, and tensile load bars at the top of the slab are negative steel.

Figure 3-35 shows average slab thickness and reinforcement for square bays between 10 and 30 feet square. Use this table as a guide for preliminary cost estimating or sizing. The values are based on typical residential floors having a total dead and live load of 100 pounds per square foot. The spacing of the reinforcement will be slightly more for interior spans. The slab thicknesses also apply to rectangular panels, although the amount of reinforcement is greater in the shorter span and less in the longer span. The maximum ratio between long and short spans is about 1.8 to 1.

Metal Pans and Concrete Joists

To reduce the amount of wood forming and shoring, you can use removable metal pans to form concrete floors and roofs. The pans have tapered sides, 20 or 30 inches wide, and 6 to 14 inches deep. The spacing between the pan edges determines the joist width.

The sides of the individual metal pans taper inward, so the joists are narrower at the bottom than at the top. The concrete topping over the pans may be 2 to 4 inches thick. This makes a concrete joist that is 8 to 18 inches deep and 4 to 8 inches wide, which creates a concrete joist floor system. The joists are cast together with the floor slab, as shown in Figure 3-36. The term "¾" clear" in the figure means the minimum concrete cover over the steel bars should be ¾ inches.

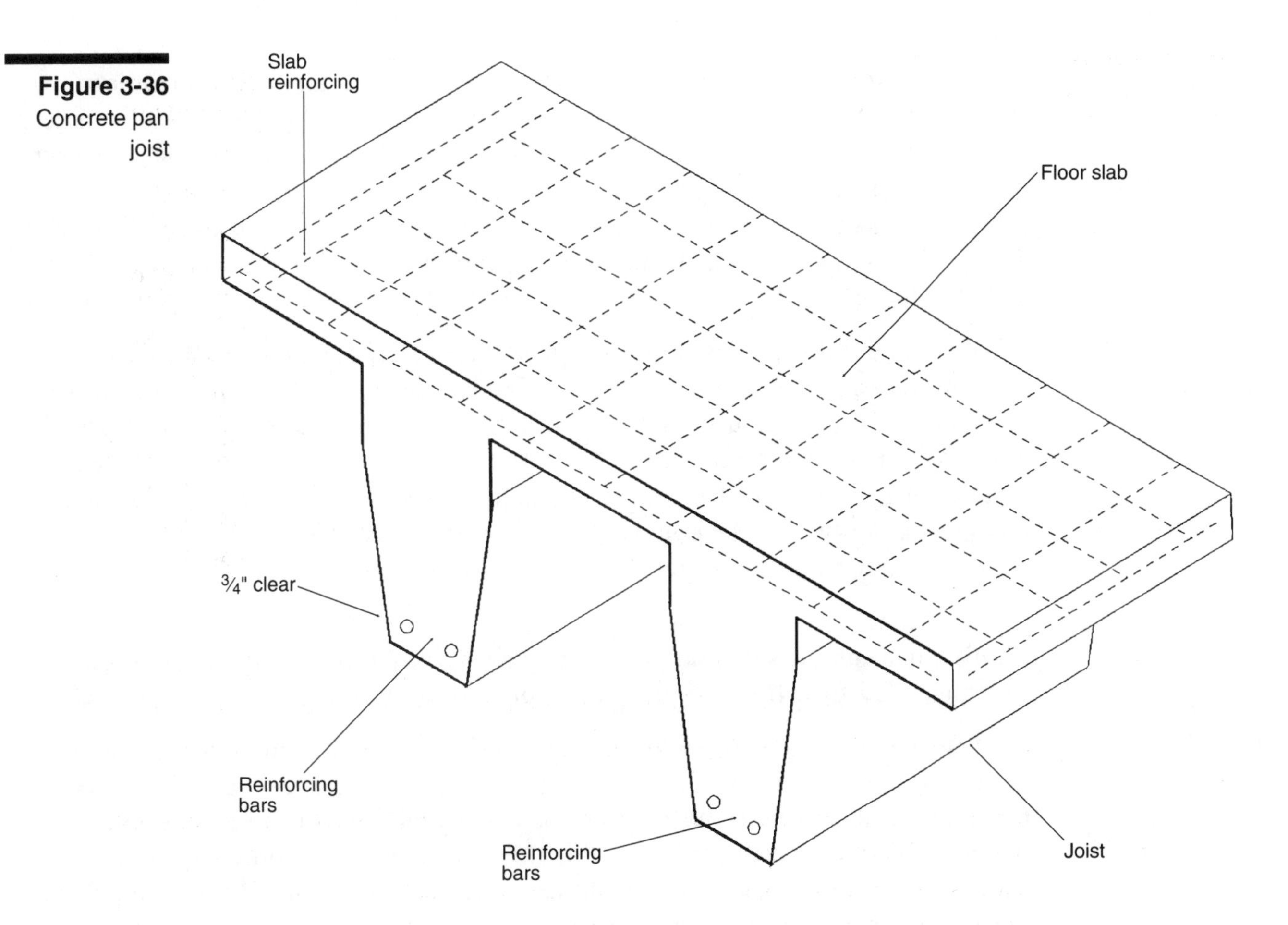

Figure 3-36
Concrete pan joist

Support the pans with beams and shoring. Set the shores at midspan to prevent deflection. Support of metal pans depends very much on the size and gauge of the steel pan and thickness of the slab. The heavier the concrete slab and joists, the closer the shores must be. You should follow the specifications of the pan manufacturer for the required shoring.

Generally, support for 20- and 30-inch wide metal pans consists of:

- 2 × 8 soffit planks which support the bottom of the pan
- 4 × 6 stringers which support the soffit planks
- 4 × 4 wood shores with Ellis clamps which support the stringers. The Ellis clamps let you adjust the shores to the required height.

As an example, these pans are usually supported at the ends, and at approximately 6- to 7-foot intervals by the stringers. The 4 × 4 posts are spaced at about 10-foot intervals.

Reshoring is usually done to facilitate maximum reuse of the formwork by utilizing the strength of the concrete before it has fully cured. Reshoring is a critical operation in formwork and should be planned in advance and approved by an engineer.

There's no single shoring procedure that's standard for all types of buildings. Each condition is different. The thickness of an elevated slab and the size of concrete beams determine the weight that must be supported. The stringers, or joists, under the plywood deck are sized according to the weight of the slab and the distance the plywood spans between supporting stringers. The stringers, in turn, are sized by their span, the weight supported, and the carrying capacity. This, in turn, depends on the height of the shores. So, as you can see, everything in a shoring system is interrelated.

The usual practice is that the contractor warehouses a stock of standard size plywood, stringers, bracing and shores. The builder plans the shoring layout for each job so that he can use the material in his inventory. Another method used by contractors is to rent prefabricated shoring material for the job. Prefabricated metal shoring consist of frames, screw jacks, cross braces, beams, purlins and the connecting hardware. The company that manufactures or rents this material will usually prepare the shoring layout for the specific job as part of the rental cost.

Most manufacturers of shoring material belong to the Scaffolding and Shoring Institute. It's a good idea to get brochures from this organization on recommended steel frame shoring erection and safety rules.

After the pans are in place, ironworkers install the reinforcing bars. Then another crew casts the concrete in, and over, the pans. Others float, screed, and trowel the topping to a smooth finish. When the concrete is almost cured, the carpenters remove the pans and supports. But the forms shouldn't be stripped off too soon. Ideally, you should leave them in place throughout the required curing period. But you usually have to strip the forms as soon as possible so you can use them someplace else. Also, if you're using some special concrete finishing operation, such as rubbing, you may have to remove the forms before the curing period is over. Whatever you do, don't remove the forms before the concrete is strong enough to carry its own weight and any other loads that may be placed on it during construction.

Screw jacks, which are a type of adjustable shores, should be released far enough so you can remove the form lumber. Use the reverse of the method you used to erect the shoring to dismantle it. Then you can move it to the next location and use it there. Frequently, you can move formwork and shoring of varying sizes from one pour to other pours without dismantling or removing it.

Usually you reshore to get the maximum reuse of the formwork. You may use the strength of the completed concrete construction below when it's fully cured and can support the new superimposed loads. Reshoring is one of the most critical operations in formwork. Be very careful when you release the adjustment screws to a point where the slab takes its actual permanent deflection. Tighten the adjustment screws until they contact the underside of the slab again. Then the reshoring below won't be carrying the load of the slab it was shoring before.

Usually you remove forms for footings, columns, and sides of beams and walls before the forms for floors and beam bottoms. It's safer to test the strength of the concrete rather than arbitrarily deciding when to remove the forms. You can figure out the relationship between the age of the concrete and its strength by testing representative samples. Under average conditions, using air-entrained concrete with a water-cement ratio of six gallons per sack, the times required to attain certain strengths for Type I-Normal cement are: 24 hours for 500 psi, 36 hours for 750 psi, 3.5 days for 1,500 psi and 5.5 days for 2000 psi. If Type III - High Early cement is used in the concrete, the times required to attain the listed strengths are: 12 hours for 500 psi, 18 hours for 750 psi, 36 hours for 1,500 psi and 2.5 days for 2,000 psi.

A minimum of 500 psi compressive strength should be attained before concrete is exposed to freezing. These times are affected by materials used, temperatures and other conditions and vary from job to job. As a rule of thumb, leave forms and shoring supporting the bottom of a slab, beam or girder for 14 days after placing the concrete. Side forms for beams, girders and columns may be removed after ten days, if arrangements are made to cure and protect the concrete that has been exposed. If the temperature is below 45 degrees F, don't remove the forms until you've tested the concrete to make sure it's thoroughly hardened.

Design the forms so they're easy to strip. Make sure you won't damage the concrete when you take off the forms. Plan on reusing the floor panels, wall panels, beam forms, and column forms, as units. Make bevel cuts and keys so you don't have to do much prying to release the forms. Right after you remove the forms, take the nails out and clean and oil the forms you're going to reuse.

You can use this system for floor or roof spans from 12 to 30 feet. Figure 3-36 shows a typical concrete joist after the steel panels have been removed. Figure 3-37 shows the pan and supporting shores. Shoring, which uses posts, struts, jacks, and bracing, makes up a form set in concrete construction. The number of form sets necessary for a multistory building depends on the weather and the shape of the members in each floor. Under ideal conditions, 1½ form sets (forms for 1½ stories) will allow work to progress at a rate of one story in a week or ten days. If the floor framing is complicated, a single form set is more economical. Where the floor area is large and the construction is uniform, less than one set of forms with additional beam bottoms and posts may be sufficient.

Metal Decking and Concrete Fill

Concrete slabs can be poured on corrugated or ribbed metal decking over steel framing. Because the corrugated metal decking can span relatively long distances between steel beams, you'll need little or no shoring. Iron crews place and weld the decking to the steel beams. They also attach a steel L-shaped angle along the periphery of the deck to act as a screed to control the thickness of the slab. Then they put steel reinforcing bars on chairs set on the decking. After the electrical conduits, pipes, and other embedded items are in place and inspected, concrete is poured 4 to 6 inches thick over the decking. Concrete finishers complete the job. Figure 3-38 is a detail of a typical spandrel beam. The concrete slab is cast on the metal deck, which is supported by steel beams. The metal decking is used as a form for an elevated slab.

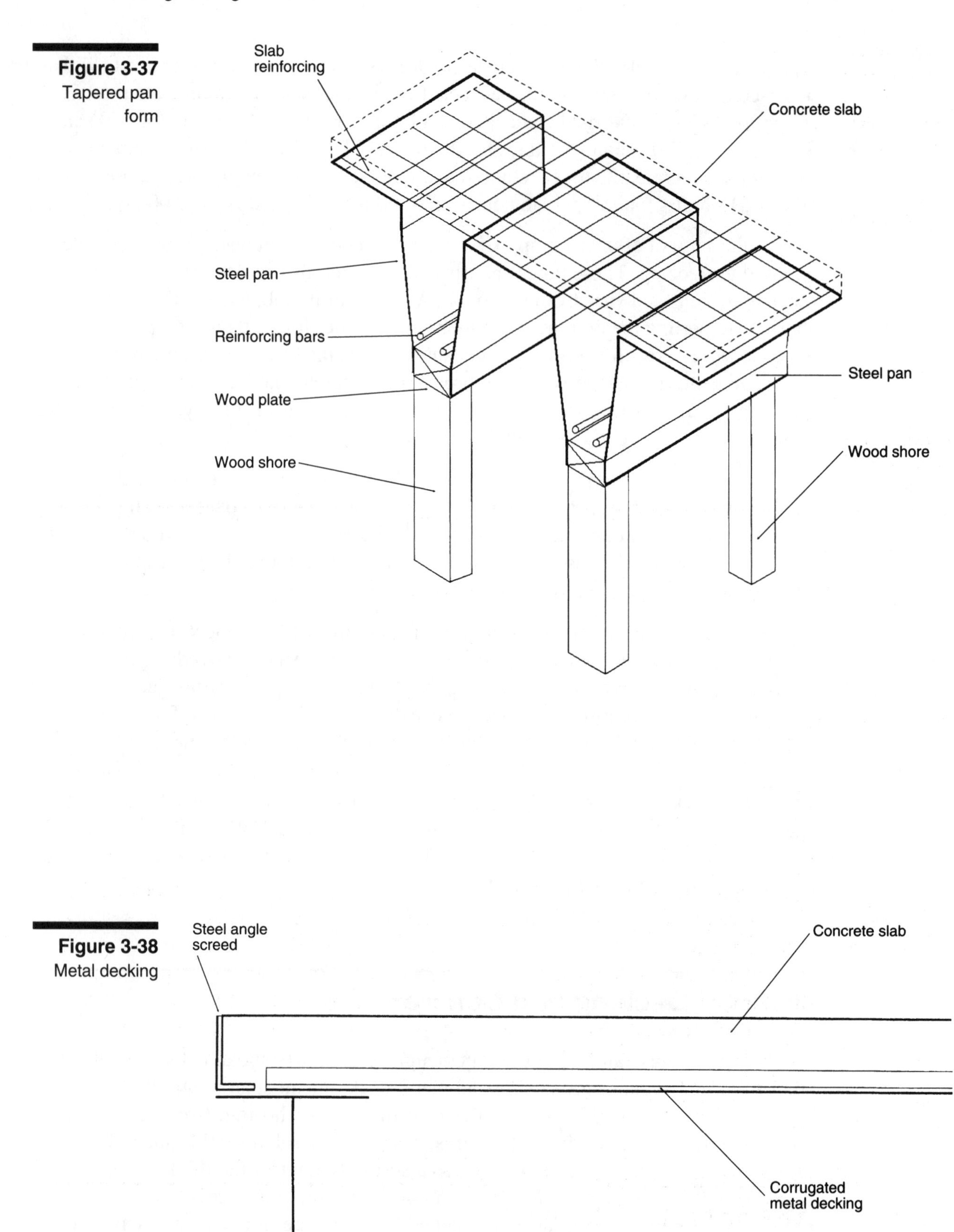

Figure 3-37 Tapered pan form

Figure 3-38 Metal decking

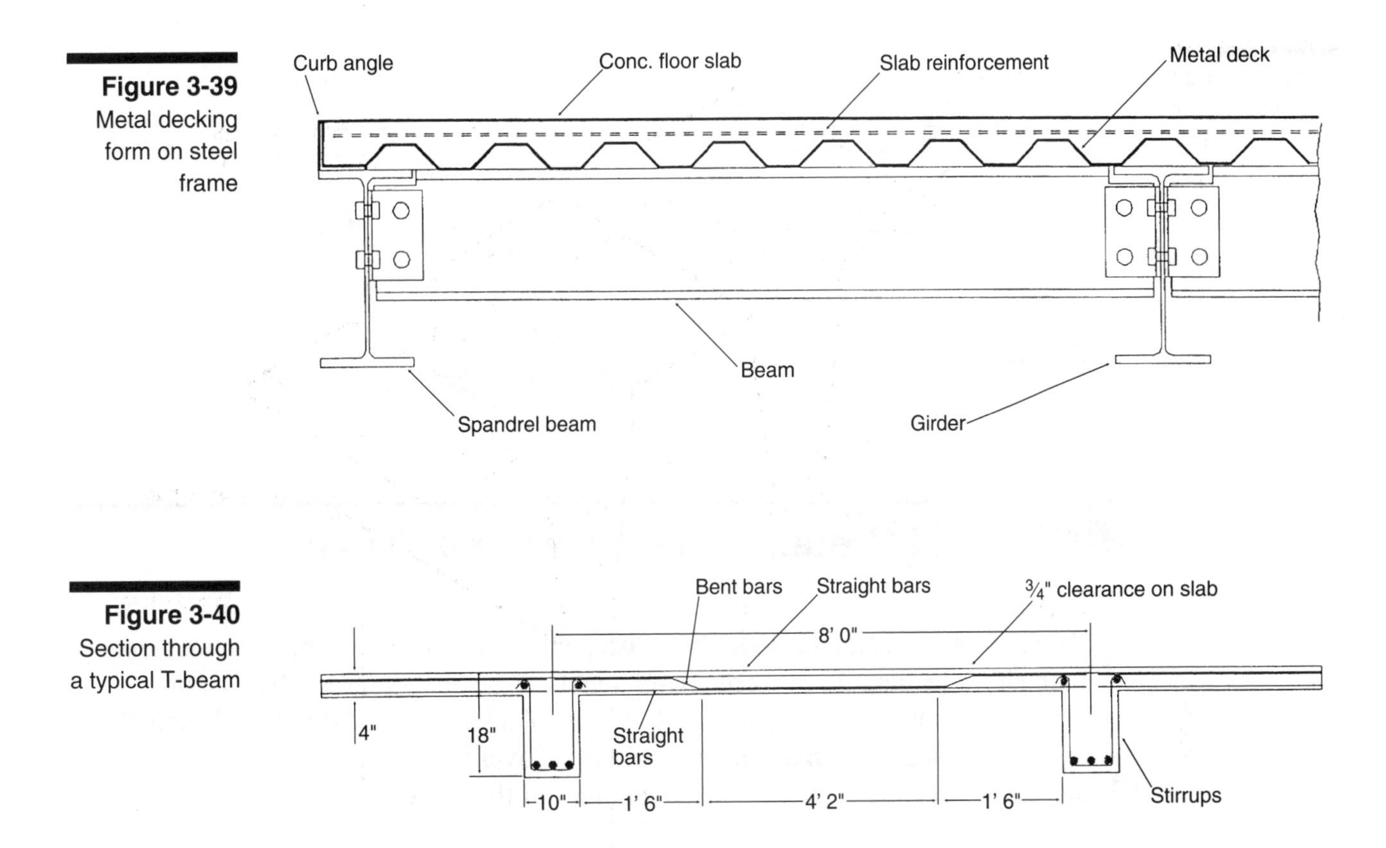

Figure 3-39 Metal decking form on steel frame

Figure 3-40 Section through a typical T-beam

Some types of corrugated and ribbed decking use steel bars or wire welded to the ribs. This way both the concrete slab and the metal deck carry the live and dead floor load. It also reduces the amount of concrete reinforcement needed. Instead of only supporting the slab as a form does, the steel deck gives the concrete tensile strength. Figure 3-39 shows a metal deck form on a steel frame.

When a deck needs a 1-, 2-, 3-, or 4-hour fire rating, you'll need to protect the underside of the steel deck. Do this with a suspended plaster ceiling, or by putting fireproof plaster directly on the bottom of the decking. Make the plaster as thick as you need to get the fire-resistance rating you want. The amount of fireproofing used in a structure is determined by the architect or engineer of the job, and should be shown on the details. The building code controls the type of construction for each type of occupancy. Each type of construction of a building requires a certain amount of fire-resistance to the structural components.

Concrete T-beams

To make T-beams, as shown in Figure 3-40, cast the slab with closely-spaced rectangular beams. Then the beams and slab work together as a single structural unit. This integrally-cast unit is stronger than if you cast the beam and slab separately.

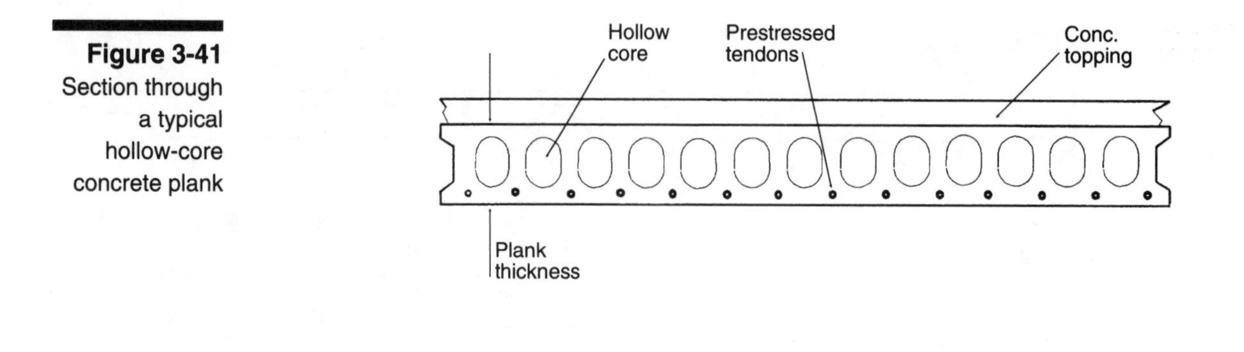

Figure 3-41 Section through a typical hollow-core concrete plank

Post and Pretensioned Slabs and Planks

For long-span concrete floor slabs, use post or pretensioned structural elements that are either rectangular, inverted U-channels, T-beams, double T-beams, or hollow-core planks. Prefabricated hollow-core planks are made 4 to 6 inches thick. They are designed to span as much as 45 feet. Cover the planks with a 2- to 3-inch-thick concrete topping. The advantages of these precast units are:

- all weather construction
- no forming or shoring
- shorter construction time
- longer spans
- low weight
- ease of erection from truck to building

Figure 3-41 shows a hollow-core precast prestressed concrete plank. Use these for floor and roof slabs. In earthquake zones, install a concrete topping to tie the individual concrete planks together.

Columns and Walls

Columns support roof and floor slabs, and transfer the building weight to the foundation. They are sometimes called posts, pillars, or supports. There are three general types of columns; long columns, short columns, and pedestals. The length of a long column is more than ten times its width. The length of a short column is more than three times, but less than ten times, its width. The height of a pedestal column is less than three times its narrowest width. When a column is part of a bearing wall, it's called a pilaster.

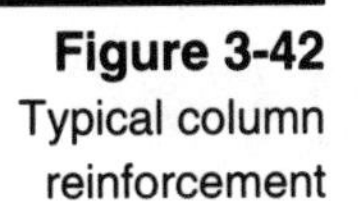

Figure 3-42
Typical column reinforcement

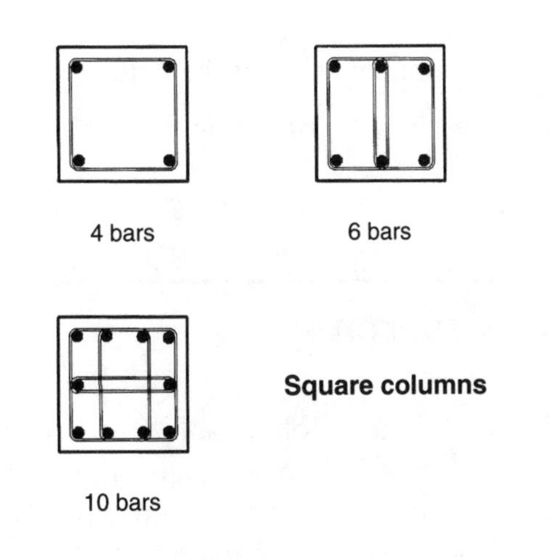

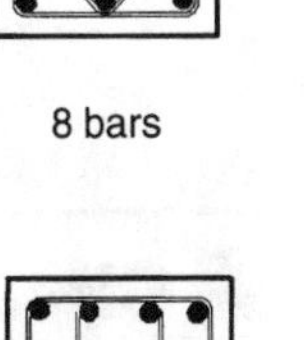

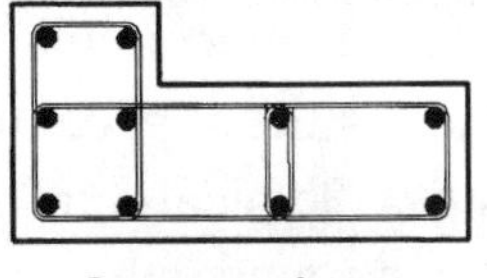

Rectangular Columns

All concrete columns have reinforcing steel. Both the steel and the concrete carry the building load. In concrete columns, the minimum area of vertical steel bars shouldn't be less than 1 percent of the area of the column. The maximum area of steel is 8 percent. For example, a 12-inch square column has an area of 144 square inches. The total cross-sectional area of the vertical steel bars must be more than 1.44 square inches but less than 11.52 square inches.

In multistory buildings, columns at the top floors carry the lightest loads; columns at the ground level carry the heaviest loads. All columns can be the same size if the columns carrying the heaviest load have more reinforcing steel. Otherwise you need to make the columns larger at the ground level to take the greater load there.

Column reinforcing steel runs both vertically and horizontally. The vertical bars are compression bars. They carry a portion of the building weight. Use steel ties to make loops around the vertical bars to keep them from buckling outward and breaking (spalling) the concrete. The size, number, spacing, and hooking of column ties is very important. These ties have to hold the vertical reinforcement in place during high winds or earthquakes. Most column failures happen when the vertical reinforcing bar buckles. Figure 3-42 shows various types of columns and reinforcement including vertical bars and ties.

Column ties in rectangular columns are usually ½ or ⅜ inches in diameter. Don't space the ties less than:

- 16 times the diameter of the vertical bars

- one-half the smallest side of the column
- 48 times the diameter of the tie

Circular Columns

It's easy to form circular columns. Fiber cylinders from 6 to 48 inches in diameter are sold as column forms. These cylinders are made of spirally-laminated plies of fiber, with exterior and interior surfaces coated with wax to make them weather- and moisture-proof. They are called sonotubes.

Place the bars in round columns in a circular pattern rather than a rectangle. Use spiral-wound No. 2 gauge wires as ties. Space these ties about 2¼ inches apart. You can also use circular loops as ties. Typical reinforcement for a 16-inch diameter column is six #6 bars; an 18-inch round column has eight #6 bars. Cover the ties with 1½ inches of concrete.

Be very careful when you reinforce columns because they are considerably more fragile than beams or slabs. In a severe earthquake, it's usually the building columns that fail.

Walls

The thickness of exterior concrete walls depends both on the load the wall has to carry and the required fire-resistance. The fire-resistance rating of a wall is based on the time it takes a fire on one side of it to ignite combustible material on the opposite side. You'll find standard test results in the Fire Resistance Directory which is published by the Underwriters Laboratories Inc. Fire-resistant concrete walls are rated from 1 to 4 hours, depending on the thickness of the wall. Ratings for solid concrete walls are:

- 4 hours for a 6½-inch wall
- 3 hours for a 6-inch wall
- 2 hours for a 5-inch wall
- 1 hour for a 3½-inch wall (Table No. 43-B UBC)

Usually you use a single layer of horizontal and vertical bars as steel reinforcing in the center of 6-inch-thick walls. Use two layers of bars, one at each face, in thicker walls. As a rule of thumb, reinforcement for any concrete wall shouldn't be less than #4 bars at 12 inches on center, both horizontally and vertically. Figure 3-43 shows the typical reinforcement for concrete walls 6 to 12 inches thick.

Figure 3-43 Typical concrete wall reinforcement

Wall thickness (inches)	Size and spacing of bars in each direction
6	#4 @ 13" o.c. center wall
8	#4 @ 18" o.c. each face
10	#4 @ 15" o.c. each face
12	#4 @ 12" o.c. each face

Precast Walls

Precast concrete construction is an economical way to build walls. Concrete wall panels are usually cast on-site or off-site at a manufacturing plant. The most popular form of site precast walls are tilted up into place. The building floor slab is usually used as the casting platform. If the floor has too many obstructions, such as trenches, pits, and curbs, build a temporary concrete slab nearby. Here's the general procedure for tilt-up construction:

- Finish the building floor slab carefully. Smooth all the joints in the slab to make a good flat place for casting wall panels.
- Set up the wood forms on the platform. Spray the forms and slab with an antibonding agent. This is very important. It's no fun at all to discover that the floor slab and the wall you are to lift into place are one solid unit.
- Put the reinforcing steel, anchor bolts, inserts, and all hardware you need in the proper position in the panels. Inserts are used for lifting and bracing the panels.
- Pour the concrete in the panels and float and screed it level to the top of the forms. Trowel the surface to a smooth finish. Add any texture finish at this time if the plans call for it.
- Spray a curing agent on the concrete to keep it from drying out too fast.
- When the concrete is fully cured, use a mobile crane to raise the panels and put them on the footings or floor slab in the final upright position. Working room for the crane is important. Be sure you have enough space to maneuver the crane around to lift and set each concrete panel.
- Brace the panels to the floor slab.
- Finally, join the concrete panels to cast-in-place concrete pilasters or weld them to steel columns.

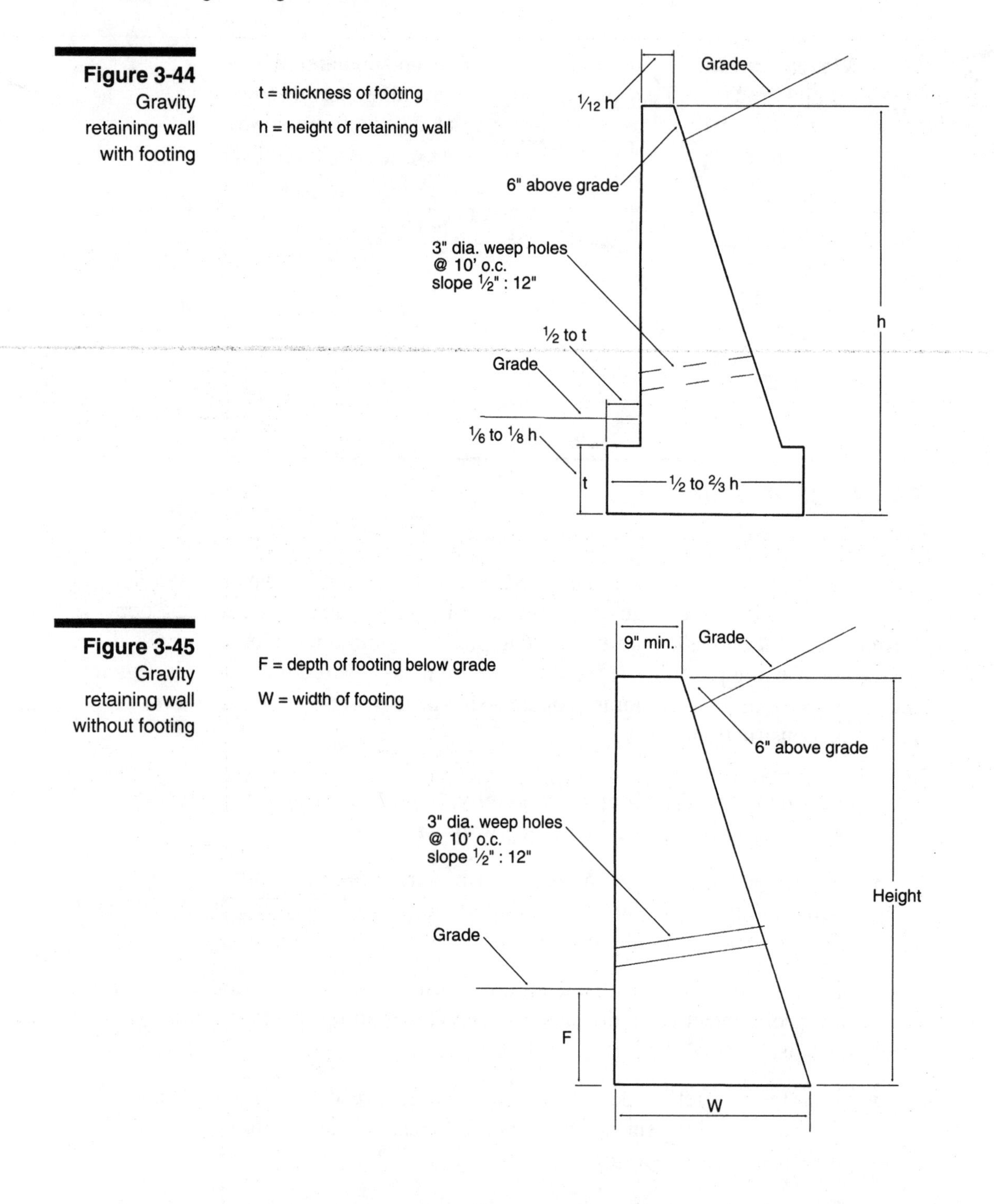

Figure 3-44 Gravity retaining wall with footing

Figure 3-45 Gravity retaining wall without footing

Retaining Walls

Retaining walls hold soil in place like a dam holds water. Generally there are two types of retaining walls — gravity and cantilevered. The gravity-type wall uses the weight of the concrete and the friction at the bottom of the wall to hold the soil behind it. Some gravity walls have large footings to increase the contact area with the soil. Figures 3-44 and 3-45 show two types of gravity retaining walls.

Figure 3-46
Typical gravity retaining walls

Height (feet)	Depth (inches)	Width (feet)	Volume (C.Y./lineal foot)
3.5	9	2.06	0.182
4.0	10	2.25	0.222
4.5	11	2.44	0.266
5.0	12	2.63	0.313
5.5	13	2.81	0.363
6.0	14	3.00	0.417
6.5	15	3.19	0.474
7.0	16	3.38	0.535
7.5	17	3.56	0.600
8.0	18	3.75	0.667

Note that there are weep holes through the wall to keep water from building up behind it. To help you figure what size gravity retaining wall to use, Figure 3-46 lists recommended dimensions and volumes for various wall heights.

The cantilever-type retaining wall is the most common. It uses less concrete but more reinforcing steel. The wall is shaped to resist the weight of the soil behind the wall. Cantilever retaining walls are T- or L-shaped. Both have a stem, or wall, and a footing. The thickness of the stem increases from top to bottom. The thicker portion resists the higher bending stress caused by the soil pressure against the wall. The maximum stress in a cantilever retaining wall is proportional to the height of wall. This is because the lowest portion of the wall resists the greatest overturning force from the retained soil. Because the wall usually bends away from the loaded side, the stress is called a bending stress. The top of the stem is usually 6 to 8 inches thick.

Design a retaining wall based on the soil pressure behind the wall. Normal dry earth puts a pressure of about 30 pounds per square foot (psf) for every foot of wall below the grade at the top of the wall. This is also called equivalent fluid pressure. A stiff cohesive clay has less than 30 psf per foot of wall below the surface of the clay. Very fluid soil, such as mud, has a pressure almost the same as water, or 62.4 pounds per square foot of wall below the surface of the soil.

If a wall must hold back a possible landslide, some codes may require that you design for an equivalent fluid pressure of as much as 125 psf per foot of wall below the surface. Any wall you design should have a safety factor of at least 150 percent. This means that the wall should be able to resist 50 percent more soil pressure than actually exists on it.

Any extra vertical load above the top of the wall caused by a sloped backfill, vehicular traffic, or the weight of an adjacent building is called a *surcharge*. This type of load puts extra horizontal pressure on a retaining wall. You can make up for a

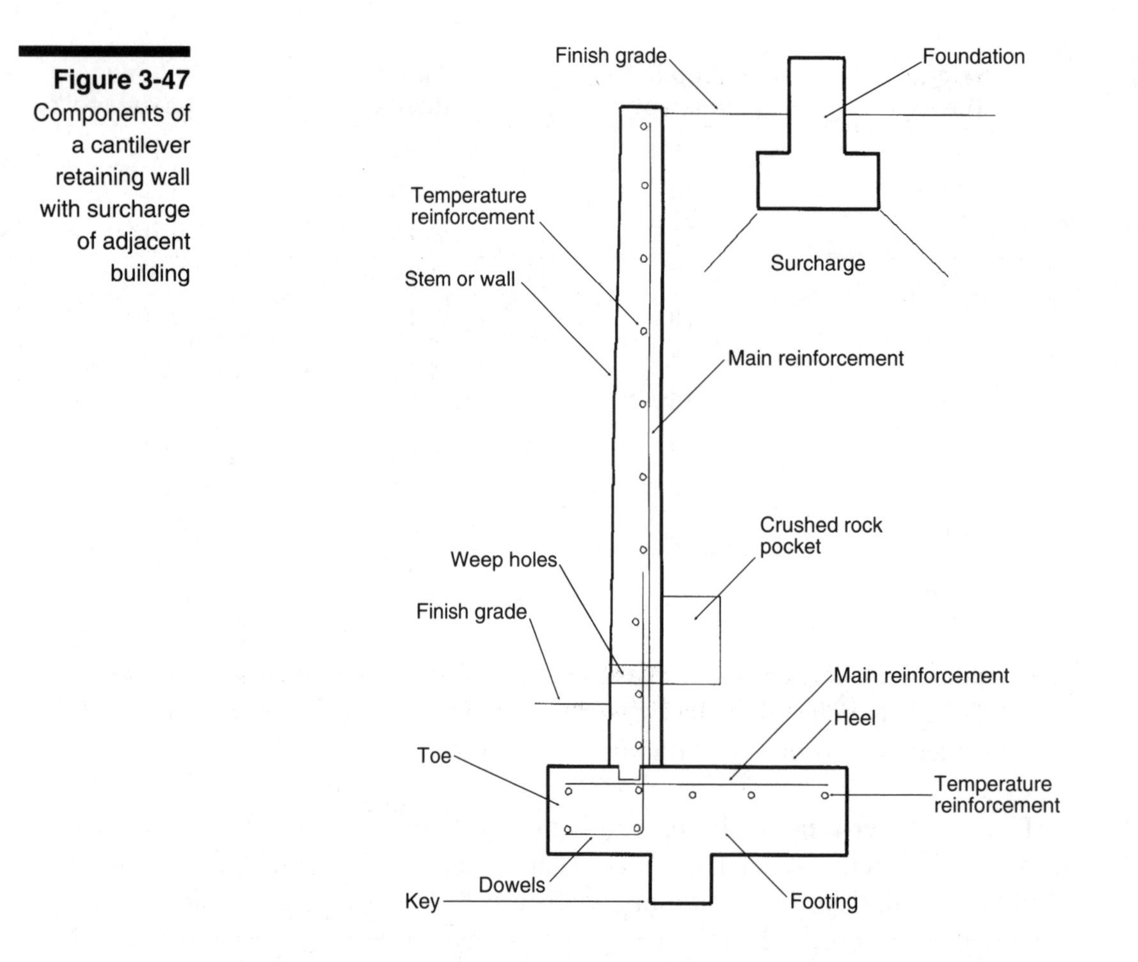

Figure 3-47 Components of a cantilever retaining wall with surcharge of adjacent building

highway loading surcharge by assuming an extra 3 feet of soil height behind the retaining wall. Figure 3-47 shows the main parts of a typical cantilever retaining wall with a surcharge.

To picture the footings of a retaining wall, think of the wall as facing away from the embankment behind it. The part of the footing that is under the retained earth is called the heel. The opposite part of the footing is called the toe. Some footings also have a part which sticks down, called a key. Use a deep footing and key to help keep a wall from sliding. Use a bigger heel and toe to keep a wall from falling over.

Most of the steel reinforcement in a retaining wall should be at the back side of the stem and bottom of the footing. Use at least 3 inches of concrete between the reinforcing bars and the soil when you cast the concrete against soil. Use horizontal reinforcing in the wall and footing to keep it from cracking as the concrete shrinks or expands. This also holds the vertical bars in place. Figures 3-48 and 3-49 will help you size retaining walls of various heights and loading conditions. All horizontal bars in these walls are #4 bars ($\frac{1}{2}$ inch) 13 inches o.c.

For a retaining wall without surcharge, Figure 3-48A shows the required thickness and reinforcement in the wall, Figure 3-48B the required shape and reinforcement in the footing, and Figure 3-48C the dimensions and reinforcement.

Figure 3-48A Typical retaining wall without surcharge (wall reinforcement)

Height (feet)	Wall thickness (inches)	X bars	Z bars
4	8	#4 @ 13" o.c.	—
5	8	#4 @ 13" o.c.	—
6	8	#4 @ 13" o.c.	—
7	8	#4 @ 13" o.c.	—
8	8	#4 @ 11" o.c.	#4 @ 13" o.c.
9	8	#5 @ 11" o.c.	#4 @ 13" o.c.
10	10	#5 @ 12" o.c.	#5 @ 14" o.c.
11	10	#6 @ 12" o.c.	#5 @ 14" o.c.
12	10	#6 @ 9" o.c.	#5 @ 14" o.c.
13	12	#6 @ 9" o.c.	#5 @ 12" o.c.

Figure 3-48B Typical retaining wall footing without surcharge (footing reinforcement)

Height (feet)	Depth (inches)	Base (feet-inches)	Toe (inches)	Rebars (y bars)	Rebars (t bars)
4	12	2'-3"	6	#4 @ 13"	3-#4
5	12	2'-8"	6	#4 @ 13"	4-#4
6	12	3'-0"	6	#4 @ 13"	4-#4
7	12	3'-6"	9	#4 @ 13"	5-#4
8	12	4'-0"	12	#4 @ 11"	5-#4
9	12	4'-4"	12	#5 @ 11"	6-#4
10	16	4'-9"	12	#5 @ 12"	6-#5
11	16	5'-3"	12	#5 @ 12"	6-#5
12	16	5'-9"	12	#5 @ 9"	7-#5
13	16	6'-3"	15	#5 @ 9"	7-#5

Figure 3-48C Concrete cantilever retaining wall without surcharge

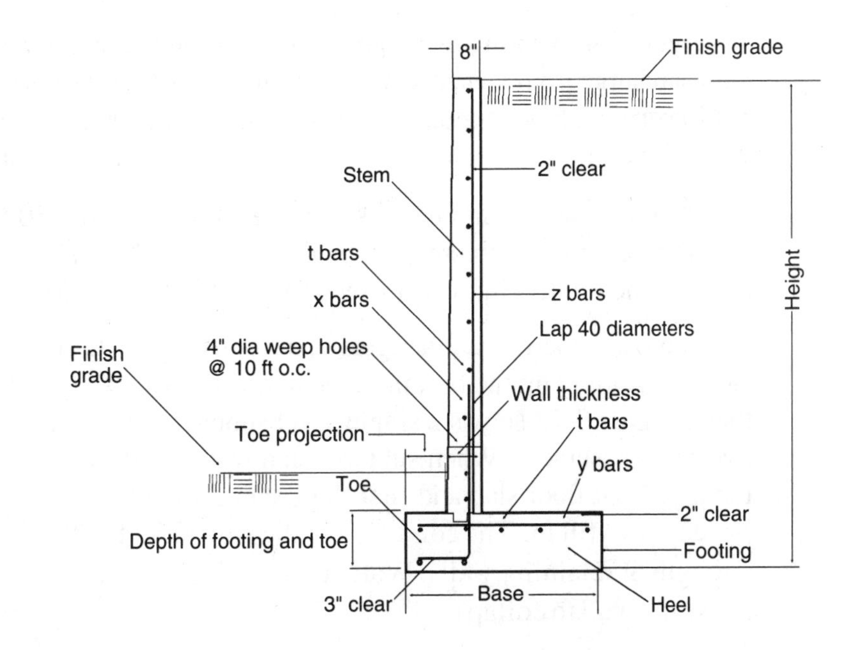

Figure 3-49A Typical retaining wall with surcharge (wall reinforcement)

Height (feet)	Wall thickness (inches)	X bars	Z bars
4	8	#4 @ 13" o.c.	—
5	8	#4 @ 11" o.c.	#4 @ 11" o.c.
6	8	#5 @ 11" o.c.	#4 @ 11" o.c.
7	10	#5 @ 11" o.c.	#5 @ 11" o.c.
8	10	#6 @ 12" o.c.	#5 @ 12" o.c.
9	10	#6 @ 9" o.c.	#5 @ 9" o.c.
10	12	#6 @ 9" o.c.	#5 @ 9" o.c.
11	16	#6 @ 11" o.c.	#6 @ 11" o.c.
12	16	#6 @ 14" o.c.	#6 @ 14" o.c.
13	16	#6 @ 13" o.c.	#6 @ 13" o.c.

Figure 3-49B Typical retaining wall with surcharge (footing reinforcement)

Height (feet)	Depth (inches)	Base (feet-inches)	Toe (inches)	Rebars (y bars)	Rebars (t bars)
4	12	3'-6"	9	#4 @ 13"	5-#4
5	12	4'-0"	12	#4 @ 11"	5-#4
6	12	4'-4"	12	#5 @ 11"	6-#4
7	16	4'-9"	12	#5 @ 11"	6-#5
8	16	5'-3"	12	#5 @ 11"	6-#5
9	16	5'-9"	12	#5 @ 11"	7-#7
10	16	6'-3"	15	#5 @ 9"	7-#5
11	20	7'-0"	18	#6 @ 12"	7-#6
12	20	7'-4"	18	#6 @ 12"	8-#6
13	20	7'-8"	24	#6 @ 10"	8-#6

Figure 3-49A shows the required thickness and reinforcement for a retaining wall with surcharge from vehicular traffic. Figure 3-49B shows the required shape and reinforcement of the footing for this type of wall. Figure 3-49C shows the height and width dimensions of a surcharge wall, and where the x and z bars are in this type of wall.

Retaining walls can fail for several reasons. Figure 3-50 shows how a wall will fail if there's too much pressure behind it. Figure 3-51 shows how a wall failed because the reinforcing wasn't securely fastened to the footing.

In most subterranean garages, you can use the exterior walls as retaining walls. Besides holding up the building, these walls must also hold back the earth around them. Figure 3-52 shows a single-level subterranean wall. Figure 3-53 shows a two-level basement. When subterranean walls span from basement to first floor, both the basement floor slab and first floor slab should be in place and cured before you put any backfill in. The condition of the soil under the footing has an effect on the strength of retaining walls. Water under the footing will soften the soil and may cause the wall to collapse.

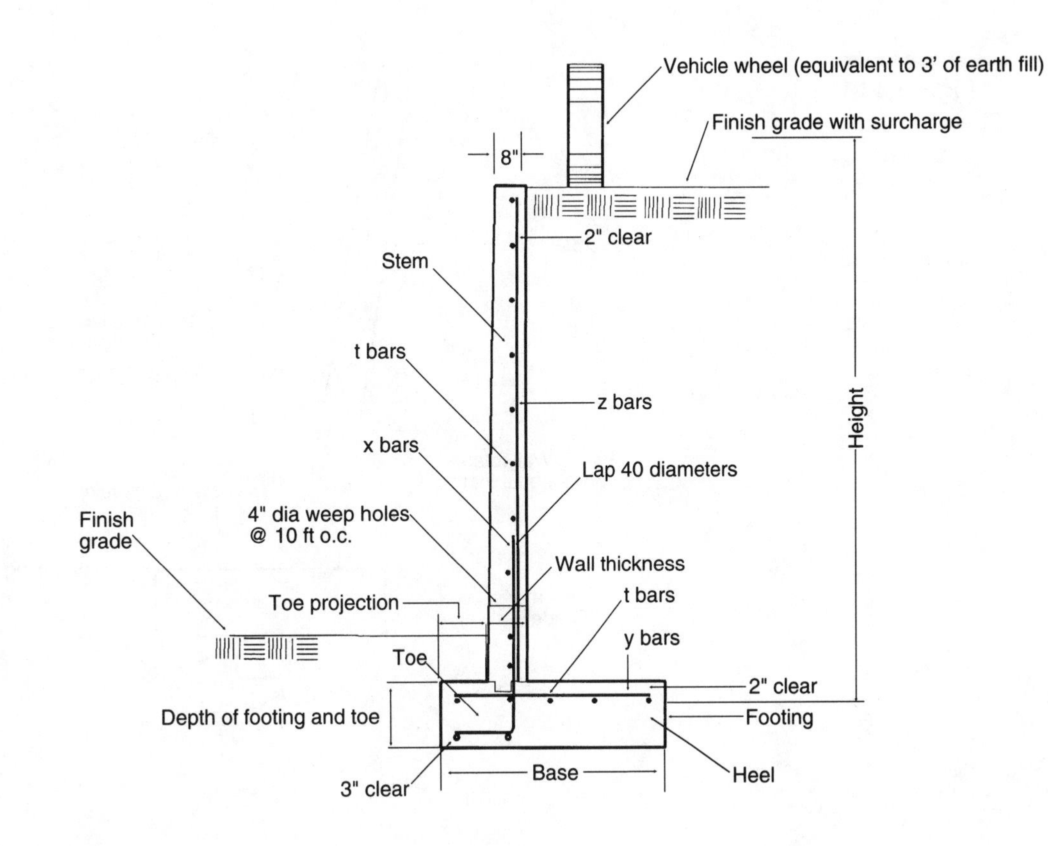

Figure 3-49C Concrete cantilever retaining wall with vehicular surcharge

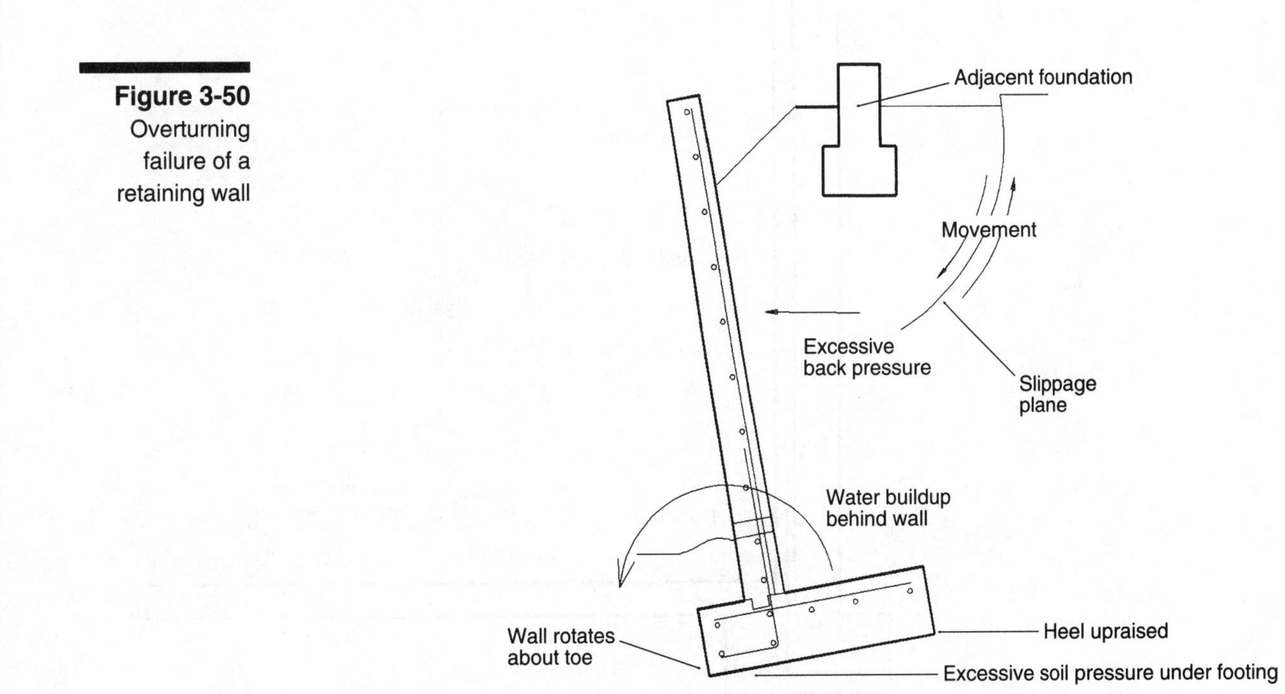

Figure 3-50 Overturning failure of a retaining wall

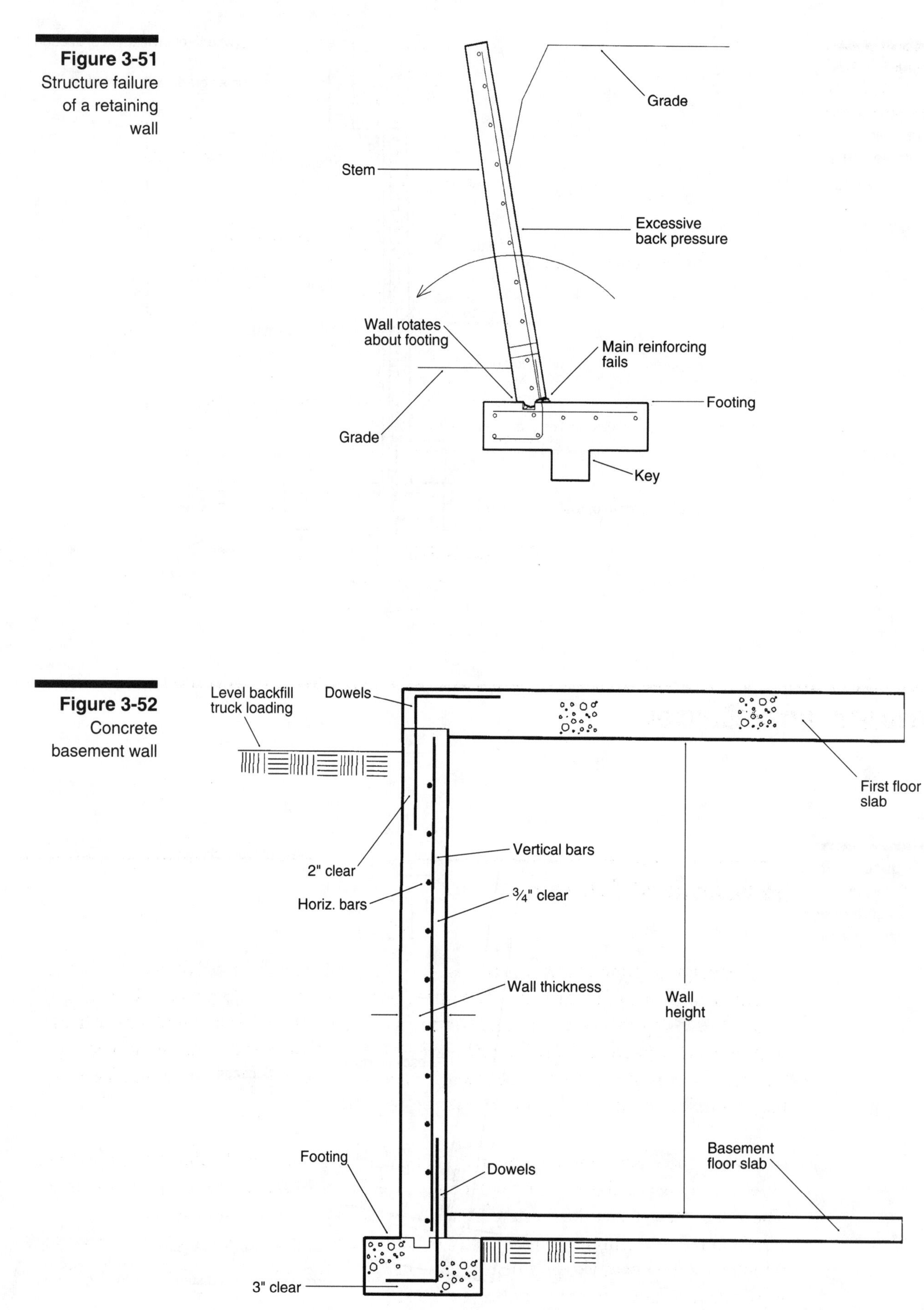

Figure 3-51 Structure failure of a retaining wall

Figure 3-52 Concrete basement wall

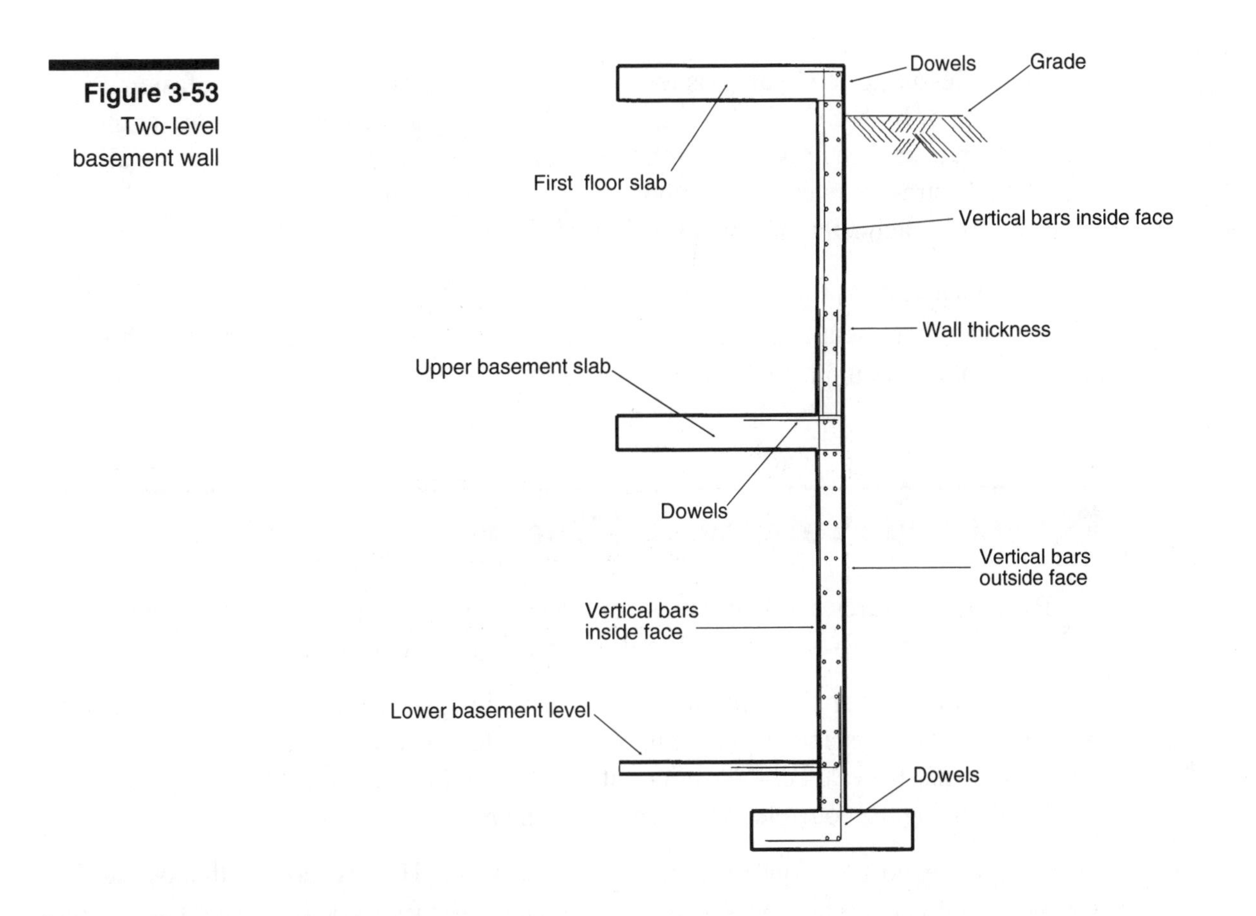

Figure 3-53 Two-level basement wall

Beams and Girders

Standard Beams

Rectangular concrete beams are used to carry floor loads in buildings. Weight on a beam causes it to flex or bend. The upper part of the beam compresses. Concrete is good at resisting this compression. But the lower part of the beam stretches, or tries to pull apart. Concrete doesn't stretch, so it may crack or break along the lower surface. Steel reinforcing bars help keep a beam from cracking or breaking. You can use a smaller size beam if you add compression steel to it.

When there isn't enough concrete in the part of a beam that's under compressive stresses, the beam size must be enlarged or steel bars added to the compression area. These bars help resist the compressive forces and are called compression reinforcement. Compression bars may be at the top or bottom of the beam depending which way the beam is being bent. The vertical reinforcement in concrete columns is also considered as compression reinforcement.

A simple-supported beam has tensile stress only at its bottom. If you fasten a beam at one or both ends so that the end can't rotate, the tensile stresses are at the top of the beam near the supports. Continuous beams, which span many supports, also have tensile stresses over each support. So you need to reinforce a continuous beam at its top over each support, and at its bottom between supports.

Stirrups are loops that hold the horizontal steel bars in place. Stirrups are $\frac{3}{8}$- or $\frac{1}{2}$-inch plain bars bent in a U or rectangular shape. Space them at intervals to hold the horizontal bars in place.

Post and Pretensioned Beams

Parking structures are sometimes constructed of pretensioned and post-tensioned building elements. These may be Single-T and Double-T beams.

Pretensioned concrete beams are made by stretching steel tendons over a casting bed within the forms. Concrete is poured into the forms to cover the stretched tendons. After the concrete has fully cured, the tendons at each end of the beam are cut, relaxing the steel and compressing the concrete in the beam. That increases beam strength.

Here's how post-tensioned concrete beams are made. Hollow tubes with loose steel tendons (cable) or specially-wrapped wires are laid in the forms before concrete is poured. When the concrete is cured, the steel cable is stretched and anchored to the ends of the beams. This is another way to compress concrete in the beam and make it stronger.

Miscellaneous Structures

Think of concrete stairs and landings as inclined slabs with steps. Size and reinforce them the same as one-way slabs. Design most commercial stairways to hold a 100 psf live load. You can design residential stairways for a 40 psf load.

Figure 3-54 shows various types of concrete stairs. The minimum thickness of a stair ranges from 3 inches to 1 foot, depending on its span. Figure 3-55 shows the approximate thickness and reinforcement for different spans of concrete stairs. The span is the distance between the centers of the supporting beams.

Summary

Concrete is the most permanent of construction materials. If it's designed and installed correctly, it'll last practically forever. On the downside, errors in concrete work are very difficult to correct. Concrete is heavy and vulnerable to earthquake forces. It's brittle and can crack if steel reinforcing isn't placed in it correctly.

Figure 3-54
Concrete stair design

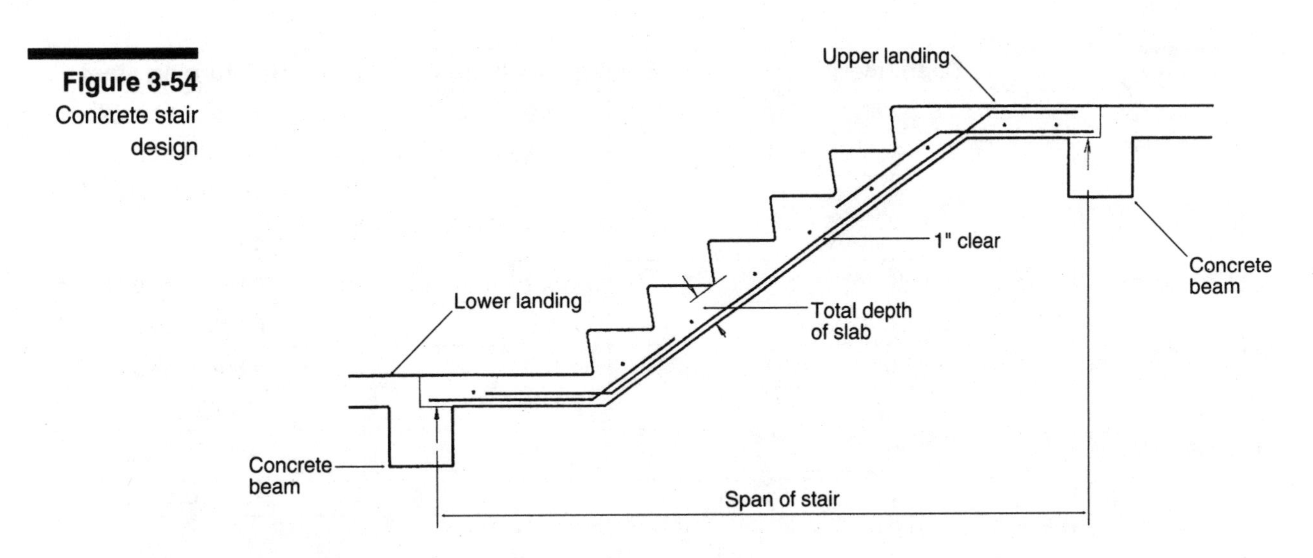

With flight and both landings supported by beams

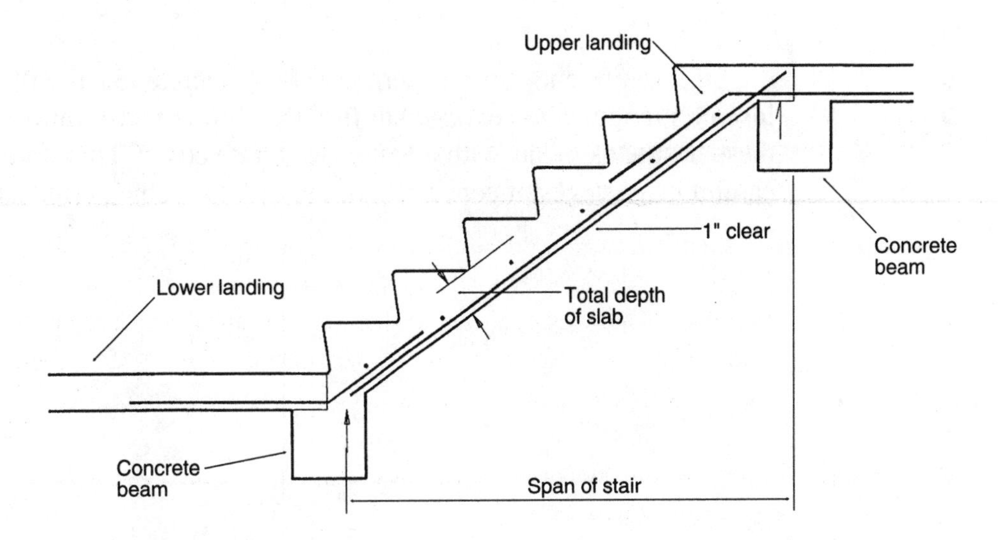

With flight only supported by beams

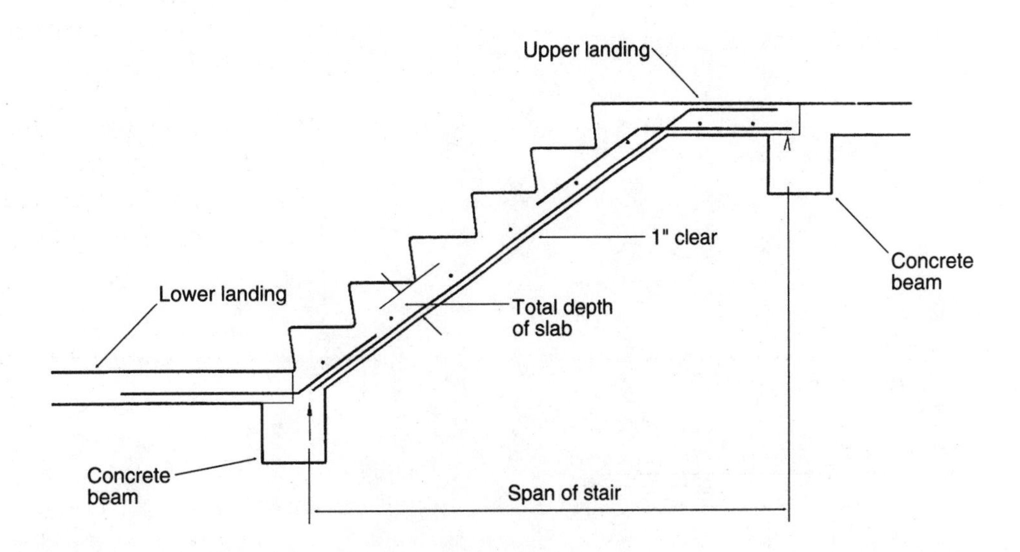

With flight and upper landing supported by beams

Figure 3-55
Typical concrete stair reinforcement

Span (feet)	Thickness (inches)	Reinforcing steel
4	3.0	#3 @ 7.5" o.c.
5	3.5	#3 @ 6.5" o.c.
6	4.0	#3 @ 5.5" o.c.
7	4.5	#3 @ 4.5" o.c.
8	5.0	#4 @ 7.5" o.c.
9	5.5	#4 @ 6.5" o.c.
10	6.0	#4 @ 5.5" o.c.
11	6.5	#4 @ 5" o.c.
12	7.0	#5 @ 7" o.c.
13	7.5	#5 @ 6.5" o.c.

Steel reinforcing isn't as permanent as concrete. Steel will rust when wet, and lose its strength when exposed to fire. Reinforcing steel must be protected against these hostile elements with adequate concrete cover. Therefore, you have to be careful in all stages of concrete work to build a stable permanent structure.

CHAPTER 4

Wood Basics

North America is blessed with abundant forests. That's probably why lumber is the most common structural material used by builders. Wood framing is preferred because it's:

- inexpensive
- strong
- easy to work with
- relatively long-lasting

Lumber Basics

This chapter deals mainly with wood framing. Finish woodwork, such as siding, trim, and door and window frames which usually are not considered part of a building's structure, will be covered briefly. Finish woodwork is also called architectural millwork. It includes casings, moldings, and similar decorative features that generally are used to cover joints between construction materials. Baseboards and base strips are used to cover the joint between a wall and the flooring. Casings close the joints around doors and windows. This work usually is done by finish carpenters rather than rough carpenters who install framing.

The most common types of wood used for millwork are white pine and sugar pine. These species are relatively soft and clear of splits, knots, and other defects. Much finish woodwork has been replaced with plastic, cast gypsum, and fiberglass.

Wood flooring may be considered as a part of finish carpentry, but it's usually installed by a flooring contractor. Most hardwood flooring was made of solid oak or maple hardwood planking and strips. However, now solid hardwood flooring is often replaced by plastic tiles covered with a thin veneer of hardwood encased with acrylic plastic.

Hardwood and Softwood

There are two major categories of wood — hardwood and softwood. These names don't necessarily mean that hardwood is hard and softwood is soft. For example, balsa wood comes from a hardwood tree. The names tell us more about the type of tree than the wood itself.

Hardwood comes from broad-leafed trees that lose their leaves during the winter months. The wood is generally heavy and close-grained. Hardwood, such as oak and maple, is used mainly for flooring, molding, paneling and furniture. It's far too expensive to use for most structural work, though you might use hardwood blocks for high bearing loads.

Nearly all framing and sheathing is done with softwood lumber and plywood. Softwood comes from trees with needle-like or scale-like leaves that stay on the tree all year. The most popular species of softwood are Douglas fir-larch, western hemlock, white fir, and Sitka spruce. Pound for pound, Douglas fir is one of the strongest woods. It resists warping, cupping, and twisting and is available at building material dealers all over the United States.

Two other types of softwood commonly used are redwood and cedar. These species are popular because they resist decay from fungi and termites. Redwood and cedar also are sawn into thin boards that can be used as siding.

Sapwood and Heartwood

Construction lumber is cut from both the sapwood and heartwood of a tree. Figure 4-1 shows the general location of sapwood, heartwood, and pith within a log.

Sapwood, also called early wood, comes from the outer layers of growth located between the bark and the heartwood. It contains the sap, or living part, of the tree. It's usually lighter in color and has more moisture than heartwood.

The heartwood, also called summer wood, comes from the center part, or heart of the tree. Sapwood surrounds heartwood. Pith is the small center core of the heartwood. Heartwood lumber is more desirable for construction than sapwood because it's more durable, and more resistant to decay than sapwood. It's also more expensive than sapwood. If sapwood is treated with a preservative, it can be as durable as heartwood.

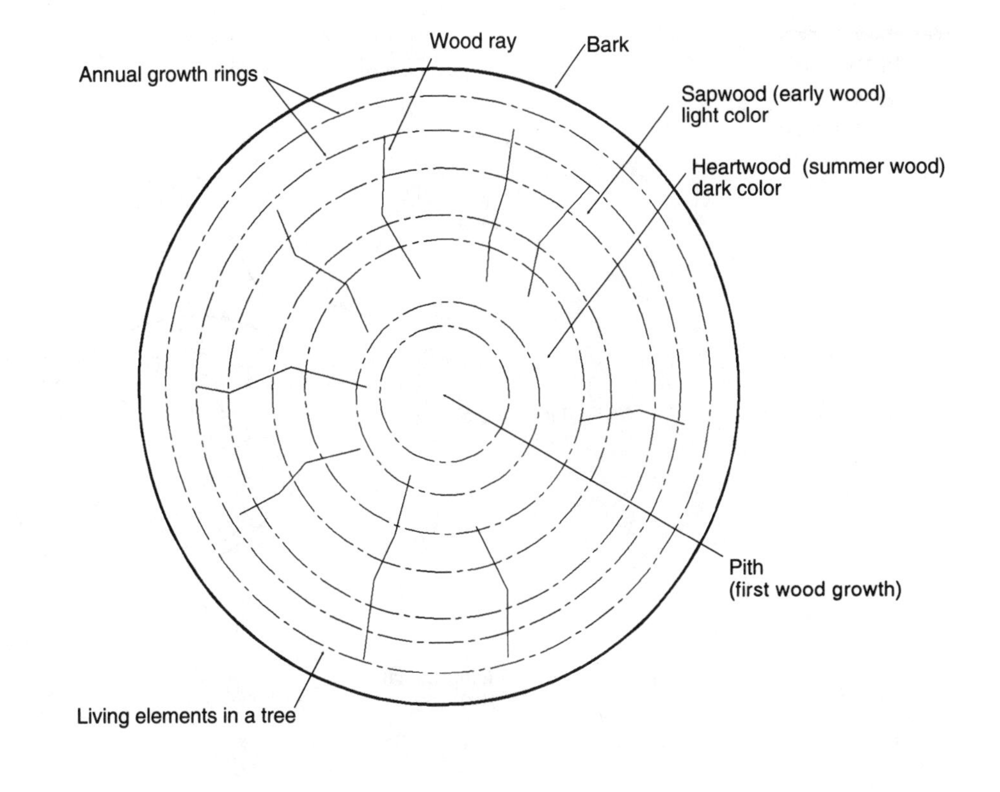

Figure 4-1
Heartwood and sapwood

Grading Lumber

Even if lumber is cut from the same tree, its quality will vary depending on which part of the tree it was cut from. Lumber grading is based on a combination of many characteristics, such as:

- density
- grain spacing and direction
- number, size, and location of knots and checks
- strength

Figure 4-2 shows the grain directions, growth rings, and wood rays in a log. This figure also shows the direction of saw cuts relative to the grain of the wood. The grain of the wood is parallel to the axis of the log. When a cut is tangential it is tangent to the growth rings. When a cut is radial, it means the cut goes through the center of the log. Look ahead to Figures 4-8 and 4-9 which show cuts relative to the growth rings.

Lumber is classified in many ways. There are several lumber industry associations and each has established its own grading rules. These rules describe the uses, sizes, and physical properties of lumber produced by member mills. The following associations publish wood grading rules for United States mills:

- West Coast Lumberman's Association (WCLA)
- Western Wood Products Association (WWPA)
- West Coast Lumber Inspection Bureau (WCIB)

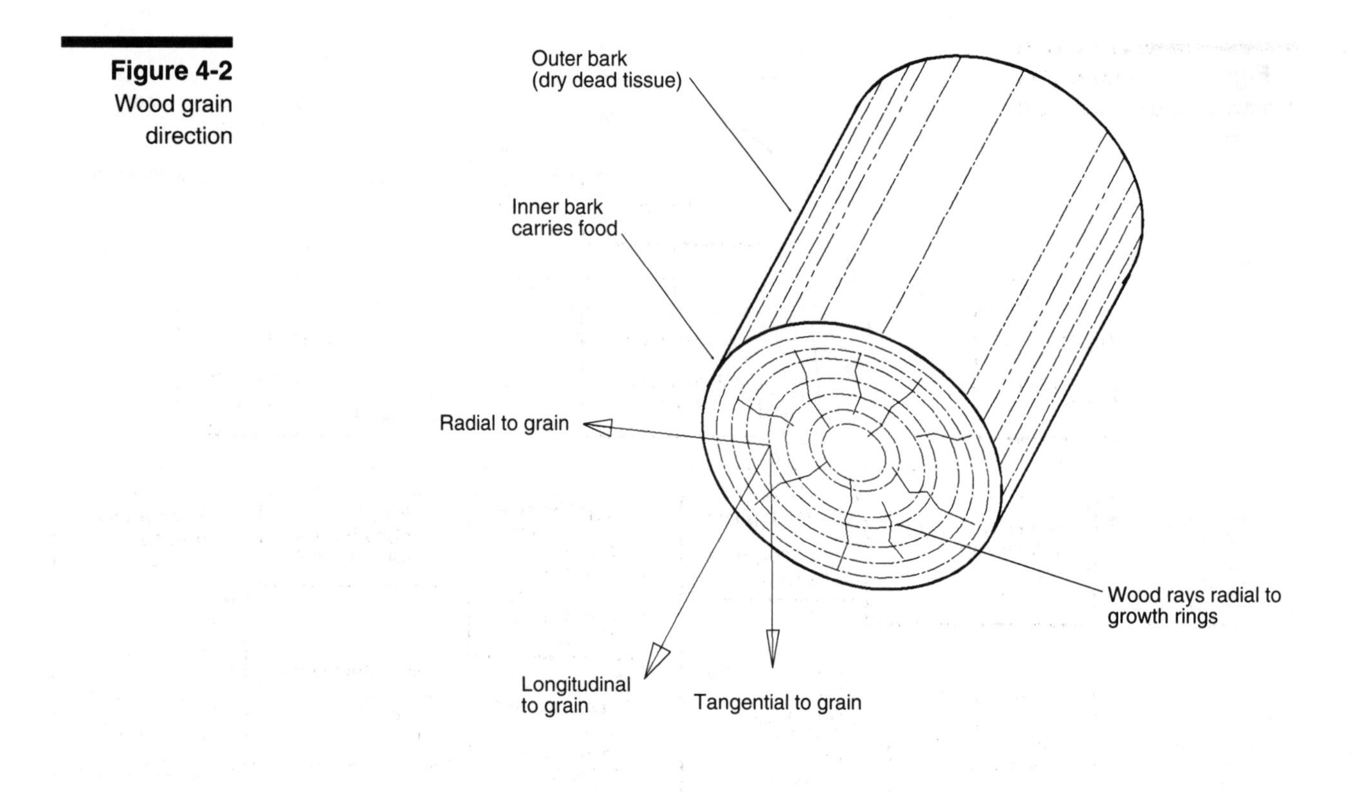

Figure 4-2
Wood grain direction

- American Institute of Timber Construction (AITC)
- American Plywood Association (APA)
- California Redwood Association (CRA)
- National Cedar Association (NCA)

Figure 4-3 shows one method of grading lumber.

Building codes use some of the wood industry rules. The *Uniform Building Code* (UBC) has a visual grading system for lumber. For example, the UBC classifies Douglas fir-larch into the following sizes, uses, and grades:

Lumber 2 to 4 inches thick and 2 to 4 inches wide is classified as:

- Dense Select Structural
- Select Structural
- Dense No. 1
- No. 1
- Dense No. 2
- No. 2
- No. 3
- Appearance
- Stud

Lumber 2 to 4 inches thick and 4 inches wide is classified as:

- Construction

Figure 4-3
Lumber grade chart

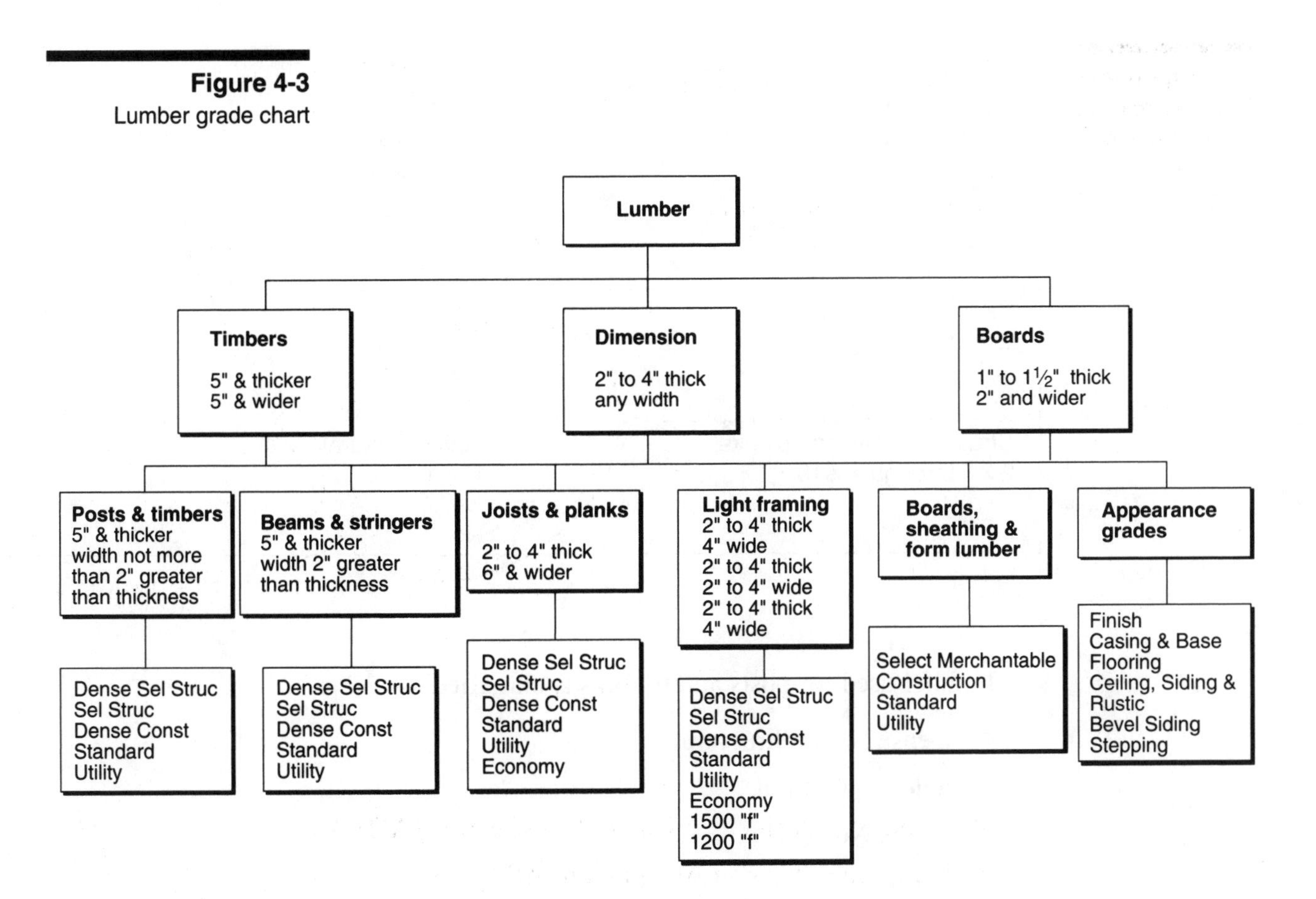

- Standard
- Utility

Lumber 2 to 4 inches thick and 5 inches and wider, is classified as:

- Dense Select Structural
- Select Structural
- Dense No. 1
- No. 1
- Dense No. 2
- No. 2
- No. 3 and Studs
- Appearance

Lumber used for beams and stringers is classified as:

- Dense Select Structural
- Select Structural
- Dense No. 1
- No. 1

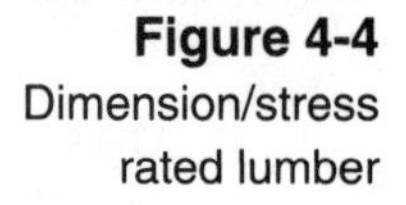

Figure 4-4
Dimension/stress rated lumber

Use	Grade
Light framing, 2 x 2 through 4 x 4	Construction Grade Standard Grade Utility Grade
Studs 2 x 2 through 4 x 6 10 ft and shorter	Stud Grade
Structural light framing 2 x 2 through 4 x 4	Select Structural No. 1 No. 2 No. 3
Structural joists and planks 2 x 5 through 4 x 16	Select Structural No. 1 No. 2 No. 3

Lumber used for posts and timbers is classified as:

- Dense Select Structural
- Select Structural
- Dense No. 1
- No. 1

Dimension Lumber, Dimension/Stress, and Construction Lumber

Lumber is also classified as Dimension and Dimension/Stress-Rated Lumber. Dimension lumber is limited to surfaced, or finished, lumber. You use it for joists, planks, rafters, studs, and light framing. It comes in pieces which range in size from 2 to 4 inches thick.

Lumber which has been tested mechanically to determine its working stresses is called Dimension/Stress-Rated Lumber. It's subdivided into the classifications shown in Figure 4-4.

Visually graded construction lumber is assigned working stresses based on how it looks. The lumber grader decides which grade applies to a piece by looking at it. Visually graded lumber classifications are shown in Figure 4-5.

Boards

Boards are lumber that is less than 2 inches thick and 1 inch or more wide. Boards less than 6 inches wide are called strips. Timber is lumber that is 5 inches or more in the least dimension. Beams, stringers, posts, caps, sill, girders, and purlins are timber.

Figure 4-5
Visually graded construction grade lumber

Use	Grade
Light framing and studs	Construction Grade Standard Grade Utility Grade Stud Grade
Beams and stringers 5" and thicker	Select Structural Grade No. 1 Grade No. 2 Grade No. 3 Grade
Posts and timbers 5" and thicker	Select Structural Grade No. 1 Grade No. 2 Grade No. 3 Grade
Beams 5" and thicker	Dense Select Structural Select Structural Dense Construction Construction
Joists and planks 2" to 4"	Dense Select Structural Select Structural Dense Construction Standard

Grade Stamps

Each piece of lumber delivered by a mill should have a grade stamp. This stamp certifies that the piece meets quality-control standards set by the lumber grading association. Figure 4-6 shows some typical grade stamps. A grade stamp shows this information:

- Mill Identification Number:
- Specie: Douglas fir, hemlock fir, etc.
- Grade: CONST, SEL STR, STUD, 3 COM, DENSE NO. 1, etc.
- Stress value: 1200f, 1500f, etc.
- Logo and Seal: WCLB, WWPA, etc.
- Seasoning: S-DRY, S-GR, MC 15, etc.

The seasoning marks printed on a grade stamp tell the moisture content of the lumber when it's surfaced. For example, S-DRY means not exceeding 19 percent moisture; MC-15 means not exceeding 15 percent moisture; and S-GRN means moisture content over 19 percent.

Figure 4-6
Typical grade stamps

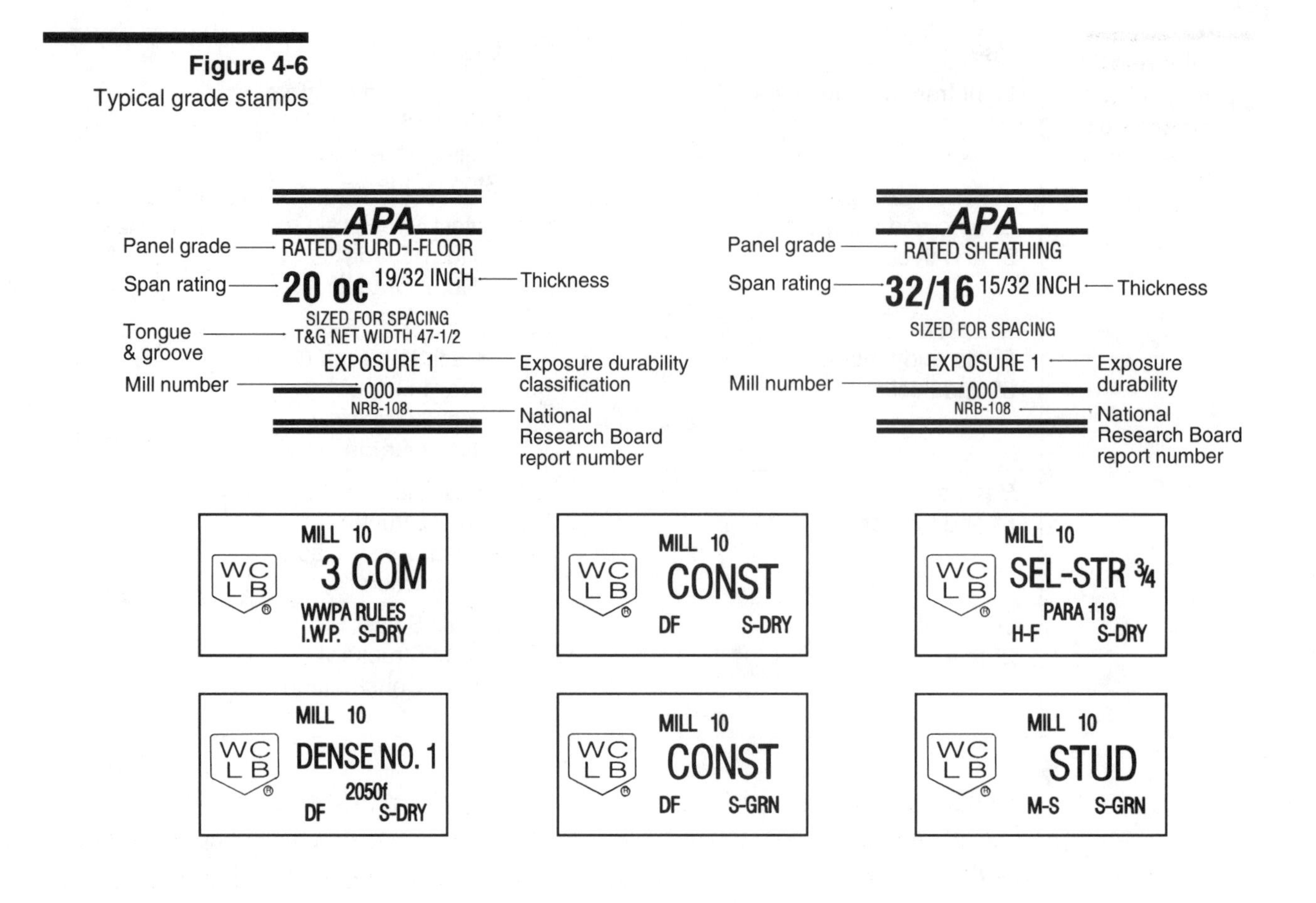

Properties of Lumber

Moisture and Defects

Wood is made up of very small hollow elongated cells. Moisture in wood is either free water or water that soaks into the cell walls. When wood contains just enough water for the cell walls, it's at its fiber saturation point. Lumber shrinks when the moisture content drops below the fiber saturation point.

If you take the free water out of wood, it doesn't do much except make the wood weigh less. But if the water goes out of the cell walls, the wood will shrink. Wood does shrink naturally as it dries. When about one-third of the moisture in the wood is gone, the wood is at a point called Equilibrium Moisture Content, or EMC. The average EMC is between 8 to 12 percent. The exact percentage depends on the moisture in the air around the lumber. The closer lumber is to the EMC, the less the shrinkage will occur as it reaches EMC.

You can take moisture out of lumber by kiln or air drying it. Both methods produce seasoned lumber. Seasoned lumber generally means that the moisture content is 19 percent or less. Lumber with higher moisture content is called green lumber. The average

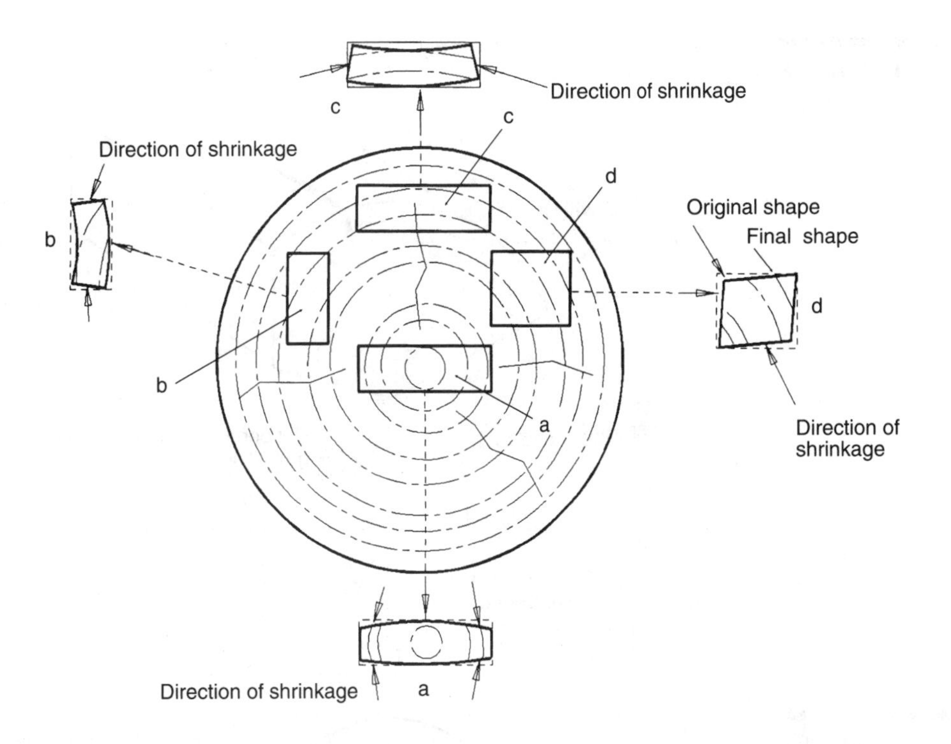

Figure 4-7
Effect of sawing location on shrinkage

shrinkage of a Douglas fir structural member from green to kiln dried is about 7.6 percent in width and 4.1 percent in thickness. This may be more than a ½-inch reduction in width for a 2 × 12 inch member. In multistory wood frame buildings, shrinkage in horizontal members can make floors uneven. In this case, it's a good idea to use seasoned horizontal members which have the lowest possible moisture content.

The amount of shrinkage in lumber also depends on what part of the tree it was cut from. The greatest shrinkage is perpendicular to the direction of the grain. It usually shrinks very little in the direction of the grain. Figure 4-7 shows the direction of shrinkage of pieces sawn from different parts of a log. When a piece shrinks, it'll cup, twist, warp, or curve depending upon the direction of the grain within the piece. Figures 4-8 and 4-9 show quarter sawn and plane, or flat, sawn lumber. The big difference between plane sawn and quarter sawn lumber is that plane sawn lumber shrinks and swells more, twists and cups more, and allows more liquid penetration than quarter sawn lumber. It also costs less than quarter sawn lumber.

You should be aware of the changes that will occur when the moisture content in the lumber you use changes. If you use green lumber that dries too fast, you'll see checks, cracks, and splits develop. Splits are separations in the wood that go across or through the annual growth rings, as shown in Figure 4-10. Any nails or bolts you drive in a split will tend to come loose. Green structural members can also warp, twist, and shrink.

Checks are separations in wood that normally occur across or through the rings of annual growth. They are usually a result of seasoning. Checks usually occur at the ends of members. End seal the fresh-cut ends of timbers to minimize end checking when the framing will be exposed to the environment. Shakes are lengthwise

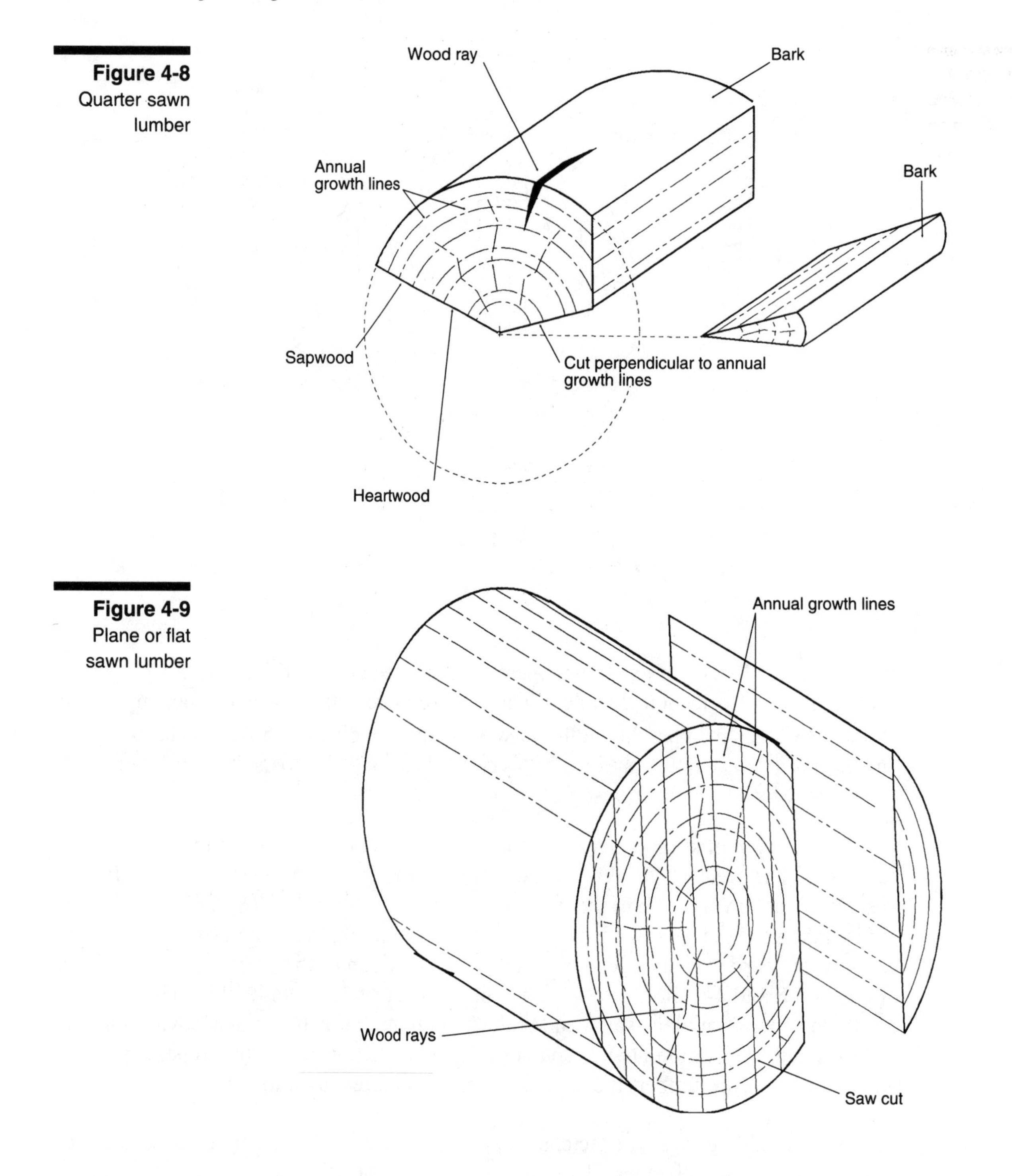

Figure 4-8
Quarter sawn lumber

Figure 4-9
Plane or flat sawn lumber

separations of wood which usually occur between, or through, the rings of annual growth. Splits are separations of the wood that happen when a wood cell tears apart. You can cut down on splitting by:

- increasing end distances of bolts and connections
- using seasoned wood
- protecting timbers against rapid seasoning
- using glued laminated timber

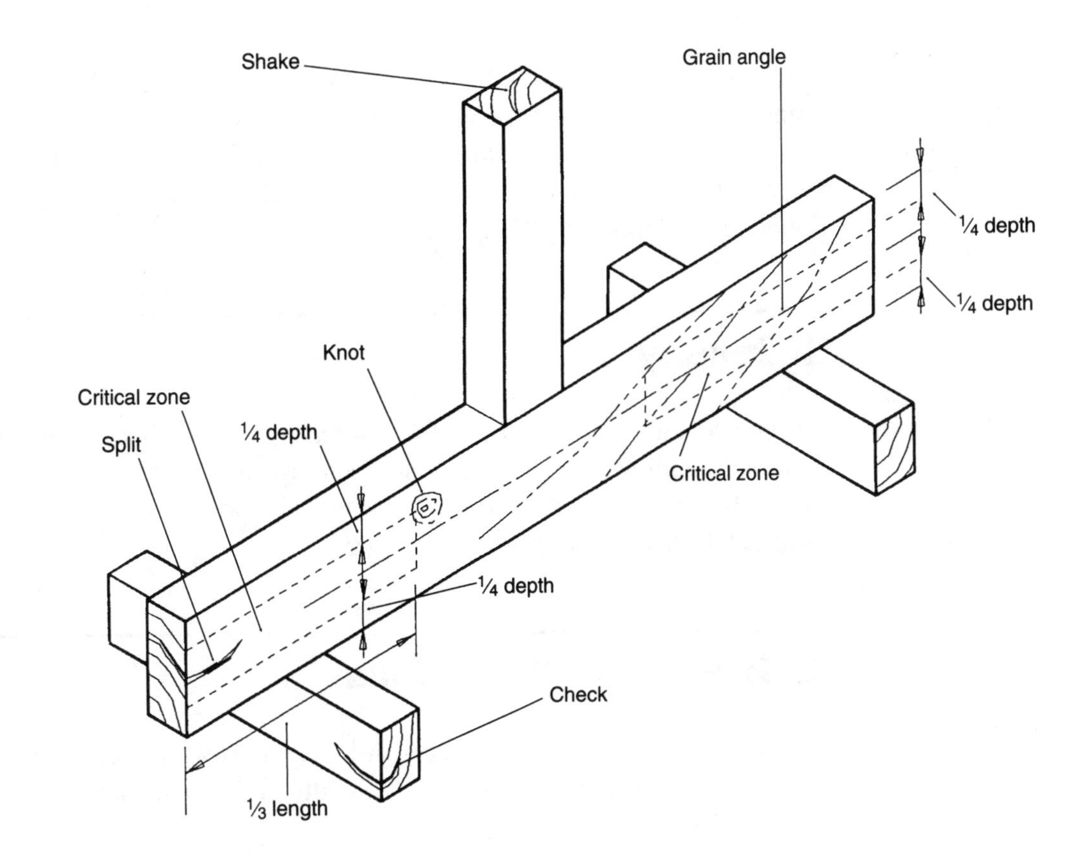

Figure 4-10
Wood defects

Sometimes it's difficult to get fully-seasoned lumber in large sizes. If you're using thick solid members, be prepared for some shrinkage. Retighten the bolts on built-up trusses or girders periodically as the members shrink.

Exposed heavy framing made of green wood is especially likely to develop problems if it dries rapidly. A large dimension timber can dry and shrink much faster on the outside surface than the interior. This can make deep cracks, splits, and warpage. If the framing will be exposed, be sure to use seasoned lumber in locations where it may dry too rapidly. Timbers that are covered with lath and plaster or building paper and wood siding are protected against drying out too rapidly.

Here's another common problem. Suppose you support the ends of green floor joists from the side of a steel member or a seasoned glue-laminated girder. Be sure to set the top of the joists above the top of the girder. If they are at the same level, the joists will shrink into the hangers and the girder will be higher than the bearing on the joists. The nails holding the flooring to the joists may pull out too. This can also make the flooring hump up over the girder, so the floor will be uneven.

If the moisture content in lumber goes up after construction, it may develop mildew, fungi, and termite damage. Wood that's in contact with, or near, soil will absorb moisture that may attract termites. All wood that touches the concrete floor slab on grade or soil, or that is less than 8 inches from finish grade, must be pressure-treated with protective chemicals to protect it from these problems. You should also treat the ends of girders that enter foundation walls. Provide at least an 18-inch clearance in the crawl space between the ground and the floor joists.

Building codes require that all wood in direct contact with masonry or concrete at a point within 48 inches of the ground must be pressure treated with an approved preservative or be made of durable wood. Typical examples of durable wood which are resistant to decay and infestation are Foundation Grade California Redwood or Foundation Lumber Western Cedar.

Density

The density of wood grain is measured by the number of annual rings per inch in a radial direction. The closer the grains are to each other in a piece of lumber, the more fibers the lumber has to resist stress. The density of the wood grain of a piece of lumber tells you how strong the lumber is.

Shape and Dimensions

Logs are sawn into lumber at a mill. The narrow lumber dimension is called the thickness. The larger dimension is the width. When used as a joist or rafter, the width is also called its depth.

You can identify lumber by its nominal size or by its net size. The net size is the actual thickness and width. The nominal size is the name given to the piece, regardless of the net size. For example, a nominal 2 × 4 stud may actually measure closer to 1½ × 3½. Net sizes have changed over the past forty years. The 2 × 4 nominal size is sometimes called the rough or unsurfaced size. Rough-sawn lumber has a rough surface. Surfacing or dressing this type of lumber reduces its thickness and depth. Some mills identify dressed lumber by the number of sides or edges that have been planed smooth. Here are some examples:

Symbol	Meaning
S1S	Surfaced one side
S2S	Surfaced two sides
S4S	Surfaced four sides
S1S1E	Surfaced one side, one edge
S1S2E	Surfaced one side, two edges

Standard nominal and dressed sizes for Dimension Lumber are:

Nominal Size (inches)	Dressed Size (inches)
1 x 2	1½ x 1½
3 x 3	2½ x 2½
4 x 4	3½ x 3½
4 x 6	3½ x 5½
4 x 8	3½ x 7½
4 x 10	3½ x 9¼
4 x 12	3½ x 11¼

Loads on Lumber

A dead or live load is a force applied to a structural member. It may be a pushing force (compression) or a pulling force (tension). Forces are usually indicated by arrows on a force diagram. Live and dead loads are static loads. Wind and earthquake loads are dynamic loads. Structural members are designed to carry 100 percent of the dead load, but the amount of live load they can carry depends on the tributary area. For example, roof live load is 20 psf if the tributary area is under 200 square feet, 16 psf if the area is between 201 and 600 square feet, and 12 psf if the area is over 600 square feet.

Because dynamic loads occur over a very short period of time, the allowable stresses in structural members are increased by one-third when resisting wind or seismic loads.

A live load is a load that's superimposed on members by the use and occupancy of a building. People, furniture, stored material, movable walls, and similar temporary loads are live loads. Live loads for buildings are listed in the UBC Table No. 23-A - Uniform and Concentrated Loads, Table No. 23-B - Special Loads, and Table No. 23-C - Minimum Roof Live Loads. These tables are shown in Figures 4-11, 4-12, and 4-13.

A dead load is the vertical load due to the weight of all permanent structural and nonstructural components of a building, such as walls, floors, roofs, and fixed service equipment.

Stresses on Lumber

The main stresses on structural lumber are flexure, tension, compression, and shear. When you put a load on a beam, as shown in Figure 4-14, the greatest tension stresses are along the face of the beam that's farthest from the load. At the same time, the greatest compressive stresses are on the edge closest to the load. Horizontal shear stresses are parallel to the grain. They are highest at the ends of the beam. Figure 4-15 shows bearing, or compressive, stresses which are perpendicular and parallel to the grain. Figure 4-16 shows the potential areas of failure on an overloaded or undersized beam or post.

The size of the beam you use to span any opening depends on the load it has to carry and the strength of the lumber. There are simple formulas for determining how much stress and deflection a wood beam can accept. Here's a list of some of the main terms, and their common abbreviations or symbols, that are used in these formulas:

A = cross sectional area of a member, in square inches.

b = breadth or width of a rectangular member, in inches, depending on its orientation. A rectangular member may be installed with its greater dimension vertical, as in a joist, or horizontal, as in a plank.

c = distance from the neutral axis, or center of a rectangular member, to the extreme fiber, in inches.

D = diameter of a round member, in inches.

Figure 4-11
Uniform and concentrated loads

TABLE NO. 23-A—UNIFORM AND CONCENTRATED LOADS

USE OR OCCUPANCY		UNIFORM LOAD[1]	CONCENTRATED LOAD
Category	**Description**		
1. Access floor systems	Office use	50	2,000[2]
	Computer use	100	2,000[2]
2. Armories		150	0
3. Assembly areas[3] and auditoriums and balconies therewith	Fixed seating areas	50	0
	Movable seating and other areas	100	0
	Stage areas and enclosed platforms	125	0
4. Cornices, marquees and residential balconies		60	0
5. Exit facilities[4]		100	0[5]
6. Garages	General storage and/or repair	100	[6]
	Private or pleasure-type motor vehicle storage	50	[6]
7. Hospitals	Wards and rooms	40	1,000[2]
8. Libraries	Reading rooms	60	1,000[2]
	Stack rooms	125	1,500[2]
9. Manufacturing	Light	75	2,000[2]
	Heavy	125	3,000[2]
10. Offices		50	2,000[2]
11. Printing plants	Press rooms	150	2,500[2]
	Composing and linotype rooms	100	2,000[2]
12. Residential[7]		40	0[5]
13. Restrooms[8]			
14. Reviewing stands, grandstands, bleachers, and folding and telescoping seating		100	0
15. Roof decks	Same as area served or for the type of occupancy accommodated		
16. Schools	Classrooms	40	1,000[2]
17. Sidewalks and driveways	Public access	250	[6]
18. Storage	Light	125	
	Heavy	250	
19. Stores	Retail	75	2,000[2]
	Wholesale	100	3,000[2]

[1]See Section 2306 for live load reductions.

[2]See Section 2304 (c), first paragraph, for area of load applications.

[3]Assembly areas include such occupancies as dance halls, drill rooms, gymnasiums, playgrounds, plazas, terraces and similar occupancies which are generally accessible to the public.

[4]Exit facilities shall include such uses as corridors serving an occupant load of 10 or more persons, exterior exit balconies, stairways, fire escapes and similar uses.

[5]Individual stair treads shall be designed to support a 300-pound concentrated load placed in a position which would cause maximum stress. Stair stringers may be designed for the uniform load set forth in the table.

[6]See Section 2304 (c), second paragraph, for concentrated loads.

[7]Residential occupancies include private dwellings, apartments and hotel guest rooms.

[8]Restroom loads shall not be less than the load for the occupancy with which they are associated, but need not exceed 50 pounds per square foot.

Figure 4-12
Special loads

TABLE NO. 23-B—SPECIAL LOADS[1]			
USE		VERTICAL LOAD	LATERAL LOAD
Category	Description	(pounds per square foot unless otherwise noted)	
1. Construction, public access at site (live load)	Walkway, see Sec. 4406	150	
	Canopy, see Sec. 4407	150	
2. Grandstands, reviewing stands, bleachers, and folding and telescoping seating (live load)	Seats and footboards	120[2]	See Footnote No. 3
3. Stage accessories (live load)	Gridirons and fly galleries	75	
	Loft block wells[4]	250	250
	Head block wells and sheave beams[4]	250	250
4. Ceiling framing (live load)	Over stages	20	
	All uses except over stages	10[5]	
5. Partitions and interior walls, see Sec. 2309 (live load)			5
6. Elevators and dumbwaiters (dead and live load)		2 × Total loads[6]	
7. Mechanical and electrical equipment (dead load)		Total loads	
8. Cranes (dead and live load)	Total load including impact increase	1.25 × Total load[7]	0.10 × Total load[8]
9. Balcony railings and guardrails	Exit facilities serving an occupant load greater than 50		50[9]
	Other		20[9]
10. Handrails		See Footnote No. 10	See Footnote No. 10
11. Storage racks	Over 8 feet high	Total loads[11]	See Table No. 23-P
12. Fire sprinkler structural support		250 pounds plus weight of water-filled pipe[12]	See Table No. 23-P
13. Explosion exposure	Hazardous occupancies, see Sec. 910		

[1]The tabulated loads are minimum loads. Where other vertical loads required by this code or required by the design would cause greater stresses, they shall be used.

[2]Pounds per lineal foot.

[3]Lateral sway bracing loads of 24 pounds per foot parallel and 10 pounds per foot perpendicular to seat and footboards.

[4]All loads are in pounds per lineal foot. Head block wells and sheave beams shall be designed for all loft block well loads tributary thereto. Sheave blocks shall be designed with a factor of safety of five.

[5]Does not apply to ceilings which have sufficient total access from below, such that access is not required within the space above the ceiling. Does not apply to ceilings if the attic areas above the ceiling are not provided with access. This live load need not be considered as acting simultaneously with other live loads imposed upon the ceiling framing or its supporting structure.

[6]Where Appendix Chapter 51 has been adopted, see reference standard cited therein for additional design requirements.

[7]The impact factors included are for cranes with steel wheels riding on steel rails. They may be modified if substantiating technical data acceptable to the building official is submitted. Live loads on crane support girders and their connections shall be taken as the maximum crane wheel loads. For pendant-operated traveling crane support girders and their connections, the impact factors shall be 1.10.

[8]This applies in the direction parallel to the runway rails (longitudinal). The factor for forces perpendicular to the rail is 0.20 × the transverse traveling loads (trolley, cab, hooks and lifted loads). Forces shall be applied at top of rail and may be distributed among rails of multiple rail cranes and shall be distributed with due regard for lateral stiffness of the structures supporting these rails.

[9]A load per lineal foot to be applied horizontally at right angles to the top rail.

[10]The mounting of handrails shall be such that the completed handrail and supporting structure are capable of withstanding a load of at least 200 pounds applied in any direction at any point on the rail. These loads shall not be assumed to act cumulatively with Item 9.

[11]Vertical members of storage racks shall be protected from impact forces of operating equipment, or racks shall be designed so that failure of one vertical member will not cause collapse of more than the bay or bays directly supported by that member.

[12]The 250-pound load is to be applied to any single fire sprinkler support point but not simultaneously to all support joints.

Figure 4-13 Minimum roof live loads

TABLE NO. 23-C—Minimum Roof Live Loads[1]						
	METHOD 1			METHOD 2		
Roof slope	TRIBUTARY LOADED AREA IN SQUARE FEET FOR ANY STRUCTURAL MEMBER			UNIFORM LOAD[2]	RATE OF REDUCTION *r* (Percent)	MAXIMUM REDUCTION *R* (Percent)
	0 to 200	201 to 600	Over 600			
1. Flat or rise less than 4 inches per foot. Arch or dome with rise less than one eighth of span	20	16	12	20	.08	40
2. Rise 4 inches per foot to less than 12 inches per foot. Arch or dome with rise one eighth of span to less than three eighths of span	16	14	12	16	.06	25
3. Rise 12 inches per foot and greater. Arch or dome with rise three eighths of span or greater	12	12	12	12	No Reductions Permitted	
4. Awnings except cloth covered[3]	5	5	5	5		
5. Greenhouses, lath houses and agricultural buildings[4]	10	10	10	10		

[1]Where snow loads occur, the roof structure shall be designed for such loads as determined by the building official. See Section 2305 (d). For special-purpose roofs, see Section 2305 (e).
[2]See Section 2306 for live load reductions. The rate of reduction *r* in Section 2306 Formula (6-1) shall be as indicated in the table. The maximum reduction *R* shall not exceed the value indicated in the table.
[3]As defined in Section 4506.
[4]See Section 2305 (e) for concentrated load requirements for greenhouse roof members.

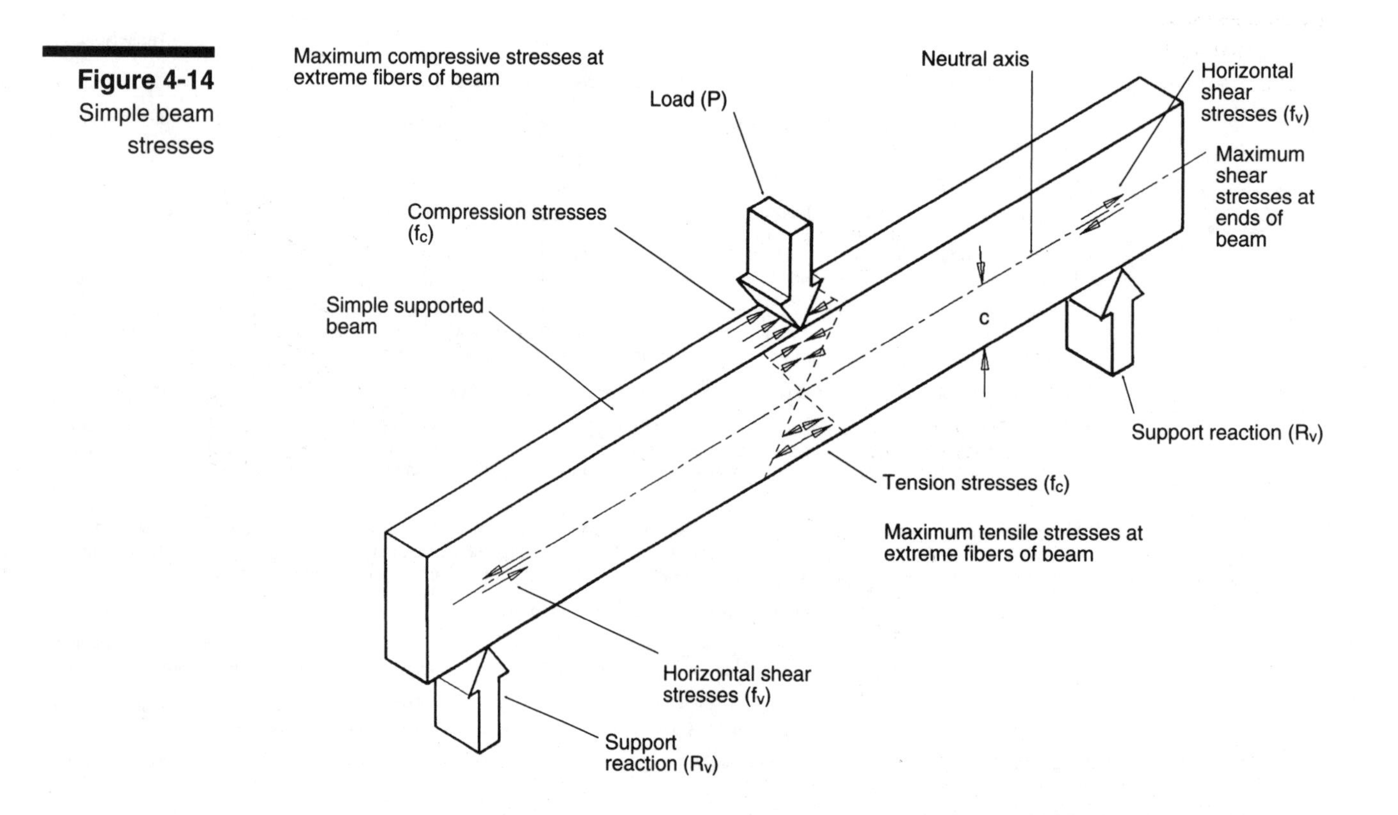

Figure 4-14 Simple beam stresses

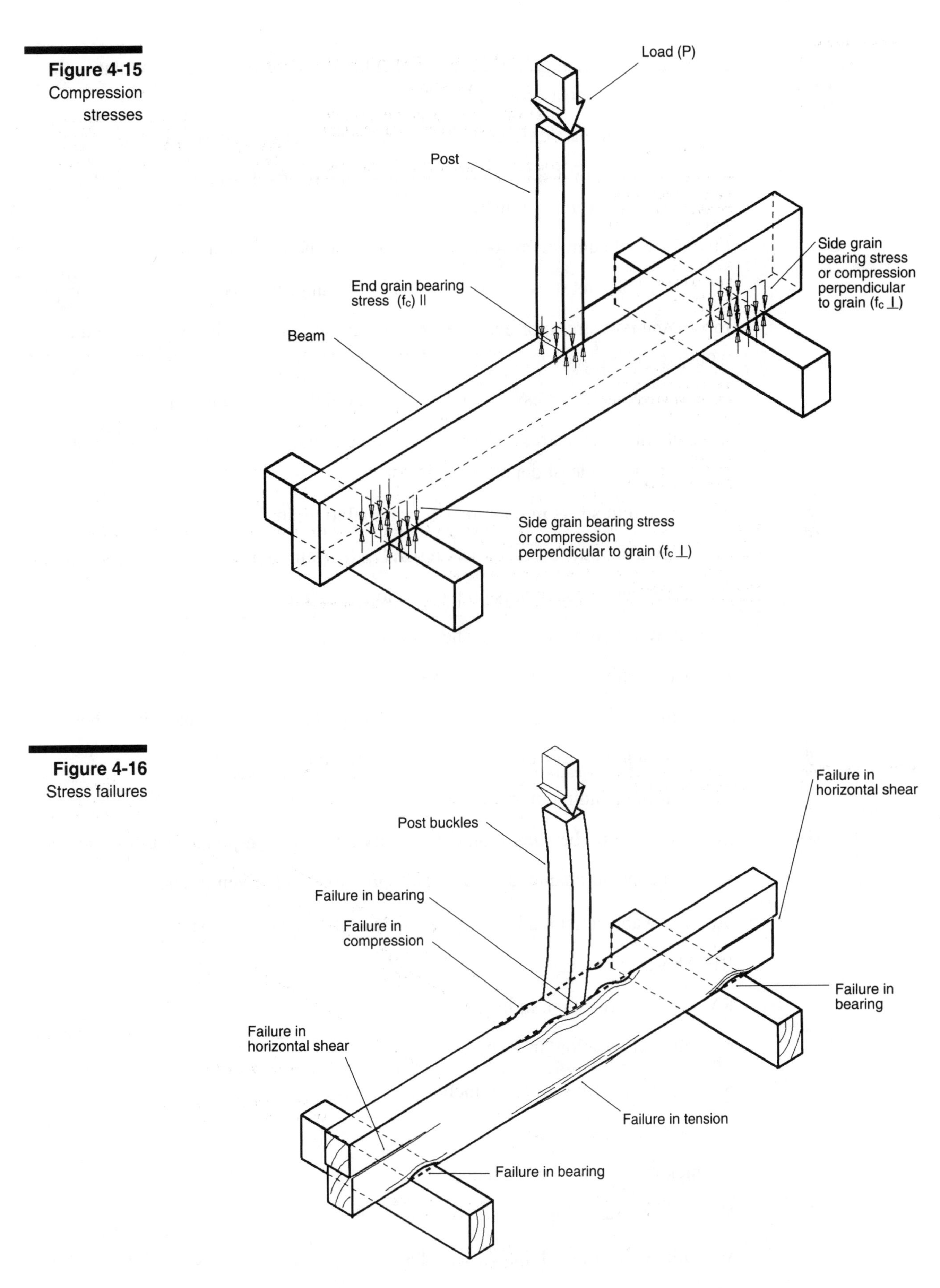

Figure 4-15
Compression stresses

Figure 4-16
Stress failures

d = depth of a rectangular member, or the least dimension of a rectangular compression member (post), in inches.

E = modulus of elasticity, or stiffness, of a construction material, in psi.

e = eccentricity, or distance a concentrated load is offset from the center of a compression member, in inches.

Fb = allowable unit stress for extreme fiber in bending, in psi.

F'b= allowable unit stress in extreme fiber in bending, adjusted for slenderness, in psi.

fb = actual unit stress for extreme fiber in bending, in psi. This is also called the working stress.

Fc = allowable unit stress in compression parallel to the grain, in psi.

F'c= allowable unit stress in compression parallel to the grain adjusted for the ratio of length to least depth, or $^{l}/_{d}$, in psi.

fc = actual unit stress in compression parallel to grain, in psi.

Ft = allowable unit stress in tension parallel to grain, in psi.

ft = actual unit stress in tension parallel to grain, in psi.

Fv = allowable unit horizontal shear stress, in psi.

fv = actual unit horizontal shear stress, in psi.

I = moment of inertia, or resistance to bending of a structural shape, in $inches^4$.

L = span of a beam, or unsupported length of a column, in feet.

M = induced bending moment, usually given in foot-pounds (ft-lbs).

m = resisting unit bending moment, usually given in inch-pounds (in-lbs).

P = total concentrated load, or axial compression load, given in lbs.

$^{P}/_{A}$ = allowable axial load per unit of cross-sectional area, in psi.

RH = horizontal reaction, in lbs.

RV = vertical reaction, in lbs.

r = radius of gyration, in inches.

S = section modulus, or $^{I}/_{c}$, in $inches^3$.

T = total axial tension load, in lbs.

t = thickness of member, in inches.

V = shear force, in lbs.

W = total uniform load, or $w \times L$, in lbs.

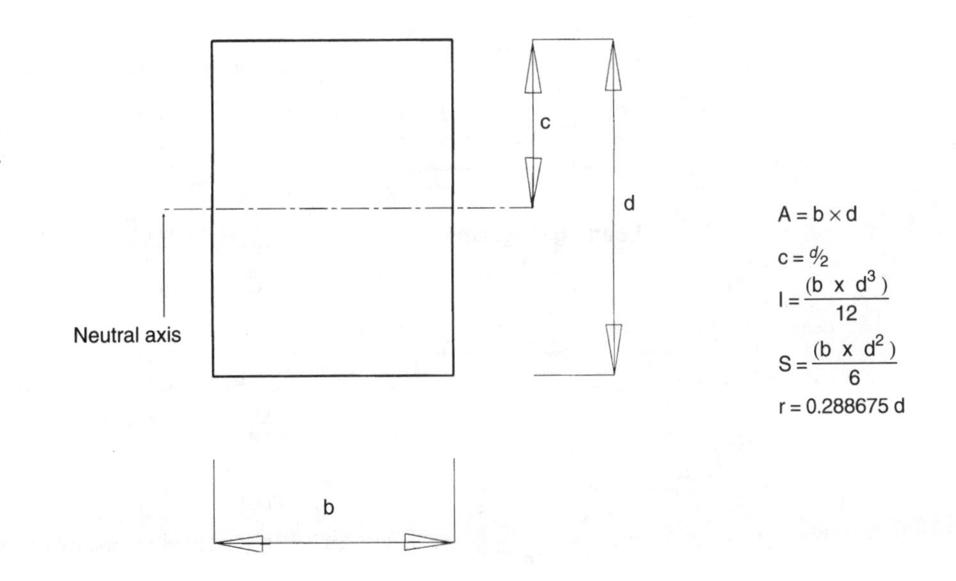

Figure 4-17 Properties of a rectangular-shape member

w = uniform load per unit of length, in pounds per linear foot, or plf.

delta allow = allowable deformation or deflection, in inches.

delta = actual deformation or deflection, in inches.

These symbols are based on the ones in the 1991 UBC, Chapter 25. Older editions of UBC and other codes and manuals may use different symbols. Be sure you read the legend or definition of the symbols used. Here are some examples of alternate symbols:

H = horizontal shear stress

q = compression stress perpendicular to grain

c = compression stress parallel to grain

Figure 4-17 shows the formulas for moment of inertia, I, section modulus, S, and radius of gyration, r, of a rectangular shape. The value of the moment of inertia of a structural member tells you how well the member can retain its shape against external bending forces which act on it. A good example is that a 2 × 12 joist, which has a high moment of inertia, resists bending better than a 12 × 2 plank, which has a much lower moment of inertia, when the plank is placed with its wider side horizontal.

To figure out what size beam you need for a particular job, you need to figure out the maximum bending moment, M_{max}, in the beam. Then you need to know the actual unit bending stress, fb, in that member resulting from that bending moment. Finally, you compare this value with the allowable unit bending stress, Fb, for the selected type of lumber. The actual unit bending stress must be less than the allowable unit bending stress. Formulas for bending and shear stresses and deflections are shown on Figures 4-18, 4-19 and 4-20. These figures show how the span, L, modulus of elasticity, E, and moment of inertia, I, affect the stresses and deflection of the member.

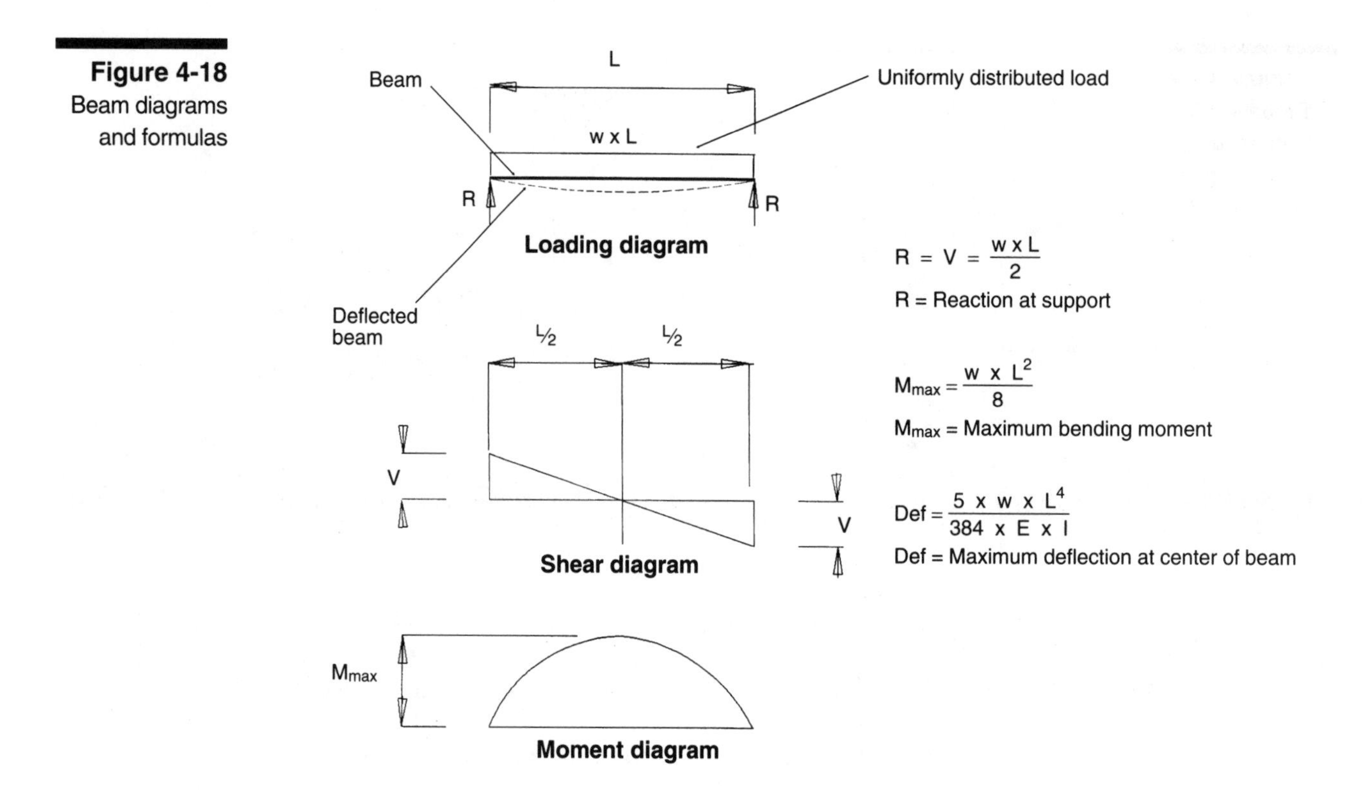

Figure 4-18 Beam diagrams and formulas

$$R = V = \frac{w \times L}{2}$$

R = Reaction at support

$$M_{max} = \frac{w \times L^2}{8}$$

M_{max} = Maximum bending moment

$$Def = \frac{5 \times w \times L^4}{384 \times E \times I}$$

Def = Maximum deflection at center of beam

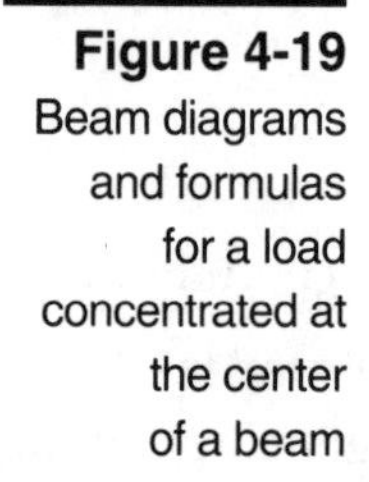

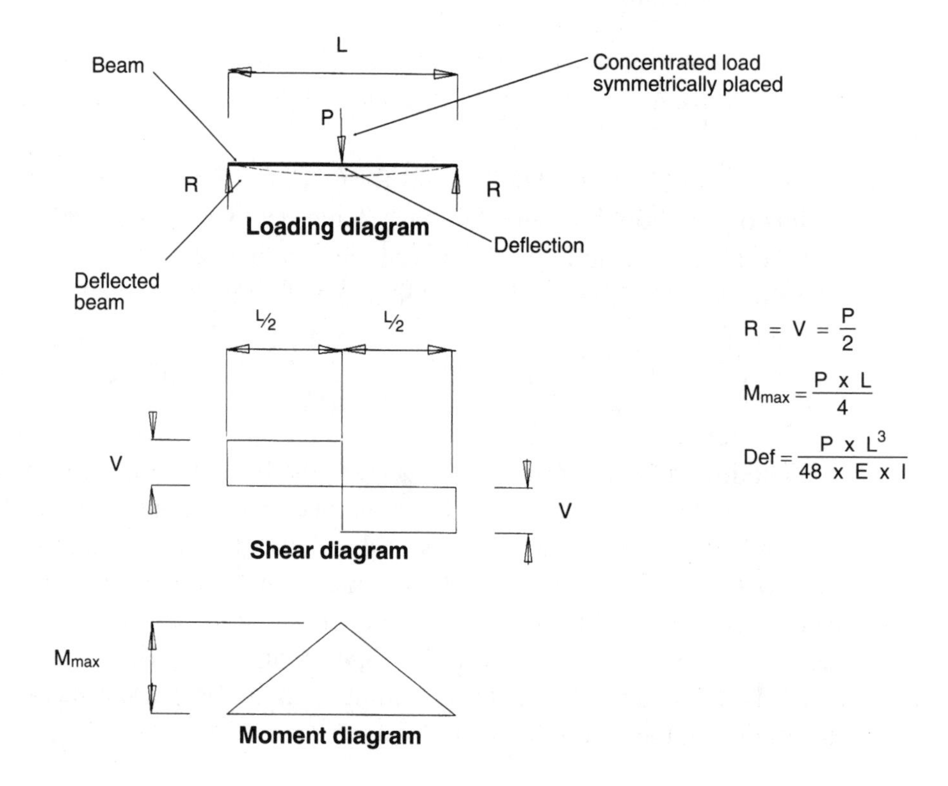

Figure 4-19 Beam diagrams and formulas for a load concentrated at the center of a beam

$$R = V = \frac{P}{2}$$

$$M_{max} = \frac{P \times L}{4}$$

$$Def = \frac{P \times L^3}{48 \times E \times I}$$

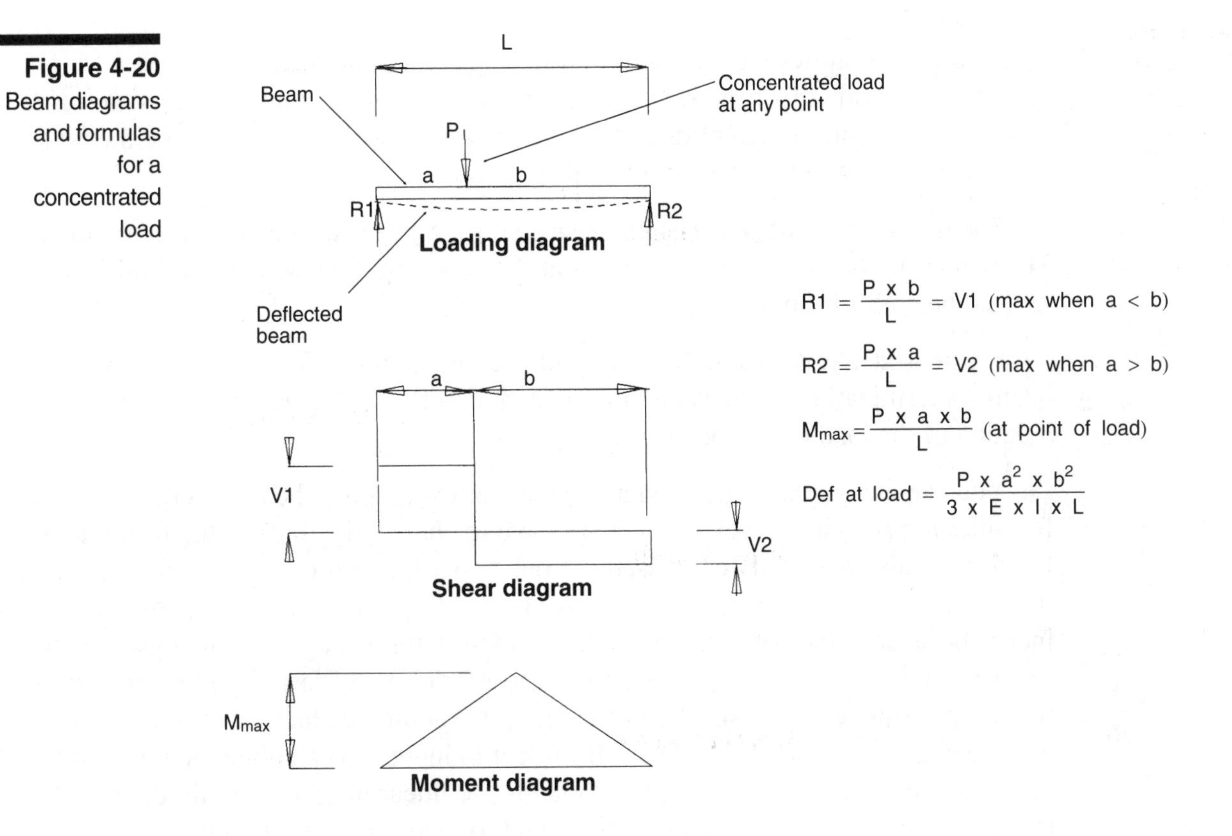

Figure 4-20 Beam diagrams and formulas for a concentrated load

Structural designers use simple diagrams to show the type of and location of loads on a beam, the amount of bending and shear stress on it, and the amount of deflection on any point of the beam under the loading condition. Figure 4-18 shows diagrams and formulas for a uniformly loaded simple beam of length L. This is the most common beam condition. The upper diagram shows the beam as a solid line and the load, w, which is usually expressed in pounds per linear foot, spread across the top of the beam. The total weight supported by the beam is w times L, or W.

The diagram at the center is called a shear diagram and illustrates the amount of shear, V, at any point of the beam. Note that the maximum shear is equal to the reaction, R, at the two supports of the beam. The sum of the reactions must equal the total load, W. When a shear diagram is drawn to scale, you can find the amount of shear at any point of the beam by measuring the perpendicular vertical distance from the point to the diagonal line. You can also determine the amount of shear at any point by use of similar triangles.

The lower diagram is called a moment diagram for the same beam. When this diagram is drawn to scale, the bending moment at any point on the beam is the perpendicular vertical distance from the point to the parabola-shaped curved line. The maximum bending moment is at the center of the beam, and zero moment is at the ends of the beam.

A deflection diagram is usually not drawn but the maximum deflection is normally calculated by the formula shown on the drawing.

Figure 4-19 shows the diagrams and formulas for a beam with a single concentrated load at its center. Note here that the maximum shear remains the same from the beam supports to midspan, and it's in the opposite direction on either side of the center of the beam.

The moment diagram is triangular-shaped for a beam with a concentrated load. The maximum bending moment occurs at the center of the beam and diminishes to zero at the ends of the beam.

Figure 4-20 shows a simple beam with a concentrated load at any point of the beam. Determination of the shear, moment, and deflection is dependent on the location of the load on the beam.

You can analyze the forces on a simple beam by adding the results of whichever formulas apply to it. However, you may have to change the units in the formulas. Uniform loads are usually expressed in pounds per linear foot, plf. Beams are usually measured in feet and inches. Formulas dealing with stresses and deflection are in inches because values of moment of inertia of structural shapes are always in $inches^4$. Section modulus is expressed in $inches^3$, and Modulus of Elasticity is expressed in pounds per square inch, psi. Therefore, all units should be changed to inches. So you have to change the length of a beam from feet to inches to calculate the uniform load on it. Convert uniform loads in plf to pounds per linear inch, pli, by dividing by 12. Change beam spans or post lengths from feet to inches by multiplying by 12. For example, a 20 foot beam with a uniform load of 60 plf would be expressed as a 240 inch beam (12×20) with a uniform load of $60 \div 12$, or 5 pli.

To demonstrate how you use some of this information, let's figure out what size wood framing you need for the detached carport shown in Figure 4-21. Assume the size of the carport is $10 \times 20 \times 8$ feet high. The preliminary design includes built-up roofing with a 90-pound cap sheet, 1-inch solid diagonal sheathing, 2×6 rafters at 24 inches o.c., 4×10 beams, and 4×4 braced posts. To confirm this design you would do the following:

1. Determine the dead load on the roof:

   Built-up roofing = 3 psf
   1-inch sheathing = 4 psf
   2×6 rafters = 2 psf
   Total dead load = $3 + 4 + 2 = 9$ psf

2. Determine the live load on the roof:

   Live load on rafters = 20 psf
   Tributary area = $10 \times 2 = 20$ sq ft
   Live loads on beams with a tributary area under 100 sq ft = 16 psf

3. Size the sheathing:

   UBC Table No. 25-R-1 states that the allowable span for ¾-inch surfaced dry roof sheathing placed diagonal to supports = 24 inches

Figure 4-21
Typical carport

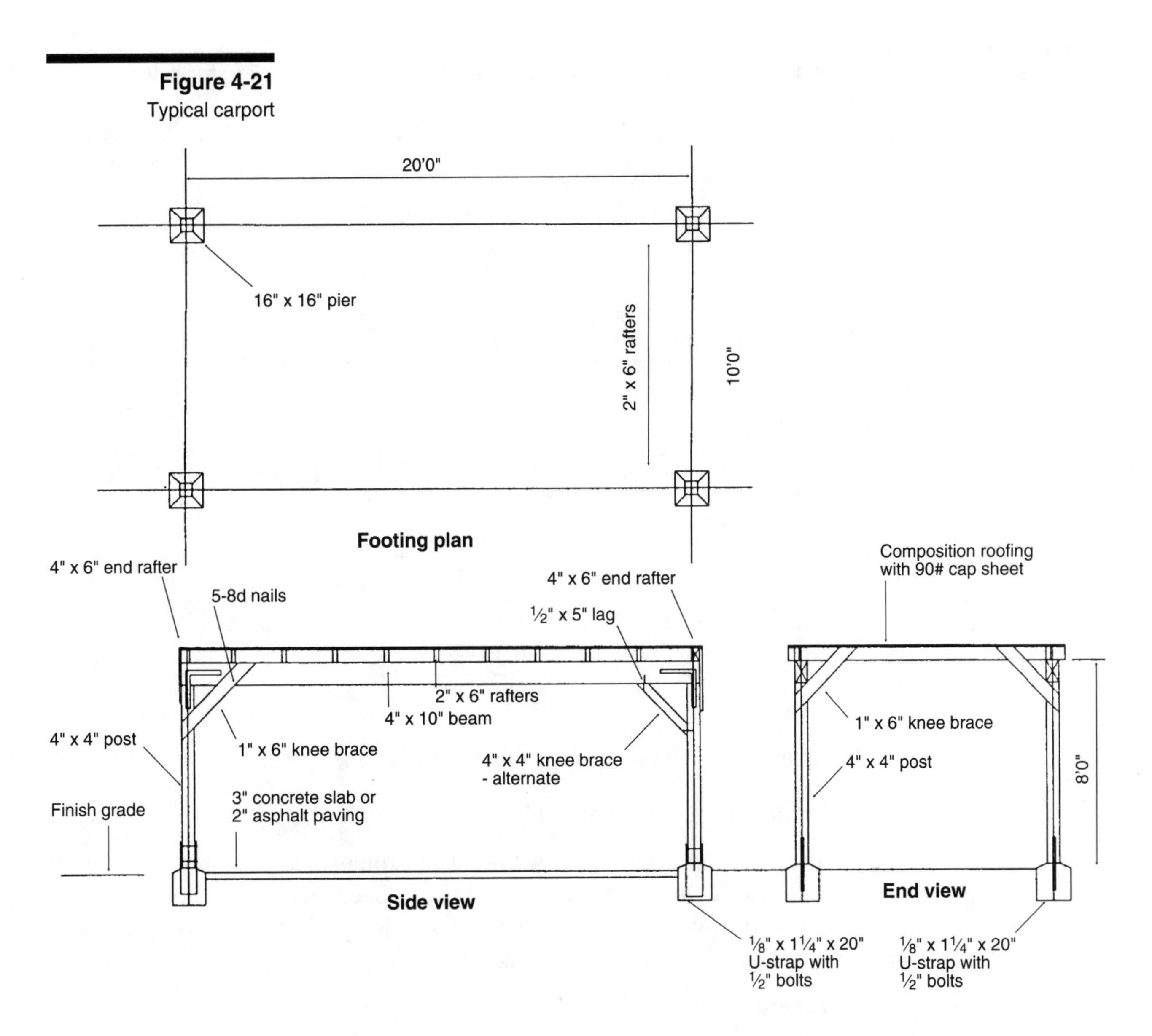

4. Size rafters:

Dead plus live load = 9 + 20 = 29 psf

Load per foot of length = w = 29 × 2 = 58 plf (You multiply by 2 because the rafters are 2 feet apart and each foot of the rafter carries 58 pounds.)

Length of rafter = L = 10 ft

$$\text{Bending moment} = m = w \times \frac{L^2}{8} = 58 \times \frac{100}{8} = 725 \text{ ft-lbs}$$

Convert to in-lbs = 725 × 12 = 8700 in-lbs

Let's see if 2 × 6 rafters will do the job:

$I = 20.80 \text{ inches}^4$ (See Figure 4-22)

$S = 7.56 \text{ inches}^3$ (See Figure 4-22)

Figure 4-22 Moment of inertia and section modulus of members

Nominal size (inches)	Moment of inertia (inches^4)	Section modulus (inches^3)
2 x 4	5.36	3.06
2 x 6	20.80	7.56
2 x 8	47.64	13.14
2 x 10	98.93	21.39
2 x 12	177.98	31.64
2 x 14	290.76	43.89
4 x 4	12.51	7.15
4 x 6	48.52	17.65
4 x 8	111.15	30.66
4 x 10	230.84	49.91
4 x 12	415.28	73.83
4 x 14	678.46	102.41
4 x 16	1034.42	135.66

Find actual maximum bending stress in extreme fibers, fb:

$$fb = m \div S = 8700 \div 7.56 = 1151 \text{ psi}$$

Rafters are repetitive members. Allowable maximum bending stress in extreme fibers, Fb, for No. 2 Douglas fir/larch is 1,450 psi. See Figure 4-23. This is more than the maximum bending stress, 1151 psi, that you calculated above, so you can use No. 2 Douglas fir/larch 2 × 6 on 24-inch centers.

5. Size beams:

Weight of 4 × 10 beam = 9.7 plf, call it 10 plf

Dead plus live load = w = (10 + (9 × 5)) + (16 × 5) = 55 + 80 = 135 plf

Length of beam = L = 20 ft = 20 × 12 = 240 inches

$$\text{Bending moment} = m = w \times \frac{L^2}{8} = 135 \times \frac{20^2}{8} = 6750 \text{ ft-lbs}$$

Convert to in-lbs = 6750 × 12 = 81,000 in-lbs

Will a 4 × 10 beam work?

I = 230.84 (Figure 4-22)

S = 49.91 (Figure 4-22)

$$fb = m \div S = 81{,}000 \div 49.91 = 1623 \text{ psi}$$

Allowable maximum unit bending stress, Fb, for Dense No. 1 DF 4 × 10 is 1,800 psi. See Figure 4-23. This is more than the 1623 psi you calculated, so the Dense No. 1 Douglas fir 4 × 10 beams will work.

Figure 4-23 Typical allowable stresses for Douglas fir-larch

Grade	Allowable bending stress		Allowable horizontal shear stress (psi)	Modulus of elasticity (psi)
	Single member (psi)	Repetitive member (psi)		
Dense sel str	2,100	2,400	95	1,900,000
Sel str	1,800	2,050	95	1,800,000
Dense no. 1	1,800	2,050	95	1,900,000
Dense no. 2	1,450	1,700	95	1,800,000
No. 2	1,250	1,450	95	1,700,000
No. 3	725	850	95	1,500,000

Figure 4-24 Capacity of typical wood posts (pounds)

Size	8 ft	10 ft	12 ft	14 ft	16 ft
4 x 4	10,000	6,330	4,400	3,230	——
6 x 6	36,300	33,600	23,300	17,100	13,100
8 x 8	67,500	67,500	67,500	59,200	45,300

6. Size posts:

Total dead plus live load on beam = W = w × L = 135 plf × 20 = 2,700 lbs.

Load carried by post = W÷2 = P = 2,700 ÷ 2 = 1,350 lbs

Length of post = 8 ft = 8 × 12 = 96 inches

Capacity of a 4 × 4 post 8 ft long is 10,000 lbs. See Figure 4-24

Will 4 × 4 posts work?

Check 4 × 4 with L = 96 inches, d = 3.5 inches, E = 1,800,000 psi by formula:

Allowable maximum compressive stress on the 4 × 4 post is:

$$\frac{P}{A} = 0.30 \times \frac{E}{(L/d)^2}$$

$= .30 \times 1{,}800{,}000 \div (96 \div 3.5)^2$

= 718 psi

Actual compressive stress = fc = $\frac{P}{A}$ =

$= 150 \div (3.5)^2 =$

= 11 psi

This is less than the allowable maximum, so you can use the 4 × 4 posts.

The formulas you use to calculate beam sizes can also be used to calculate floor and ceiling joist sizes. However, there can be more bending stress when framing members are repetitive, or run parallel and side by side, such as joists and rafters. These increased values are shown on Figure 4-23.

The critical stress in relatively short beams carrying heavy loads may be horizontal shear, fv. A horizontal shear stress is caused by an applied load tending to shear off the end of a member. This type of large girder is often used in heavy timber construction. It is also found in the main supporting beams on hillside houses. The formula for determining maximum shear stress is:

$$fv = (3/2) \times \left(\frac{V}{A}\right)$$

where:

fv = unit shearing stress, in psi
V = force tending to shear the beam, in lbs
A = cross-sectional area of beam, in sq inches

As an example, assume V = 1,100 lbs, A = 32.38 sq inches (4 × 10). Then the maximum horizontal shear is:

$$fv = (3/2) \times (1100 \div 32.38)$$
$$= 50.96 \text{ psi}$$

Loads on Posts

The carrying capacity of a wood post depends on the strength of the lumber the post is cut from, the cross-sectional area of the post, and its unsupported length. A post will tend to buckle under a heavy load. Resistance to buckling usually determines the carrying capacity of a post. The tendency to buckle depends on the stiffness of the wood post, and the ratio of its length, L, to its least side dimension, d. That's known as the "L/d" ratio. The technical name for stiffness is modulus of elasticity (E). This may range from 1,900,000 psi for dense select structural lumber, to 1,500,000 psi for No. 3 Douglas fir lumber, as shown on Figure 4-23. The formula for the allowable axial compression stress in a wood post or column, Fc, is:

$$Fc = 0.30 \times E \div (L/d)^2$$

As an example, here's how to calculate the allowable compressive stress on a 6 × 6 inch post, with net size $5\frac{1}{2} \times 5\frac{1}{2}$ inches, 10 feet long, made of No. 3 Douglas fir with a modulus of elasticity (E) of 1,500,000 psi:

$$Fc = 0.30 \times 1{,}500{,}000 \div (120 \div 5.5)^2 = 945.38 \text{ psi}$$

A nominal 6 by 6 post has an area of about 30 square inches, since the timber actually measures $5\frac{1}{2}$ by $5\frac{1}{2}$. To find the maximum allowable load on this post, multiply 30 square inches by 1,125 pounds per square inch. The answer is 33,750 pounds.

Do the calculation yourself to find that if the same post was 20 feet, or 240 inches, long it could carry only 7,093 pounds.

$$P/A = 0.3 \times E \div (240 \div 5.5)^2$$
$$= 236.3 \text{ psi}$$
$$P = 30 \times 236.3$$
$$= 7093 \text{ lbs}$$

Figure 4-24 shows typical load capacities of wood posts of various lengths.

Deflection

The size of most long beams is controlled by the expected deflection under load rather than the bending stress. Excessive deflection (sagging) looks bad and can crack plaster ceilings. The maximum allowable sag of a beam is given as a fractional part of the span. Here are some maximum allowable deflections that the *Uniform Building Code* requires for roof beams:

- ceiling without plaster - $\frac{1}{240}$ of the span for live load, and $\frac{1}{180}$ of the span for a combination of live and dead load
- plastered ceiling - $\frac{1}{360}$ of the span for live load and $\frac{1}{240}$ of the span for live plus dead load.

The allowable sag for floor beams, with or without plaster ceiling, is $\frac{1}{360}$ of the span for live load and $\frac{1}{240}$ of the span for the total of live plus dead load. The expected sag (deflection) of a simple supported beam with a uniform load is:

$$\text{def} = \frac{5 \times w \times L^4}{384 \times E \times I}$$

where:

E = the modulus of elasticity
I = the moment of inertia
w = uniform load in pounds per linear foot, plf

Figure 4-22 shows the I and S values for common-size lumber. Figure 4-23 shows the following for various grades of Douglas fir-larch wood:

allowable bending stress for a single member
allowable bending stress for repetitive members
allowable horizontal shear stress
modulus of elasticity

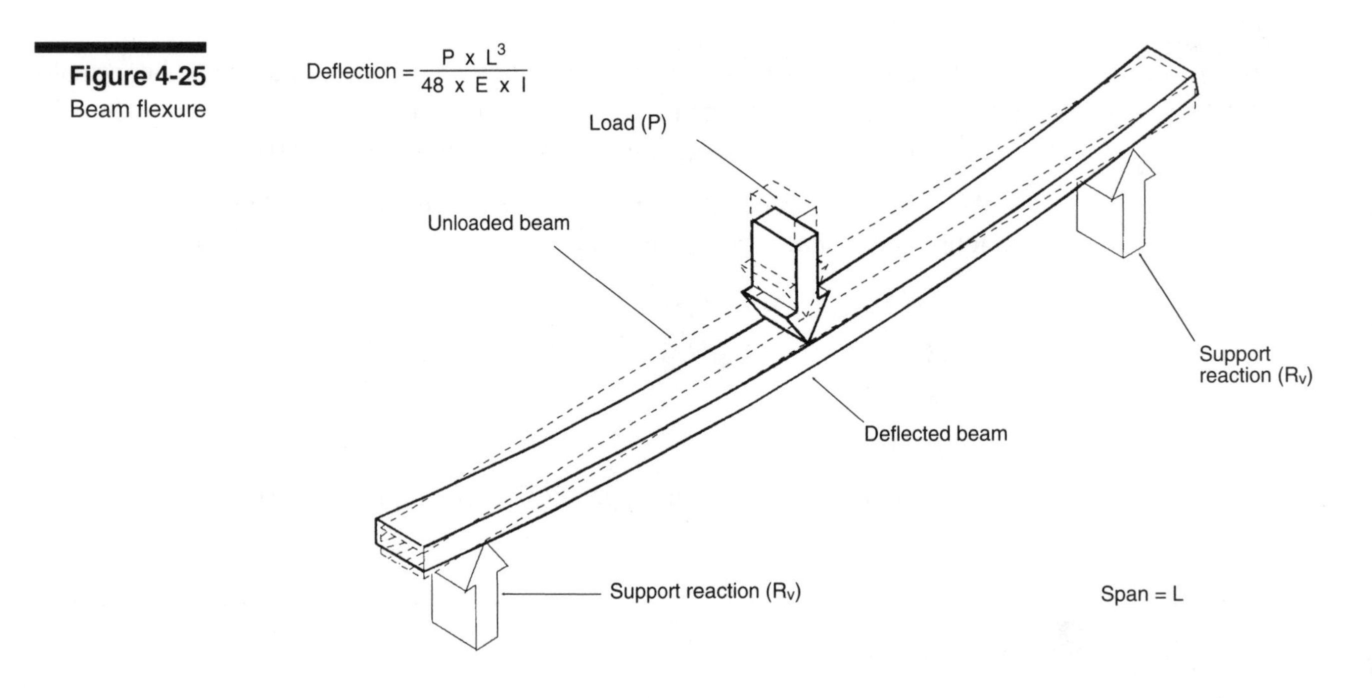

Figure 4-25
Beam flexure

The modulus of elasticity of various types of woods is given in the UBC and the *Timber Construction Manual* published by the American Institute of Timber Construction.

Any loaded wooden beam will sag. Figure 4-25 shows the flexure and deflection of a beam. Flexure is another word for the bending of a beam. Deflection is caused by a beam bending or flexing. Wood with a low modulus of elasticity will deflect more than wood that has a higher one. For example, the modulus of elasticity of Dense No. 1 Douglas fir is 1,900,000 psi. For No. 3 Douglas fir it's about 26 percent less, only 1,500,000 psi. So the deflection for a No. 3 Douglas fir beam would be 26 percent more than a same-size beam of No. 1 Douglas fir. The amount of deflection in a beam is also inversely proportional to its moment of inertia. Figure 4-22 shows that deeper beams with higher moment of inertia values are stiffer.

Wood Members

Forming

Forms for concrete foundations are made of sheathing, studs, wales, braces, stakes, shoe plates, spreaders, and tie wire. Each of these can help create a rigid and accurate concrete form. Chapter 3 contains detailed information on forming. Figure 4-26 shows parts of the formwork for a foundation grade beam. Figure 4-27 shows forming for a concrete wall.

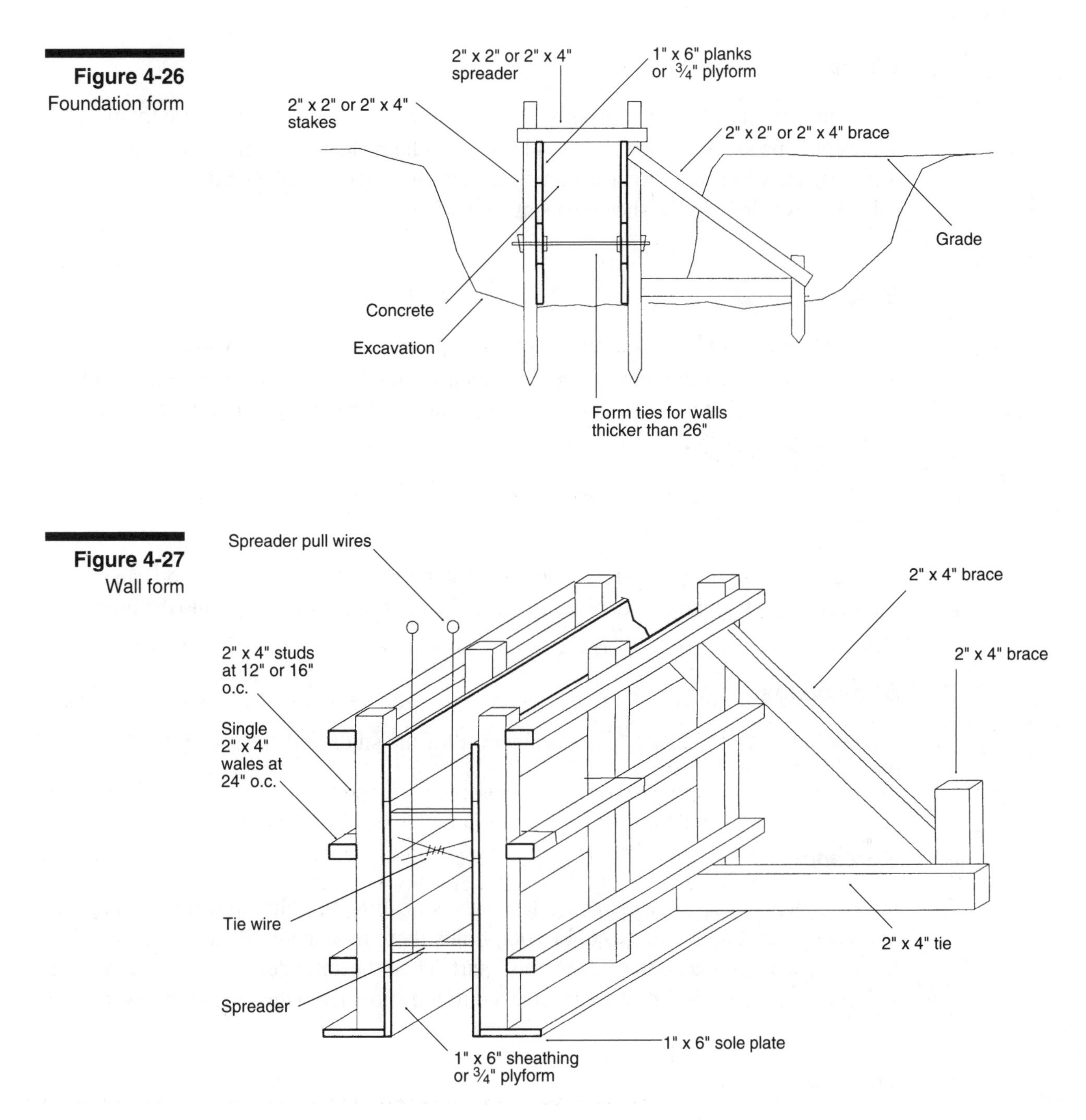

Figure 4-26
Foundation form

Figure 4-27
Wall form

Sheathing

Sheathing forms the surface of the concrete. It must be smooth and tight, and should be strong enough to hold wet concrete. Most sheathing is made of 1-inch boards (13⁄16-inch dressed) or 3⁄4-inch plywood.

Studs

Studs hold the sheathing in place and resist the pressure of the liquid concrete. Vertical form studs are normally made of 2 × 4 or 2 × 6 lumber. The studs make the wall form rigid.

Wales

If the studs are over 4 feet high, you use wales, or walers, to reinforce them. Wales may be single or double 2 × 4 lumber. Prefabricated forms are usually about 10 feet long. They use wales to fasten the form sections together with 16d double-headed nails. Wales help keep the sheathing straight and flat.

Braces

The braces also help hold the concrete in place until it sets. A typical brace is a diagonal 2 × 4 member nailed to the wale and to a stake driven into the ground. To make the form more stable, you can also nail horizontal members to the studs, braces, and stakes.

Stakes

Stakes are usually pointed 2 × 4 lumber that's driven into the ground. You can also use steel stakes with holes where nails can be driven through the stakes into the braces.

Shoe plates

The shoe plate makes a level foundation for the studs and sheathing. Nail studs to the shoe plate.

Spreaders

Spreaders are set between the sides of the sheathing to hold the form walls apart. Spreaders aren't nailed in place. Friction holds them in place until enough concrete has been poured to keep the form walls apart. Then the spreaders are pulled out of the wet concrete with wires that were attached to the spreaders when the forms were built.

Tie wires

Tie wires are made of No. 8 or 9 gauge, soft, black, annealed iron wire. You attach them to each stud at the wales on both sides of the forms. Wires are tightened by twisting them with a wedge. Tie wires help keep the forms from spreading. See Chapter 3 for a great deal more information on forming, including wales, studs and spreaders.

Horizontal Framing Members

Horizontal structural members include joists, rafters, beams, and girders. Figure 4-28 shows the principal horizontal members of a building section.

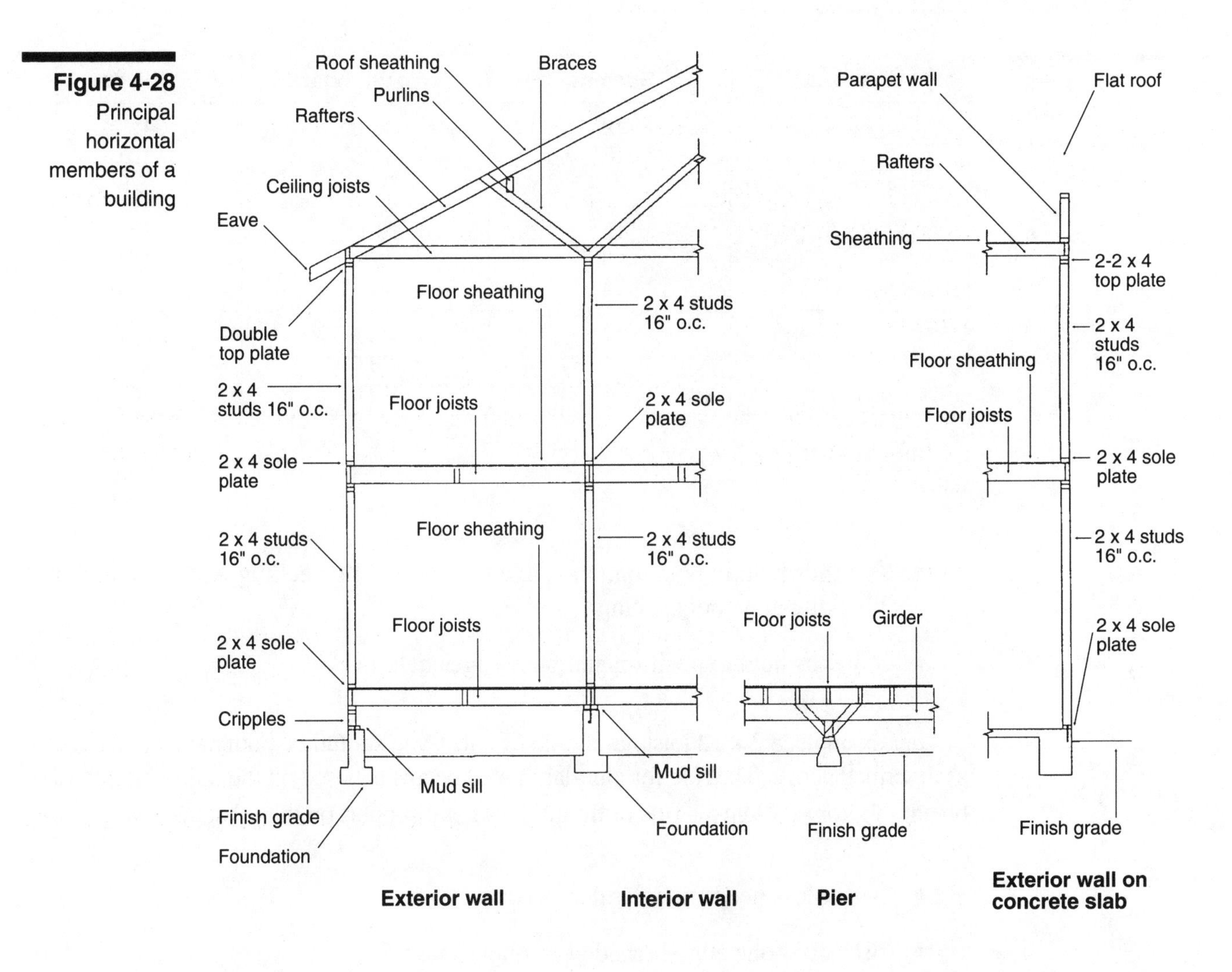

Figure 4-28 Principal horizontal members of a building

Most horizontal members are supported at each end. Joists and rafters are relatively thin and tend to twist or buckle when loaded. When the top portion of a beam is compressed from the bending, it tends to buckle and move horizontally. It's important that you provide lateral support at the top of girders, beams, joists, and rafters to prevent buckling.

You can provide lateral support at the ends of beams, at intermediate points along spans, or continuously along the tops of the members. Base the size of lateral support for beams on the ratio of the depth of the beam (d) to its thickness or breadth (b). Here's a rule of thumb for lateral support on beams, rafters, joists, and girders based on the ratio of d/b:

- 2:1 no lateral support needed
- 3:1 ends of beam held in position
- 4:1 ends of beam held in position and member held in line as in a well-bolted chord in a truss
- 5:1 ends held in position and one edge held in line by decking

Figure 4-29 Allowable spans for ceiling joists

Joist size	Spacing	Gypsum board	Plaster
2 x 4	12"	12'8"	11'0"
	16"	11'6"	10'0"
	24"	9'11"	8'9"
2 x 6	12"	19'11"	17'4"
	16"	17'9"	15'9"
	24"	14'5"	13'9"
2 x 8	12"	26'2"	22'10"
	16"	23'5"	20'9"
	24"	19'0"	18'2"

For drywall or plaster ceilings, Douglas fir-larch, No. 1 or better

- 6:1 ends held in position, one edge held in line by decking, and bridging or blocking at 8-foot spacing
- 7:1 ends held in position and both edges held in line.

For example, a 2 × 12 joist has a ratio of 6 to 1 and should be both nailed and blocked as described above. The rule for glue-laminated beams is based on the ratio of depth (d) to breadth (b), or d/b. Here's a rule of thumb for lateral support on these beams:

- 5:1 or less needs no lateral support
- 6:1 needs one edge braced at frequent intervals

Beams

Normally you use sawn wood timbers for spans of 6 to 40 feet. Use glue-laminated beams for spans from 10 to 100 feet. For a span that's less than 25 feet, solid wood beams are cheaper than glue-laminated members. For spans over 30 feet, it's more practical to use glue-laminated beams, because it's difficult to get extra long solid sawn timbers in structural grades.

Joists, girders, and lintels

Figure 4-29 shows allowable spans and spacing for ceiling joists with plastered ceilings. Figure 4-30 shows allowable spans and spacing for floor joists. Figure 4-31 shows allowable spans for floor girders which support floor joists. Figure 4-32 shows allowable spans for door and window lintels or headers. These spans and spacings are for conventional wood frame dwellings. Check your governing building code for local requirements.

Figure 4-30 Allowable spans for floor joists

Joist size	Spacing	Supporting partitions	Non-supporting partitions
2 x 6	12" 16" 24"	10'11" 9'6" 7'9"	10'11" 9'11" 8'6"
2 x 8	12" 16" 24"	14'5" 12'7" 10'3"	14'5" 13'1" 11'3"
2 x 10	12" 16" 24"	18'4" 16'1" 13'1"	18'5" 16'9" 14'4"
2 x 12	12" 16" 24"	22'4" 19'6" 15'11"	22'5" 20'4" 17'5"

For Douglas fir-larch No. 2 or better and a floor live load of 40 psf. If 50 psf, reduce spans by 10%.

Figure 4-31 Allowable spans for floor girders

Girder size	Spacing (ft)	Supporting partitions (ft)	Non-supporting partitions (ft)
4 x 4	6 8	4 3	4 3
4 x 6	6 8	7 6	7 6
4 x 8	6 8	8 6	9 7

For Douglas fir-larch, No. 2 or better

Figure 4-32 Allowable spans for lintels

Lintel size	Supporting floor, roof & ceiling	Supporting roof & ceiling only
4 x 4	3'6"	4'0"
4 x 6	5'6"	6'0"
4 x 8	7'0"	8'0"
4 x 10	9'0"	10'0"
4 x 12	10'0"	12'0"

For Douglas fir-larch, No. 2 or better. For 16 foot garage door opening in one-story attached or detached garage without ceiling, a 4 × 12 Douglas fir-larch No. 1 grade may be used

Rafters

The local building codes usually list the sizes and spacing of roof rafters. Generally the tables are based on a minimum live load of 20 psf, 16 psf, and 12 psf, depending on the tributary area of roof the individual member supports. The tributary area is the part of the loaded area that's carried by the beam being sized. The larger the tributary area, the lower the unit live load.

Rafters for flat roofs can get overloaded from buildups of rain water or snow. During rain storms, if roof drains get plugged up on a roof so water backs up several inches deep, the added weight of the water can collapse the roof. Standing water weighs about 5 pounds per square foot for each inch of depth. If you have just 2 inches of water on a 10,000 square-foot flat roof, you've got 100,000 pounds extra weight. That's about the same as 20 full-grown African elephants on the roof.

In cold climates, another type of loading problem is possible. The roof may cover both heated and unheated portions of a building. Water from melting snow on the warmer roof can freeze as it drains onto the unheated portion of the roof. The weight of this accumulating ice may be far more than the weight of any snowfall that could be expected. You should frame such portions of a roof to withstand the extra load. The UBC requires that you double the standard snow load in these critical locations.

Notching the top or bottom edge of a beam for pipes or ducts will reduce its strength. Cutting a notch at the ends of a beam will have the same effect. The safest place to cut a hole in a beam is on its neutral axis. A line parallel and midway between the top and bottom edges of a rectangular beam is called the neutral axis of the beam.

Vertical Members

Studs

In buildings up to two stories high, use 2 × 4 studs for the stud walls, as shown in Figure 4-33. In three-story buildings, the studding at the bottom of the first floor, under the second floor joists, shouldn't be less than 3 × 4 or 2 × 6, as shown in Figure 4-34. Unless laterally supported by adequate framing, the maximum allowable height of studs is:

- 10 feet for 2 × 3 studs
- 14 feet for 2 × 4 studs
- 16 feet for 3 × 4 studs
- 20 feet for 2 × 6 studs

Don't space studs more than 16 inches o.c. unless vertical supporting members in the walls are designed as posts. You can make these walls of posts that are at least 4 × 4 inches, spaced apart not more than 5'4" o.c.

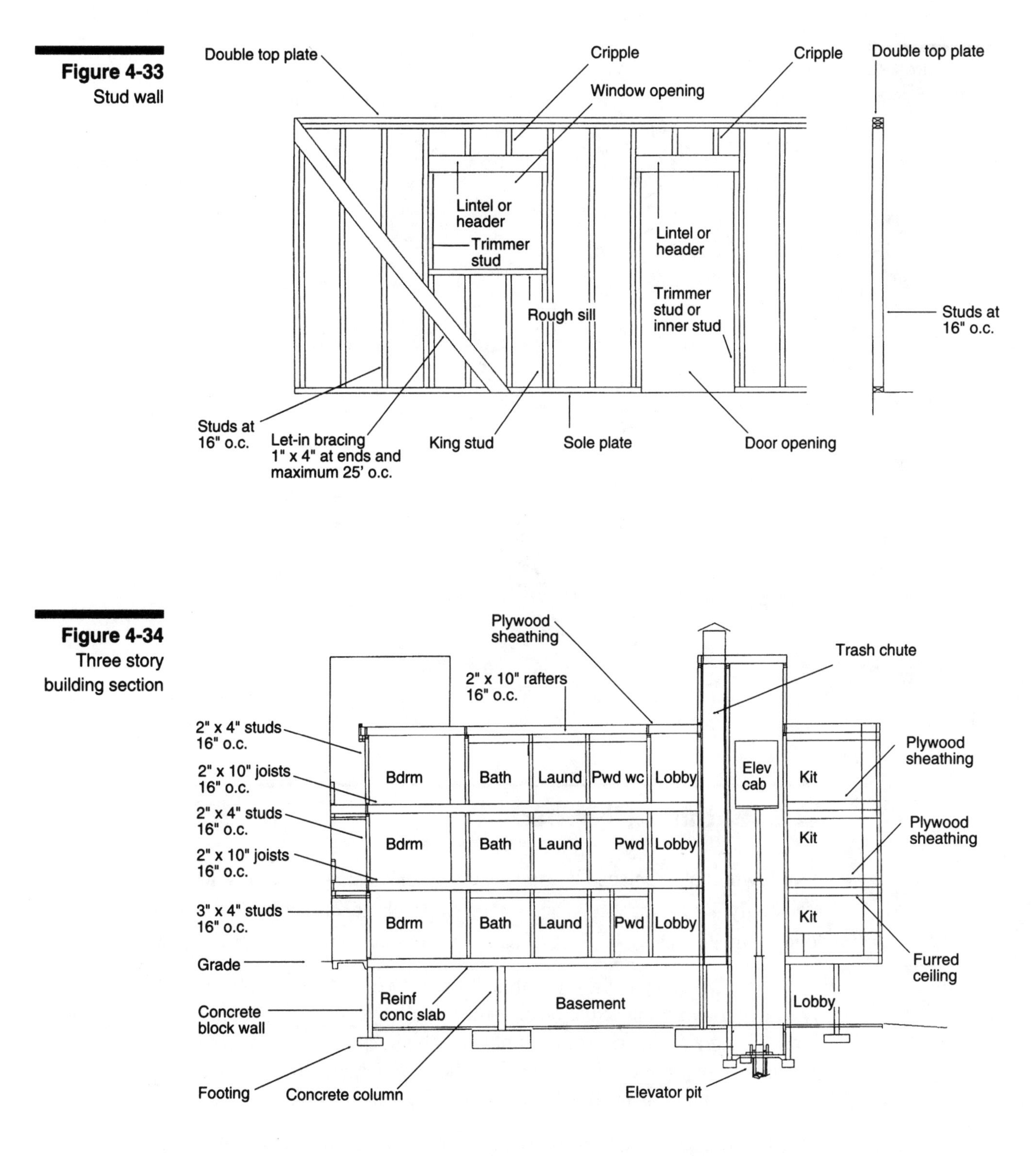

Figure 4-33 Stud wall

Figure 4-34 Three story building section

Posts

The most common size wood posts used in residential and commercial buildings are 4 × 4 and 6 × 6 inches. For special requirements you may use 8 × 8 posts. The carrying capacity of a post depends mainly on its size and length. Figure 4-24 shows the approximate capacity of wood posts of various sizes and lengths.

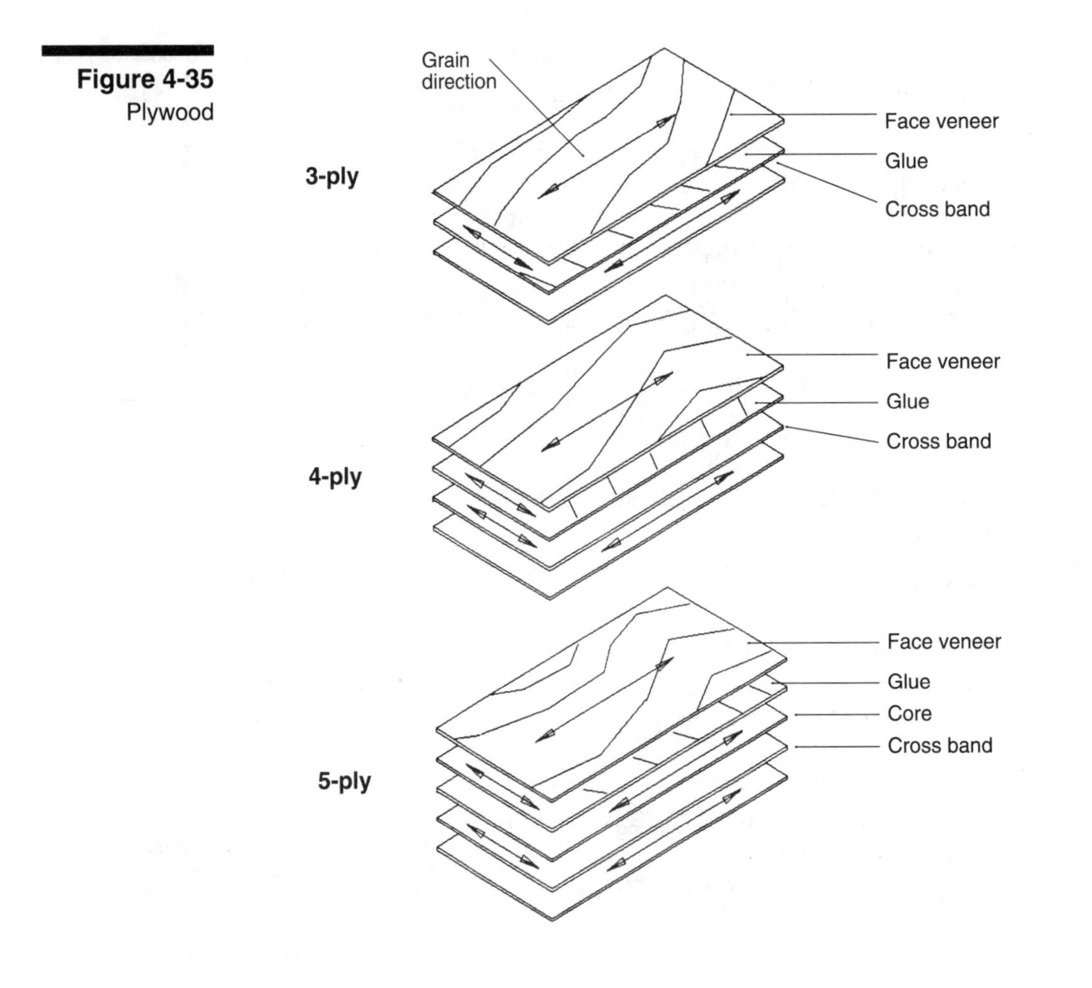

Figure 4-35
Plywood

Sheathing Plywood

Plywood has replaced board sheathing for most applications. Most lumber yards stock only 4 × 8-foot sheets, but 4 × 9-foot panels are also considered a standard size and longer panels, such as 4 × 12, are available from many mills by special order.

Most plywood panels have three, four, or five layers with glue between each layer. The direction of grain of the outside plies is always parallel to the long dimension of the panel. The grain direction of each ply is 90 degrees to the adjacent ply, except on 4-ply panels. In 4-ply panels, the two inner plies are perpendicular to the direction of the face plies. Figure 4-35 shows the parts of 3-, 4-, and 5-ply plywood panels.

Plywood panels are normally stamped with a grade mark that tells the following information:

- Grade of veneer on panel face: A, B, C, or D
- Grade of veneer on panel back: A, B, C, or D
- Identification Index: 12/0, 16/0, 20/0, 24/0, 32/16, 42/2, or 48/24
- Type of plywood: Interior or exterior
- The Product Standard followed by the manufacturer: PS-1-74

- Type of glue: Interior or exterior
- Mill No.
- APA logo

The grade of face veneer depends on the number and size of knot holes, loose knots, plugged holes, splits, and other defects. The first of the two letters indicates the face veneer, and the second letter is the back face. Generally, the grades tell the following:

- A is the highest grade and it has no open defects.
- B is a solid face, with tight knots and repairs allowed.
- C is a plugged veneer with splits limited to ⅛-inch wide and holes from ¼ inch to ½ inch in size.
- D is the lowest grade, with open spaces and no repairs allowed.

Plywood comes in various surface conditions; unsanded, touch sanded, and sanded. The net thickness varies according to the amount of sanding. Unsanded plywood is usually used for sheathing and subfloors where you're going to cover the surface.

The Identification Index is made up of two numbers. The first number shows the maximum on-center span in inches for use on roofs. The second number shows the maximum on-center span for use on floors. An Index of 32/24 means that the panel may be used on roofs with rafters no more than 32 inches center to center. The panels may also be used on floors over joists spaced a maximum of 24 inches o.c. The two major grades of plywood are Structural I and Structural II.

Glue used between the plies is either interior or exterior grade. If you use interior type panels in an exposed situation, the glue may soften and allow the plies to peel apart. In a 3- or 4-ply panel, if one of the outer plies comes loose, it would destroy the strength of the plywood.

Moisture can enter plywood through a defective roof, leaking pipes, or when you install it outdoors. You should use Exterior Type Plywood if it will be exposed to high-moisture conditions. This type has a better resistance to moisture than the interior type.

Diaphragms

Structural plywood is often used to create a diaphragm that strengthens framing members. Plywood sheathing nailed firmly to a wall adds rigidity that helps resist wind and earthquake loads. One type of diaphragm is horizontal, as in a roof sheathing or subfloor. Another type is a vertical diaphragm, as in shear walls. Figure 4-36 shows horizontal and vertical building diaphragms.

If you use plywood flooring as building diaphragms, you'll need to nail all edges of each plywood panel to framing. This means you'll have to add blocking between joists to support panel ends. Space the nails, depending on the location of the panel, in relation to the building plan. Nail the panel's intermediate supports 12 inches o.c.,

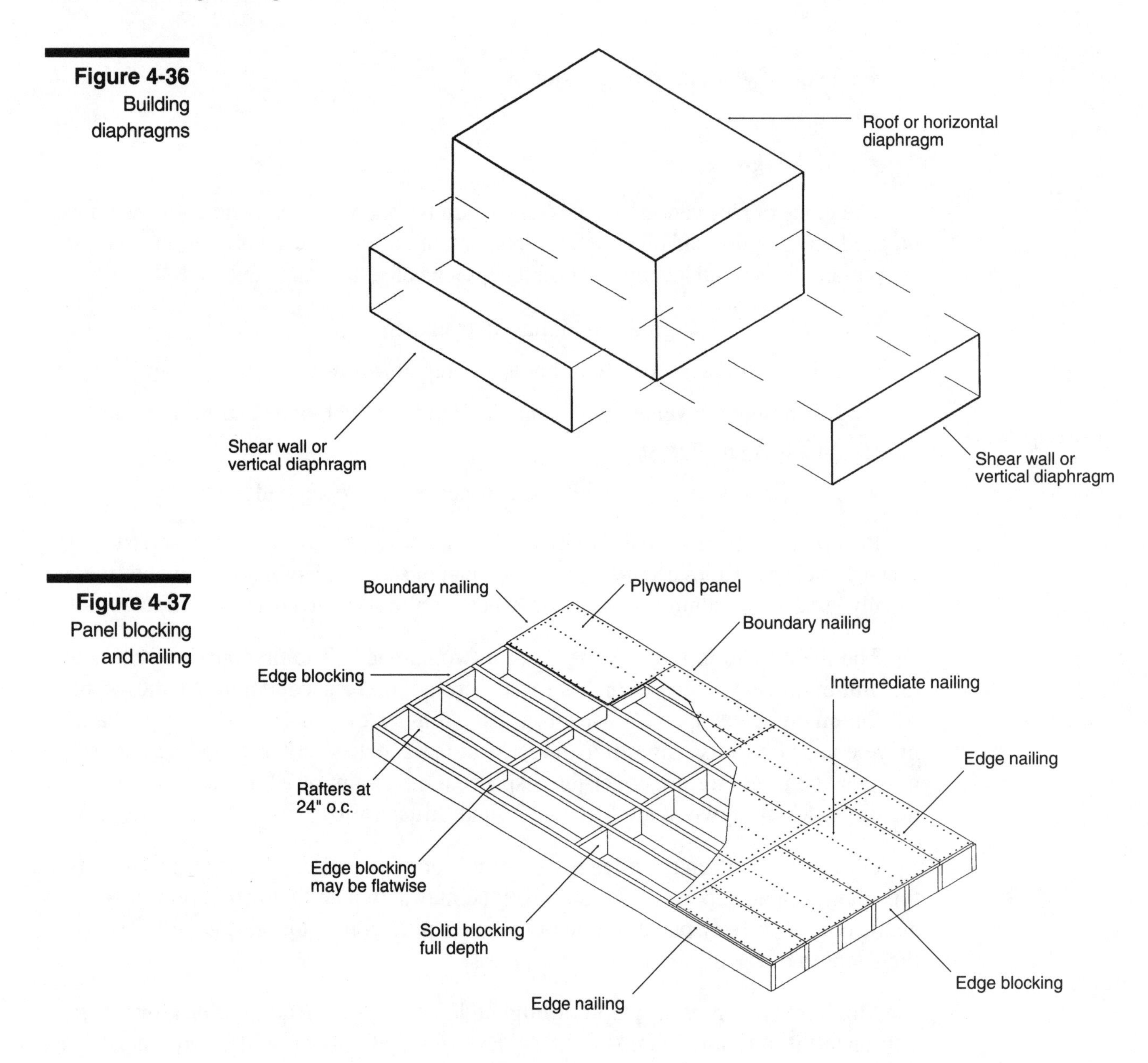

Figure 4-36 Building diaphragms

Figure 4-37 Panel blocking and nailing

as shown in Figure 4-37. Space the nails closer together at the boundaries of the diaphragm than you do near the center of the diaphragm. Spacing may vary from 3 o.c. to 12 inches o.c.

Don't butt plywood panels together tightly. They expand when they get wet. Leave an ⅛-inch gap between panels. If the panels are butted together, they may buckle when they expand.

You can use 1-inch sheathing laid diagonally instead of plywood subflooring. This type of sheathing is often used on joists that are spaced as much as 4 feet apart. Figure 4-38 shows thickness and allowable spans of plywood roof decking. The plywood panel index is shown in Figure 4-39.

Figure 4-38
Allowable span for plywood roof & floor sheathing

Thickness (inches)	Index	Roof blocked (inches)	Roof unblocked (inches)	Floor blocked (inches)
3/8	24/0	24	16	—
1/2	32/16	32	28	16
5/8	42/20	42	32	20
3/4	48/24	48	36	24
1 1/8	GR. 1 or 2	72	48	48

Figure 4-39
Plywood panel index

Panel index	Thickness (inches)	Span (inches)	Span with unsupported edges (inches)
12/0	5/16	12	12
16/0	5/16 – 3/8	16	16
20/0	5/16 – 3/8	20	20
24/0	3/8 – 1/2	24	20 - 24
32/16	1/2 – 5/8	32	28
42/20	5/8 – 3/4	42	32
48/24	3/4 – 7/8	48	36

Shear walls

A shear wall is a vertical diaphragm of a building. It's cantilevered from the foundation and loaded at the top by a horizontal diaphragm. The most common shear wall is braced with let-in 1 × 4 braces or metal straps. Unless the wall is sheathed with plywood, you should brace each end at approximately 45 degrees. Figure 4-33 shows a typical let-in wall brace in a stud wall.

Composite Members

Glue-laminated beams

Glue-laminated beams are made from selected clear lumber that is glued together and cured under pressure. The grades of the lumber making up a finished beam may be a combination of Dense Select Structural, Select Structural, Dense Construction, and Construction and Standard Grade Douglas fir.

Figure 4-40 Glued laminated beam widths

Nominal width (inches)	Net finish width (inches)
3	2¼
4	3¼
5	4¼
6	5
8	7
10	9
12	11
14	12½
16	14½

Large structural members are often made from glue-laminated wood. Glue-laminated beams have many advantages:

- good appearance
- fewer wood defects
- more uniform seasoning
- shape based on stress
- higher-quality wood used for higher stresses
- more economical and more plentiful than solid beams

Structural glue-laminated timber should be made by an approved Type I manufacturer in accordance with Product Standard PS-56. The manufacturer should furnish a signed certificate to the building department for every glue-laminated member. The certificate should have a statement saying that the member conforms to product Standard PS-56. It should also have the following:

- name and address of the approved manufacturer
- address of installation
- specie of lumber
- type of glue
- combination symbol, number of laminations, and AITC specification
- every member has to bear the manufacturer's certification

The individual laminations are from nominal 1- and 2-inch thick lumber. Typical sizes are 1 × 6, 1 × 8, 2 × 4, and 2 × 6 inches. The number of laminations varies with the size of the finished girder. Figure 4-40 shows the net width of glue-laminated members. Grades for glue-laminated lumber are: Industrial Appearance, Architectural Appearance, and Premium Appearance. The grading is based on the number and types of voids, inserts, open and loose knots, and surface condition. Use tapered and double tapered glue-laminated girders for roof spans of 25 to 100 feet. Use curved glue-laminated girders for spans of 25 to 100 feet.

Based on the *National Construction Estimator*, the material cost of glued laminated girders is between $2,100 and $2,400 per MBF (thousand board feet), and 6 × 12 solid sawn beams cost about half that, or $1230 per MBF. So use ordinary timber if you can, because glued laminated isn't economical. For spans over 25 feet in length, it may be difficult to get heavy timbers so you'll have to consider using glued laminated girders instead.

Glued laminated girders do have many advantages over solid sawn timbers. They resist shrinkage and warpage, and have fewer cracks, knots, and other defects. Also the allowable bending stress in extreme fibers is higher than in cut timbers because only the outer laminations need be of high grade lumber. Glued laminated girders also come in much greater lengths than cut timbers. Glued laminated girders are smaller in section than an equivalent solid sawn section, as shown below:

Solid Sawn Section, inches	Equivalent Glued Laminated Section, inches
3 x 8	3⅛ x 6
3 x 10	3⅛ x 7½
3 x 12	3⅛ x 9
3 x 14	3⅛ x 10½
3 x 16	3⅛ x 12
4 x 10	3⅛ x 9
4 x 12	3⅛ x 10½
4 x 14	3⅛ x 12
4 x 16	3⅛ x 15
6 x 10	5⅛ x 9
6 x 12	5⅛ x 10½
6 x 14	5⅛ x 12
6 x 16	5⅛ x 13½

The solid sawn sizes are based on the use of No. 1 or select structural grade of Douglas fir-larch. The glued laminated girders have the corresponding grade in other laminations.

Typical examples of glued laminated girders supporting a 10 psf dead load and 20 psf live load are:

Span, feet	Spacing, feet	Girder Size, inches
10	8	3⅛ x 6
20	8	3⅛ x 12
30	8	3⅛ x 18
40	12	5⅛ x 22½
50	12	5⅛ x 28½
60	12	6¾ x 30
70	12	6¾ x 36
80	12	6¾ x 40
90	12	8¾ x 42
100	12	8¾ x 46½

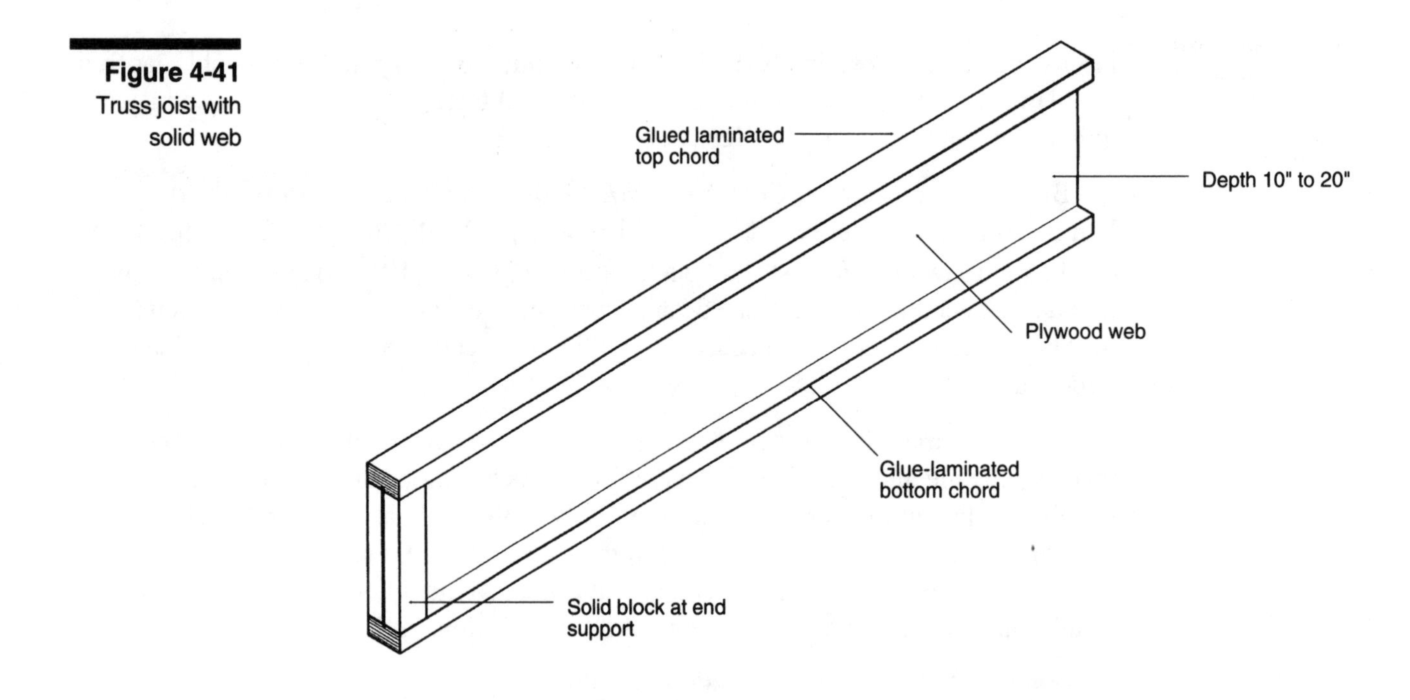

Figure 4-41
Truss joist with solid web

This table is based on straight, simply-supported laminated timber beams. The roof should have a minimum slope of ¼ inch per foot to eliminate water ponding. The total load includes the beam weight.

Truss joists

Two types of truss joists are used for floor and roof framing. One type is shaped like an I-beam, and it's called TJI. It has a laminated wood top and bottom chord and a solid plywood web all glued together. See Figure 4-41. The other type is made of a Warren truss shape with wood top and bottom chords and steel tube web members. See Figure 4-42. The following table shows the approximate size of TJI Series truss joists used as roof rafters. The members are spaced 2 feet apart. The dead load is 10 psf and the live load is 20 psf.

Span, feet	Size, Truss Joist
12	10" TJI
14	10" TJI
16	10" TJI
18	10" TJI
20	10" TJ
22	12" TJI
24	12" TJ
26	14" TJI
28	16" TJ
30	16" TJI
32	18" TJI
34	20" TJI

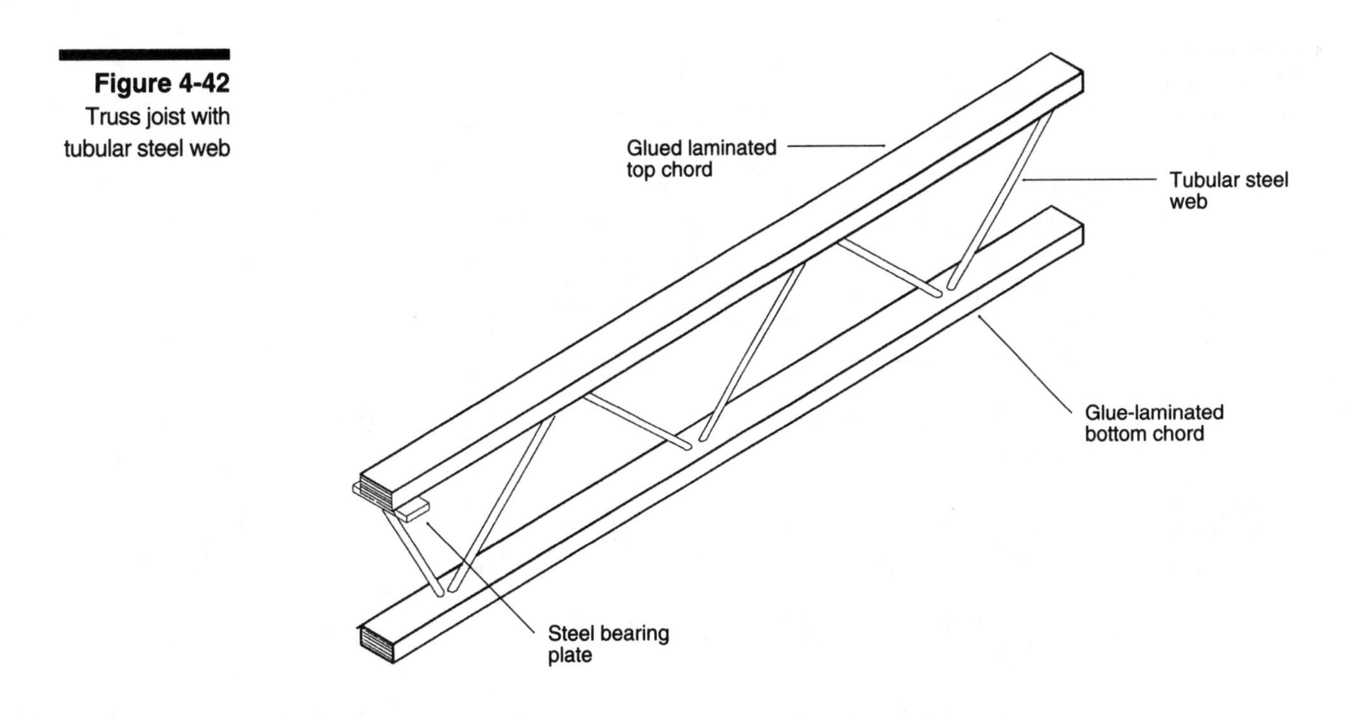

Figure 4-42
Truss joist with tubular steel web

The following table provides approximate size of TJI Series truss joists used as floor joists. The members are spaced 2 feet apart. The dead load is 10 psf and the live load is 40 psf.

Span, feet	Size, Truss Joist
12	10" TJI
14	10" TJI
16	10" TJI
18	10" TJI
20	10" TJI
22	12" TJI
24	12" TJI
26	14" TJI
28	16" TJI
30	18" TJI
32	20" TJI

Truss rafters or gang-nailed trusses

Prefabricated trussed rafters, or gang-nailed trusses, are commonly used for rafter and ceiling assemblies. Figure 4-43 shows the main parts of this type of truss. Slopes of the top chord are made in 2½:12, 3:12, 4:12, and 5:12 pitches. The truss span may be from 16 feet to 46 feet. The top chord may overhang, forming roof eaves. You'll have to use a crane to install these trusses.

Truss rafters are pre-engineered by a fabricator. Figure 4-44 shows some common truss rafters. These trusses don't require engineering design by a job engineer or a contractor. The fabricator submits his bid based on the shape and span of the roof framing shown on the architectural plans. Most prefabricated truss rafters are made up of 2 × 4 or 2 × 6 lumber connected by gang nail plates. A gang nail plate is made from sheet steel

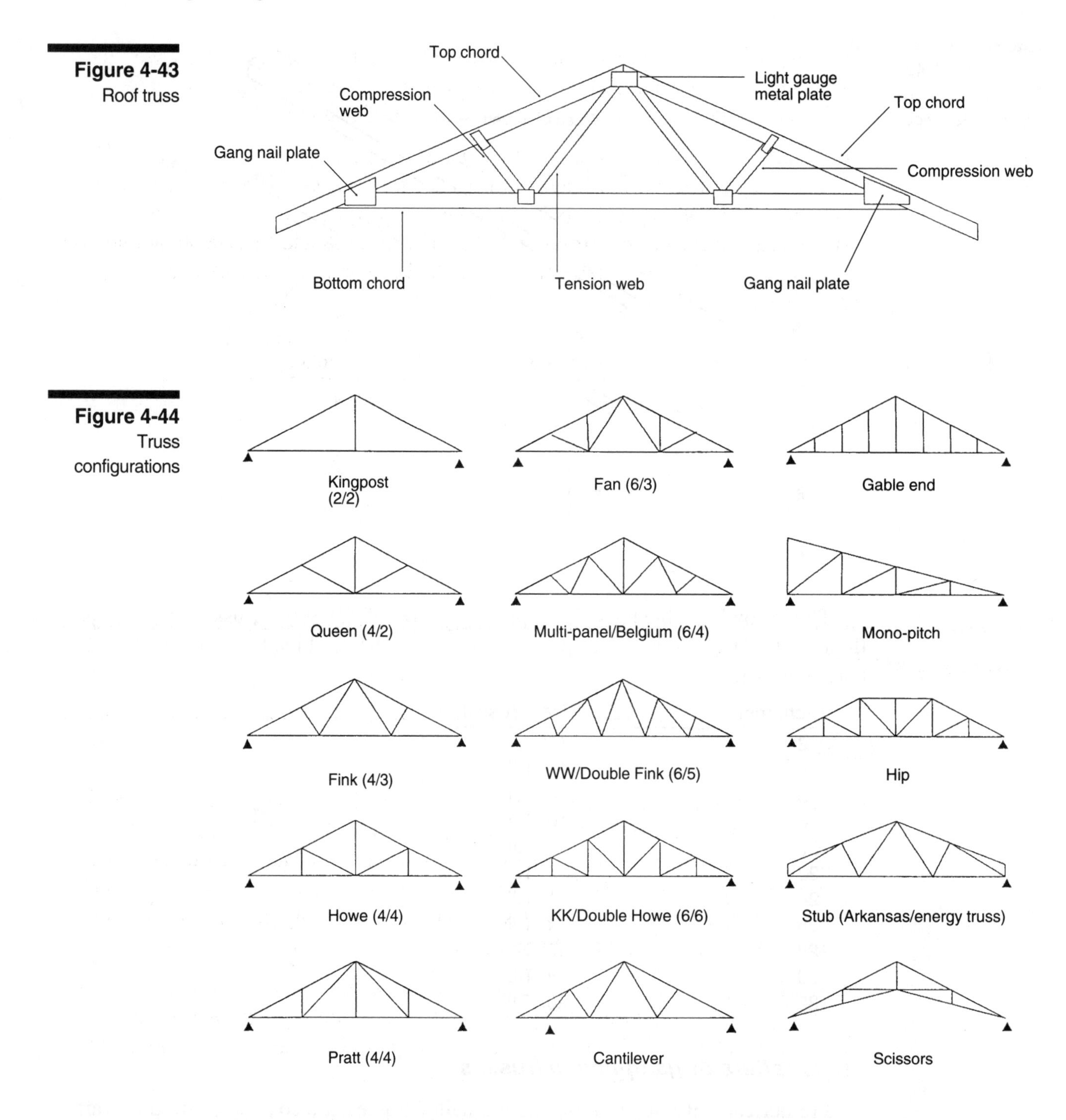

Figure 4-43 Roof truss

Figure 4-44 Truss configurations

which has been punched to form many sharp points on both faces of the plate. The gang nail is driven across the joint between wood members, thereby, securely connecting the members together. The trusses are spaced 16 or 24 inches o.c.

If you can't inspect all the parts of an assembled prefabricated wood truss carefully, make sure it's made by an approved manufacturer. A building inspector must be able to see all parts of the roof framing system before it's covered. Since the design and details of a truss rafter don't appear on architectural or structural drawings, an inspector has no way of knowing whether the trusses comply with the code or not. So, the manufacturer must verify that the truss was approved by a qualified testing agency and that it meets code requirements.

Prefabricated roof trusses are made with flat or parallel chords in lengths of 50 to 150 feet. They also come in triangular or pitched shapes, 50 to 90 feet in length, as shown in Figure 4-44. Here's a table showing the approximate size of L Series truss joists that can be used as roof trusses. The heel of the truss and truss depth at midspan are as shown on table. The members are spaced 2 feet apart. The dead load is 10 psf and the live load is 20 psf.

Span (feet)	Heel (inches)	Truss depth (inches)
20	18	22
25	18	22
30	18	22
35	18	30
40	18	46
45	18	50
50	26	50

Connections

Nails

Nails are classified according to use and form. Nails are designated by the term "penny" which is abbreviated "d." The term penny came from the price of 100 of a particular size nail in the 1400s. It applies to the length of the nail regardless of the wire gauge. Nails come in sizes 2d to 60d, or from 1- to 6-inch lengths. Figure 4-45 shows the length, diameter, and steel wire gauge of commonly-used nails. The most common nails you use are made from steel wire.

A nail should be at least three times as long as the thickness of the wood it's supposed to hold. Two-thirds of the length of the nail should go into the second piece. Drive nails at a slight angle toward each other to keep them from pulling out. Avoid driving nails parallel to the grain as they don't hold well.

The normal nail to use for house construction is the common wire nail. Box nails are similar to common nails but the wire sizes are one or two numbers smaller for a given length of box nail than for a common nail. Scaffold or form nails are temporary nails which have two heads (one above the other). Drive these nails to the first head to get the proper depth. Then use the second head for pulling the nail out when you're done. The upper head projects above the surface of the wood to make it easy to pull out.

Nails are manufactured with different finishes, including:

- bright smooth wire nails, made from 6d to 60d sizes
- cement-coated nails, which are surfaced with a resin to increase friction

Figure 4-45
Nail sizes

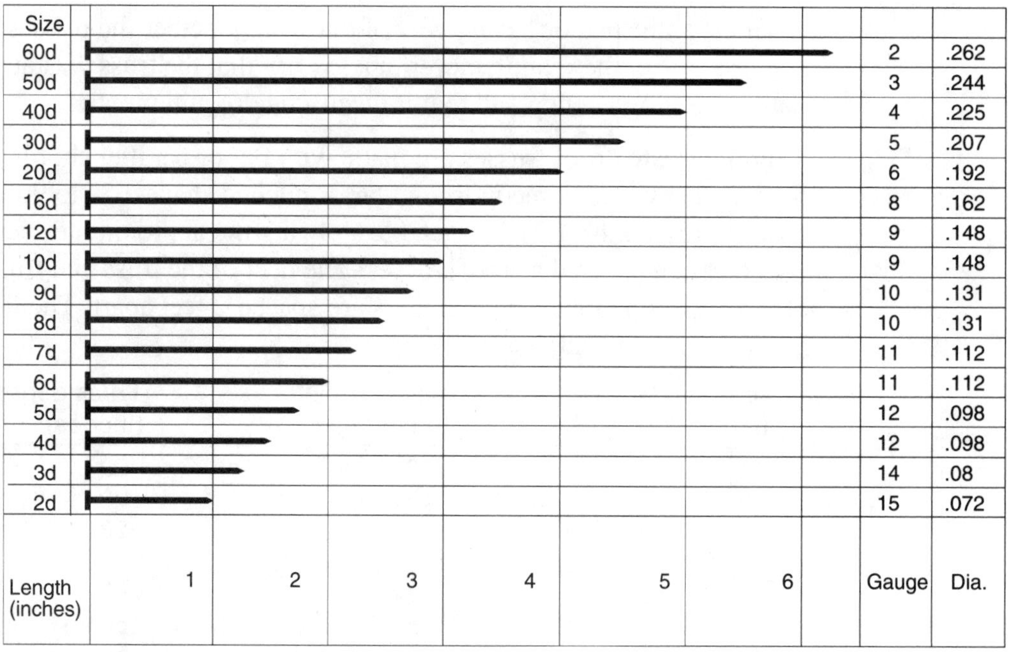

Size	Gauge	Dia.
60d	2	.262
50d	3	.244
40d	4	.225
30d	5	.207
20d	6	.192
16d	8	.162
12d	9	.148
10d	9	.148
9d	10	.131
8d	10	.131
7d	11	.112
6d	11	.112
5d	12	.098
4d	12	.098
3d	14	.08
2d	15	.072

Length (inches): 1 2 3 4 5 6

- zinc-coated (galvanized) or plastic-coated nails, which resist corrosion
- chemically-etched nails, to keep them from corroding. The surface of iron nails are etched so the zinc used in galvanizing will stick better to the metal.
- annularly and helically threaded nails, to keep them from pulling out

Most (but not all) nails are made from steel. Use nails made from copper and aluminum alloys, or stainless steel nails where you may have a problem with corrosion. The minimum number and size of nails you use for connecting wood framing members are:

Framing member	Number and size of nails
joist to sill or girder, toenail	three 8d
bridging to joist, toenail	two 8d
sole plate to joist	16d @ 16 inches o.c.
top plate to stud	two 16d
double stud, face nail	16d @ 24 inches o.c.
double top plate, face nail	16d @ 24 inches o.c.
top plate laps and intersections	two 16d
continuous header, two pieces	16d @ 16 inches o.c.
ceiling joist to plate, toenail	three 8d
ceiling joist parallel to rafter	three 16d
rafter to plate, toenail	three 8d
1 inch brace to each stud and plate	two 8d
built-up corner studs	16d @ 24 inches o.c.

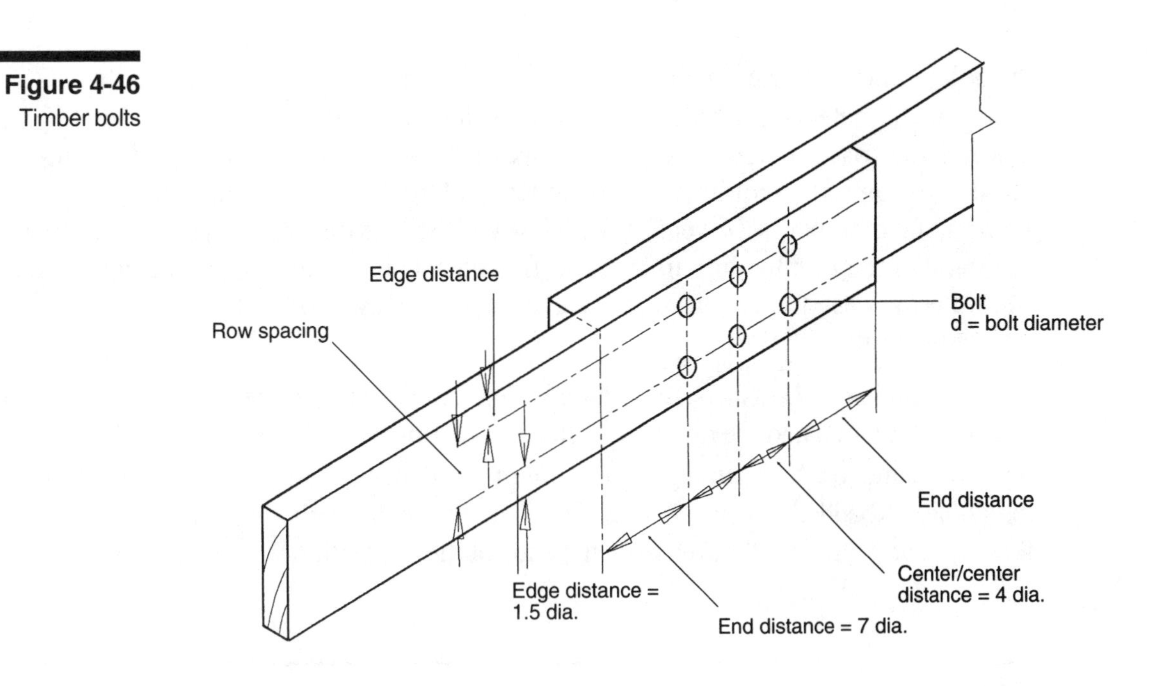

Figure 4-46
Timber bolts

To nail plywood to supports use:

- 6d nails for 5⁄16-inch-thick plywood
- 8d nails for 3⁄8-inch-thick plywood
- 10d nails for 1⁄2-inch-thick plywood

Bolts

Use bolts when you need extra strength or when your work must be disassembled. Usually you use washers and nuts with bolts. Use bolts to fasten wood members together, side grain to side grain, as shown in Figure 4-46. The most common sizes to use are 1⁄2- and 3⁄4-inch diameter bolts. Use bolts to fasten metal to wood side grain. It's a good idea to make prebored holes that are 1⁄16 inch smaller in diameter than the bolt.

Be careful that you don't split the wood when you put in any bolts. When the stress from the bolt is perpendicular, rather than parallel to the grain of the wood, you can put the bolt closer to the edge. See Figure 4-46 for bolt spacing when load is parallel to grain.

Usually you space bolts four times the bolt diameter. The minimum spacing is usually two inches. The end distance space is seven times the bolt diameter for members in compression, and four times the bolt diameter for members in tension. Edge distance is usually four times the bolt diameter at the edge the load acts on. Edge distance should be 1½ times the bolt diameter on the opposite edge. Row spacing is generally 1½ times the bolt diameter. To obtain full strength of a bolted connection, the spacing of bolts along the grain should not be less than four times the bolt diameter. For perpendicular to grain loading, bolt spacing need only be 2½ times the diameter of the bolt provided steel side plates are used. For parallel to grain loading, the spacing across the grain between rows of bolts is determined by the number of rows, the required edge distance, and the net section requirement. The net section of a member is the section which gives the maximum stress in the member. For parallel to grain loading in Douglas fir, the net area remaining at the net section should be at least 80 percent of the total area in bearing under all bolts in the particularjoint.

Use standard cut washers, or metal plates or straps, between the wood and bolt head and between wood and the nut to distribute the bearing stress. Use anchor bolts to attach a building frame to a foundation. The minimum anchorage is ½-inch diameter by 10-inch long bolts embedded 7 inches into the concrete. Space the anchor bolts a maximum of 6 feet apart, and within 12 inches from the ends of the mud sill.

Screws

Wood screws have more holding power than nails. Use them to fasten pieces, 1 inch or less, to heavier members, side grain to side grain, or side grain to end grain. A screw should go in two-thirds of its total length, with one-third of the length in the first member.

Screws are usually made of unhardened steel, stainless steel, aluminum, or brass. The most common are steel, bright-finished or blued, zinc, cadmium, or chrome plated. Wood screws are threaded from the point to about two-thirds the length of the screw. The head of a screw may be flat, round, oval, slotted, or Phillips.

The size of a wood screw is given by its length and gauge number. Each gauge number determines the diameter of the shank. A gauge may come in several lengths. Here are some screw numbers, or gauges, and their related shank diameters in decimals of an inch:

Gauge	Shank Diameter (inch)
4	.112
5	.125
6	.138
7	.151
8	.164
9	.177
10	.190
11	.203
12	.216
14	.242
16	.268
18	.294
20	.320
24	.372

Lag Screws

Lag screws, or lag bolts, are longer and larger in diameter than wood screws. Use them to fasten 1- and 2-inch pieces to heavier members and to attach metal shapes, such as angles, channels, and straps, to heavy wood members.

A lag screw should go into wood at least seven times its shank diameter. You should prebore holes $\frac{7}{8}$ of the diameter of the screw at the root of the threads. Usually you'll use a wrench to tighten lag screws.

Shear Plates

Over 60 types of timber connectors are used to strengthen the joint where wood meets wood or other structural material. These come in sizes from 2 to 8 inches and can be fastened with either $\frac{1}{2}$-inch or $\frac{3}{4}$-inch bolts.

Split rings are made of low-carbon steel in $2\frac{1}{2}$-inch and 4-inch diameters. Use them between two timber faces in cut grooves. The depth of the grooves should be half the depth of the ring.

Tooth rings are corrugated and toothed from 16 gauge low carbon steel. Use them between timbers in relatively light construction. Drive the teeth of the ring into the wood by tightening the bolts.

Use clamping plates, or gang nail plates, to connect wood chords to the web members on prefabricated wood trusses.

Post Anchors and Hangers

Figure 4-47 shows a U-strap post anchor used to support and anchor wood posts to a foundation. This is also shown on Figure 4-21 where the U-strap holds the post to the foundation. Figure 4-48 shows an angle and bearing plate post anchor that you can use to support and anchor wood posts to concrete foundations. Figure 4-49 shows two common types of joist or beam hangers.

Workmanship

Rough Framing

After the rough framing work is done, you should check the quality of the finished work. At the end of this chapter is a checklist of some of the more important items. Make as many copies as you need to use on your jobs.

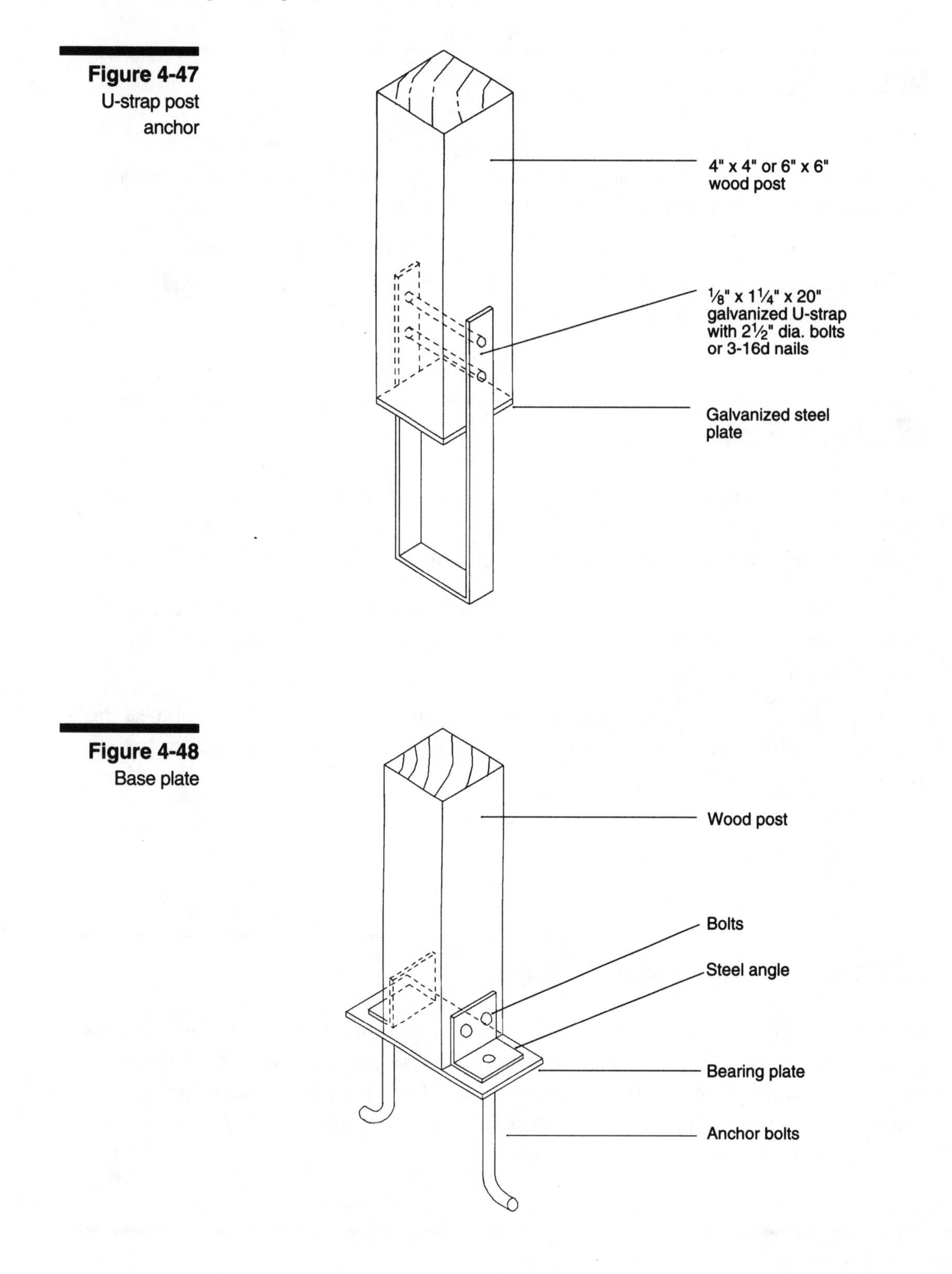

Figure 4-47
U-strap post anchor

Figure 4-48
Base plate

Problems Due to Poor Workmanship

A common problem in multistory wood-frame buildings is uneven floors. This condition may be due to careless measurement of vertical members, shrinkage of green horizontal members, or construction over an uneven concrete slab.

Figure 4-49
Hangers

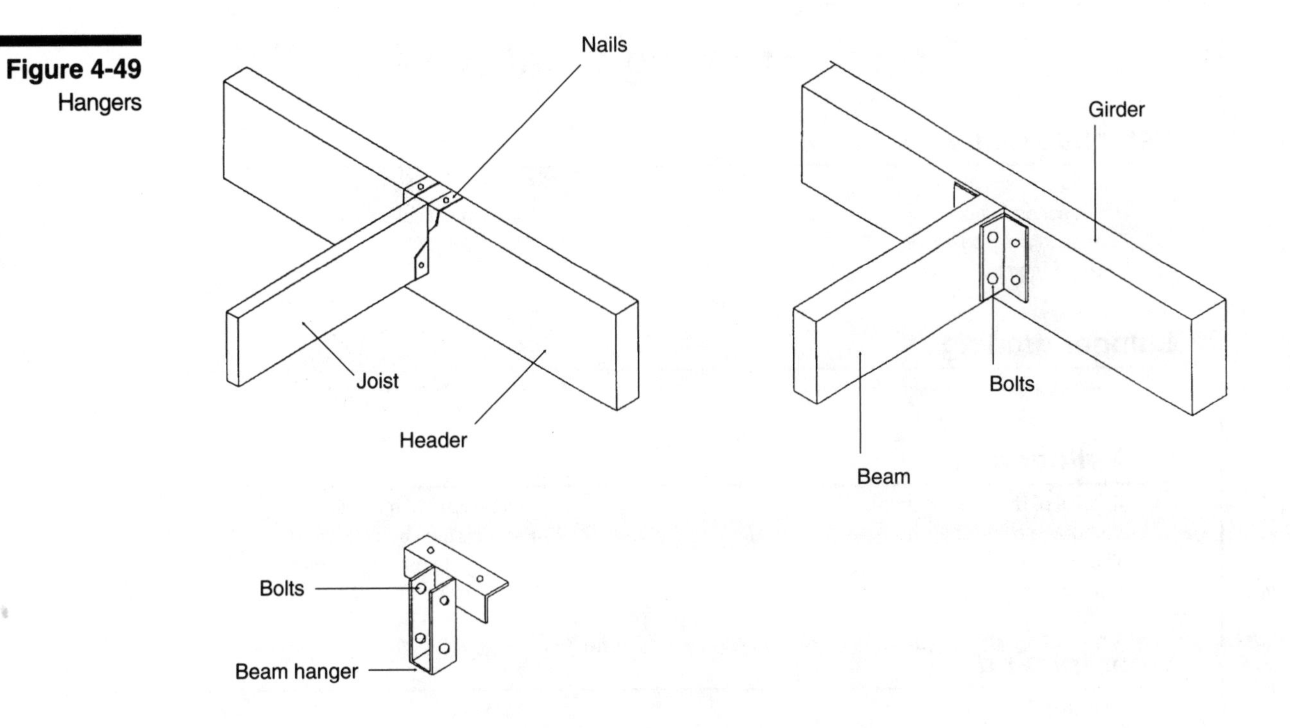

Most structural concrete slabs of underground garages are constructed with an upward camber between supports. That creates a curved floor slab for first floor framing. Studs for that floor should be cut to adjust for the curve in the floor slab. Use survey equipment to find the difference in floor elevation and then cut studs so the second floor is joists are horizontal.

Rough Framing Quality Checklist

Defective lumber

- ☐ Checks and splits
- ☐ Unseasoned lumber
- ☐ Termite infestation
- ☐ Improper notching or cutting
- ☐ Fungus and dry rot
- ☐ Knots in critical locations
- ☐ Warped members

Lumber grading

- ☐ Grade stamps per plans and codes
- ☐ Grade stamps not concealed

Underfloor framing

- ☐ Mudsills of redwood or cedar
- ☐ Anchorage of mudsills
- ☐ Blocking at joist ends
- ☐ Bracing of cripples
- ☐ Full bearing of mudsills
- ☐ Blocking between floor joists
- ☐ Bracing of posts under girders
- ☐ Preservative treatment of wood in contact with concrete

Floor framing

- ☐ Joist size
- ☐ Double joists under bearing walls and partitions
- ☐ Moisture barrier of embedded beams
- ☐ Joist span
- ☐ Nailing to top plates

Wall framing

- ☐ Bottom plate full bearing
- ☐ Lateral bracing, by let-in, cut-in or metal straps
- ☐ Trimmers, size
- ☐ Post size, and end connections
- ☐ Top plate lapped and nailed
- ☐ Header size, span, and end bearing
- ☐ Firestopping between studs
- ☐ Fireproofing treatment

Ceiling framing

- ☐ Joist size
- ☐ Joist spacing
- ☐ Joist span
- ☐ Nailing to top plates

Roof framing

- ☐ Bracing size, spacing, and angle
- ☐ Rafters size, spacing, span, framed opposite each other, butted to ridge board
- ☐ Ridge board size
- ☐ Purlin size and spacing

Sheathing

- ☐ Roof sheathing bearing, spacing, nailing
- ☐ Flooring bearing spacing, nailing

Composite beams and girders

- ☐ Glue-laminated beams by approved fabricator
- ☐ Lumber grade, moisture content, surfaces, adhesive
- ☐ Identification mark, size, span, laminations
- ☐ Truss joists by approved fabricator, size, span, lumber grade, moisture content, and adhesive

Connections

- ☐ Nails: number, size at connections, edge distance
- ☐ Shear plates: size, number, approval
- ☐ Hangers: clips, straps, approval
- ☐ Bolts: sizes, washers and nuts, number, spacing, edge distance, angle to grain, tightness

CHAPTER 5

Steel

Steel is the most versatile material used in engineered buildings. It's made in uniform quality and dimension and its properties are well understood. Freezing and thawing don't affect steel. And it can be treated so it won't rust easily. Steel's major limitation is that it loses strength at high temperatures.

Every contractor who handles commercial and industrial construction should be familiar with the basics of steel framing. This includes the design, fabrication, and erection procedures of structural steel. This chapter will help you understand the important principles of steel construction.

Steel Construction

One- to three-story buildings often have some structural steel components. Steel beams and columns are commonly used in two-story residences. If bedrooms are built over a two-car garage, the garage probably includes a steel beam and a pair of columns. Many multiple-family residential complexes, such as town houses and apartments, have steel-framed parking structures.

Two categories of steel are used in buildings. The first is structural steel for the building frame. The other, nonstructural steel, is classified as miscellaneous metal, or miscellaneous iron. Nonstructural steel includes items such as metal railings, corner guards, and sight screens. Structural steel parts are listed in the *Code of Standard Practice*, which is published by the American Institute of Steel Construction (AISC). Usually, anything in the *Code of Standard Practice* will be sold by steel fabricators. According to this *Code,* the term "Structural Steel," when used in construction documents, consists of only the items on the list on the next page.

- Anchors for structural steel
- Bases of structural steel
- Beams
- Bearing plates for structural steel
- Columns
- Crane rails, stops, splices, bolts, and clamps
- Door frames constituting part of a structural frame
- Expansion joints connected to a steel frame
- Fasteners for connecting structural steel items
- Permanent shop bolts
- Shop bolts for shipment
- Field bolts for permanent connections
- Floor plates (checkered or smooth) connected to structural steel frame
- Girders of structural steel
- Girts
- Grillage beams and girders of structural steel
- Hangers of structural steel, if attached to structural steel framing
- Leveling plates
- Lintels shown on the framing plans, or otherwise enumerated or scheduled
- Machinery foundations of rolled steel section and/or plate attached to structural steel framing and shown on structural steel framing plans
- Marquee framing of structural steel
- Monorail beams of structural shapes when attached to a structural frame
- Permanent pins
- Purlins
- Separators, angles, tees, clips, and other detail fittings essential to a structural steel frame
- Shear connectors
- Steel cables forming a permanent part of a structural frame
- Struts
- Supports of steel structural shapes for piping, conveyors, and similar structures
- Suspended ceiling supports of steel structural shapes 3 inches or greater in depth
- Tie, hanger, and sag rods forming part of a structural steel frame
- Trusses

All steel, iron, or other metal items not listed above are not considered to be structural steel. Both structural and nonstructural steel are installed by steel erectors.

Code Requirements

Certain types of buildings must use steel framing or other noncombustible materials. For example, the *Uniform Building Code* (UBC) says that residential buildings over three stories must be built of noncombustible materials. According to the UBC, noncombustible materials include:

- Materials which will not ignite and burn when subjected to fire, such as concrete, masonry, or steel.
- Materials having a structural base of noncombustible material with a surfacing material under ⅛-inch thick having a flame-spread of 50 or less.

The flame-spread rating of a building material is based on a comparison between the flame propagation of the material and a standard sample of wood as tested under laboratory conditions (1991 UBC Standard 17.6).

The building code classifies buildings according to their fire-resistive (F.R.) rating. These classifications are Type I-F.R., Type II-F.R., Type II-One-Hour, Type II-N (not fire-resistive), Type III-One-Hour, Type III-N, Type IV-H.T. (Heavy Timber), Type V-One-Hour, and Type V-N.

Type I-F.R. is the most fire-resistive and Type V-N is the least fire-resistive. Fire resistance is measured in hours. Any height building can qualify as Type I-F.R. if it has a three-hour fire-resistive structural frame. A Type II-F.R. building must have a two-hour fire-resistive structural frame. It may be up to 12 stories in height. Type III-N and Type IV-N residential buildings can't be more than three stories in height.

Steel loses strength when it's heated. To protect it against fire, layers of plaster, concrete, or gypsum board are installed over it. The thickness of the cover depends on how much fire-resistance you need.

Many commercial buildings, factories, and warehouses which have open and clear spaces are built with steel columns, open-web joists, tapered girders, and trusses. Figure 5-1 shows an example using long-span trusses to support the second floor of a two-story warehouse. A rigid frame supports the roof. Figure 5-2 shows the X-bracing system for this building. Figures 5-3, 5-4, and 5-5 show alternative bracing methods for the same warehouse building. Together, these figures show the main structural members of a steel-framed building.

Typical Steel-frame Buildings

Figure 5-6 shows a typical framing plan for a six-story commercial building. Figure 5-7 shows the framing elevation for the same six-story building. This building has steel columns placed in 30 foot by 20 foot bays. It also has 24-inch-deep wide-flange girders, weighing 62 pounds per foot. These girders are spaced 20 feet apart and span 30 feet between columns. The girders support 14-inch-deep wide-flange beams that are spaced 7.5 feet apart and span 20 feet. Roof framing members are shown on the framing plan in parentheses. These consist of W24X44

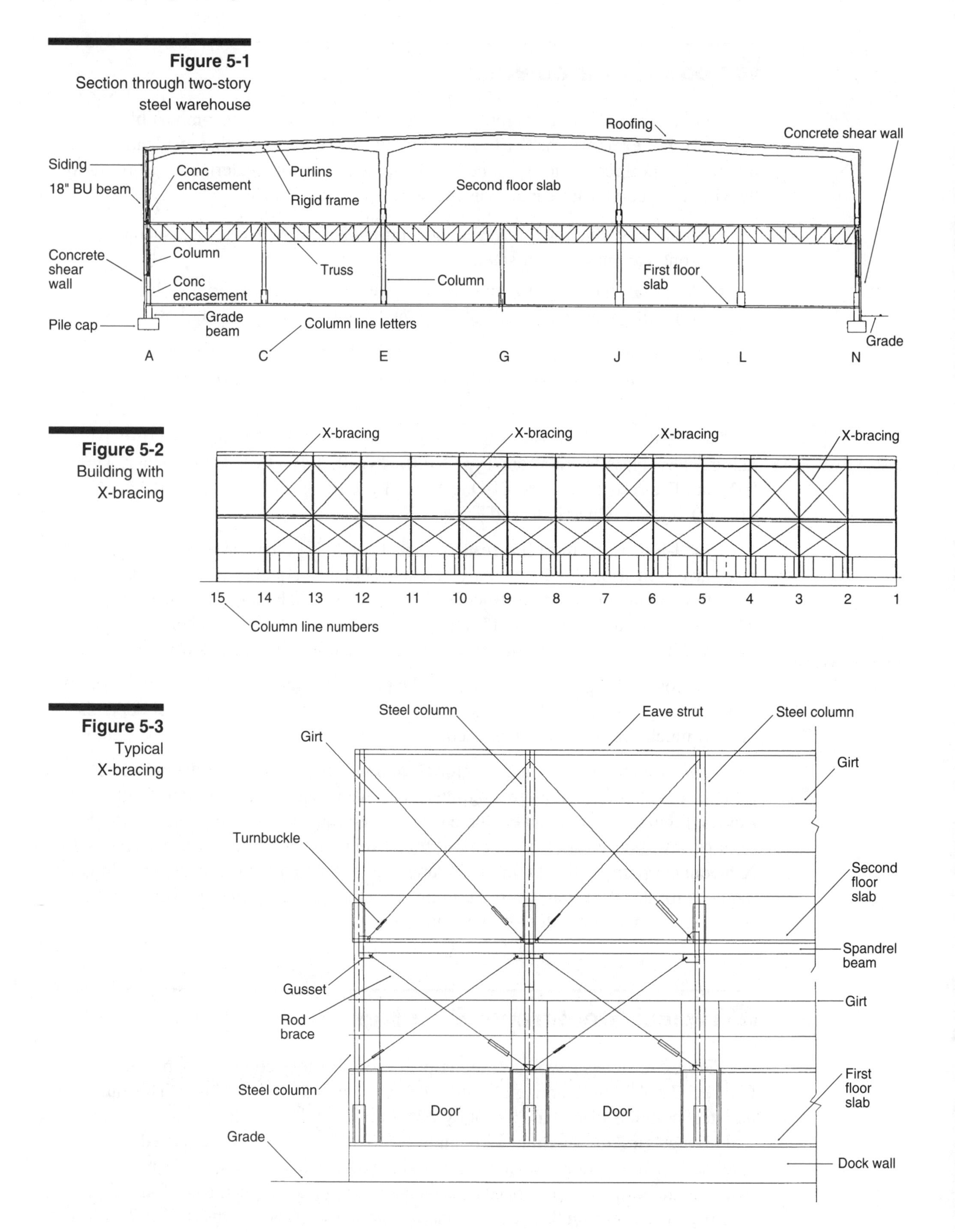

Figure 5-1
Section through two-story steel warehouse

Figure 5-2
Building with X-bracing

Figure 5-3
Typical X-bracing

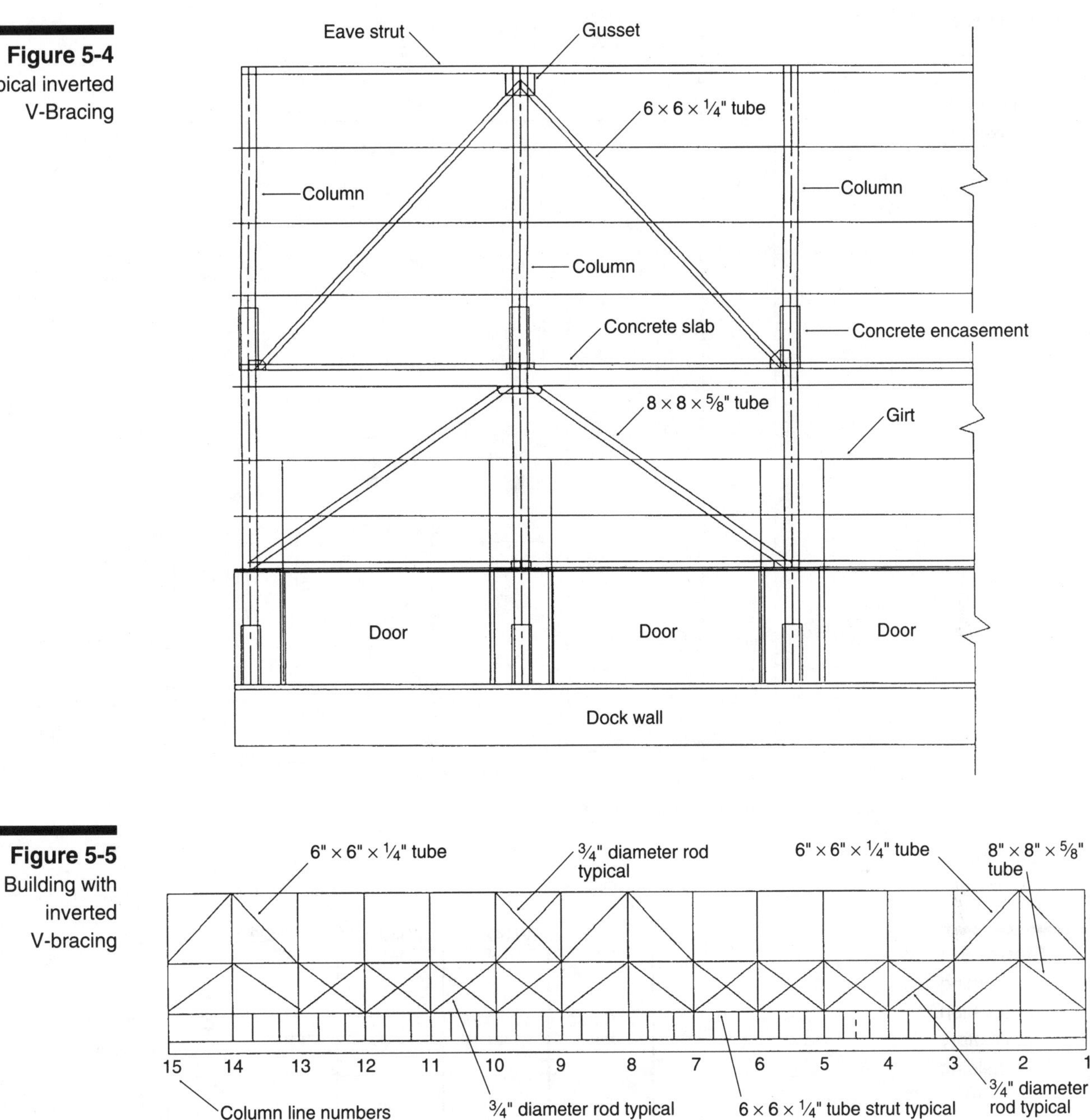

Figure 5-4
Typical inverted V-Bracing

Figure 5-5
Building with inverted V-bracing

girders and W14X22 beams. The term W24X44 describes a particular steel section. The "W" means the section is a wide-flange section. The number "24" stands for the nominal depth of the section in inches. The number "44" says that the section weighs 44 pounds per linear foot.

The abbreviation "Do" in Figure 5-7 stands for "Ditto" which means that the same type, size, and weight section is used on both sides of the members that are marked "Do." For more information on the names of steel sections, see the section on Steel Shapes.

An apartment or condominium building, four stories or more high, will usually be built of steel columns, beams, and steel decking. Decking supports the upper concrete slabs and the suspended ceilings. Exterior walls may be of masonry, precast concrete

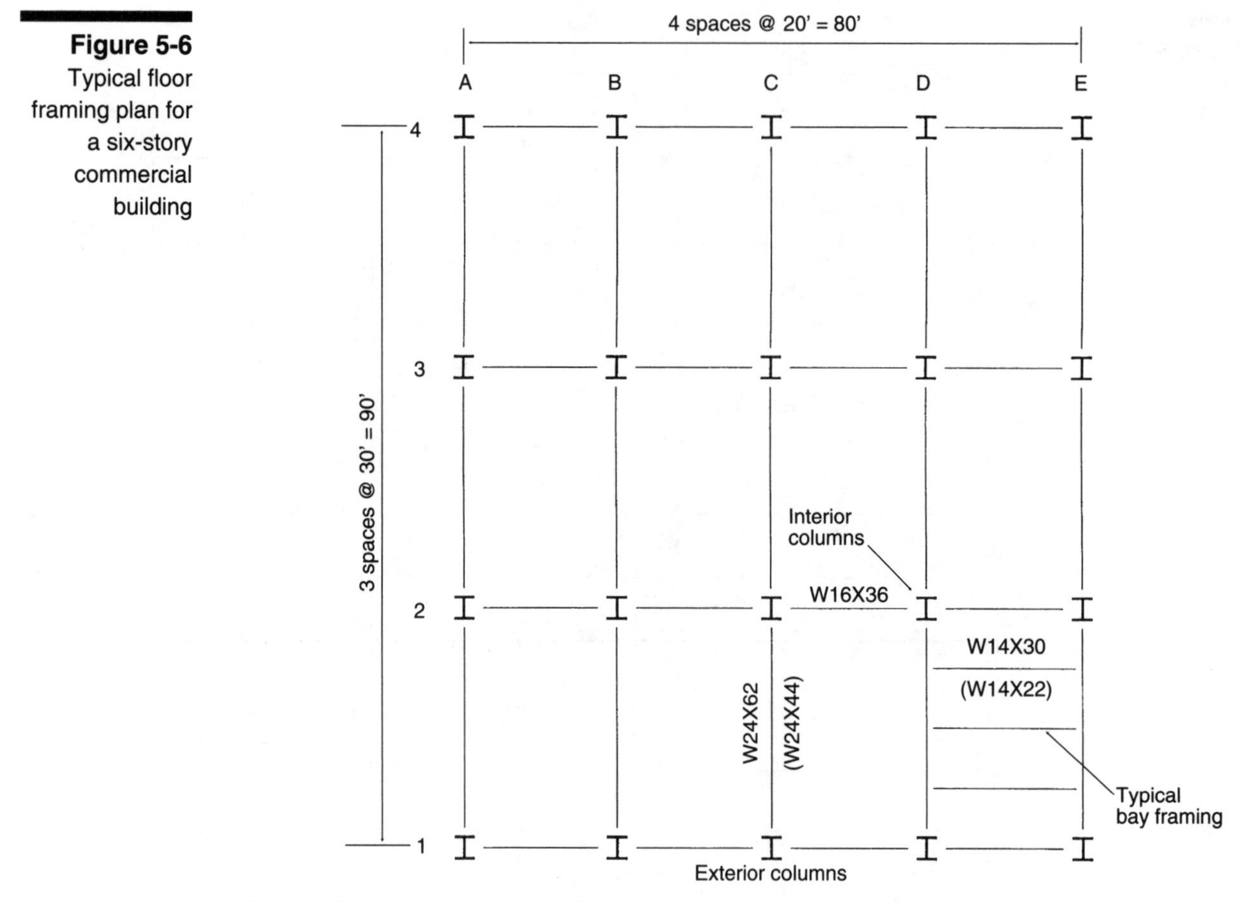

Figure 5-6 Typical floor framing plan for a six-story commercial building

Figure 5-7 Typical framing elevation multi-story building

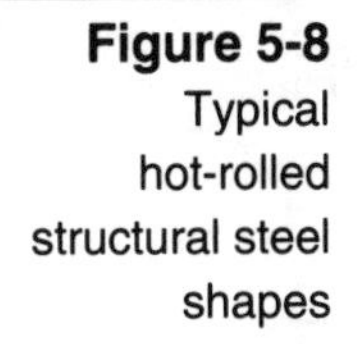

Figure 5-8
Typical hot-rolled structural steel shapes

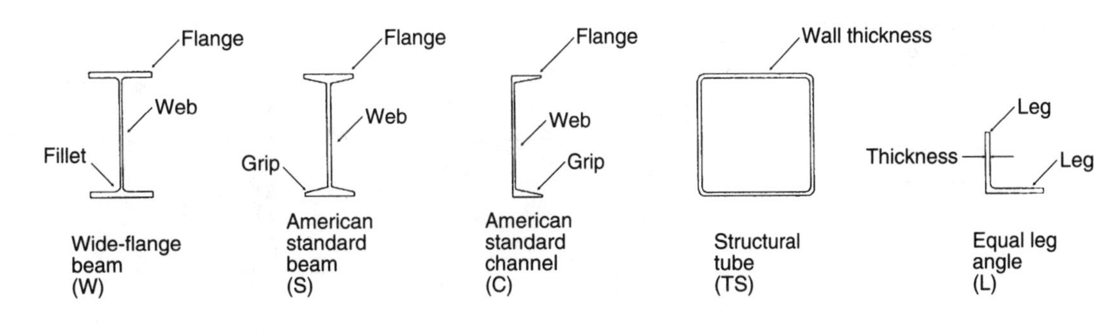

curtain walls, or insulated metal panels. A smaller building such as a two-story home or townhouse with rooms over a two-car garage may require steel only in the columns and beams that support the upper floor.

Pre-engineered steel buildings are popular for industrial applications. More recently, pre-engineered all-metal residential and commercial buildings have become available. No matter what the application, pre-engineered steel buildings use steel for the entire enclosure, not just the framing. Insulated metal siding and roofing are formed and factory-painted in a wide range of patterns and colors. Metal roofing simulates shingles, tiles, and copper mansard sheets. Butler is one manufacturer of pre-engineered steel buildings.

Steel Shapes

Steel products formed into structural members by rolling mills are called hot-rolled shapes, or sections. Figure 5-8 shows the most common hot-rolled shapes. Each shape is designed for a particular use. The steel industry has a specific name and symbol for each shape. The *Manual of Steel Construction*, published by AISC, gives the dimensions and structural properties of each shape.

Every steel shape is listed in the AISC manual two ways: (1) dimensions and properties for designing and (2) approximate dimensions for detailing, as shown on Figure 5-9. Designers use the following dimensions and critical properties of each shape:

- Section Designation, with letter indicating type of shape and nominal depth in inches
- Weight in pounds per foot
- Cross-sectional area (A) in inches^2
- Depth in inches (d)
- Width and thickness of flanges (b_f, t_f) in inches
- Web thickness (t_w) in inches
- Moment of inertia about the x-axis, or horizontal axis (I_x), in inches^4
- Section modulus about the x-axis (S_x) in inches^3
- Radius of gyration about the x-axis (r_x) in inches

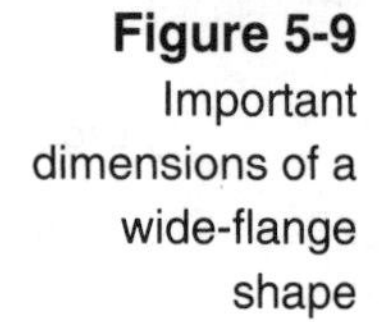

Figure 5-9 Important dimensions of a wide-flange shape

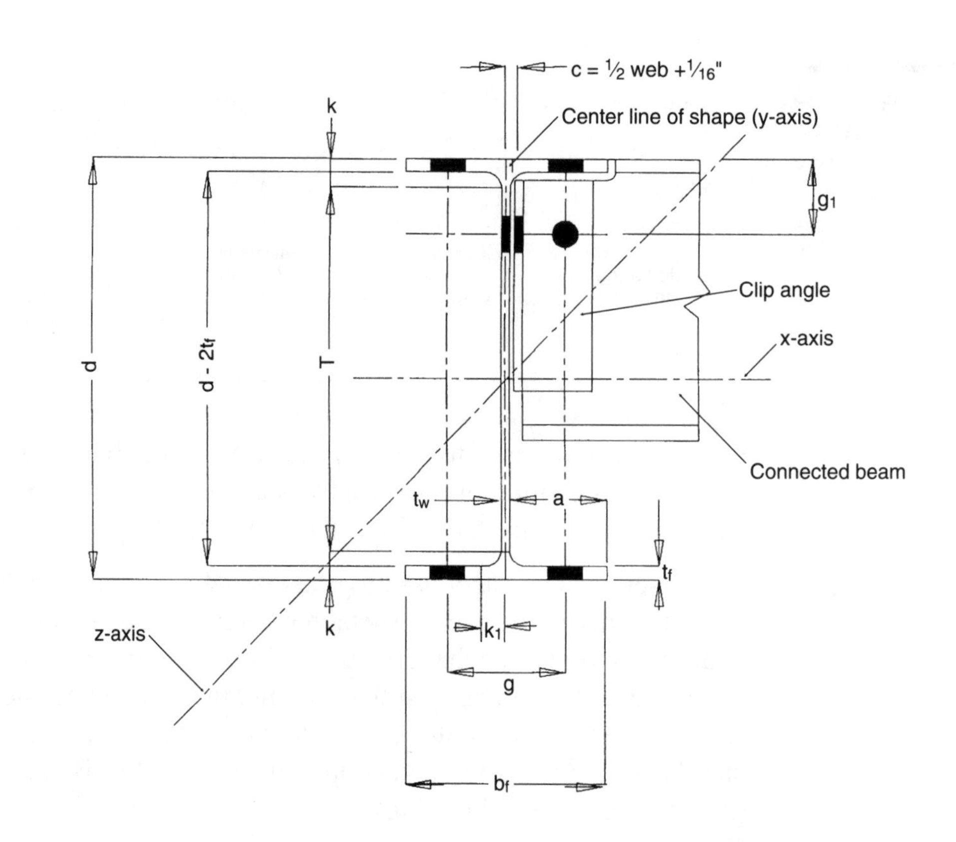

- Moment of inertia about the y-axis, or vertical axis (I_y), in inches^4
- Section modulus about the y-axis (S_y) in inches^3
- Radius of gyration about the y-axis (r_y) in inches
- Radius of gyration about the z-axis, or diagonal axis (r_z), in inches

Steel detailers use the following data, with all distances expressed in inches:

- Section Designation, with letter indicating type of shape
- Weight of shape in pounds per foot
- Full depth of section (d)
- Width (b_f) and thickness (t_f) of flange
- Web thickness (t_w)
- Half web thickness ($t_w/2$)
- Clear distance between flanges ($d\text{-}2t_f$)
- Distance from the web to the edge of the flange (a)
- Distance between the fillets of the flange (T)
- Distance between the outside face of the flange and the edge of the fillet at the web (k)
- Distance between the center of the web and the edge of the fillet at the flange (k_1)
- Usual gauge of bolt holes in the flange (g)

- Gauge to top hole (g_1)
- Distance to face of connecting member, i.e. one-half of the web thickness plus $\frac{1}{16}$ inch ($c = t_w/2 + \frac{1}{16}$ inch)

The symbols representing these dimensions are shown on Figure 5-9.

Quality and Strength

Structural steel can be classified by its chemical composition. There are two general types: carbon steel and alloy steel. Carbon steel is used mainly in construction. There's no minimum standard for how much of any other metals such as cobalt, chromium, molybdenum, or nickel there is in carbon steel. Alloy steel, such as stainless steel, is classified by the amount of alloy metals it contains. Alloy steel is used in construction where corrosion may be a problem.

When a steel mill forms steel shapes, it creates a Certified Mill Test Report which describes the chemical composition of the steel and the results of tests performed on samples. If there's a question about the quality or strength of the steel, the test report will be used to identify any possible problem. Building departments may not let you reuse steel if you don't have the certified test reports from the mill that rolled the steel. Some authorities require that you have laboratory tests performed on coupons (specimens) cut from the reused steel. Other agencies may allow you to install reused steel, but you can only design it to use 50 percent of its original strength.

The American Society of Testing and Materials (ASTM) publishes specifications for structural steel. These specifications control the properties of all types of structural steel and how it's manufactured. Names, such as ASTM A36, A242, and A441, are terms they use for different types of steel. A36 steel has a minimum yield strength of 36,000 psi, A242 has 50,000 psi, and A441 has 46,000 psi. In the 1960s, the most common steel used in construction was ASTM A7 steel. In the United States, ASTM A7 has been replaced by ASTM A36 steel, and it's now the standard grade used in building construction. ASTM A36 steel (also known as carbon steel) is stronger than ASTM A7 steel.

Following the ASTM A36 requirements, the steel supplier should mark high-strength steel before he delivers it to the fabricator or the site where it'll be used. If the supplier didn't mark some high-strength steel you get, don't use it until it's been tested and marked with the fabricator's identification mark. During fabrication, each piece of high-strength steel should have a fabricator's or an original supplier's identification mark. The fabricator's identification marking system should be on record so that you, the building department, or the inspector can check it before fabrication. The building code says that all steel that is stronger than ASTM A36 steel must be individually marked by the fabricator. If there's any uncertainty about the grade of some steel, the fabricator may have to provide an affidavit or laboratory test report that shows the actual grade.

In special cases, where higher-strength steels are required, you can use ASTM A242 and A441 steel. Both are high-strength, low-alloy steel. ASTM A442 is also rust resistant, so you can use it in exposed locations. ASTM A53, A120, and A500 steels are used to make steel pipes and rectangular structural steel tubes.

Hot-Rolled Steel Shapes

The AISC uses a simple system to name all types of hot-rolled shapes. For example, a steel angle with 6 × 6 × ½-inch thick legs that's 6'2" long, is:

L 6 × 6 × ½ × 6'2"

where

L means this is angle steel

6 × 6 is the length of each leg in inches

½ is the thickness of each leg in inches

6'2" is the length of the piece in feet and inches

When an angle has unequal legs, the longer leg is noted first, as in L 6 × 4 × ½ × 6'2". Figure 5-8 shows a typical equal leg angle.

An American Standard Beam, sometimes called an I-beam, 12 inches deep and weighing 31.8 pounds per foot, is:

S 12 × 31.8

See Figure 5-8 for a typical American Standard Beam

The most common hot-rolled steel shapes and their symbols are:

- Plate (PL)
- Flat Bar (FLT)
- Square Bar (BAR)
- Wide-flange Beam (W), the basic shape used for beams and columns
- American Standard Beam (S), used for beams with narrow flanges
- Miscellaneous Beam (M), a lightweight beam
- American Standard Channel (C), used for lightweight beams, door frames, purlins and girts
- Junior Beam (JR), a lightweight beam similar to the I-beam and used in secondary members
- Round, or rod (ROD), used for cross bracing in roofs and walls
- Angle (L), used in secondary members

- Structural Tube, (TS), square or rectangular, slit from flat rolled steel, formed, fusion welded, and cut to length in one continuous operation. This shape is commonly used for columns.
- Standard Pipe (STD PIPE), Extra Strong Pipe (X-STRONG PIPE), and Double Extra Strong Pipe (XX-STRONG PIPE), similar to Schedule 40, 80, and 120 Steel Pipe, and used for columns
- Structural Tees cut from Wide-Flange shapes (WT), American Standard Shapes (ST), and Miscellaneous Shapes (MT)

Figure 5-8 shows cross sections of some of these shapes.

Beams are rolled into shapes that are from 4 to 36 inches deep. For each depth, a different weight and strength beam can be rolled by making a beam's flange and web thicker. For example, there may be several other standard sections of an American Standard 12-inch deep beam which have different weights and strengths. S12X31.8, S12X35, S12X40, and S12X50 are all 12-inch deep American Standard shapes.

There are so many different size beams, it's not practical for warehouses to stock all sizes and weights. They usually keep just the lightest beam of each size in stock. Consequently, designers usually select the lightest beam of each size that will meet job requirements. If you have to special order steel from the rolling mill, it'll cost more and take longer. Often it's cheaper to order the next larger size in that shape rather than the next heavier shape of the same size.

Cold-Formed Shapes

Cold-formed structural shapes are made from sheet steel in various thicknesses and strengths. They're formed in fabricating shops or at the building site using portable rolling machines. The cold-formed shapes include angles, channels, and G shapes. Angles are used for trim. Channels are used for studs, joists, girts, and purlins. G shapes are used for eave struts. These shapes are made from higher-strength steel and weigh less than hot-rolled shapes of the same load-carrying capacity.

Cold-formed structural shapes are used as secondary members in most pre-engineered metal buildings. Secondary members aren't part of the main frame but carry loads to the primary members. Secondary members include purlins, girts, struts, diagonal bracing, wind bents, knee braces, headers, jambs, and other miscellaneous framing.

Cold-formed shapes are cheaper than hot-rolled shapes used for the same purpose. It's usually simpler to form and cut secondary members at the site. And it's easy to make the metal-to-metal connections between the cold-formed shapes using self-drilling screws which don't need any drilling or punching. A self-drilling screw is a fastener which combines the functions of drilling and tapping. It's commonly used for attaching wall and roof panels to girts and purlins. A self-tapping screw is a fastener which taps its own threads in a predrilled hole.

Metal Studs

When the building code requires that a building be entirely noncombustible because of its height or occupancy, steel studs will probably be used for the interior and exterior walls. There are other advantages of steel-framed walls over wood-framed walls besides being noncombustible. Steel framing is free of weakness caused by knots or termites and it won't warp, crack, or shrink from drying action. Steel stud walls are also lighter than masonry walls. This results in a lighter structural system which is less vulnerable to earthquake forces.

A steel stud is a vertical wall member that you can attach exterior or interior covering or collateral material to. The stud may be loadbearing or nonloadbearing. Steel studs are usually channel-shaped sections roll-formed from corrosion-resistant steel. The most common steel used has a yield strength of 40,000 psi (40 ksi). The channels are made in many sizes, such as 3⅝, 4, 6, 7¼, 8, 9¼, 11½ and 13½ inches. The first three sizes are commonly used for studs while the others are used for ceiling or floor joists. The channels may be made in various thicknesses — 20, 18, 16, and 14 gauge.

Selection of the proper size and gauge of steel stud depends upon the location and loading condition:

- Bending stress for exterior walls exposed to wind
- Axial stress for loadbearing walls
- Combined bending and axial stress for exterior loadbearing walls

You can use 18 gauge steel studs, spaced 16 inches on center, for nonbearing walls to the following heights:

- 3⅝- or 4-inch studs can be used for 10 foot height without bracing and 20 foot with bracing at midheight for interior walls.
- 3⅝-inch studs can be used up to 9 feet in height, 4-inch studs up to 10 feet in height, and 6-inch studs up to 14 feet in height for exterior walls with a 20 psf wind load.

A steel stud wall is assembled flat on the floor the same way as a wood stud wall. After door and window openings are framed in, the wall is lifted into position and temporarily braced. Without bracing, the construction loads may cause the studs to fail from the weight of framing on upper floors, materials stored on upper floors, or concrete or other floor decking.

Permanent wall bracing is usually made of 1½-inch cold-formed channels placed horizontally at midheight or at third points. In areas prone to earthquake, stud walls are also diagonally braced.

Stud flanges are screw-attached or welded to top and bottom runner flanges. Where cutouts occur at the stud/runner intersection, a 1 × 1½-inch end stiffener is screw-attached, or additional welds are placed in the runner.

Figure 5-10
Metal stud wall detail

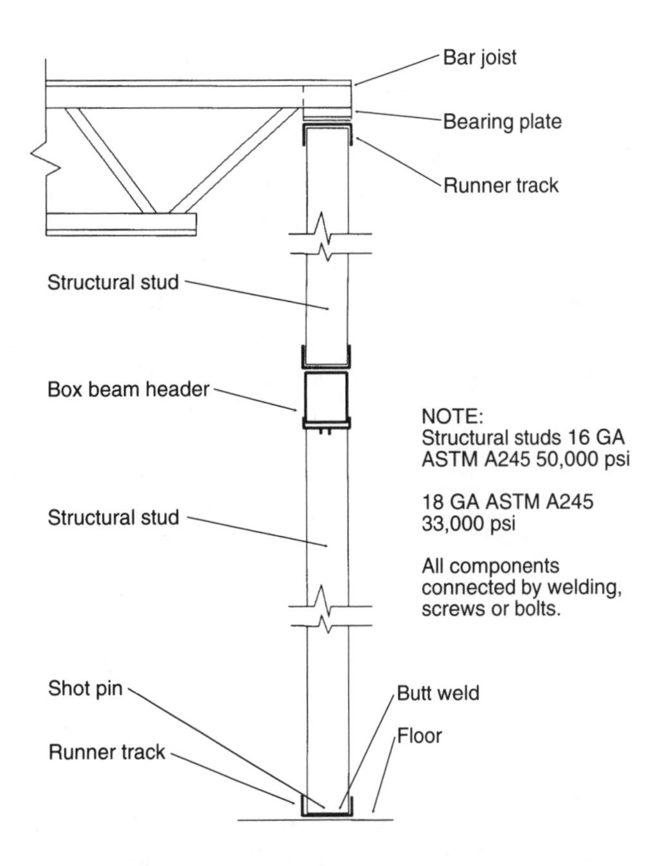

The runner is anchored to concrete floors with power-driven fasteners, and to plywood floor with 1$^{15}/_{16}$-inch Type S-12 pilot-point bugle head nails. Each stud has slotted holes to allow passage of electric conduits.

In summary, a completed metal stud wall has a base runner, studs at 12, 16, or 24 inches on center, and a top runner track. Headers for doors and windows are formed by a combination of channels. It's best to add channel bridging on walls more than 10 feet high. Channel bridging is cold-formed channels placed horizontally between studs. You must use one row of bridging when studs are 10 to 14 feet high. You'll need two rows of bridging for walls over 14 feet high. Use self-drilling or self-tapping screws to connect studs and channels.

Figures 5-10 and 5-11 show a typical framing system for an exterior or interior wall. Figure 5-10 is a sectional detail through the wall and shows the wall supporting bar joist ceiling members. Figure 5-11 is an elevation of a metal stud wall.

Structural Steel Design Drawings

The plans for structural steel framing are called Structural Drawings. Engineers create these drawings after calculating loads for the structure. Figures 5-6 and 5-7 show simplified versions of typical design drawings. Complete framing plans and elevations usually have an anchor bolt layout plan, connection details, general notes, title block information, and other pertinent data. These drawings show the shape, size, weight and

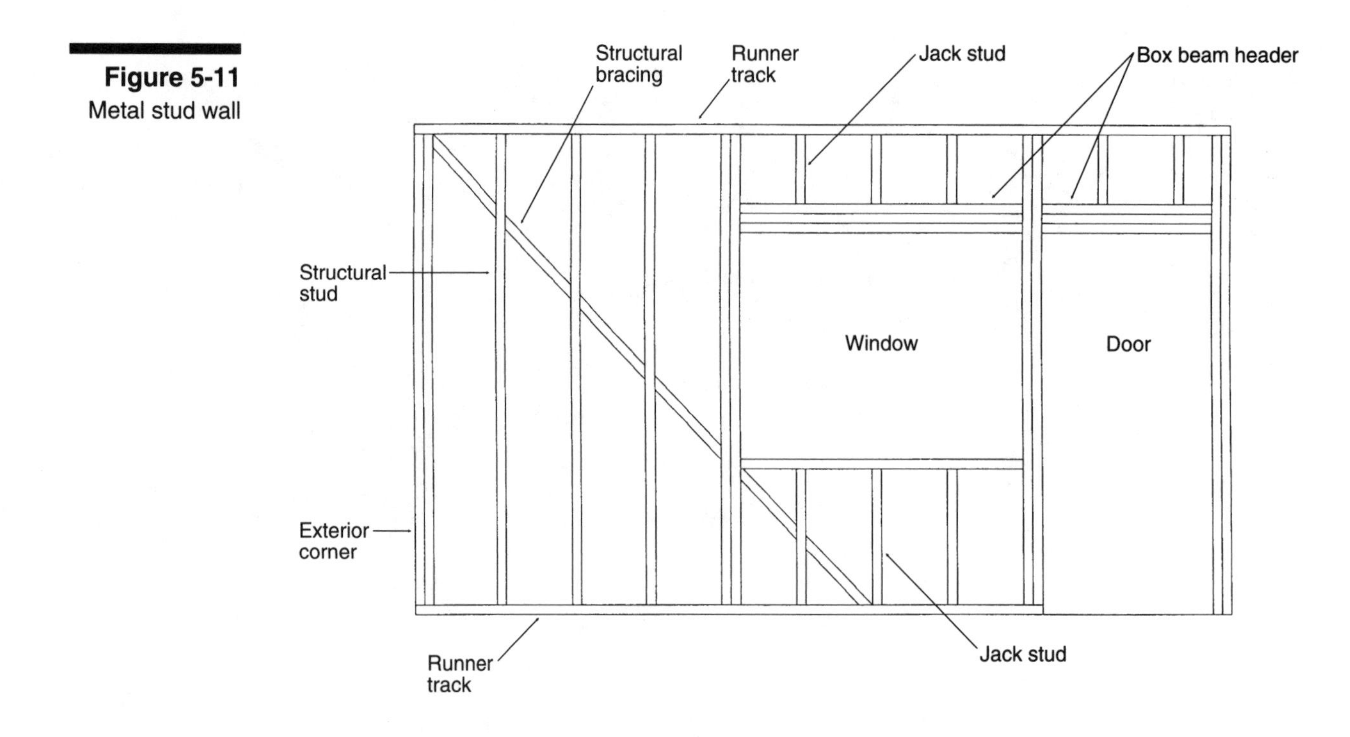

Figure 5-11
Metal stud wall

relative position of steel members so the steel fabricator can produce the steel that's needed. Floor levels and column centers are usually stated to the nearest $\frac{1}{16}$ inch. The following information should be on every set of final structural plans:

- A North arrow on each sheet
- Notes covering connections, fasteners, cleaning, and painting on each sheet
- All revisions made to the plans after the contract was signed, flagged with a triangle, marked with a revision number, dated, and described in the revision block
- Size and location of each framing member
- Dimensions to working lines or working points

A working line in a framing elevation may be the top of a beam, centerline of a column, or the neutral axis of an angle. In framing plans, the working lines are usually the centerlines of wide-flange beams and columns, back side of channels and angles, and centerlines of rods. Figure 5-12 shows the gauge and axis lines of some structural shapes. The working point is the intersection of two or more working lines. Gauge lines show the standard location of bolts in structural members. Common gauges for angles are shown in Figure 5-13.

There are some special rules for drafting structural steel drawings. The top surface of beams, centerline of columns, neutral axis of angles, and backs of channels are drawn with single continuous lines. Each single line has a small section which shows the full depth orientation. These small sections are shown in Figure 5-7. A space or gap at the end of a line which represents a member means that the member is

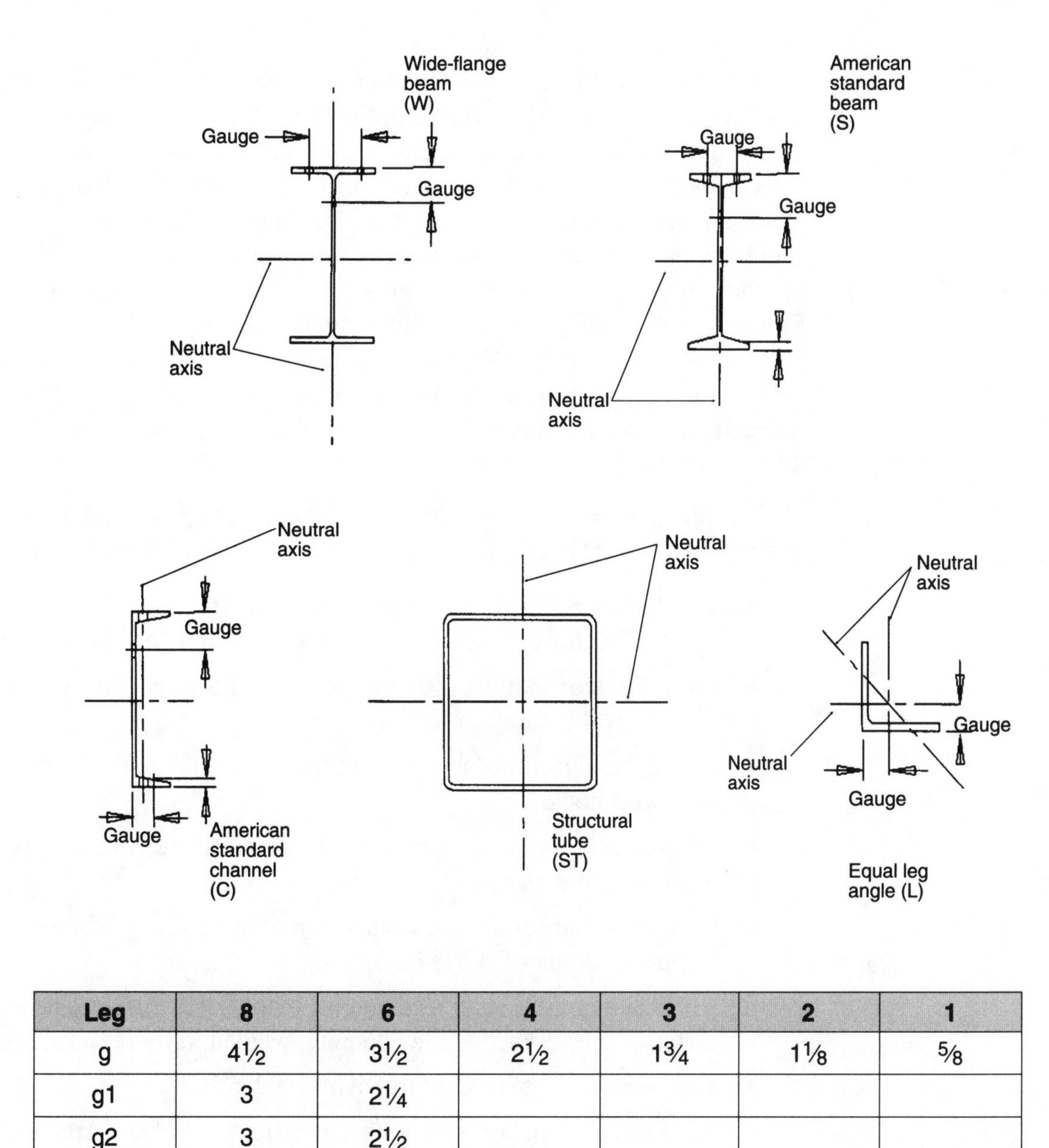

Figure 5-12 Gauges of structural shapes

Figure 5-13 Usual gauges for angles (inches)

Leg	8	6	4	3	2	1
g	4½	3½	2½	1¾	1⅛	⅝
g1	3	2¼				
g2	3	2½				

a separate piece of steel. Otherwise, the fabricator may think that a long line represents a single member rather than two members. Figure 5-7 shows that the columns are made of three separate pieces of steel spliced together.

Although members are shown with single lines, be sure to consider the depth and width of each member to avoid interference. When two beams cross each other, be aware of the depth of the upper member. If there are unusual or complicated connections on a plan, the design drawings should include details that give more information about them.

There should be a description of the steel shape required for each steel member. Where many members have the same description, just the outer ones need to be fully identified. Others can be marked "ditto" (or "do") as shown in Figure 5-7.

All framing plans should clearly state the elevation of the top of steel (T.O.S.). The T.O.S. is usually the elevation of the top surface of the major framing members at each story. This elevation is the measured vertical distance from the bottom of the column base plates to the top of main beams of floor considered. All secondary members that are either above or below the T.O.S. elevation at each floor level should be marked as plus or minus dimensions. For example, the elevation of a platform that's 6 inches above the main floor beams of a story would be noted as +6". Similarly, a monorail which is 12 inches below the top of floor beams would be noted as -12". In other words, each floor has its own datum elevation which is called T.O.S. The elevations used by the steel detailer will not necessarily correspond with the architect's elevation of a floor or story. Remember that the steel detailer only sees the steel framing of a building.

The plans should also include notes about the grade of steel, type of connections, and any shop paint. Here's a typical set of notes for structural steel specifications:

- The detailing, fabrication, and erection of structural steel shall conform to the latest AISC Specifications Serial Designation A36.
- The fabricator shall furnish three sets of completed, checked shop drawings for approval before fabrication.
- All field connections are ¾-inch bolts with 13⁄16-inch diameter holes unless otherwise noted.
- All connections using high-strength bolts shall be friction-type or bearing-type with regular-head bolts.
- Material manufacture and installation of high-strength bolts shall conform to the latest edition of ASTM A325.
- All steel members shall receive one shop coat of rust-resistant paint, except surfaces to be embedded in concrete, welded, or to receive high-strength bolts.
- The steel fabricator shall furnish all field bolts.
- All welding shall conform to the specifications of the American Welding Society.
- All field welding shall be done by certified welders and shall be continuously inspected by qualified inspectors.

Of course, these notes will vary with each job. The reference to the AISC specifications makes it unnecessary for the designer to describe and illustrate the details and connections that are standard practice.

If a brand name product is specified, make sure it's readily available, or get permission to use a substitute of equal quality.

Any steel you use that will be supplied by someone other than the structural steel fabricator should be clearly marked on the plans. Examples are: "masonry anchors by mason," "stairs by miscellaneous iron contractors," or "bolts by elevator supplier." If miscellaneous iron items are shown on the structural drawings, label them clearly. Any reference to the AISC manual should identify the edition. The phrase "latest version" leaves room for confusion.

Basic Steel Design

Properties of Shapes

The most important properties of steel shapes are:

- moment of inertia (I)
- section modulus (S)
- radius of gyration (r)

The moment of inertia (I) of a steel member is a physical property which helps define its stiffness, resistance to being bent, and its strength and deflection characteristics.

The section modulus (S) of a member is a property that determines the load it can carry. You find out the section modulus by dividing the moment of inertia by the distance (c) between the neutral axis and the outside surface of the shape, as shown below:

$$S = I / c$$

The radius of gyration (r) of a member is a measure of its resistance to buckling. You calculate it by dividing the moment of inertia by the cross-sectional area of the shape, and taking the square root of your answer as shown below:

$$r = \sqrt{I/A}$$

Here's how you can calculate these properties for a simple rectangular beam:

$$I = bd^3 / 12$$

$$S = bd^2 / 6$$

$$r = \sqrt{I/A}$$

where

b = width of the beam, in inches

d = depth of the beam, in inches

A = cross-sectional area of the beam, in square inches

You don't have to calculate these values for most structural shapes because they're listed in the *Manual of Steel Construction* under the heading "Dimensions and Properties."

Stresses

Tension, compression, shear, and bending stresses are the fundamental types of stresses. Tension happens when a rod or bolt resists a pull at each of its ends. The tension load may be the weight of an object suspended by a rod. The tensile stress (f_t) in a member is the tension load (T), in pounds, divided by its net cross-sectional area (A), in square inches, or

$$f_t = T / A$$

Compression (P) happens when a column or strut resists an inward push at each of its ends. The compressive stress (f_c) on a member is the compressive force (P), in pounds, divided by its cross-sectional area (A), in square inches, or

$$f_c = P / A$$

A compression force on a column may be the weight of a roof supported by the columns.

Shear (V) happens when two forces, almost in the same plane, act toward each other on a structural member. This is similar to what happens when you cut a sheet of steel with a tin snipper. The shear stress (f_v) on a member is the shear force (V), in pounds, divided by its net cross-sectional area (A), in square inches, or

$$f_v = V / A$$

A shear force on a bolt can be the outward pull of two plates connected by the bolt. A single shear stress is found by dividing the pull on the end of the two plates, expressed in pounds, by the cross-sectional area of the bolt through the plane of the threads. Double shear occurs in the bolt when a single plate is bolted to two plates and there is a tendency to cut the bolt in two planes.

A bending moment (M) on a member causes a bending stress (fb) in that member. A bending moment, sometimes called a moment, occurs when the forces on a beam tend to cause the beam to bend, or rotate about a point or axis. A moment at any given point of a member is the force applied to the member multiplied by its distance to the point. The distance, also called the lever arm, is the distance measured by a line drawn perpendicular to the line of action of the force to the point.

If the force tends to rotate the structural member in a clockwise direction, it's called a positive moment. If it tends to rotate the member in a counterclockwise direction, it's called a negative moment. A bending moment is always expressed as a force and distance, such as foot-pounds, inch-pounds, or foot-kips (one kip = 1000 pounds).

A simple example of a bending moment occurs in a horizontal beam of length (L) supported at each end and carrying a single concentrated load (P) at midspan. The upward force at each support is called the reaction (R). It's equal to half of the concentrated load (P/2). The bending moment at midspan is the reaction, in pounds, times half the span in feet, or

$$\begin{aligned} M &= P \times L / 4 \\ &= R \times L / 2 \end{aligned}$$

The bending stress (f_b) on a member will vary with the bending moment (M). You should calculate the bending stress where the moment is the greatest. The most commonly used formulas for determining bending stress (f_b) are:

$$f_b = M \times c / I \text{ or } M \times S$$

where

c = Distance from neutral axis to outside surface, or extreme fibers of the member, which is usually half the depth of the member

I = Moment of inertia

S = Section modulus

M = Maximum bending moment in inch-pounds

Stress-Strain Curves

A stress-strain curve for a type of steel tells you its physical properties. Figure 5-14 tells the physical characteristics of A36 steel. Stress is measured along the vertical axis in 1000 pounds per square inch (ksi). Strain, or elongation, is measured along the horizontal axis in inches per inch (in/in). Strain is the change in length per unit length. It's the deformation of a body as forces act on it. The values for stress and strain were obtained by placing a specimen rod made of A36 steel in a testing machine and recording the amount of elongation in the rod with each incremental pull until the rod fails. Stress is the pull (kips) on the specimen divided by its cross-sectional area (kips per square inch). The values are plotted on the graph.

The straight portion of the curve is the range in which the stress is proportional to the strain. The slope of this line represents the modulus of elasticity (E), or Young's modulus, of the steel specimen. It's also called the elastic range, since the specimen acts like a rubber band in this range. When the stress is reduced to zero, the specimen returns to its original length. The modulus of elasticity is also called the ratio of stress to strain, or:

modulus of elasticity (E) = stress / strain

The value "E" for A36 steel is about 29,000,000 psi.

The point where the curve changes direction suddenly is called the elastic limit, and the stress at that point is called the yield stress, or yield strength. This is the maximum stress that can be put on the steel without permanently deforming it. Beyond the yield strength, the values for stress vs. strain follow an erratic course and may eventually drop before the steel fails completely. If the stress is relieved after the yield point is reached, there will be permanent deformation, or set, in the material. The allowable stress, or safe working stress, as required by the *Uniform Building Code*, is based on a percentage of the yield stress of a specific material. For example, the yield stress of A36 steel is about 36 ksi and the allowable stress is only 22 ksi, or about 60 percent of yield strength.

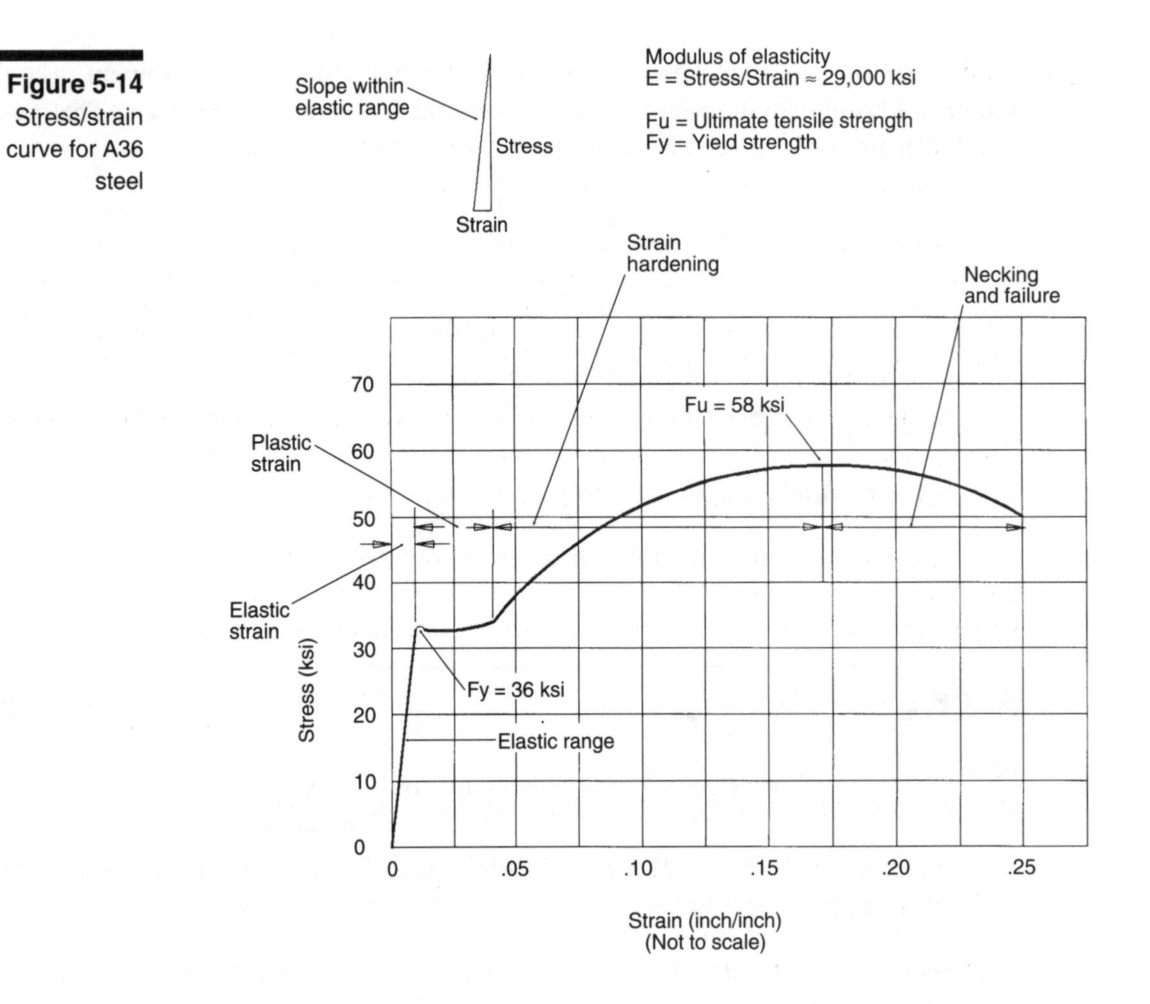

Figure 5-14 Stress/strain curve for A36 steel

The stress at the highest point on the curve is called the ultimate stress, or ultimate strength. The ultimate strength of a material is rarely used in design because by the time it happens the material is too deformed to be useful. It's used as a factor in earthquake design, however, because steel becomes ductile at stresses greater than the yield stress and can better absorb some of the shock of an earthquake.

Types of Loading

The structural frame of a building should be able to withstand all possible loading conditions. Typical building loads are: dead loads, floor live loads, impact loads, roof live loads, seismic loads, wind loads, crane loads, collateral loads, and auxiliary loads. There may be a combination of loads, such as dead load plus live load, dead load plus wind load, and dead load plus seismic load. When you design a warehouse or storage area, you should combine the dead load with the seismic load and a portion of the live load.

Design loads are minimum live loads specified in the building codes published by federal, state, county, or city agencies, or in owner's specifications. The floor live load is based on the type of occupancy. For example, residential buildings require a minimum floor live load of 40 psf, office buildings require 50 psf, and retail stores require 75 psf. When the code specifies a minimum unit live load on a floor, people and furnishings are included. Remember that these are minimum live loads. Special cases, such as office file rooms, may require higher live loads.

A roof live load includes temporary loading during construction or maintenance. In certain localities, snow loads are part of roof live loads. You should contact the local building department to determine how much snow load you should design for. Snow, which is very heavy, can build up in roof valleys and on unheated areas.

The dead load of a building is the weight of all permanent construction, such as floors, roofs, framing, and enclosure. Roof-mounted mechanical equipment can also be considered dead load, as it's permanently in place.

Impact loads are assumed dynamic loads resulting from the motion of machinery, elevators, cranes, vehicles, and other similar moving forces. For example, in elevator supporting members, you double the dead and live loads of the elevator cab.

Auxiliary loads are certain dynamic live loads, other than the basic design loads which the building must safely withstand. Typical examples are cranes, material handling systems, and rotating or reciprocating machines. Reciprocating machines, such as air compressors, can shake their supports apart if the machine malfunctions.

Collateral loads are dead loads other than the building dead load. They include sprinklers, mechanical systems, electrical systems, and ceilings.

Seismic loads are lateral loads acting in any horizontal direction on a building due to the action of earthquakes. These loads are a function of an earthquake's magnitude (Richter Magnitude), the soil type under the building, the shape and height of the building, the building's dead load, and the flexibility of its structural frame.

Wind loads are loads caused by the wind blowing from any horizontal direction. A wind load is a function of the velocity of the wind, and the height and location of the building part exposed to the wind. Wind loads can be either inward or outward, causing pressure or suction.

A load may be either concentrated or uniformly distributed on structural members. A concentrated load is a point load. A uniformly-distributed load is assumed to be evenly distributed over an area or along the length of a structural member.

The loads described above cause a force on a structural member which changes or tends to change its state of rest or motion. A force must be defined as to amount, direction, and position. A force may be expressed in pounds, kips, or other similar units, and may act in the following ways:

- A compressive force is a push acting on a member, such as a column, tending to compress the member.
- A tension force is a pull acting on a member, like a rod brace, tending to make the member longer.

- A shear force is sliding pressure acting on a body, like a bolt holding two plates together, which tends to slide one portion of the bolt against the other portion of the bolt. Shear can be described as the force tending to make two contacting parts slide upon each other in opposite directions parallel to their plane of contact.
- A torsion force is a turning force acting on a body, like a handle on a shaft, which tends to twist the shaft.

A uniformly-distributed load is the load per linear foot of beam, or per square foot of the floor or roof area, expressed in pounds per square foot (psf). For example, if a 10-foot-long beam is carrying a 100 square foot portion of floor area which is loaded at 50 psf, the total weight carried by the beam is 100 × 50, or 5,000 pounds. The uniform load on the beam is 5,000 / 10 or 500 pounds per linear foot (plf).

Beams and Girders

Simple and Fixed-End Beams

Figure 5-15 shows several types of beams. The ends of a simple supported beam are free to rotate. The ends of a fixed-end beam can't rotate. A cantilever beam is fixed at the support end and free at the unsupported end. Figure 5-16 shows the allowable uniform loads, in kips, on laterally-supported beams. The loads shown are the total weight that the beams can carry. Laterally-supported means that the beams are restrained from moving sideways, as with a floor or roof deck. A uniformly-loaded beam has the load spread evenly over the beam.

Steel shapes, such as standard and wide-flange beams, can be reinforced by welding cover plates to the top and bottom flange. These members are then called built-up beams. The cover plates increase the moment of inertia of the section and, thereby, improve the beam's strength. When you use built-up members instead of heavier rolled shapes, be sure to compare the savings in weight with the extra cost of shop labor.

Welded beams, or welded plate girders, are made of three plates welded together to form an I-shaped section having a web and two flanges. A prismatic welded beam has both flanges parallel about its longitudinal axis. A tapered welded girder is a built-up plate girder consisting of flanges welded to a variable depth web. These girders are usually used for roofs with clear spans up to 100 feet. The lower flange is horizontal, while the top flange provides the necessary roof slope for drainage. The depth of the web increases from the ends to the center of the girder. The flange thickness may also be increased at the center portion of the span. This provides a greater moment of inertia at the center of the beam where the bending moment is the

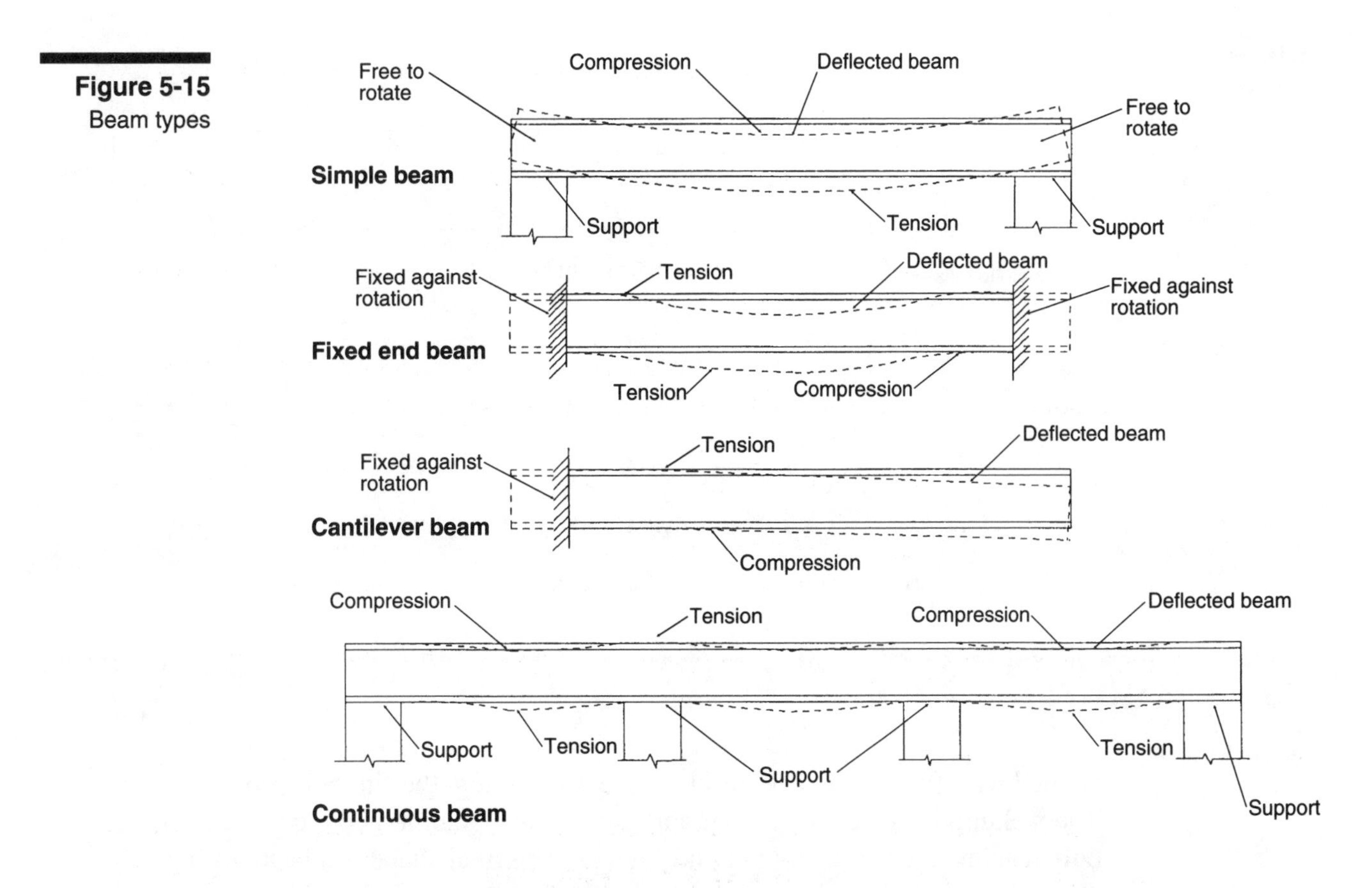

Figure 5-15 Beam types

Figure 5-16 Allowable uniform loads on laterally supported beams (kips)

Span (feet)	W8X17	W10X21	W12X27	W14X30	W16X36	W18X50
10	18.8	29	45	56	75	119
12	15.7	24	38	46	63	99
14	13.4	21	32	40	54	85
16	11.7	17.9	28	35	47	74
18		15.9	25	31	42	66
20		14.3	23	28	38	59
22			21	25	34	54
24			19	24	31	49
26				21	29	46
28				19.9	27	42
30				18.6	25	40

maximum. The web of a tapered girder is usually stiffened at the heel and midspan by ¼-inch-thick plates. Angle knee braces also stiffen the lower chord. These stiffeners and braces keep the girder from buckling.

Castellated beams are also called open-web beams. Figure 5-17 shows how a parent beam is made deeper and stronger. These are fabricated by splitting a standard wide-flange shape with an oxyacetylene cutting machine. After the two halves of the original beam are cut, they are placed with high point to high point. They are then

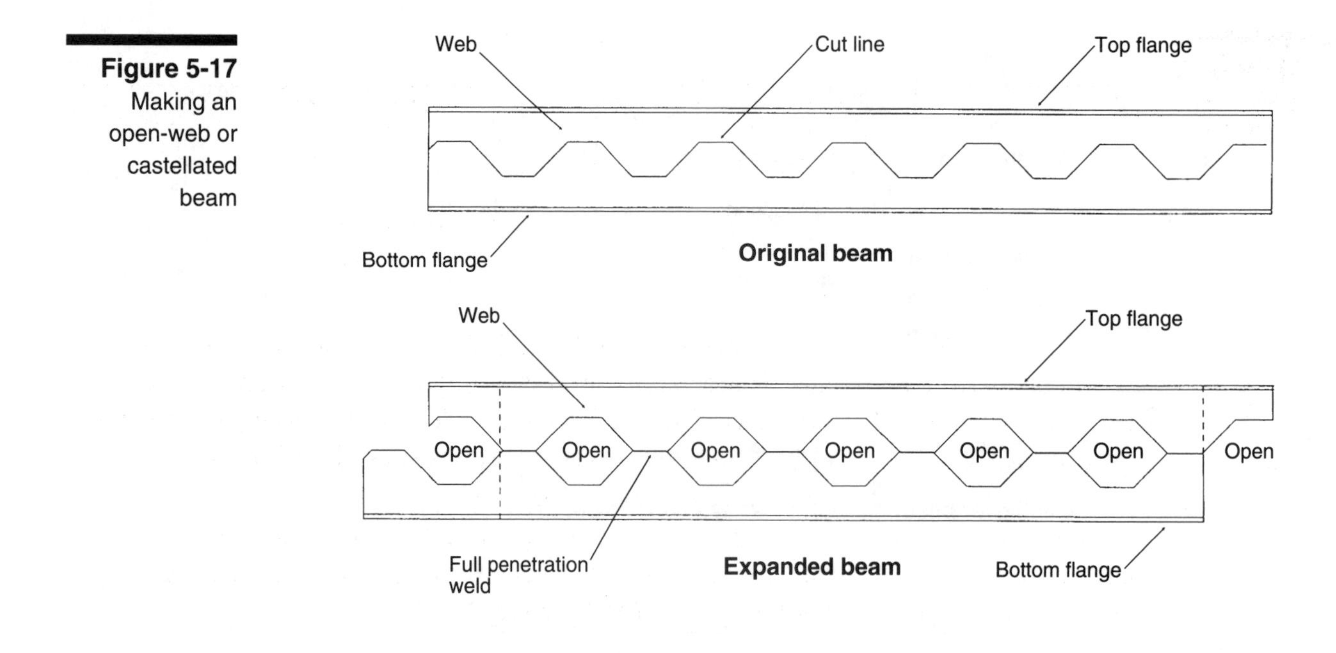

Figure 5-17 Making an open-web or castellated beam

welded with full penetration welds while pressed together in their proper position. The welding is made with a continuous automatic, submerged arc process on one side only, against a grooved copper chilling bar. The final shape is a beam of higher strength than the original beam. It has a row of hexagonal holes in the web so you can put pipes and conduits through it. When you use open-web beams as joists, tie the top and bottom flanges together by angle cross bridging to keep them from buckling. Use open-web beams only on uniformly-distributed loads.

Other built-up members are combinations of channels and angles mainly used as eave struts in an all-steel building. An eave strut is a horizontal member connected to the top of columns at the eave line of a building. The channel supports the weight of the siding while the attached angle supports the roofing. The combination also forms a strut for the bracing system.

Composite beams have steel studs welded to the top flange. The studs, called shear connectors, are embedded in the concrete slab that is supported by the beams. The advantage is that the slab works with the beam to support the floor load, so you can use a smaller size beam to support the floor. These beams are common in multistory buildings with steel frame and concrete floor slabs.

Open-Web Joists

Although open-web joists and long-span joists aren't usually considered to be structural steel, they're often used in residential and commercial buildings. These members are shop-fabricated lightweight trusses. Open-web joists are made by manufacturers but usually installed by a steel erector.

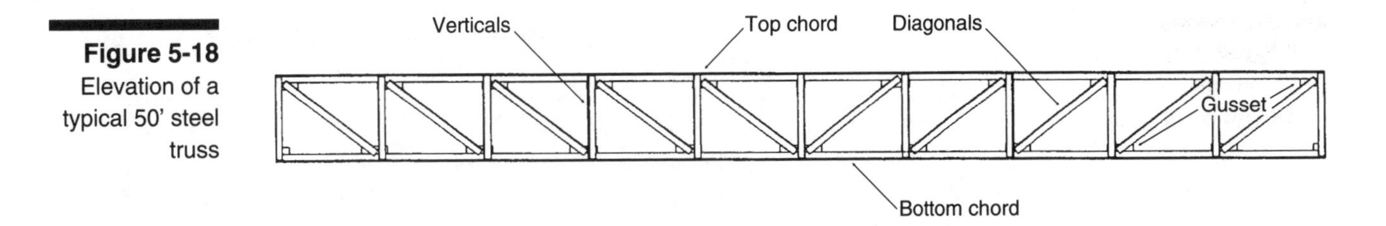

Figure 5-18 Elevation of a typical 50' steel truss

You can use them for short, medium, and long spans. They can function as joists or girders between columns. Made mainly of high-strength steel, they come in several series; for example, J, H, and LH. The J series has a yield strength of 36,000 psi. The H series has a yield strength of 50,000 psi. The LH series is used for long spans.

Open-web steel joists, also called bar joists, are used for floor and roof construction. As floor joists, the top and bottom chords are parallel, while in roof construction the upper chord is pitched for drainage.

The top chords are usually made of two angles placed back-to-back. The lower chord may be of two angles or two rods. Web members are bent rods welded to the chords.

Cross bridging or straight bridging between open-web joists is made of steel rods or angles. They support the top chords against lateral movement during the construction period, and hold the joists in place. Without bridging the joists might rotate or buckle. Cross bridging is usually placed at 8 foot intervals along the axis of the joists.

Trusses

Where you have a long span that has to carry a heavy load, it's a good idea to use a steel truss. This has a top and bottom chord made from structural tees (ST) or double angles. The diagonal and vertical web member is made of angles that are bolted or welded to the chords through gusset plates. Steel trusses can span from 20 to 200 feet. Figure 5-18 shows a steel truss.

Typical ratios of truss depth to span for parallel chord trusses are:

Depth (ft)	Span (ft)
2	20 to 30
3	35 to 45
4	50 to 60
5	65 to 75
6	80 to 90
7	95 to 110
8	120 to 200

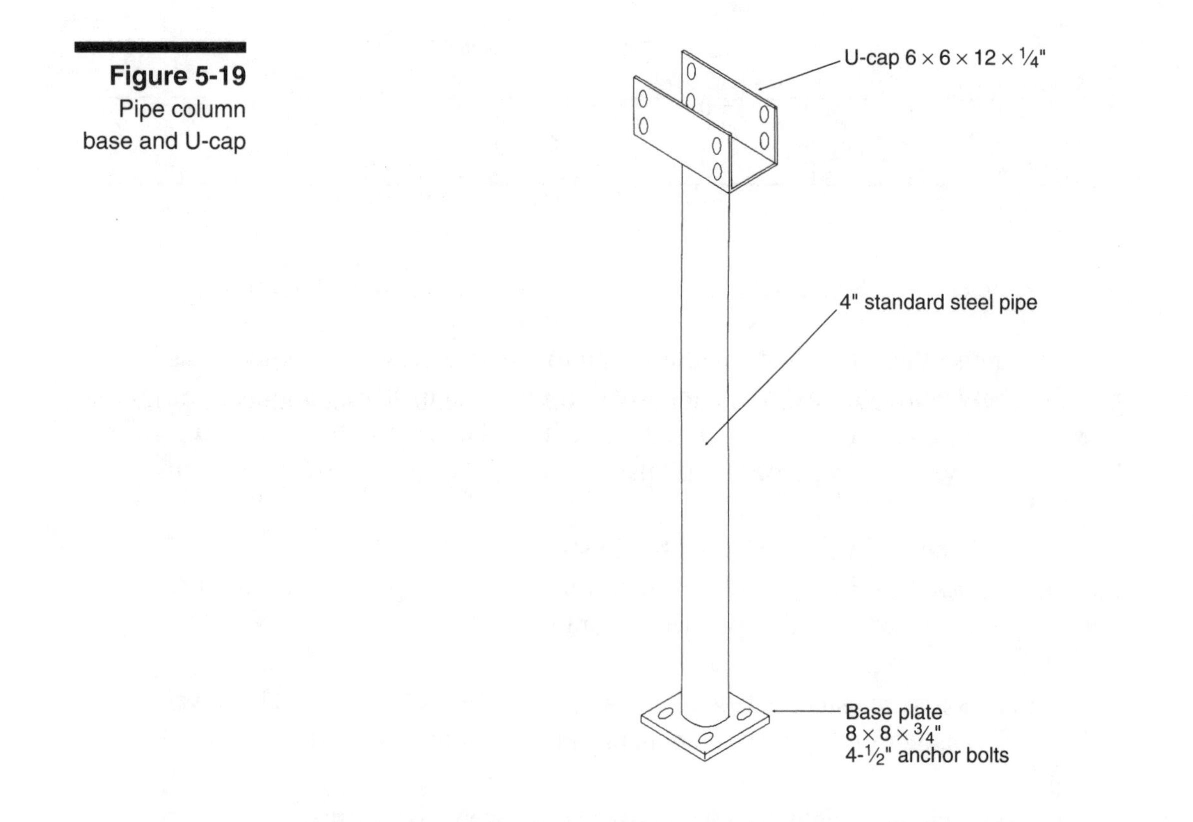

Figure 5-19 Pipe column base and U-cap

Columns

Types of Columns

Generally there are three shapes that are used for columns; wide-flange, pipes, and structural tubes. Most standard steel pipes are made of A53 steel and are similar to Schedule 40 pipe. Extra-Strong and Double Extra-Strong pipes have thicker walls and can carry greater loads. Figure 5-19 shows a typical steel pipe column with end connections.

Most columns used to support upper floors in multistory residential building are made of standard steel pipe columns or square structural tubing. These have an advantage over the wide-flange steel columns because they are equally strong in both directions. Also, they have higher load-carrying capacity pound for pound, less painting surface, and neater appearance. They don't have pockets or grooves that collect moisture.

A typical steel pipe column is fabricated with a 5/8- or 3/4-inch thick steel base plate with two 1-inch or four 5/8-inch diameter holes for anchor bolts. The standard cap is a 1/2-inch thick steel plate with four 13/16-inch diameter holes that are used to attach it to a steel or wood beam. Both plates are welded to the pipe with a continuous fillet weld.

You can normally get quicker delivery on 4- and 6-inch steel pipe and structural tees because the steel supplier usually has them in stock. Otherwise, you might have to make a special order to the steel mill and the fabricator must wait for the steel sections to be rolled and delivered to the supplier.

Loads on a Column

If possible, apply loads to a column concentrically or as close as possible to the center line of the column. An eccentrically-loaded column can buckle if it's not designed to resist the bending moment caused by an off-center load.

The strength of a column depends on its cross-sectional area (A), unsupported length (L) and its least radius of gyration (r). Some shapes, like the wide-flange ones, have two radii of gyration, one about the x-axis and the other about the y-axis. The least radius of elasticity is the smaller one. Pipe columns have one radius of gyration.

The ratio of L/r should be less than 120 for maximum stability against slight eccentricity of the load on a column. If L/r is less than 120, then the safe load (P) on a pipe column, is:

$$P = (17{,}000 - 0.485\,(L/r)^2)\,A$$

where

P = safe carrying load in pounds

L = unsupported length in inches

r = radius of gyration in inches

A = area of column in square inches

The safe concentric load for a square or rectangular column based on a round column of equivalent diameter (D) is:

$$D = A_S / (P \times t)$$

where

A_S = area of the steel in the square or rectangular column, in square inches

t = thickness of the wall of the square or rectangular column, in inches

After you use this formula to figure out what D is, you can use the pipe diameter that's closest to (but not smaller than) that value of D. Figure 5-20A lists the capacity of columns of varying sizes and lengths. Figure 5-20B lists the capacity of wide-flange shapes that are used as columns.

Column Connections

One-story columns are usually fabricated with a base plate and a connection at the top to support a beam. If the column supports a wood beam, weld a U-cap plate to the top of the column, as shown in Figure 5-19. The base plate spreads the load over the surface of the concrete foundation. Usually you anchor the base plate with two or four anchor bolts and fill the space between the base plate and the concrete with nonshrinking grout.

Figure 5-20A Allowable concentric load on standard steel pipe columns (kips)

Pipe size	10 (ft)	12 (ft)	14 (ft)	16 (ft)	18 (ft)	20 (ft)
3½	35	30	25	19	15	10
4	44	39	34	28	23	19
5	64	60	56	51	45	39
6	87	83	79	75	69	64
8	136	133	129	125	120	115

Figure 5-20B Allowable concentric load on wide-flange columns (kips)

Column shape	10 (ft)	12 (ft)	14 (ft)	16 (ft)	18 (ft)	20 (ft)
W6X15.5	63	56	48	15		
W8X24	101	93	90	71	59	49
W8X31	139	132	124	115	98	92
W10X33	147	139	125	119	100	92
W12X40	178	169	157	144	129	113

Make field splices of the columns in multistory buildings to hold the upper and lower sections in alignment. Figure 5-21 shows this type of connection. The direct load from the upper section is transmitted to the section below. For this reason, column ends are milled smooth and square. The sections are held together by splice plates you attach to the outside surface of the flanges. Use filler plates to make up for any difference in depth between the upper and lower sections. Columns should go about 3 feet above the second level. This makes connecting the column splice easier. Have the splice plates shop welded or bolted to the lower section. Weld or bolt the upper section to the lower section in the field.

There are many ways to connect a beam to a column. It depends on the type and amount of load. A light load may need only a single plate connection. A heavier load may need a double angle. A moment connection may need a seat to hold the beam in place until field welding is all done. A moment connection, also called a rigid connection, is a connection between two structural members that holds them at a particular angle. A moment connection is also a connection that maintains continuity in a line of beams.

Use leveling plates to set the column base plates at the proper level. You can either shim them in place or support them with threaded nuts to the anchor bolts. Use a template when you place anchor bolts in concrete to accurately set the anchor bolts and establish the proper elevation.

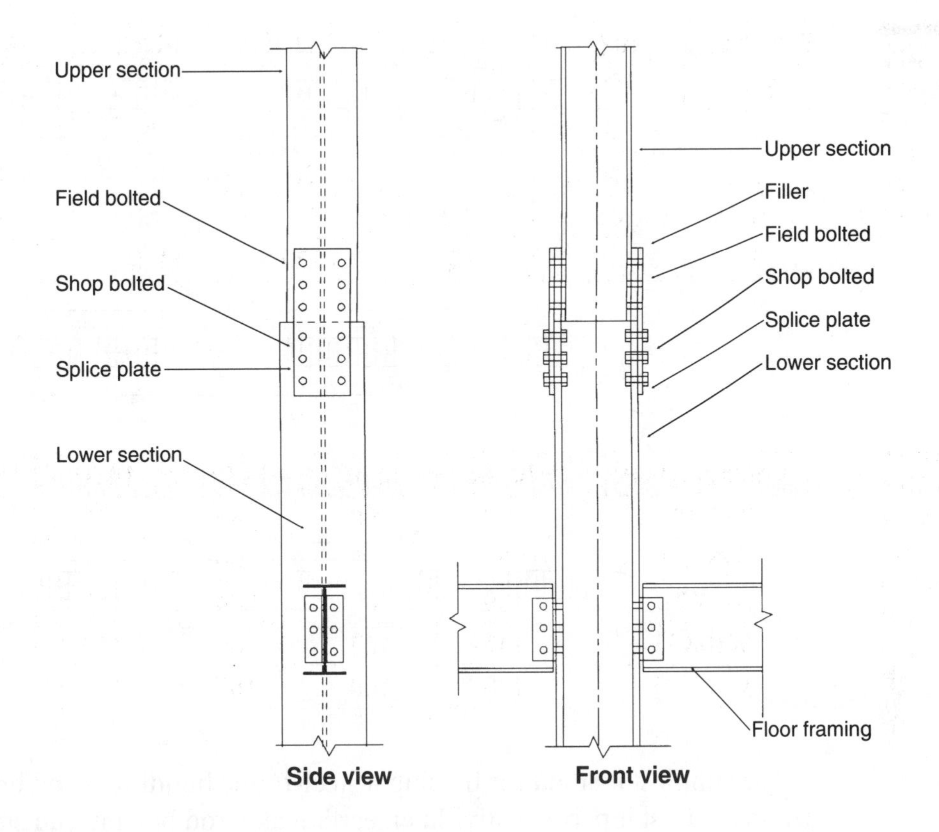

Figure 5-21 Typical bolted column splice

Bracing

Rod Bracing

You can connect lightly loaded rod bracing with threaded nuts and beveled washers. Rod bracing is usually installed in an X pattern. Figure 5-3 shows a typical X-braced building bay.

Clevises and turnbuckles are devices used to attach and tighten rod bracing. Steel bracing rods, usually ½ to 1¼-inch diameter, are fabricated with each end threaded for a distance of 4 inches. The clevis is made by bending a ¼ to ⅝-inch thick by 2½ to 4-inch wide steel bar into the desired U-shape. A hole is made at midpoint to attach the bracing rod, and there are two holes at the ends to attach it to the steel frame. The rod is attached to the clevis with a pair of nuts.

Use a turnbuckle when you connect two rods if you can't tighten a single rod enough by the nuts. A turnbuckle is made of an elongated cast steel fitting with threaded holes at each end. You screw the ends of the two rods into the turnbuckle and then rotate the turnbuckle by pulling the two rods inward, which tightens the brace. Figure 5-22 shows a turnbuckle with four types of end fittings.

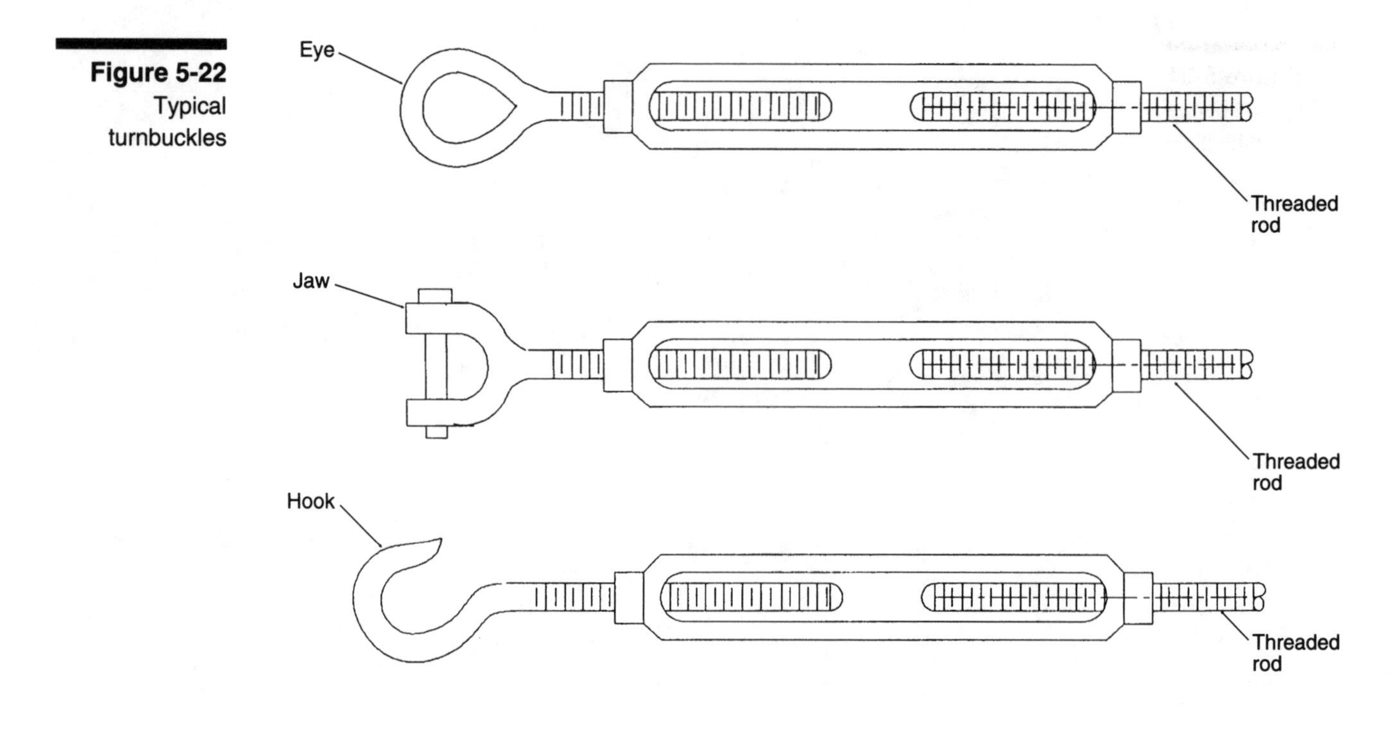

Figure 5-22 Typical turnbuckles

The simplest method for bracing a steel-frame building is rod bracing, but you need to readjust it periodically. In an earthquake, rod bracing can stretch so much that it must be replaced.

Angle Bracing

Angle bracing is more rigid and stable than rod bracing. You make this type of connection by bolting to gusset plates that are welded to adjoining columns or beams. To tighten angle bracing, force them into position with drift pins (tapered steel pins that align holes in steel members). Usually you install angle bracing as a single diagonal or in an X pattern.

Structural Tube Bracing

The most rigid type of bracing uses structural tubes. Figure 5-4, on page 215, shows an inverted V brace. Some ways to install structural tube bracing are:

- Chevron bracing — a pair of braces located either above or below a beam terminate at a single point within the clear beam span.
- Diagonal bracing — a type of bracing that diagonally connects joints at different levels.
- K-bracing — a pair of braces located on one side of a column terminate at a single point within the clear column height.

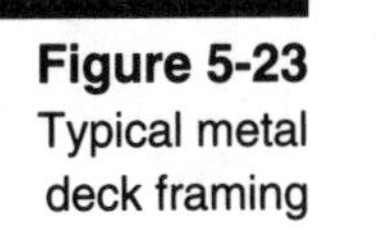

Figure 5-23
Typical metal deck framing

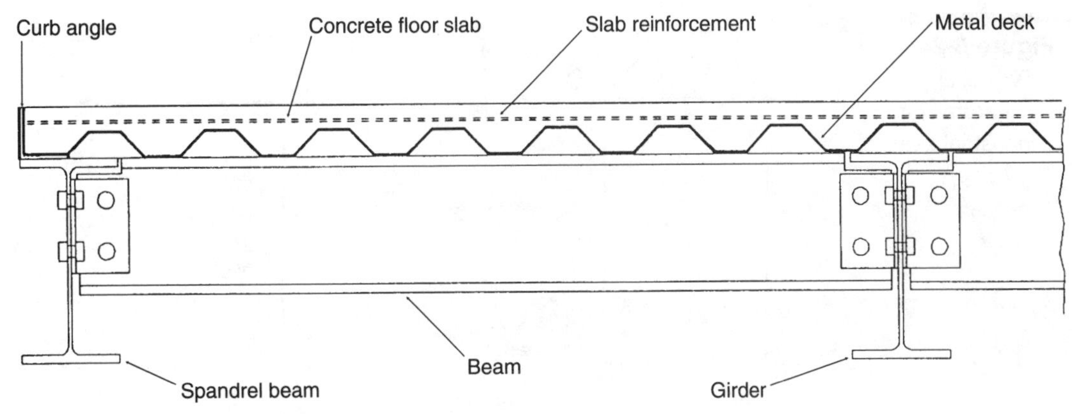

- V-bracing — a form of chevron bracing that intersects a beam from above.
- Inverted bracing — a form of chevron bracing that intersects a beam from below.
- X-bracing — a pair of diagonal braces cross near the middle of the bracing members.

Decking and Siding

Floor Decking

Steel decking comes in single ribbed sheets or combinations of a ribbed and flat sheet. The gauge of the metal and the depth and shape of the corrugation determine the strength of the decking. Decking is also used in combination with shear connectors to the steel beams forming a composite construction. Steel decking should be strong enough to carry the weight of wet concrete on it without excessive deflection. The maximum deflection is about $1/360$ of the span of the decking. If the steel decking you use is strong enough, you won't need any temporary shoring for concrete slabs. Figure 5-23 shows one type of metal decking.

Steel decking is used for roofs, floors, form decks, long span and composite floor construction. Typical thicknesses are 16 and 12 gauge.

Siding

You can make building enclosures of insulated metal panels or siding. These are made of cold-formed steel or aluminum sheets. They come in several configurations, gauges, and finishes. A panel may be up to 60 feet long, so you won't need any horizontal joints. You usually attach siding to the framing with self-tapping or self-drilling metal screws.

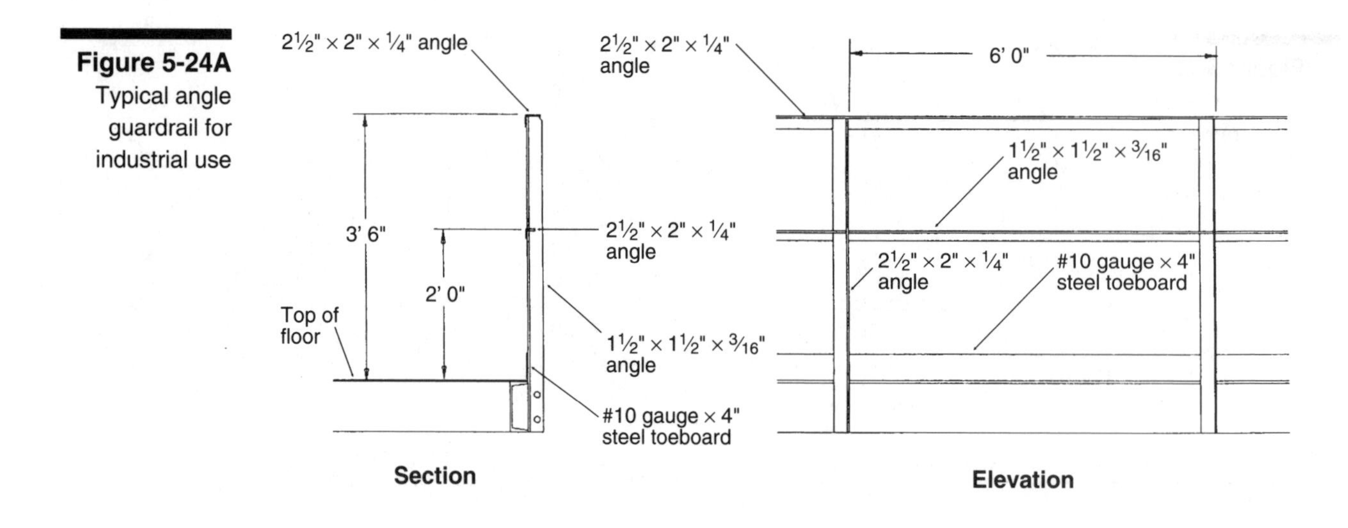

Figure 5-24A Typical angle guardrail for industrial use

Miscellaneous Iron

Metal Stairs, Guardrails, and Ladders

Metal stairs are common in most residential and commercial buildings. A typical stair consists of two stringers, treads, landing and railing. Stringers are usually made of 9- or 10-inch deep steel channels with flanges pointed outward. The stringers support the steel treads by bolts or welding. The treads may be ¼-inch thick bent checker plate, called safety tread, when used for emergency purposes. More commonly, treads are made of steel pans filled with concrete. Some residential stairs are made with precast concrete treads with embedded plates welded to the stringers.

Guardrails for stairs and balconies are made of either aluminum or steel tubing. Figure 5-24A shows a typical steel angle guardrail used in an industrial building. Figure 5-24B shows an aluminum guardrail common for residential buildings. The parts of a guardrail are:

- top rail
- bottom rail
- posts
- balustrades

Recent building code requirements for residential buildings limit the opening between the balustrades to less than 4 inches to prevent a child's head from getting caught between the balustrades. Vertical balustrades are considered safer than horizontal rails because they are harder to climb.

Roof access ladders are common for buildings with roof-mounted mechanical equipment. A typical ladder is made of ⅜ by 3-inch steel side rails and ¾-inch diameter steel rungs. The side rails are 16 inches apart and the rungs are spaced 12 inches apart.

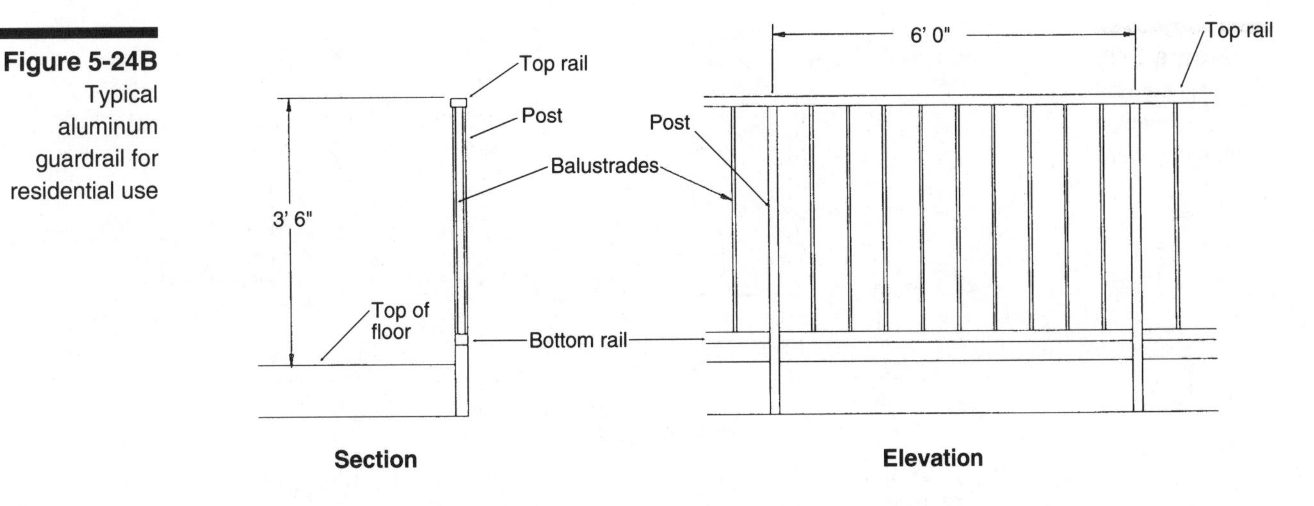

Figure 5-24B Typical aluminum guardrail for residential use

Other examples of miscellaneous steel are:

- post bases
- post caps, tees and ells
- connections for glu-lam girders: saddles, hinges and knee braces
- joist hangers and anchors
- framing anchors
- tie straps
- threaded rods and turnbuckles
- trench and pit frame and covers, safety plate or grating
- access openings for roofs
- curb guard angles and anchor bolts for trenches, pits and other exposed concrete corners

Trade practice designates sheet steel as flat steel less than $3/16$-inch thick. Sheet steel may be hot-rolled or cold-rolled. You can identify hot-rolled steel by the presence of mill scale on the surface. This is common for all hot-rolled steel. Cold-rolled steel has a bright oiled finish. Plates are flat steel $3/16$-inch or thicker.

Connections

Types of Connections

The most common types of connections are:

- Single and double clip angles
- Angle seats

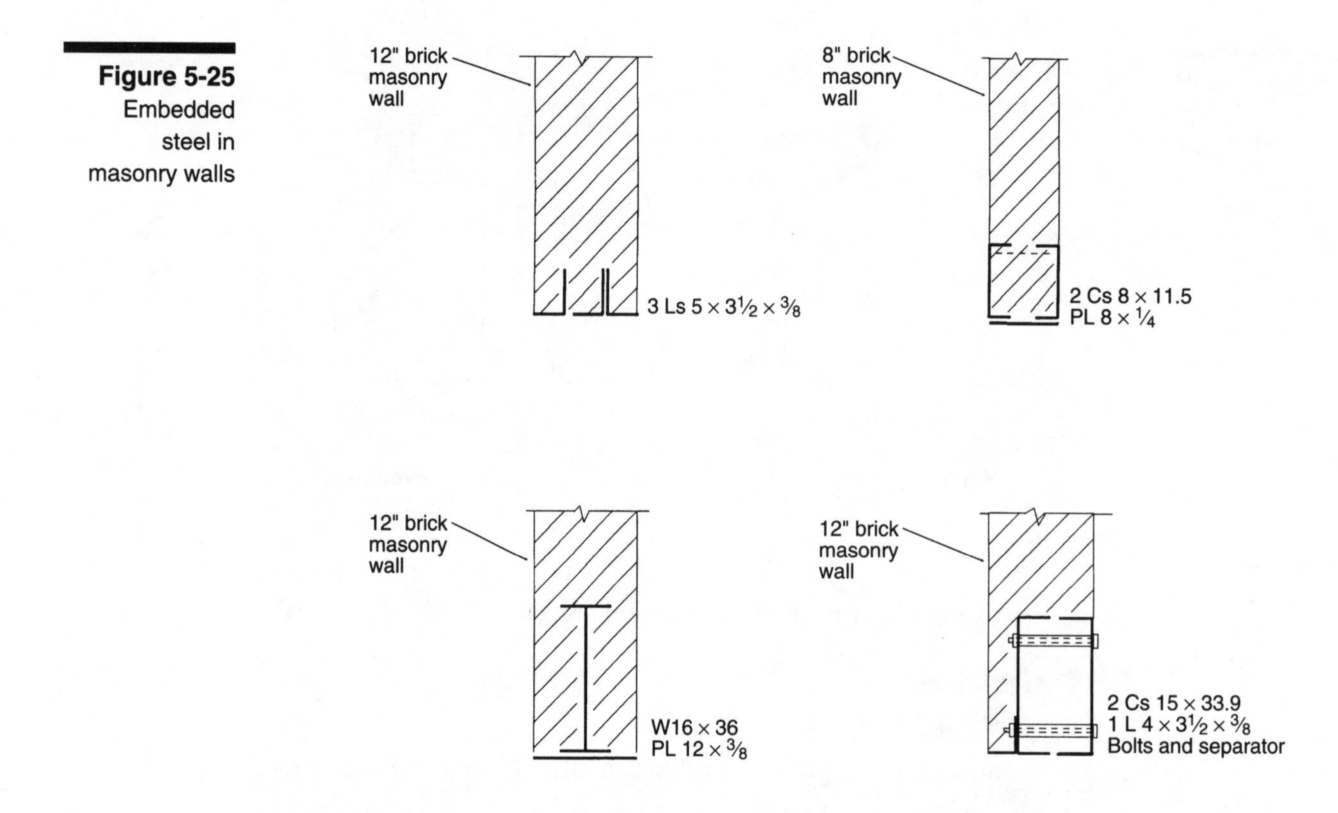

Figure 5-25 Embedded steel in masonry walls

- Pinned, rigid, or moment-resisting connections
- Embedded connections. See Figure 5-25 for typical embedded steel in masonry walls

Fasteners used in metal framing, siding, and roofing include self-tapping screws, machine and high-strength bolts, and welding. Self-tapping screws have hexagonal heads so you can power-drive them in. They are chrome or cadmium coated to help keep them from rusting. Seal the holes you make for self-tapping screws with a neoprene washer attached to the head of the screw.

Bolts

The major portion of the cost of steel fabrication and erection is in the connections. You'll save money if the connections are easy to make. For lightly-loaded members, use a single plate connection, as shown in Figure 5-26A, instead of a conventional two-angle connection, as shown in Figure 5-26B. A double-angle connection may be six times stronger than you need.

If you have to use bolts of varying sizes or off-normal gauge line, they'll cost more than bolts of the same size that are on gauge. If you use more of the same diameter and type bolts, there's less chance of a mix-up in the shop or field. This also makes the drilling and punching operations easier. If you have a member bolted and welded at the shop, it'll cost more to fabricate.

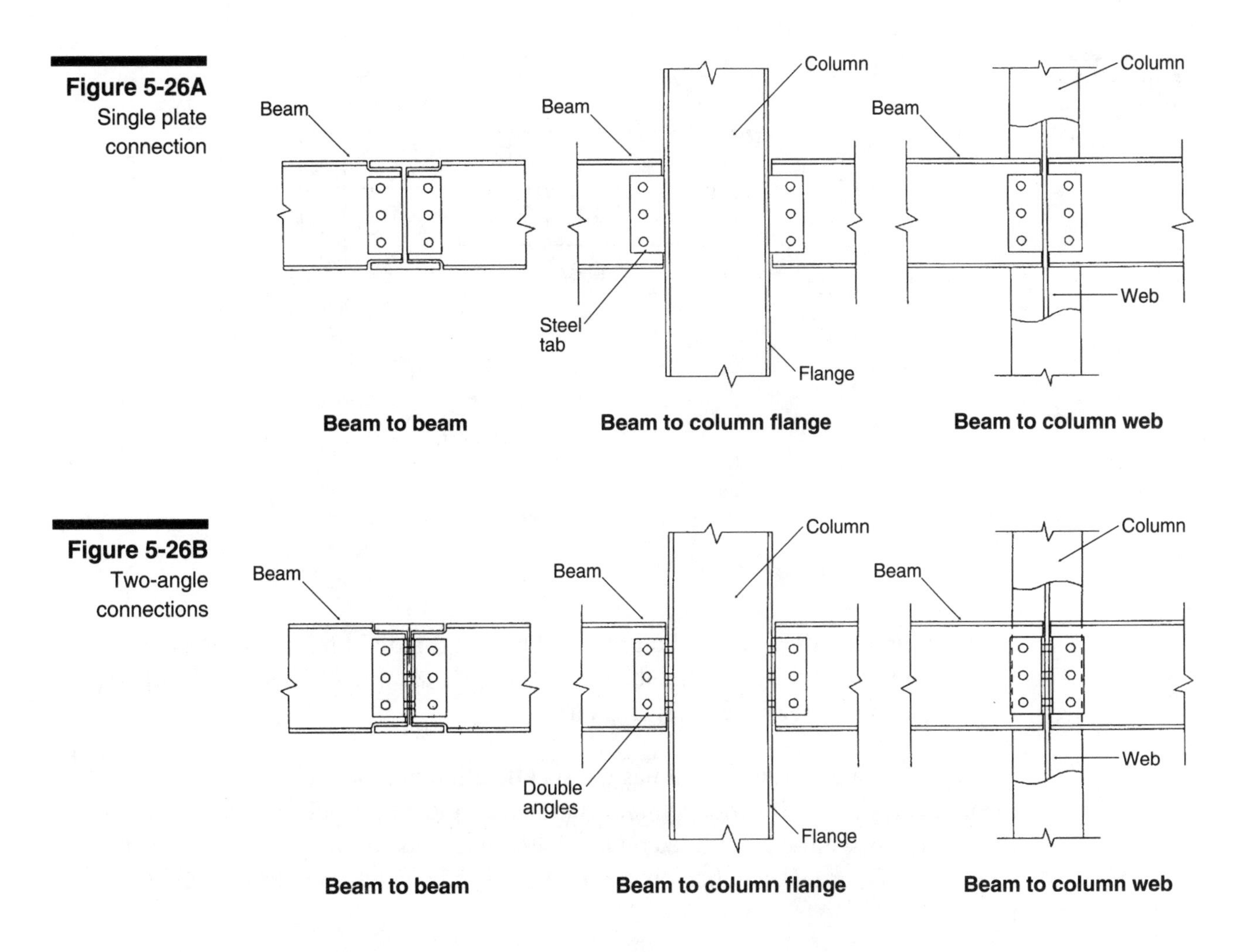

Figure 5-26A Single plate connection

Figure 5-26B Two-angle connections

Common A307 bolts, also called unfinished, machine, mild steel, or rough bolts, are the cheapest. Usually you use them on secondary connections in the shop or field. Figure 5-27 shows the important parts of a machine bolt.

You should use high-strength bolts on primary and critical connections. Primary connections are those holding the main, or primary, framing members together, such as columns, girders, and beams. Secondary connections are connections between members which carry loads to the primary members. These include purlins, girts, struts, diagonal bracing, wind bents, knee braces, headers, jambs, sag members and other miscellaneous framing.

High-strength bolts are bolts made of steel having a tensile strength in excess of 100,000 pounds per square inch. Tensile strength is the longitudinal pulling stress a material can bear without tearing apart. Some high-strength bolts are identified as ASTM A325 and A449. The A325 high-strength bolt is made of carbon steel and the A449 is made of alloy steel. Each is made in three types. The A325 is made from medium carbon steel, low carbon martensite steel, and weathering steel. The A449 bolt is made in Type 1 alloy steel, Type 2 low carbon martensite, and Type 3 weathering steel.

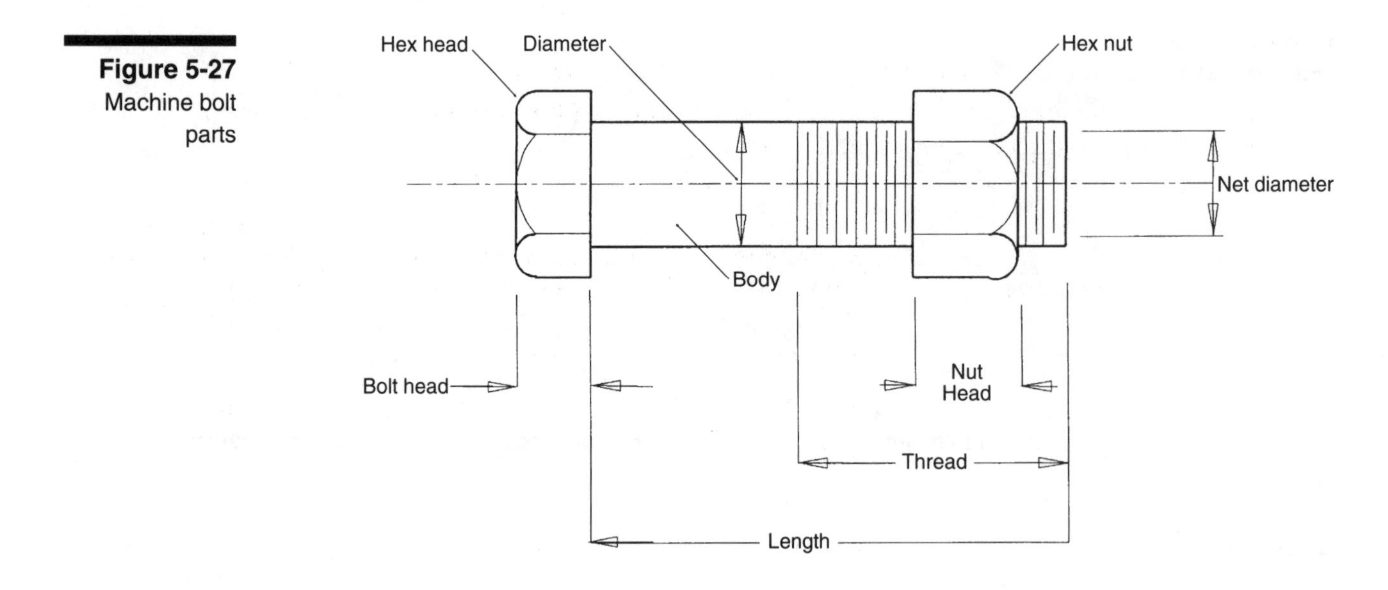

Figure 5-27 Machine bolt parts

You can identify an A325 bolt by the three short radial lines on its head, the specification number A325, and the manufacturer's identification symbol. An A325 high-strength nut has three short arcs on it.

An A490 high-strength bolt has the specification number A490 and the manufacturer's identification symbol stamped on the bolt head. The manufacturer's identification symbol is stamped on the A490 nut. Nuts for high-strength bolts have a *1, 2, 3, 2H, D*, or *DH* stamped on them. These symbols describe the specific type of nut.

High-strength bolts are either friction or bearing type. Friction type bolts depend on the surface friction between the connected parts to make them hold. They are better for dynamic loading, load reversal, fatigue loading, or oversize holes. Dynamic loading is impact loading such as an earthquake or elevator support. Load reversal occurs when a structure shakes or oscillates back and forth. Fatigue loading is a repeated variation of live load. These are classified by the number of cycles of tension and compression stress reversals in 25 years. Bearing type bolts depend on the bearing pressure between the bolt shank and the surface of the connected parts to make them hold.

Don't paint adjoining surfaces when using friction type high-strength bolts. Bearing type bolts are more common to use as they're more economical, can take higher stresses, and have fewer restrictive paint requirements.

Don't make any bolt holes more than 1⁄16 inch larger than the bolt size. Deburr holes for high-strength bolts. When bolt holes don't match, don't indiscriminately make them bigger with a cutting torch. Have an engineer redesign the connection.

Tension in a bolt is related to the torque you use to tighten the bolt. For example, if you apply 320 pound-feet to a 3⁄4-inch diameter bolt, you'll produce about 25,600 pounds of tension. You can use a calibrated manual torque wrench or a power torque wrench to put a specific amount of tension on a high-strength bolt to tighten it. The amount of tension you need will depend on the size of the bolt. Here's a typical table showing bolt tension and torque:

Bolt size (in)	Recommended bolt tension* (lbs)	Required minimum bolt tension** (lbs)	Approximate torque for minimum bolt tension*** (lb ft)
½	12,500	10,850	90
⅝	20,000	17,250	180
¾	29,000	25,600	320
⅞	37,000	32,400	470
1	49,000	42,500	710

* Approximately 15 percent in excess of the required minimum bolt tension.
** Equal to 90 percent of the minimum proof load of bolt (ASTM 325). There is no maximum bolt tension.
*** Equal to 0.0167 lb ft per inch diameter per pound tension for nonlubricated bolts and nuts. Values given are experimental approximations. If you're measuring torque rather than tension, determine the torque-tension ratio by actual condition of the application.

The manufacturers of high-strength bolts test their bolts and they can provide tables for you on the results of these tests.

There are several methods you can use to install high-strength bolts:

- Tension-indicator method
- Turn-of-the-nut method
- Load-indicating washer method
- Torque-control or twist-off nut method
- Turn-of-the-bolt method

You can use the tension-indicator method with a torque-tension table and several types of tools. Some of these tools are:

- manual torque wrench with adjustable presetting
- torque-safe wrench which slips when you reach the preset torque
- electronic torque wrench with a built-in computer. Torque is measured by an electrical signal generated by two pairs of battery or AC-powered strain gauges.
- large face dial indicating torque wrench

When you use the turn-of-the-nut method, you turn the nut a specific number of times from a snug-tight position, corresponding to a few blows of an impact wrench or the full effort of a man using an ordinary spud wrench. An impact wrench is an electric or pneumatic device used to tighten nuts on bolts. To tighten high-strength bolts, one man holds the bolt head with a spud wrench while the other tightens the nut with the impact wrench.

When you use the load-indicator washer method, the load-indicator washer will show you when you've squeezed the nubs on the washer enough to put the tension you need on the bolt to tighten it.

When you use the torque-control method, you use a torque wrench. A torque wrench is a tool with an adjustable mechanism for measuring and controlling the amount of torque or turning force it produces. If you use a manual torque wrench, you read the torque on the wrench dial. If you use a power torque wrench, you should find out from the manufacturer how to calibrate it. With this method, when you give the bolt the proper torque, the splined end will separate from the bolt shank.

When you use the turn-of-the-nut method, you tighten the bolt to a snug-tight fit and then rotate the nut or bolt another fraction of a turn. How much of a fraction will depend on the diameter and length of the bolt. For example, if the bolt is less than 4 times its diameter, make a ⅓ turn; between 4 and 8 times its diameter, make a ½ turn; over 8 times its diameter, make a ⅔ turn.

When you use the twist-off nut method, you can control the proper tension on a specially-manufactured bolt by twisting off the projecting shank of the bolt. When you reach the desired torque, the splined end of the bolt will separate from the bolt shank.

Although the AISC doesn't say that you have to use hardened washers, you'll get a more accurate torque reading if you do. You should use hardened washers for A490 high-strength bolts, but they're not required for A325 high-strength bolts. Don't use lock washers on high-strength bolts.

The strength of a bolted connection depends on whether the bolt has a single or double shear force acting on it. Bolts are stronger against double shear forces than single shear forces. An example of single shear force on a bolt is where the bolt holds two steel plates together that are being pulled apart in opposite directions. If the bolt fails due to this shear force, it'll be cut in two as shown in Figure 5-28A. An example of a double shear force is where there are two plates on one side and a single plate on the other that are being pulled apart in opposite directions. If the bolt fails here, it'll be cut in two places, as shown in Figure 5-28B.

Figure 5-28C shows how a bolted connection fails due to a bearing force. The metal of the plate in contact with the bolt is squeezed by the high bearing force. If the bolt fails, it'll be deformed from a circle to an oval. Figures 5-28D and E show tension and friction failures in a bolted connection.

The minimum thickness required to develop full shear strength of bolts is shown in Figure 5-29.

■ Welding

The most common welds are:

- Fillet weld
- Square groove butt joint, with one side, two sides, or a backing strip

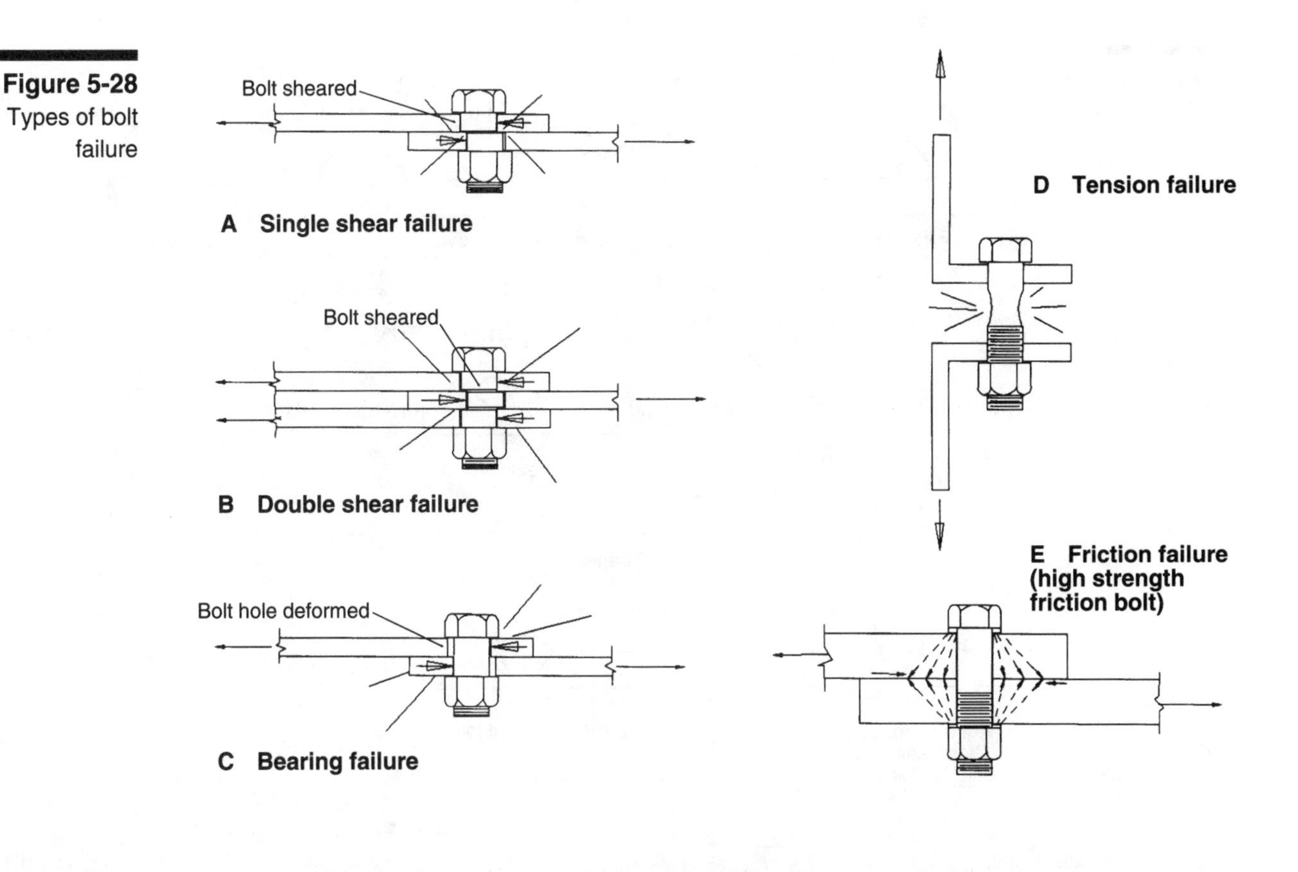

Figure 5-28 Types of bolt failure

Figure 5-29 Minimum plate thickness to develop full bolt strength

Bolt size	Single shear	Double shear
3/4"	.294	.471
7/8"	.343	.549
1"	.393	.628

- V-groove butt joint, with or without a backing strip
- Single bevel groove weld
- Double bevel groove weld
- Single J-groove weld
- Double J-groove weld
- Single U-groove butt weld
- Double U-groove butt weld

Symbols and illustrations of these welds are shown on Figure 5-30. A fillet weld has a triangular cross section and is usually used to connect two surfaces that are 90 degrees from each other. A square groove butt joint weld holds the squared edges of two plates together. The weld can be applied from one side or both sides of the plates. The backing strip allows the weld to fuse both edges of the plates together fully. The V-groove butt weld is similar to the square groove butt weld except that the edges of both plates have been beveled by cutting or grinding. The other welds are variations of the groove welds.

Figure 5-30
Typical welds

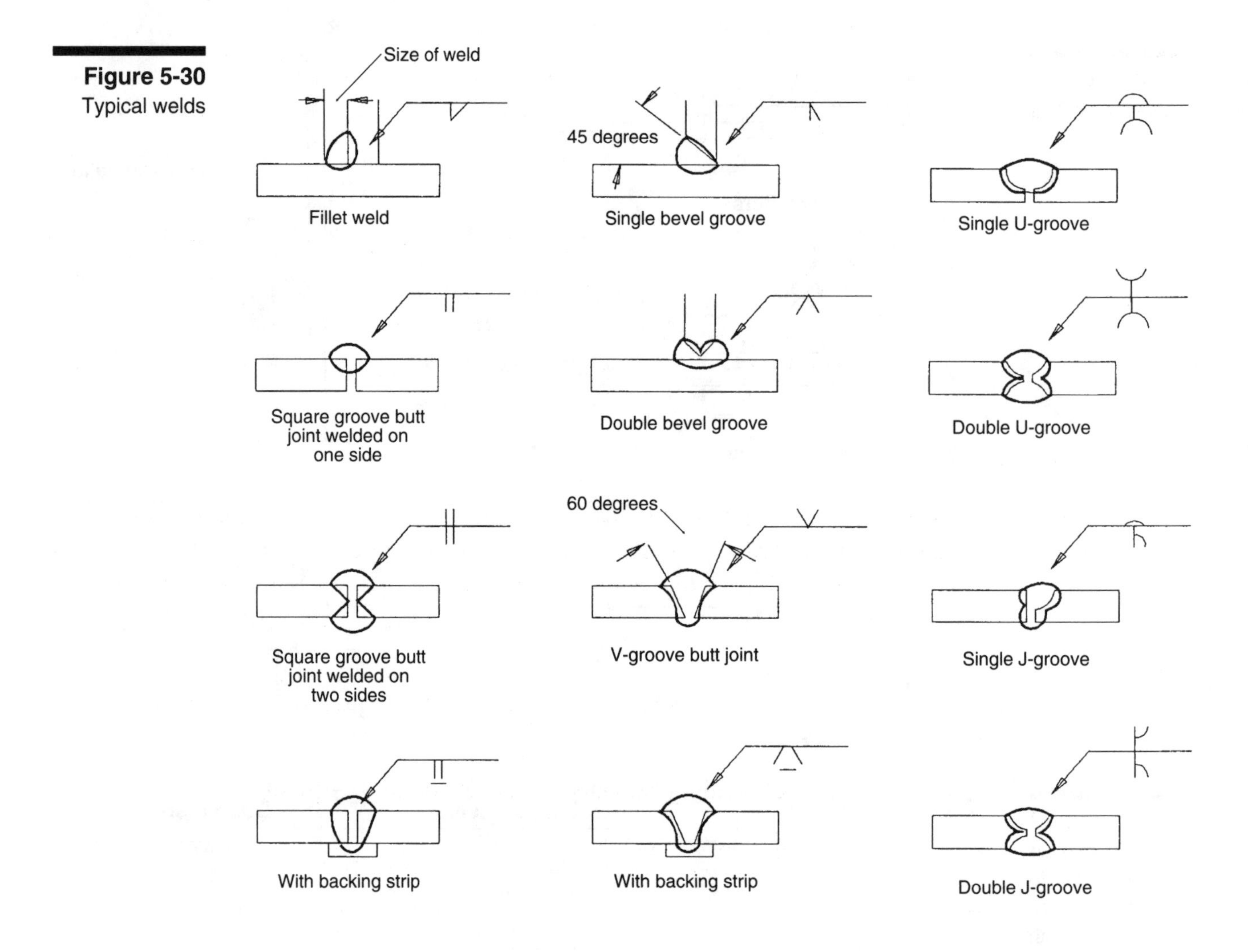

You'll recognize these welds on design drawings by the special symbols the steel industry uses for them. Each weld symbol tells the location, size, type, and whether it's a field or shop weld.

Figure 5-31 shows the standard location of the most common elements of a welding symbol. The horizontal line is the basic reference line. It's always connected to a diagonal leader and an arrow which points to the joint to be welded. The arrow may be on either side of the reference line. If the weld will be done in the field, there will be a small flag where the reference line and the leader meet. If the weld is done in the shop, there's no flag. A small circle where the reference line and the leader meet means that the weld should go all around the joint. A tail at the opposite end of the reference line means a note or specification has been added about the weld.

The basic weld symbol is near the center of the reference line. It may be above, below, or on both sides of the reference line. When it's above the line, make the weld on the side of the joint that's opposite the leader arrow. When the symbol is below, make the weld on the same side of the joint as the leader arrow. When the symbol is above and below the reference line, make the weld on both sides of the joint.

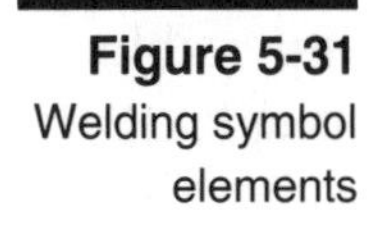

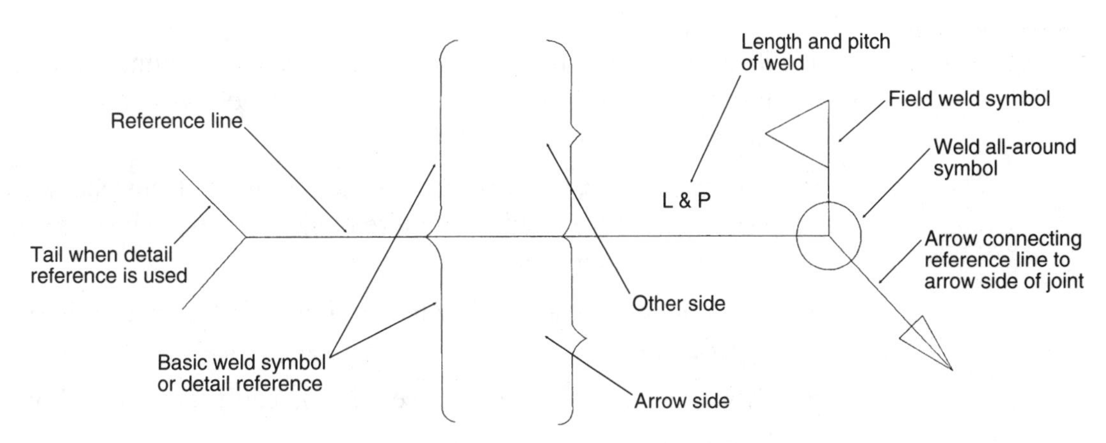

Figure 5-31
Welding symbol elements

The size, weld symbol, length and pitch (distance between centers) of weld should be read, in order, from left to right along the reference line. The perpendicular line of the fillet weld, V-groove, U-groove, and J-groove weld should always be on the left side.

For an explanation of other information that can be shown on a weld symbol, see the *Manual of Steel Construction* of the AISC or the manual of the American Welding Society (AWS).

The total cost of welding a connection is the combined cost of:

1) pounds of weld material applied
2) size of weld rod
3) amperage used
4) weld position
5) arc time
6) time to move in and set up the welding machine
7) building the scaffolding
8) use of a welder-helper
9) removal of welding machine and scaffolding

Make sure the designer hasn't overdesigned the welds. Oversize welds that you don't need just add to the cost of fabrication and erection. Here are some basic rules for economical welding design:

- Don't specify more than a minimum size weld.
- State how much weld is needed.
- Don't state "weld all around" when it's not needed.
- If possible, make fillet welds 5/16 inch or less, or a size that can be made in one pass.

- Specify manual intermittent welds instead of continuous welds, if possible. Intermittent welds cost less when their center-to-center distance is at least twice the length of the weld.
- Where possible, specify partial penetration welds instead of full penetration welds. A groove weld costs more than a fillet weld because of the joint preparation and positioning.
- You only need seal welds where you must have an airtight or watertight connection.
- Don't specify an all-around weld if you can't reach the whole perimeter of the weld.
- Overhead or vertical field welds are more difficult than horizontal and downward welds.
- Note on the plans any welds that need ultrasonic inspection. Ultrasonic inspection is usually required for full penetration welds.

Failures

The most important parts of a steel-frame building are the connections. Most failures in steel framing happen in the connections, not in any undersized beams or columns. When a beam is stressed beyond its design capability, it'll deflect, twist, or buckle. Usually you'll see some warning signs before a member reaches the point of collapse. When a connection fails, however, there's usually little warning before failure. Stresses in a bolted or welded connection can reach, and pass, the yield point of the connection with no warning sign beforehand. You may not find out about a failure in a connection until after that portion of the structure has collapsed. Here's a checklist of things to look for to prevent failures in steel connections:

- loose bolts in high-strength friction-type connections
- flawed or burnt bolt holes
- unburred holes for high-strength bolts
- oversize bolt holes
- unequal stressing in bolts due to poor arrangement. Position bolts symmetrically around the line of force of a connection. If the center of gravity of a bolt formation isn't on the line of force, torque will tend to rotate the connection and increase the shear stress on some of the bolts.
- unequal stressing in gusset plates due to poor layout. Lay out a group of bolts in a gusset plate so the center of gravity of the group is at the intersection of the two lines of force. This is usually where the neutral axis of each of the two members meet.
- excessive bolt flexure. If the shank of the bolt bends between the head and the nut, there is too much bolt flexure on the connection.

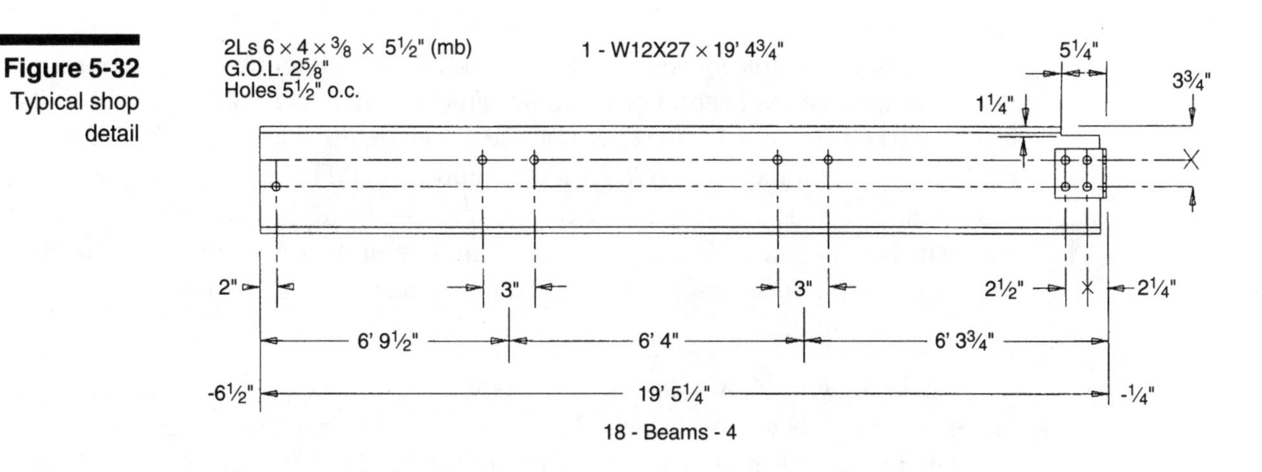

Figure 5-32 Typical shop detail

- excessive bolt tension. If there's too much tension on the connection, the nuts come loose.
- reverse stresses in the connection. If there are reverse stresses on the connection, which may be due to cycles of tension and compression or changes in direction of the shear forces, the nuts come loose.
- secondary stresses due to eccentricity of connection
- used or nonconforming steel installed
- anchor bolts that are incorrect size, length, plumb, or incorrectly embedded
- machine bolt holes that are incorrect number, alignment, location, size, arrangement, or tolerance
- high-strength bolts that are incorrect size, length, type; incorrect washers and nuts used.

Most building departments require that all structural welding and high-strength bolts be inspected continuously during installation by a special inspector. The inspector may require that some of the shop welds be cut to expose the workmanship. The inspector may also require that a few of the high-strength bolts be removed so he can check the condition of the steel surface and the holes.

Fabrication

Shop Drawings

Although many detailing shops use computer-aided drafting methods, the final shop drawing still looks like a manually-drawn sheet. Figure 5-32 shows how to fabricate a single beam.

This detail contains symbols and abbreviations which tell the fabricator certain information about the beam. For example, Figure 5-32 tells the fabricator to make 18 beams marked Beam-4. It also says that the gauge on the outstanding leg of the clip angle at the end of the beam (*GOL* on the figure) should be 2⅝ inches. The *mb* on the figure stands for the clip angle. The dimensions shown at the ends of the beam (6½ and ¼ inches) are the distances to the working line of the two connecting members. The two working lines are 20 feet apart. Twenty feet is the sum of 6½" plus 19'5¼" plus ¼".

A shop drawing shows details of each individual piece of steel, whether it's a major column or beam, or a small clip angle. It shows the cutoff length of the member, all cutouts, holes for bolts, and surfaces that will be milled or painted. A shop drawing also has a bill of materials that lists every item of steel fabricated and shown on that particular sheet.

The shop detail follows the design drawings but allows for fit-up, dimension, clearances, cutouts, matching holes, and all bevel and diagonal dimensions. Fit-up of a structural frame is the initial loose connection of the members. During fit-up, the bolts are usually tightened one turn until the building frame has been plumbed and in alignment. After the joint is bolted up, the fitting-up bolts should be loosened and retightened by one turn of the nut.

Usually, beams or other members are shipped with some markings on them. One set of markings uses a prefix number for the detail sheet number, a letter for the type of the member, and a number for which place the member has on the sheet. For example, on a shipping piece number 2B21, the B stands for a beam, the 2 before the B says it's on sheet 2, and the 21 says it's the twenty-first member on the sheet. A piece marked 3G10 is a girder that's the tenth detail on sheet 3.

Symbols and abbreviations

On a shop drawing:

- a heavy solid continuous line means an edge that you can see on an object
- a dashed line means an edge that you can't see on an object
- a fine continuous line means a dimension or an extension to a dimension
- a dash-dot line means a centerline
- an irregularly-drawn line means a break line

A drawing also shows how an object looks when you look at it from different angles. These are called orthographic views. Usually they look at the object from the top, front, bottom, left end, and right end. To show more about the inside of an object, a view may show how it would look if you cut through it and looked at the cut edge.

If there are any questions that the owner, design engineer, or general contractor need to answer, they should be circled and marked with something like "clarify" or "verify."

Approval of shop drawings

It's a good idea for the fabricator to have the owner, designer, and builder approve the shop drawings before steel fabrication begins. This is the last chance to make sure there are no misunderstandings about what the drawings mean. And it's the last chance to find any design and drafting errors in dimensioning and clearances.

Erection drawings

The shop detailer also makes erection drawings. These are similar to design drawings. A mark on each member, such as B-2, G-4, or C-7, identifies the member for fabrication and erection. On the erection plans, the mark is noted on the same end of the beam as it's painted on the fabricated piece. This helps the erector install the member properly. Usually the structural engineer gives the shop detailer transparent or reproducible prints of the steel framing drawings to use for the erection plans.

Bill of materials

A bill of materials is a list of steel that the fabricator must buy to fabricate each structural member. The base price of steel varies by its weight and size. Beams may be ordered cut to length or in full lengths. Warehouse steel is steel stored in the distributor's warehouse. It can be ordered and delivered rapidly to the fabricator. Mill steel has to be ordered from the rolling mill. It may cost less than warehouse steel but may also take longer to get. It's cheapest if the fabricator buys the steel in standard lengths of 20, 40, or 60 feet. Steel that's cut to length by the warehouse includes a cutting charge. If the warehouse keeps the remnants, there's another charge because remnants must be stored and they're harder to sell. When ordering steel shapes from a supplier, there are several factors to consider. The steel supplier will provide a price list of all types of steel shapes. Each item contains a base price per 100 pounds, or cents per pound, subject to quantity extras.

Certified Steel Fabricating Shop

All steel fabricating not done at the job site should be done by a certified steel fabricating shop. Then you won't need to have a special inspector around while you weld or bolt the connections in the shop. Most building codes have this requirement.

Shop equipment

A typical steel fabricating shop may have the following equipment:

- shear — a guillotine-type machine used to cut plate
- cutting torch — a gas flame used to cut steel plate
- rolls — used for curving bars, angles, and plates
- center punch — used for starting bolt holes
- detail punch — used for bolt holes

- multiple spindle drill — used for groups of bolt holes
- mill — used to smooth and even column end surfaces
- press brake — to make angles in wide sheets and plates

Shop procedure

As a builder, you should know what the steel fabricator does. Here's a brief description of the steps that happen after the subcontract between the general contractor and the steel subcontractor is signed, and the steel subcontractor gets the design drawings.

Engineering work:

1) Prepare advance billing for ordering material, especially items that take a long time to get
2) Prepare erection plans using the design drawings if possible
3) Prepare a system of marks, index sheets, and sheet numbers of shop drawings
4) Prepare typical details, layouts, and calculation sheets for determining the lengths of members
5) Check detail drawings
6) Write the bill of materials
7) Get approval of the drawings from the customer
8) Itemize field fasteners and weld electrodes
9) Calculate material weights
10) Prepare shipping bills for the finished parts

Shop work:

1) Handle and cut steel
2) Make templates
3) Lay out steel
4) Punch and drill bolt holes
5) Straighten steel members
6) Bend and roll curved plates or shapes
7) Fit-up temporary bolted pieces for shipping
8) Fasten, bolt, and weld
9) Finish
10) Inspect
11) Clean and paint
12) Ship to job site

If the steel won't be exposed, it may not have to be painted at the shop. If it will be exposed, you can have it painted at the shop to give it temporary protection. If you want permanent protection, give it a field finish painting. It's a good idea to galvanize steel that's exposed to the weather. Try not to weld galvanized steel at the building site. Not only will you have to grind the zinc coating off the places you're going to weld, but the welding will give off toxic fumes.

Erection

A basic steel erection crew may have an ironworker foreman, an ironworker, an operating engineer for heavy equipment, an oiler, and a laborer. You should follow these steps in the erection of a steel-frame structure:

- Brace the frame of the steel structure so that the structure remains true and plumb at all points.
- As the work progresses, securely bolt or weld the structure to carry the dead loads, wind loads, and erection stresses.
- Don't permanently bolt or weld a structure until it's been properly aligned.
- Wire brush and clean all areas that will be field welded to remove all shop paint and reduce any film to the minimum.
- Leave the bracing in place as long as required for safety.

You shouldn't remove temporary bracing until you've completed the structural frame and it's been accepted for plumbness, level, and alignment. You don't need the temporary supports in a self-supporting structure after the self-supporting elements are finally fastened. You can remove the temporary supports of a nonself-supporting structure when the necessary nonstructural elements, such as masonry walls, are complete. Don't remove temporary supports without the consent of the erector.

Erecting a pre-engineered metal building is similar to putting up a wood building. You cut the secondary framing members, roofing, and siding in the field. You use portable shears, punches, and drills to fit-up the building. Use portable rolls to form the roofing and siding from flat sheets at the site. Portable rolls are machines used on the job site to cold form roofing and siding. Use shearing devices to cut steel, aluminum, fiberglass, and plastic formed panels. To decrease erection time, use high-speed pneumatic screw-nailers to fasten heavy-gauge sheet metal. Use cordless drills and screwdrivers with detachable energy packs instead of manual tools.

It's very important that you place the foundation anchor bolts accurately. Make holes in the base plates at least 3⁄16 inches oversize. Use leveling nuts on pairs of anchor bolts to help keep the column base plate at the proper elevation. Don't use leveling nuts on base plates with four or more anchor bolts. This can make an uneven restraint on the plate.

Give the inspector access to all parts of the work as it's being done. The *Uniform Building Code* says that all field welding and installation of high-strength bolts must be continuously inspected by a special inspector. The inspector may want to check the mill and test reports to confirm the quality and grade of steel.

Some common problems in steel erection are:

- Bolt holes that don't fit. You'll have to dismantle parts of the frame and refabricate them.
- Improper field welding with inclusions, voids, cracks in welds. Inclusion is foreign material within the weld, such as a nail or bolt.
- Field bolting of high-strength bolts with inadequate torque, painted surfaces, and unburred holes.
- Missing bolts.

Summary

Some designers tried to reduce steel weight to lower costs with little or no regard to the effect on shop costs. As the price of material decreases, the cost of shop labor may increase. For example, it may be more economical to slightly increase the weight of a member to avoid shop fabrication of stiffener plates.

You can also save by using single clip angles instead of double clip angles for beam connections. For example, a W24X55 beam 30 feet long with seven rows of bolts in a single clip angle as opposed to four rows of bolts in a double clip angle would be a savings of $10 per beam for fabrication and $15 per beam for erection. That's $25 per beam.

Here are some other tips for reducing costs:

- The cost of steel construction is approximately one-third for material, one-third for fabrication and one-third for erection. Since this relationship varies, you should keep up with the current prices to determine the best way to save costs.
- Use all of the allowable stress increases permitted by the building code for short-term loads such as earthquake, wind and live loads.
- Contact your steel supplier to avoid specifying sections that are not readily available and seldom rolled.
- Use composite design where possible. Don't give the concrete slab a free ride on a steel frame. Make the concrete carry its share of the load.
- Select the most economical bay dimensions. A length-to-width ratio of 1.25 to 1.5 and a bay area of 1000 square feet is recommended. The interior members should span the long direction.
- Use oversize or slotted holes where possible to facilitate fit-up and erection time.

- Avoid unnecessary stiffeners.
- Space the floor beams to avoid shoring of the decking.
- Avoid unnecessary full penetration welds.
- Use the same size section rather than varying them for a small savings in weight.
- Detail all nonstandard connections. Don't leave this up to the shop detailer.
- Don't release incomplete plans or plans with lots of addenda.
- In multistory buildings, set column splices about 3 feet above floor level to allow attachment to safety cables.
- Flag all revisions on drawings.
- Always make tolerances compatible with the type of construction. Remember masonry work has greater tolerance than steel.
- Where possible, use single pass fillet welds, $\frac{1}{4}$ or $\frac{5}{16}$ inch.
- Use uniform-sized bolts.
- Use uniform spacing of anchor bolts.
- When designing a steel structure, pretend that you have to do the shop detailing yourself. Keep it simple.

Remember that the main purpose in structural design is to select members that will safely and economically carry anticipated loads. The responsibility for the design depends on who prepared the drawings and specifications. When you, the builder or the owner, provides the design, plans, and specifications, the fabricator and erector aren't responsible for the suitability, adequacy, or legality of the design. Also, the fabricator isn't responsible for the practicability or safety of erection if the structure is erected by others. If you want the fabricator or erector to prepare the design, plans, and specifications or to assume any responsibility for the suitability, adequacy or legality of the design, you should clearly state so in the contract documents (Ref. 1.51 Standard Specifications AISC).

CHAPTER 6

Masonry

Building with masonry has many advantages. It's durable, fire- and heat-resistant, sound insulating, and hard to chip. It goes up quickly and easily because no forms are needed and you can mix the mortar and grout on the job. That makes masonry the lowest-cost material for many applications. Also, masonry needs little or no maintenance. Architects and designers like masonry because it offers so many choices of texture, color, style and pattern.

There are, however, a few disadvantages to masonry construction. Long masonry walls tend to crack when they expand or contract with temperature changes. Vertical expansion joints in a long wall help avoid this problem. Masonry is also a heavy material that needs a stronger building frame in earthquake-prone areas. The horizontal shaking of an earthquake will tend to bend a wall — but masonry walls don't bend. Although masonry walls have high compressive strength, they're weak in flexing. So, in earthquake zones, steel reinforcement is required in all walls.

Masonry Basics

Masonry units are used in:

- loadbearing walls, both below and above grade
- nonbearing walls, both interior and exterior
- fire and curtain walls
- retaining and garden walls
- piers and columns
- chimneys

Types of Masonry

Modern building codes classify the different types of masonry as:

- stabilized unburned clay masonry, or adobe block masonry
- stone masonry
- cavity wall masonry
- hollow unit masonry, also called hollow concrete unit masonry, hollow load-bearing concrete unit masonry, and concrete block masonry
- solid brick masonry
- grouted brick masonry
- reinforced grouted brick masonry
- reinforced hollow unit masonry

Solid brick masonry is made from fired clay or shale brick. There are two basic types of clay brick — building brick and face brick. Building bricks come in various sizes and textures, and many shades of red. Face bricks are made with a controlled mixture of clays and shales. There are minerals in these bricks which give them color when they're fired in a kiln. Face brick must meet a stricter tolerance for color, size, warpage, and resistance to chipping than building brick.

Bricks are also made from a mixture of sand and lime, which are hardened under pressure and heat. Another type of brick is made of cement and aggregate, similar to concrete.

Hollow masonry units are usually concrete block. They are made of cement, sand, and fine gravel aggregate. Porous aggregate is used to make lightweight block. The aggregate may be expanded shale, slag, pumice, or similar material.

Making Masonry

Clay bricks are made by mixing clay in pug mills and then extruding the mud under vacuum. A pug mill is a special mixer used to knead ground clay, shale, additive, and water to make a plastic mix before extruding into brick-sized columns of clay. The extrusion is then cut by wire into required sizes. The units are roasted in tunnel-driers or fired in field kilns at 1950 to 2100 degrees F. Tunnel-driers make a more uniform quality brick.

The surface texture of clay brick may be matte face, rug face, or wire-cut face. Matte face texture is made with light vertical scratches on two sides and two ends of the brick. The other two surfaces have a wire-cut texture. A rug face has a ruffled texture that's produced by making heavy scratches on the brick surface and rolling the excess material back on the face. This texture is on one side and one end of the brick. The wire-cut face texture is on two faces of brick units. Wire-cut faces have exposed aggregate and a ruffled surface. This is caused by the wire cutting through the ribbon of extruded clay.

Another type of clay brick is cored brick. This brick has holes made in its body while the clay is still plastic. The holes make the brick lighter, and give it more sound and thermal insulation.

Solid and hollow concrete units are made in power-tamping machines. The wet concrete is tamped into a mold. The mold is stripped off after the concrete has hardened. The concrete blocks are then cured in wet steam at 125 degrees F for about 15 hours — long enough for them to reach about 70 percent of the 28-day strength. The full compressive strength of concrete is attained after a 28-day curing period. Compression tests are then made on 3 × 6-inch cylindrical or 2 × 2 × 2-inch cube-shape specimens. Concrete blocks that will be exposed to the weather should be made with a concrete mix that has at least six sacks of cement per cubic yard of concrete.

Grout poured into concrete block cores is similar to concrete, except the aggregate is much smaller. Pea gravel is commonly used. This makes the mix fluid so the grout can flow around the reinforcing steel and into the cavities in the masonry. It's more important that grout be fluid than plastic. Grout should be fluid enough for pumping. Grout is usually delivered to the job by truck and pumped by hose to the walls.

Mortar

It's important that mortar that is to be worked with a trowel in layers and that must adhere to the surface of masonry be plastic. Lime is a good plasticizing agent.

The most important properties of mortar are its plasticity and ability to form a strong and durable bond between the masonry units. Other critical factors are workability, compressive and tensile strength, durability, and minimal volume change during climatic changes.

Mortar used to bond courses of masonry is made of a mixture of cement, sand, water, and a plasticizing agent. The cement in mortar may be portland cement or a masonry cement with a plasticizer in it. Use low alkali cement to reduce the possibility of efflorescence (a coat of white salt that may form on the surface of masonry). Use clean sharp sand in mortar. But don't use too much sand. If you do, the mortar will tend to drop off the trowel, and once in place, it won't weather well.

There are various types of mortar: S, M, N, and O. Use Type S for severe exposure masonry work. Use Type M mortar for moderate exposure. Use Type N for no exposure, and Type O for nonstructural work. You must use specific proportions of cement, lime, and sand with each of these types of mortar. The proportions for these types of mortar are listed on the following page.

Type S mortar

1 part portland cement
½ part hydrated lime
4½ parts sand

or

1 part Type II masonry cement
4½ parts sand

Type M mortar

1 part portland cement
¼ part hydrated lime
3 parts sand

or

1 part Type II masonry cement
6 parts sand

Type N mortar

1 part portland cement
1 part hydrated lime
6 parts sand

or

1 part Type II masonry cement
3 parts sand

Type O mortar

1 part portland cement
2 parts hydrated lime
9 parts sand

or

1 part Type I or II masonry cement
3 parts sand

You can convert these to weight proportions by multiplying the unit volume by the weight per cubic foot of the material. As a rule of thumb, use the following densities of the materials:

Masonry cement	94 pounds per cubic foot
Portland cement	94 pounds per cubic foot
Hydrated lime	40 pounds per cubic foot
Mortar sand	85 pounds per cubic foot

For Type S mortar

1 part cement	$= 1 \times 94$	= 94 pounds
½ part lime	$= \frac{1}{2} \times 40$	= 20 pounds
4½ parts sand	$= 4\frac{1}{2} \times 85$	= 382.5 pounds

You'll find a more detailed description of mortar in ASTM Specifications C270.

Quality and Strength

Like other construction materials, the structural properties of masonry are determined by laboratory tests. Grades are established by mutual agreement between the industry and engineering professions. For example, the ASTM Specification C62 gives the standards for building brick, and ASTM Specification C216 gives the requirements for face brick.

Compressive strength is the most important property of masonry. To get flexural strength, you have to use a combination of masonry and steel reinforcement. Masonry alone has very little tensile or flexural strength. The shear strength of masonry depends mainly on the mortar joint. If you use a weak mortar, you'll get an unstable wall.

There are two categories of masonry structural strength: ultimate strength and allowable strength. The ultimate strength, or f'm, is the load in psi that caused the material to fail when it was tested in a laboratory. The allowable strength, or fm, is the maximum stress permitted by code or standard practice. It allows for a factor of safety, varying quality of material, tolerance in construction, and other factors. Typical ultimate and allowable strengths of masonry materials are shown in Figure 6-1.

For Grade 60 reinforcing steel, the yield strength is 60,000 psi and the allowable tensile strength is 24,000 psi. You'll find the accepted quality and strength of masonry materials described in detail in these ASTM Specifications:

Common brick, Grade MW	ASTM C62
Facing brick, Grade MW	ASTM C216
Sand	ASTM C144
Portland cement, Type I & II	ASTM C150
Plastic cement, Type I & II	ASTM C150
Hydrated lime, Type S	ASTM C207
Reinforcing steel	ASTM A15, A305

Figure 6-1 Typical ultimate and allowable strength for masonry

Material	Ultimate f'm (psi)	Allowable fm (psi)
Grout		2,000
Mortar		1,500
Solid clay	14,000 10,000 6,000 3,000 2,500	5,300 4,000 2,600 1,800 1,500
Concrete block: Type N mortar		1,350
Concrete block: Grouted solid		1,500
Concrete brick	3,500	2,100

Shapes and Sizes of Clay Masonry

Bricks come in many different sizes and types. Most buildings use standard or common bricks. In the United States, common bricks are about 2¼ inches deep, 3¾ inches wide, and 8 inches long. This may vary in different locations, and from manufacturer to manufacturer. Oversize bricks are 3¼ inches high, 3¼ inches wide, and 10 inches long. Imperial bricks are 3 × 6 × 16 inches.

Another popular size is the modular brick. It's called modular because it's based on the modular measure of 4 inches. Modular bricks are about 3⅜ inches high, 3 inches wide, and 11⅜ inches long. When you add the mortar joint, the unit becomes 4 × 12 inches. You can combine these bricks in various ways to make up a wall based on the 4-inch module.

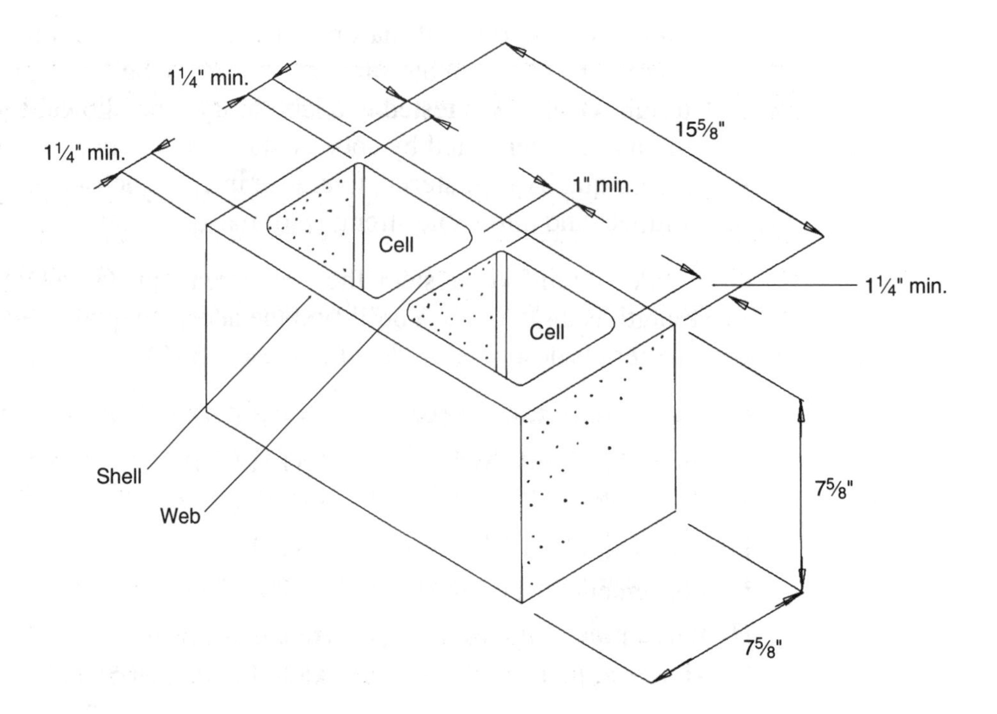

Figure 6-2 Standard concrete block

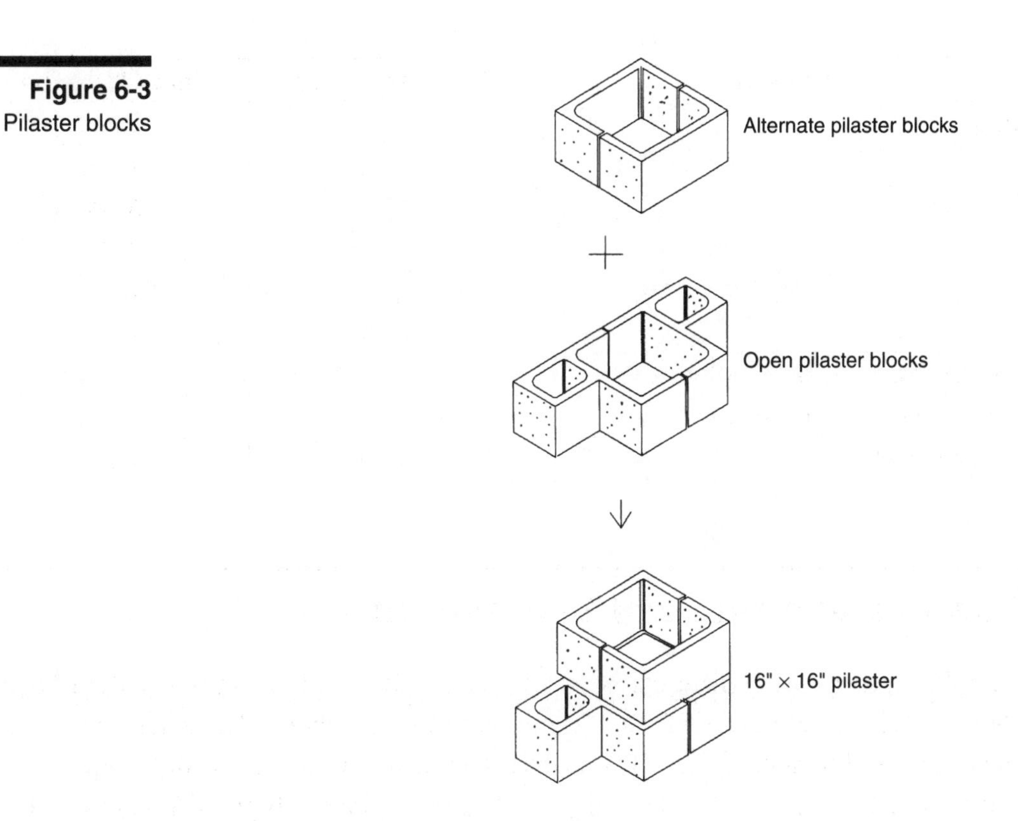

Figure 6-3
Pilaster blocks

For special architectural treatment, you can use English, Roman, and Norman bricks. English bricks are $3 \times 4\frac{1}{2} \times 9$ inches. Roman bricks are $1\frac{1}{2} \times 4 \times 12$ inches and Norman bricks are $2\frac{3}{4} \times 4 \times 12$ inches. The actual dimension of brick varies because of shrinkage during firing.

Shapes and Sizes of Concrete Masonry

Concrete masonry units are made for many purposes. Concrete blocks are used for interior and exterior loadbearing and nonbearing walls, partitions, and backing for veneer. The weight, color, and texture of concrete block depend on the type of aggregate used to make it. Here are the names and uses of some different types of concrete block:

- Standard units are used for walls. Figure 6-2 shows a typical $8 \times 8 \times$ 16-inch unit. Similar blocks are 4 inches high and 6 or 12 inches thick. These are nominal sizes and include $\frac{3}{8}$-inch mortar thickness. The figure shows the net sizes.
- Cap and Paving units are used at the top of walls, and as pavers.
- Pilaster units are used for construction of pilasters, as shown in Figure 6-3. The open pilaster block is also called a "banjo block."

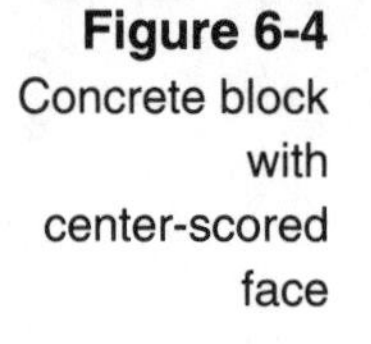

Figure 6-4
Concrete block with center-scored face

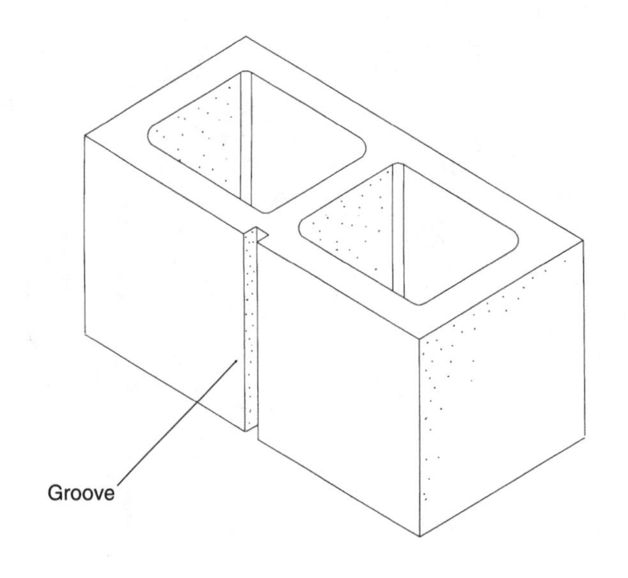

Figure 6-5
Concrete block with offset face

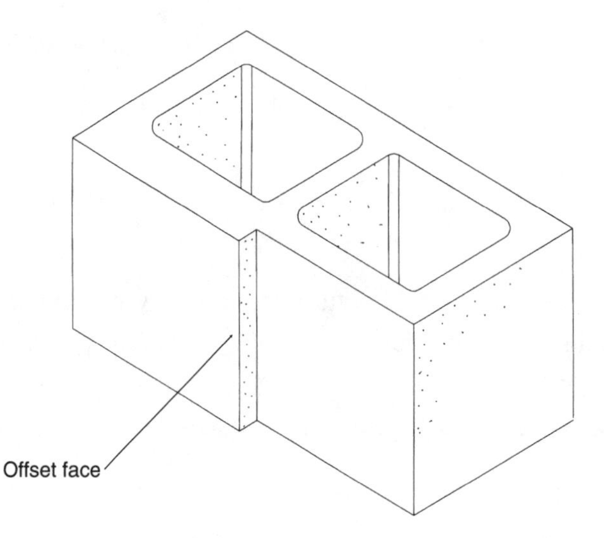

- Split faced units are used because they look like undressed stone, which is stone that hasn't been smoothed. It has a rough quarried finish.
- Center Scored units create a stacked bond wall pattern. This type of block is shown in Figure 6-4.
- Slumped units are used to look like hand-formed clay block.
- Offset Face units make a wall pattern with many shadows, as shown in Figure 6-5.
- Lintel Blocks, also called channel blocks, are used as headers over door and window openings. Figure 6-6 shows a lintel block. The channel shape permits placing horizontal bars over the opening.
- Open End Blocks, shown in Figure 6-7, are used in filler walls when the blocks can't be threaded over existing vertical bars. Filler walls are used to fill the space between steel or concrete columns and spandrel beams in exterior walls.

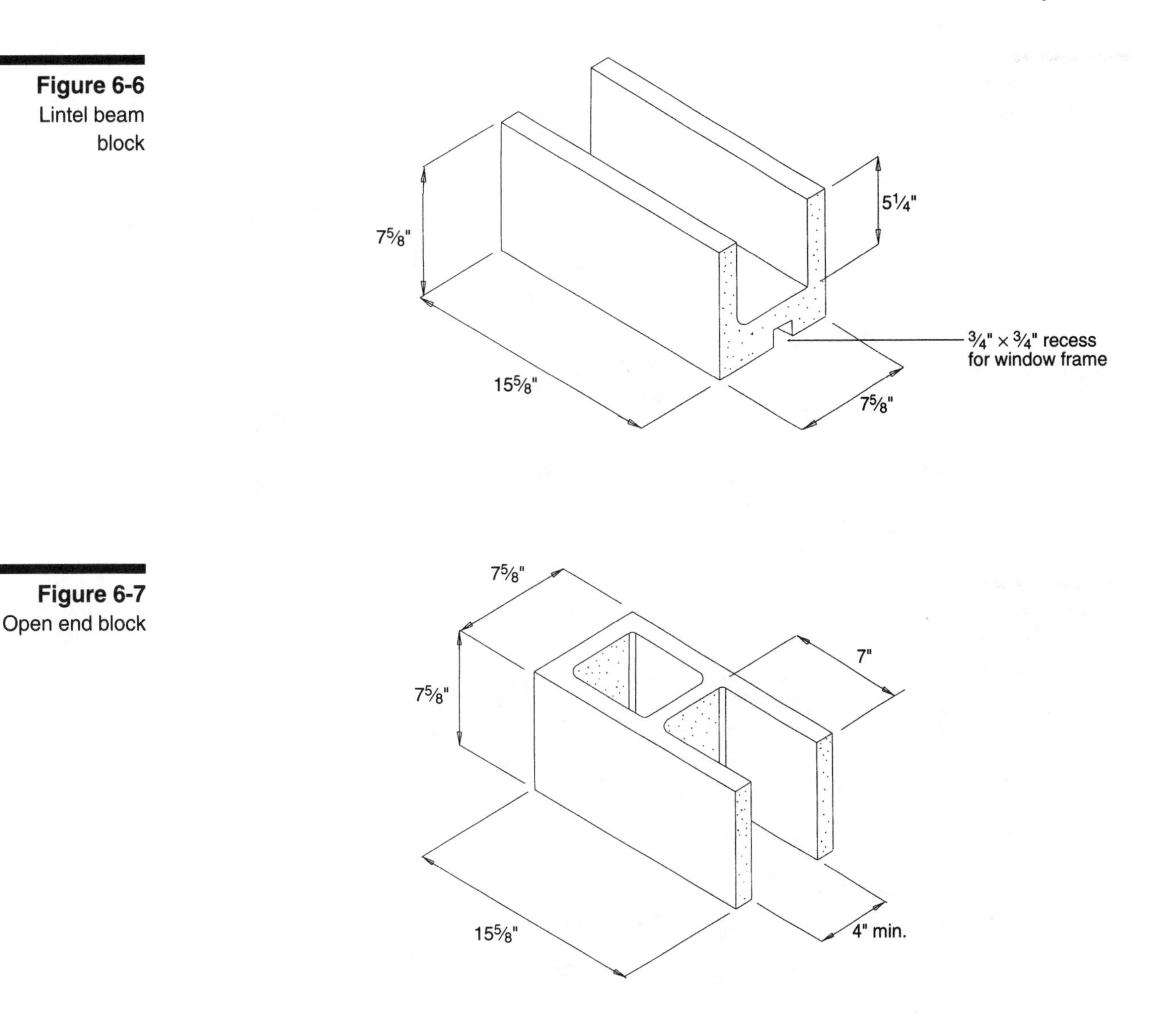

Figure 6-6 Lintel beam block

Figure 6-7 Open end block

- Sash Blocks are used in window jambs. Figure 6-8 shows this block, and the groove the window frame fits into.

- Bond Beam Blocks are used to form a horizontal beam within a block wall. This block is shown in Figure 6-9. The webs have been cut so horizontal reinforcing bars can be put in.

There are many concrete blocks other than these two-cell concrete blocks. Figures 6-10 and 6-11 show examples of other types of hollow concrete units.

Wall heights and lengths in concrete unit masonry and modular brick construction should be in multiples of 4 inches. Then you can use standard masonry units both horizontally and vertically with a minimum of block or brick cutting. Window and door openings should also be in multiples of 4 inches.

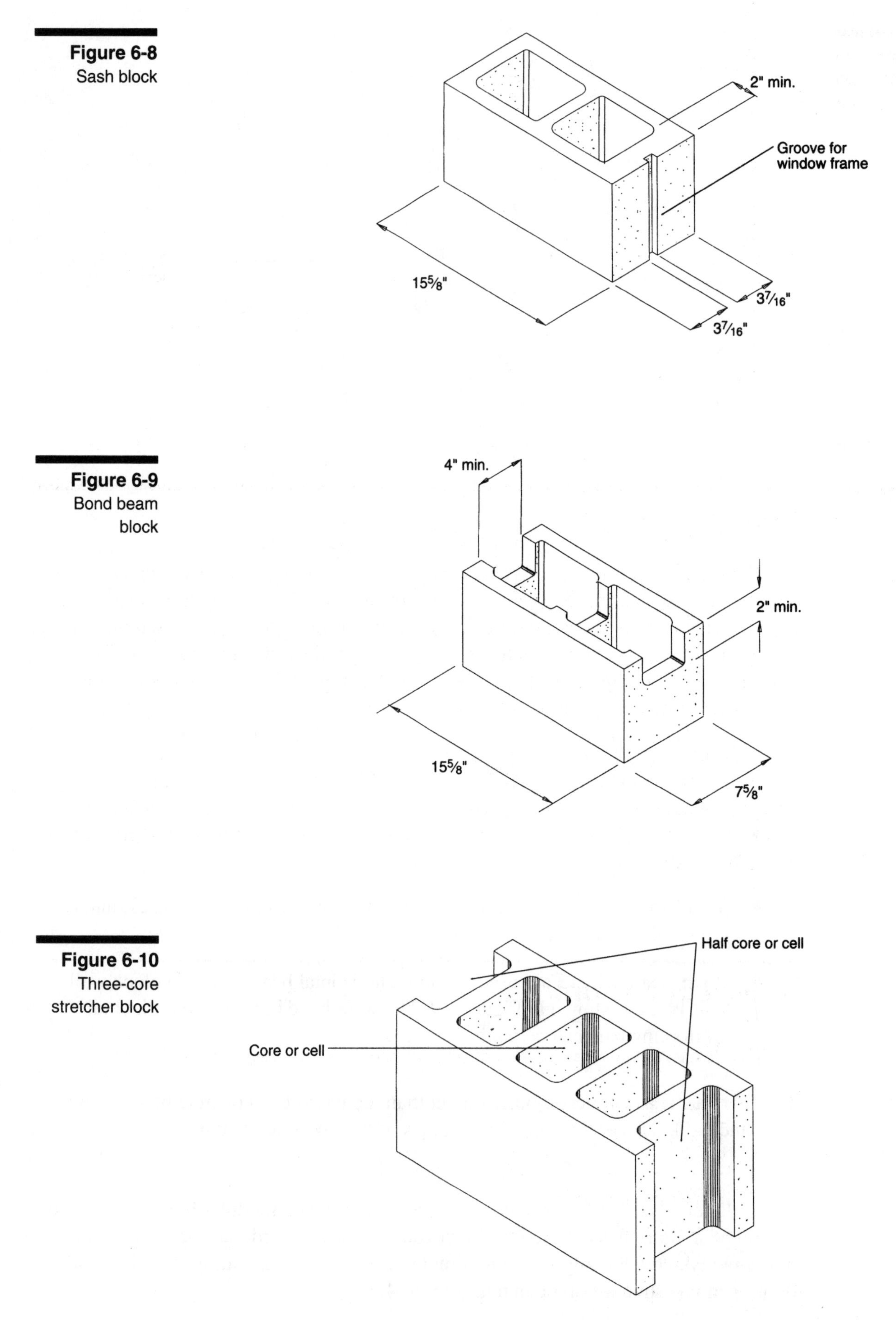

Figure 6-8
Sash block

Figure 6-9
Bond beam block

Figure 6-10
Three-core stretcher block

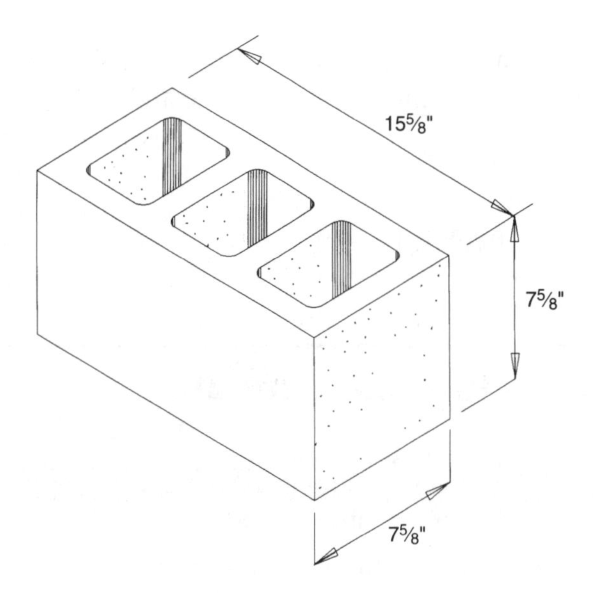

Figure 6-11 Three-core pier block

Basic Masonry Design

There are two basic types of design for masonry buildings, those which are designed to resist earthquakes and those which are not. Masonry buildings have been built for centuries in nonearthquake areas by rule-of-thumb formulas based on trial and error. The thickness of a wall can be based on wall height, and walls can be designed to carry only vertical loads. Reinforcement is added to the walls mainly to prevent cracking due to temperature changes.

In the early 1900s, when earthquakes struck in many areas of California, earthquake-resistant design in brick buildings became very important. Since then, design standards have been upgraded and revised after every major earthquake. If you live in a seismic zone, check with your building official before beginning any masonry construction.

Design Factors

The design of masonry construction uses these factors, all expressed in psi:

f'm = Ultimate compressive stress
fm = Allowable compressive stress
u = Bond stress between grout and reinforcing bars
fa = Computed axial unit stress
Fa = Allowable axial unit stress
fb = Computed unit bending stress
Fb = Allowable unit bending stress

A basic formula for finding the allowable stress in a masonry structure is that the ratio of the computed stress to the allowable stress is less than 1, or expressed as:

$fa/Fa < 1$

and

$fb/Fb < 1$

Code Requirements

Some of the codes and standards regulating masonry construction are:

- Uniform Building Code
- Uniform Building Code Standards
- ASTM Quality Control Standards
- Masonry Institute of America
- BOCA Building Code
- Standard Building Code

Building codes usually specify when masonry work must be inspected. There are generally two types of inspections, required or called, and special. For required inspections, you shouldn't cover the reinforcing steel or structural framework of any part of a building before you get the approval of the building inspector.

If full allowable masonry stresses are used in the design of a building, it must have a special inspection. Without continuous special inspection, only half of allowable masonry stresses may be used to design a building. In special inspection, the owner must pay a special inspector to check the following types of work during construction:

- sampling and placing of masonry units
- placement of reinforcement
- all grouting operations
- preparation of masonry wall test prisms. Test prisms are usually 2 × 2 × 2-inch cubes of mortar or 3 × 6-inch cylinders made at the job site which are cured and tested for compression in a testing laboratory.

Certain types of buildings must have special fire-resistant exterior walls. For example, in Construction Type I, II, III, and IV buildings, exterior walls must have 4-hour fire resistance. Look back to Chapter 5 for additional description of types of construction. You can obtain a 4-hour wall by using one of the following:

- 8-inch thick concrete block wall with all cells filled with grout
- 9-inch thick solid brick wall
- 9-inch thick grouted brick wall

Interior fire division or separation walls also require fire resistant properties. The required fire rating depends on the type of occupancy the wall separates.

Masonry Drawings and Specifications

You should prepare plans for clay brick construction with wall thicknesses that are combinations of the width and length of the bricks. Lay out the foundation plans carefully to show the location of steel dowels projecting from the foundation. Dimension all door openings. Make sure the length and height of walls are increments of 4 inches.

The architectural drawings should show the bond pattern and joint tooling. Architectural drawings show the final appearance of masonry work. This can include outside dimensions, type of finish and mortar joint, and door and window openings. Structural drawings show the structural components of masonry work, such as location, type and lapping of steel reinforcement, anchors, and grouting.

Specifications for masonry work are generally divided into these major subjects:

- Scope of Work — This usually includes all masonry, placing of reinforcing, ties, and anchor bolts, and installing door and window frames.
- Work Not Included — This usually is all concrete work, anchor slots in concrete, furnishing and fabricating steel reinforcement, wire mesh backing for veneer, welding, shoring, and bracing for openings and bond beams.
- Materials — This includes mortar, portland cement, hydrated lime, sand, water, brick, face brick, common brick, concrete masonry units, wall coping, wall reinforcing, anchors, expansion joint filler, and control joint material.
- Delivery and Storage of Material
- Laying Walls, Brick, Concrete Masonry Units
- Built-in Work
- Pointing
- Cleaning

Walls

Base the thickness of a masonry wall on the size of the masonry unit and mortar joint you use to build the wall. It's a bad idea to cut hollow concrete units. For example, a solid brick wall made of common brick should be about $8\frac{1}{4}$ inches thick. This is controlled by two widths of $3\frac{7}{8}$ inches plus the $\frac{1}{2}$-inch mortar joint, or length of the header brick ($3\frac{7}{8}$ inches + $3\frac{7}{8}$ inches + $\frac{1}{2}$ inch). See Figure 6-12.

The thickness of a reinforced grouted brick masonry wall of common brick with a 2-inch-wide center core for horizontal and vertical reinforcement would be about $9\frac{3}{4}$ inches ($3\frac{7}{8}$ inches + $3\frac{7}{8}$ inches + 2 inches).

Don't start any masonry walls if the horizontal or vertical alignment of the foundation is 1 inch or more out of plumb or line. Figure 6-13 shows the sequence you should use in a grouted brick wall.

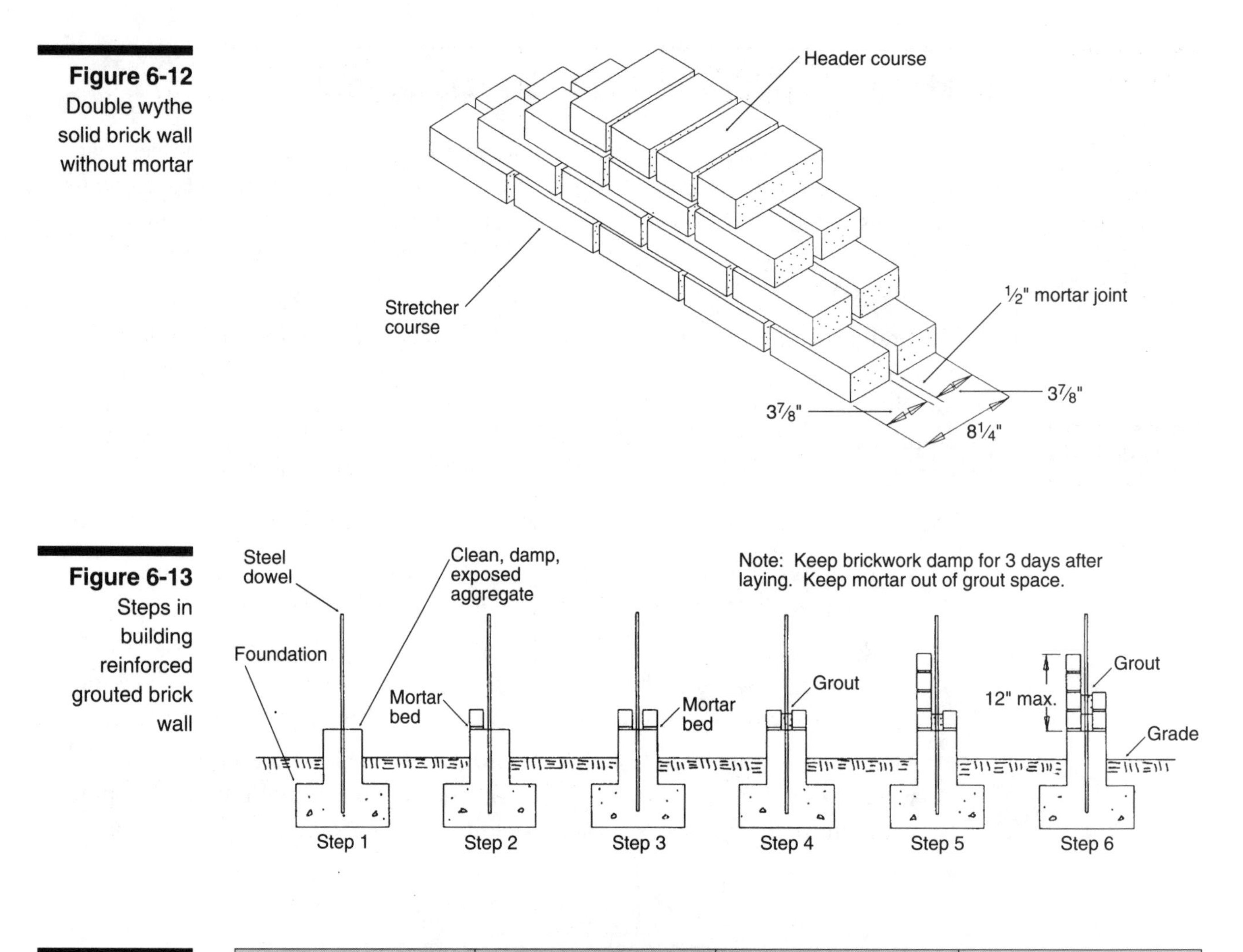

Figure 6-12 Double wythe solid brick wall without mortar

Figure 6-13 Steps in building reinforced grouted brick wall

Figure 6-14 Bearing masonry walls with only vertical load and vertical reinforcing (.002 × ⅓ gross wall area)

Thickness, wall (in.)	Spacing #5 bars (in.)	Spacing #4 bars (in.)	Spacing #3 bars (in.)
8	48	34	22
9	48	32	18
10	45	30	16
12	—	48	27
14	—	42	24

The minimum amount of vertical steel reinforcement in a reinforced grouted brick masonry wall is 0.002 × ⅓ of the gross area of the wall. The gross area is the total cross-sectional area including the grout-filled cells. The minimum horizontal steel reinforcement is 0.002 × ⅔ of the gross area of the wall. Other examples of minimum steel reinforcement are shown in Figures 6-14 and 6-15. Figure 6-16 shows the typical reinforcement needed in seismic zones, which are areas of various earthquake probability. Masonry work should be reinforced in all seismic zones. Seismic Zone 0 is the least vulnerable to earthquakes and Seismic Zone 4 has the greatest probability for earthquakes.

Figure 6-15 Bearing walls with only vertical load and horizontal reinforcement (.002 × ⅔ gross area of wall)

Wall thickness (inches)	Total reinforcing (square inches per foot wall height)
8	.13
9	.15
10	.16
12	.20
14	.23

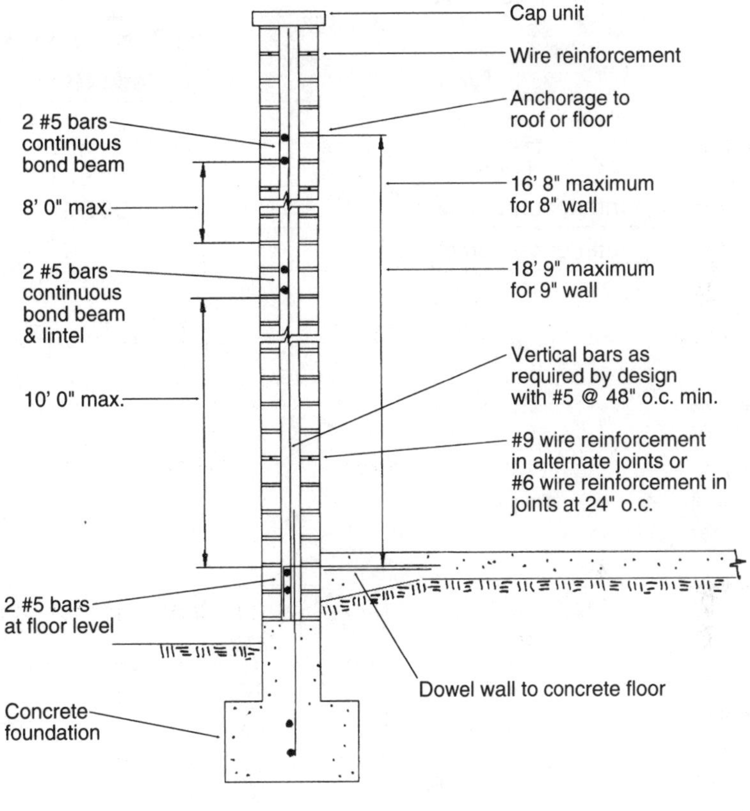

Figure 6-16 Reinforced grouted brick wall

Cap unit
Wire reinforcement
Anchorage to roof or floor
2 #5 bars continuous bond beam
8' 0" max.
16' 8" maximum for 8" wall
2 #5 bars continuous bond beam & lintel
18' 9" maximum for 9" wall
10' 0" max.
Vertical bars as required by design with #5 @ 48" o.c. min.
#9 wire reinforcement in alternate joints or #6 wire reinforcement in joints at 24" o.c.
2 #5 bars at floor level
Dowel wall to concrete floor
Concrete foundation

The minimum thickness or maximum height of a masonry wall is generally based on the ratio of the height to the thickness of the wall. Figure 6-17 is a rule of thumb for different types of walls. Height, H, and thickness, t, are in inches. For example, a 10-foot high (120 inches) solid masonry wall, has a maximum H/t ratio of 20. Therefore, the minimum thickness of the wall is 120 divided by 20, or 6 inches. This can be expressed as:

$H/t = 20$

or

$t = H/20$

$= 120 / 20$

$= 6$ inches

Figure 6-17 Ratio of height (H) to thickness (t)

Bearing Walls		
Type	**Ratio H/t**	**Thickness (t)**
Unburned clay	10	16
Stone wall	14	16
Cavity wall	18	8
Hollow unit	18	8
Solid wall	20	8
Grouted wall	20	6
Reinf. grouted wall	25	6
Reinf. hollow unit	25	4
Non Bearing Walls		
Type	**Ratio H/t**	**Thickness (t)**
Exterior unreinforced	20	2
Exterior reinforced	30	2
Interior unreinforced	36	2
Interior reinforced	48	2

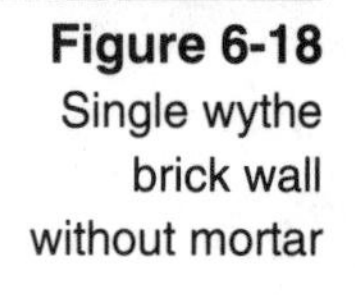

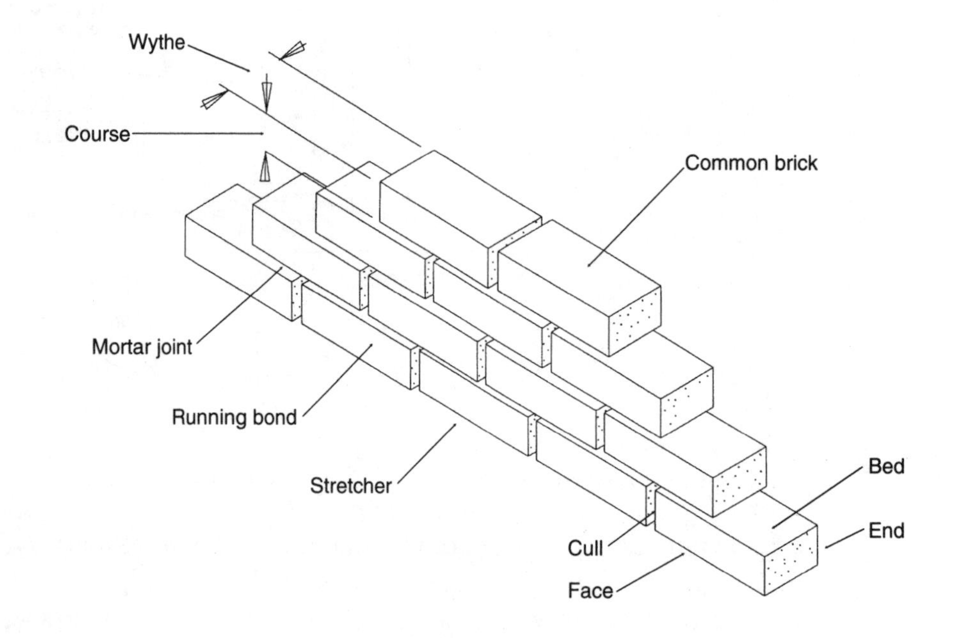

Figure 6-18 Single wythe brick wall without mortar

Solid Masonry

A traditional solid brick wall is made of two or more wythes. A wythe is one brick thick. Two wythes are two bricks thick. Figure 6-18 shows the simplest type of wall. It also shows the various parts of a brick wall, such as the wythe, course, stretcher, running bond, bed, face, and end, or cull. Figure 6-19 shows another way

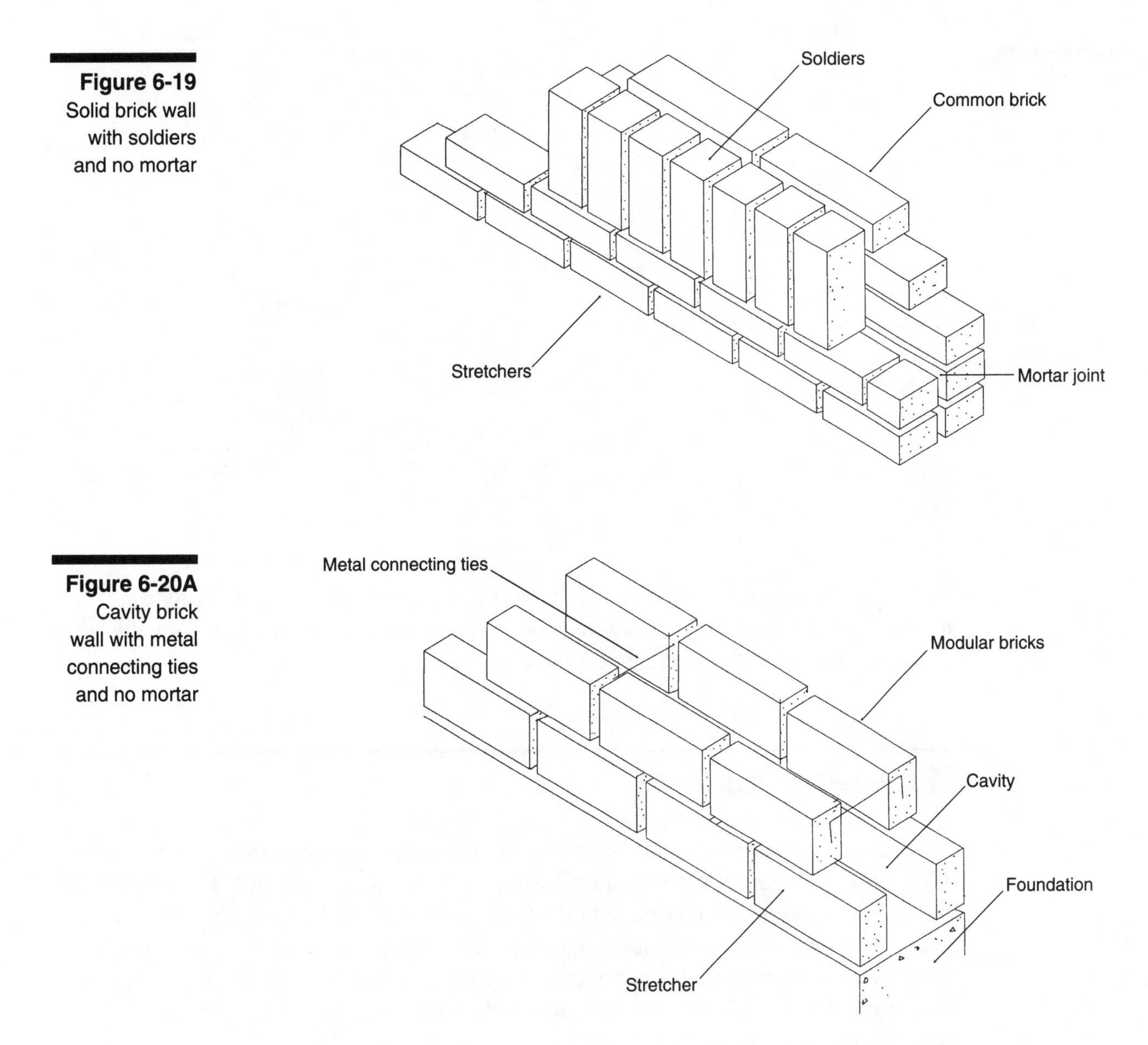

Figure 6-19 Solid brick wall with soldiers and no mortar

Figure 6-20A Cavity brick wall with metal connecting ties and no mortar

you can build a solid masonry wall. The bricks that are set vertically are called soldiers. Usually, you can only build this type of wall to a limited height. Any exterior nonbearing wall shouldn't be more than 20 times as high as it is thick.

Cavity Walls

Cavity walls have two or more wythes. Each wythe is separated by a space that's 1 to 4 inches wide. You anchor the wythes mechanically with metal ties that are embedded in the horizontal joints of the masonry. Figure 6-20A shows a type of mechanical tie. Figure 6-20B shows another type of cavity brick wall that's not reinforced. The bull headers tie the two wythes together. The exterior wythe is called

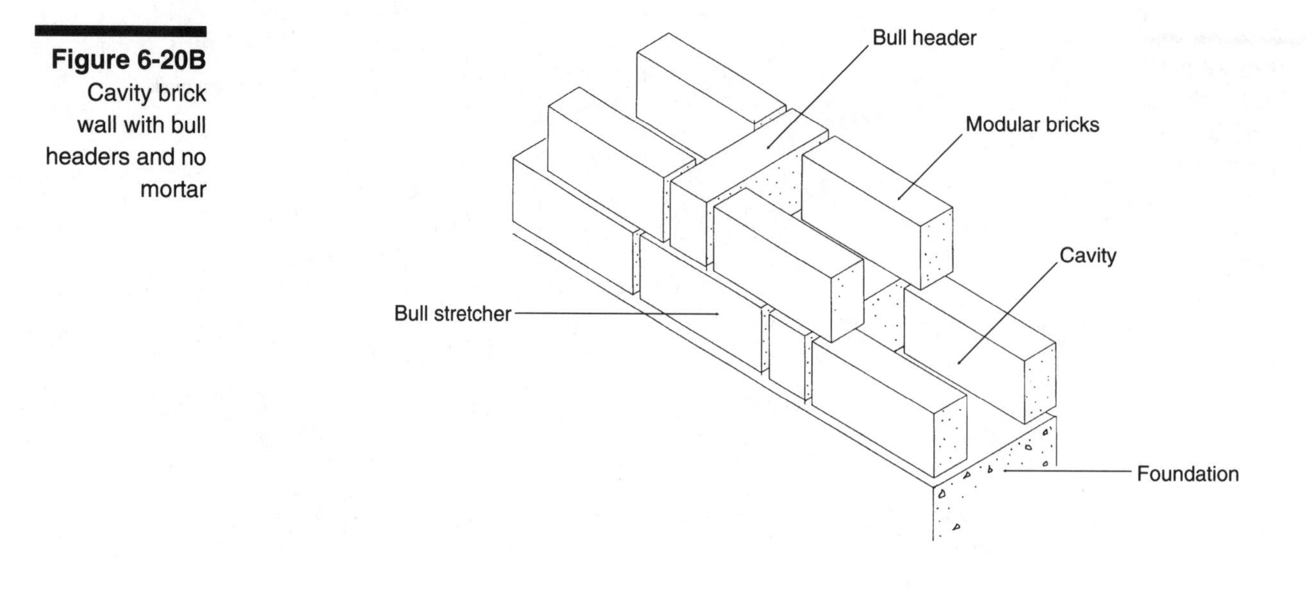

Figure 6-20B Cavity brick wall with bull headers and no mortar

the facing. The interior wythe is called the backing. The facing and backing must be at least 3 inches thick. The cavity must be at least 1 inch wide. Most cavity walls are now tied with metal ties instead of bull headers.

Grouted Masonry

Grouted masonry walls are similar to cavity walls except that the cavities are filled with grout. Grout may be placed with each course of brick laid in grouted brick masonry, or at greater intervals in hollow unit concrete masonry. A lift is the vertical distance that grout is continuously placed to. Low-lift grouting in grouted brick masonry requires that the lift not exceed six times the width of the grout space, with a maximum of 8 inches until the top is reached. High-lift grouting requires the placement to be continuous in 6-foot maximum lifts.

Low-lift grouting in hollow-unit masonry requires that the lift not exceed 8 feet, but if it exceeds 4 feet, cleanouts at the base of the wall must be provided. High-lift grouting in hollow-unit masonry may be used for any height wall, but the grout must be placed in 4-foot lifts. Continuous inspection is required during grouting.

A grouted masonry wall may also have horizontal and vertical steel reinforcement, which is put in before the grout is added.

Hollow Unit Masonry

Hollow unit masonry (concrete block) walls are either grouted or not grouted. If a wall isn't grouted, it depends on the mortar for its strength. If it *is* grouted, the hollow cells with steel reinforcing bars are filled with grout. This forms a series of reinforced concrete studs within the wall, which gives it much greater strength.

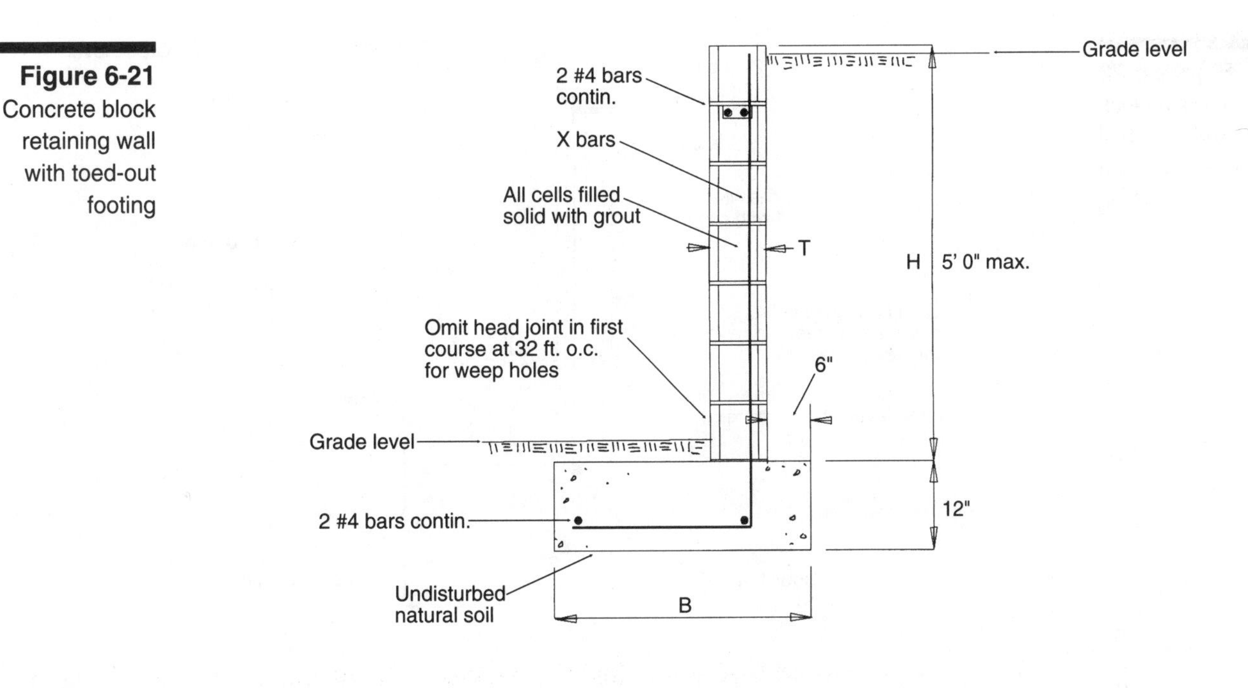

Figure 6-21 Concrete block retaining wall with toed-out footing

Retaining Walls

Concrete block makes an economical retaining wall. Retaining walls withstand the pressure of the retained soil and superimposed loads, or surcharge. You should always provide weep holes or open joints near the bottom of a retaining wall to drain the retained soil. The horizontal pressure of the retained soil is called "equivalent fluid pressure." This means that the soil is equivalent to water weighing 30 pounds per cubic foot instead of 62.4 pounds per cubic foot. For most drained soils, you design for a minimum fluid pressure of 30 pounds per square foot per foot of height.

The retaining walls shown in Figures 6-21, 6-22, 6-23, and 6-24 show the size and reinforcement for retaining walls of varying heights and footing arrangements. Figure 6-21 shows a toed-out footing with the heel of the footing near the wall, or stem. Figure 6-22 shows a toed-in footing with the toe near the wall. Figure 6-23 is similar to Figure 6-21 except it's higher than 5 feet. The wall is made up of 12- and 8-inch concrete blocks. Figure 6-24 is similar to Figure 6-22 except it's also higher than 5 feet.

Figures 6-25 and 6-26 give the various dimensions and reinforcement for these four types of retaining walls. These figures are based on a soil bearing value of 1,000 pounds per square foot and a level grade at the top of the retaining walls. The concrete in the footing has a compressive strength of 2,000 psi and the concrete blocks are Grade "A," per ASTM Specification C-90. Most building departments hand out standard details for walls up to 4 feet in height. For walls higher than 4 feet, you can buy publications that show details of retaining walls from 4 feet to 8 feet high. One of these is a booklet called the *Concrete Masonry Design Manual.* It's published by the Concrete Masonry Association of California.

You can use these publications to make preliminary designs and cost estimates but the plans you submit for a building permit may require an engineer's stamp.

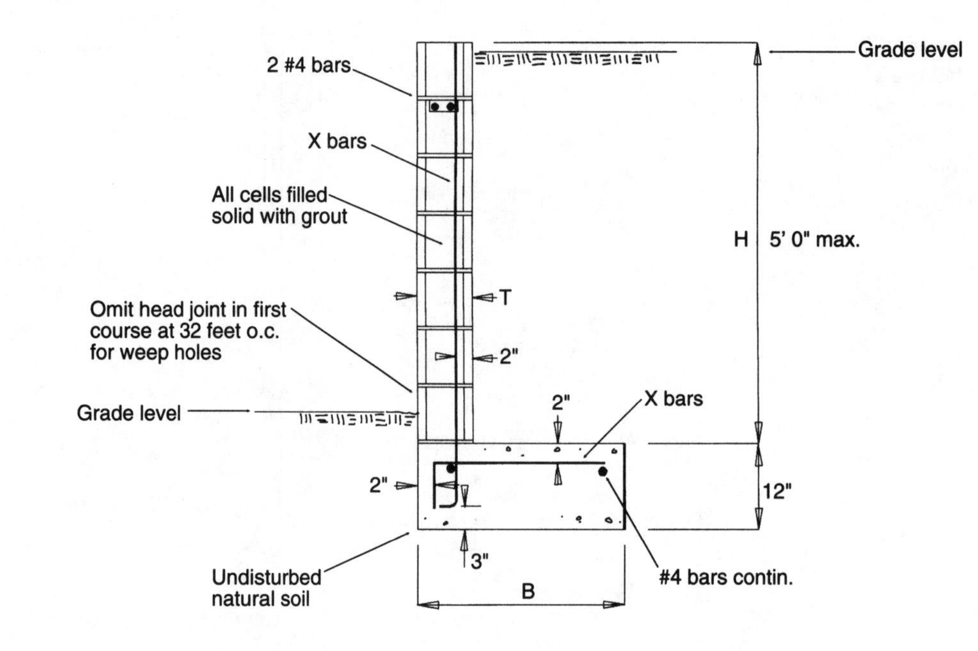

Figure 6-22 Concrete block retaining wall with toed-in footing

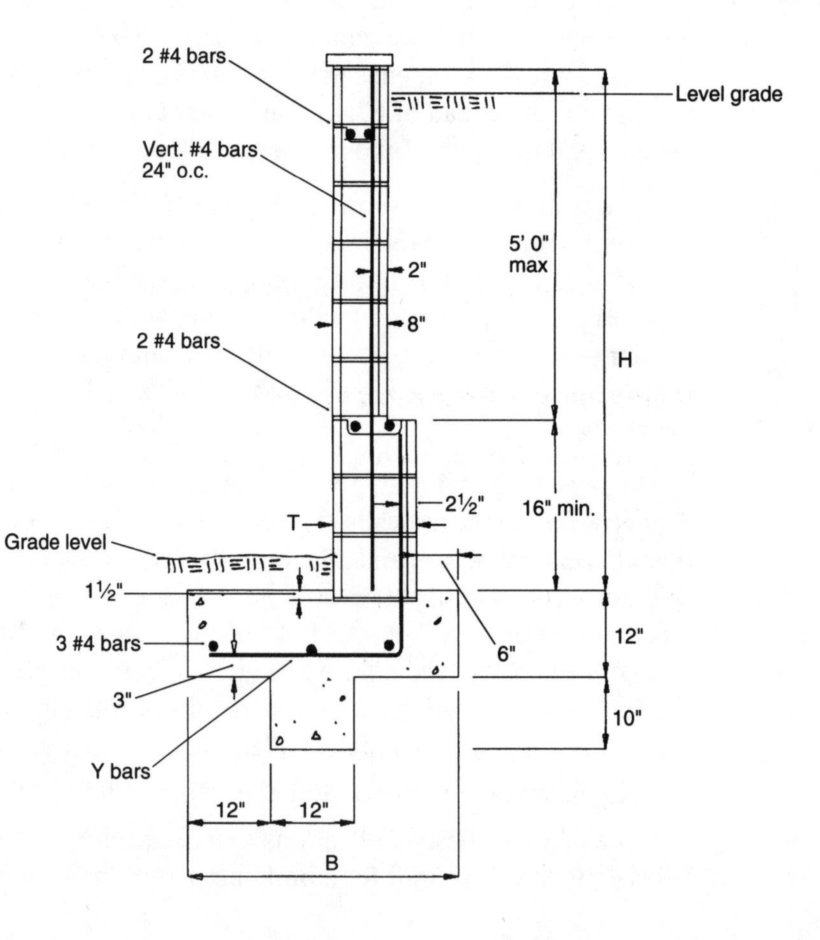

Figure 6-23 Concrete block retaining wall over 5 feet high with toed-out footing

Figure 6-24
Concrete block retaining wall over 5 feet high with toed-in footing

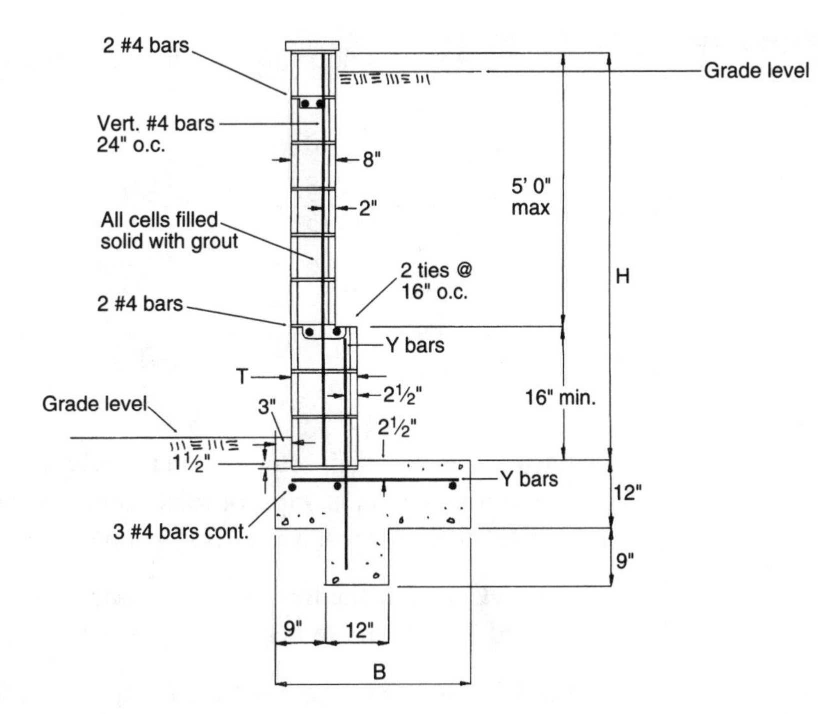

Figure 6-25
Concrete block retaining wall - toed-out footing

H (feet)	T (inches)	B (feet-inch)	X bars	Y bars
3	6	1' 9"	#3 @ 32" o.c.	
4	8	2' 2"	#4 @ 48" o.c.	
5	8	2' 9"	#4 @ 24" o.c.	
6	12	3' 3"		#4 @ 24" o.c.
7	12	3' 10"		#4 @ 16" o.c.
8	12	4' 6"		#5 @ 16" o.c.

For walls over 8 feet in height, you should hire an engineer.

Designing a Retaining Wall

When you design a retaining wall, you should follow these steps:

Step 1 - List the material specifications. This includes lightweight or regular weight block, block size, allowable stresses in blocks and reinforcing steel, soil pressure, and type of inspection.

Step 2 - Select tentative size and spacing of steel reinforcement from the design manual.

Step 3 - Find total horizontal pressure on the wall, per foot of width, based on an equivalent fluid pressure of 30 psf per foot of height. This is also called the total horizontal force on the wall. When there's a push or pull on a unit area, such as a square foot or a square inch, you usually talk about "total horizontal pressure" in psi

Figure 6-26
Concrete block retaining wall - heel-out footing

H (feet)	T (inches)	B (feet-inch)	X bars	Y bars
3	6	1' 10"	#3 @ 32" o.c.	
4	8	2' 6"	#4 @ 48" o.c.	
5	8	3' 0"	#4 @ 24" o.c.	
6	12	3' 8"		#4 @ 24" o.c.
7	12	4' 8"		#4 @ 16" o.c.
8	12	5' 3"		#5 @ 16" o.c.

or psf. But when the push or pull is on the whole wall, it's called the "total horizontal force", and it's in pounds, kips, or tons. This is because, for design purposes, you consider this force for just a one-foot width of wall.

Step 4 - Calculate the maximum bending moment, per foot of width, at the foot of the wall and the maximum bending stress in the wall.

Step 5 - Determine whether the computed stresses on block and reinforcement are within those allowable by code. If not, increase the bar size or reduce the spacing.

Step 6 - Find out whether the soil pressure at the toe of the footing is less than that permitted by code. If not, increase the size of the footing.

Step 7 - Check whether the wall will overturn or not by calculating the clockwise and counterclockwise forces about the toe of the footing. There should be at least a 150 percent safety factor against overturning. If there isn't, increase the size of the footing.

Now let's design a retaining wall that's 5 feet high with no surcharge. We'll assume that the soil under the footing has an allowable bearing pressure of 1000 psf.

For a trial design we'll use Grade A concrete blocks, ASTM C-90, and grout made of 1 part cement, 3 parts sand, and 2 parts pea gravel. The mortar is made of 1 part cement, $\frac{1}{2}$ part lime putty and $4\frac{1}{2}$ parts sand. The allowable bending stress for this material is 225 psi. We'll also assume that the wall will be 8 inches thick, and that the footing is 2'9" wide, 12 inches thick, and projects 1'7" from the face of the wall. For reinforcing, we'll use #4 steel bars 24 inches o.c.

For Steps 3 and 4, it helps to make a sketch of the outline of the trial design to scale. Figure 6-27 is a scale drawing of this trial design. It also shows some of the basic calculations for designing the wall.

In the figure, the triangle to the right of the wall is called a force diagram. It's used to represent some of the forces on the wall. The height of the wall, 5 feet, is the height of the force diagram triangle. The pressure at the top of the wall and the triangle is zero, but this increases at the rate of 30 psf for every foot of wall. The pressure at the base of the wall and the triangle is 30 psf × 5 feet, or 150 psf. Now, the total horizontal force on the wall is the area of the force diagram triangle, or $\frac{1}{2}$ × 150 psf × 5 feet. This is 375 lbs.

Figure 6-27 Calculations for a 5-foot retaining wall

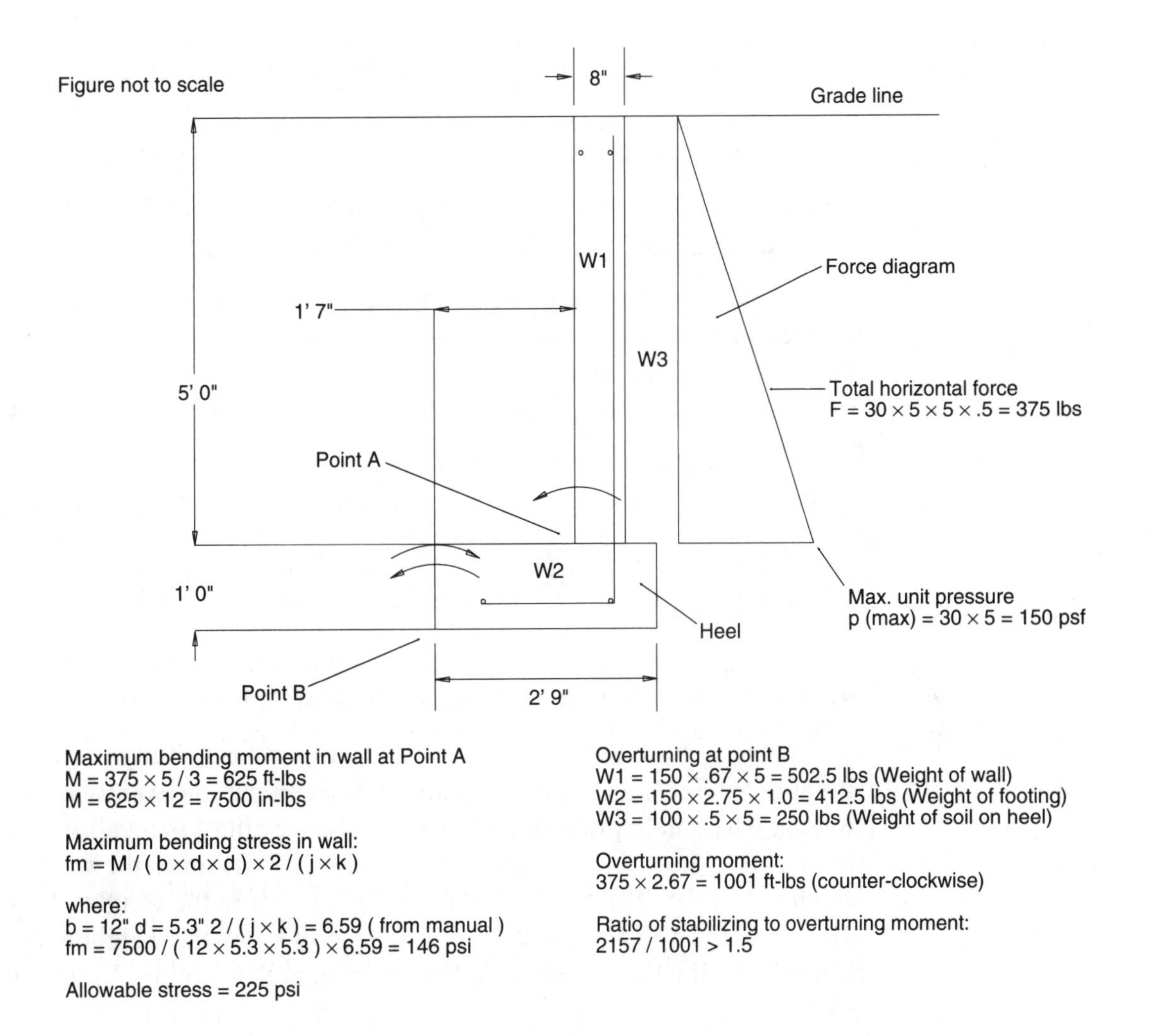

Now we're ready for Step 4. For design purposes, we assume that the total horizontal force on the wall acts one-third of the way up, or 1.67 feet up from the bottom. Then the bending moment on the wall at Point A in Figure 6-27 is the total horizontal pressure times the distance from Point A to the place where the horizontal force acts on the wall, or:

$$= 375 \text{ lbs} \times 1.67 \text{ ft}$$
$$= 625 \text{ ft-lbs}$$

Converting to in-lbs:

$$= 625 \times 12 \text{ in}$$
$$= 7500 \text{ in-lbs}$$

To figure out the maximum bending stress, you use this formula:

$$fm = M / (b \times d \times d) \times 2 / (j \times k)$$

where

M = 7500 in-lbs (maximum bending moment calculated previously)
b = 12 in (for a 1-foot width of wall)
d = 5.3 in (the distance from the face of the wall to the reinforcing bars)

The values of j and k in the formula depend on particular properties of concrete and steel, and the percentage of reinforcing steel to concrete in the wall. The tables in the design manual you use should give you the appropriate values of j and k. For the trial design we used the *Concrete Masonry Design Manual*, which gave a value of 6.59 for the quantity 2 / (j × k).

So:

fb = 7500 in-lbs / (12 in × 5.3 in × 5.3 in) × 6.59
= 146 psi

We find that the trial wall is adequate, since the maximum bending stress (146 psi) is less than the allowable bending stress of 225 psi.

Now let's go to Step 7 and check the wall for the stability of the toe of the footing (Point B on Figure 6-27). First you need to figure out the weight of the wall and footing, which help prevent overturning. The weight of the wall is its height (5 feet) times its thickness (0.67 feet) times its width (1 foot) times the weight of concrete, which is 150 pcf. This comes out to 502.5 lbs. Figure the weight of the footing the same way. It's 412.5 pounds. Now you need to know the weight of the 6-inch-thick column of soil which is supported by the heel of the footing. You figure it the same way as the footing, except the soil weighs 100 pcf, so you'll get (5 × 0.5 × 1 × 100), or 250 lbs.

Now calculate the overturning moment. The horizontal force on the wall acts at the same height on the wall as the center of the force diagram triangle. This is ⅓ of the way up the 5 foot wall, or 1.67 feet. Then you add another foot for the height of the footing to get 2.67 feet. The overturning moment is 375 lbs × 2.67 ft, or 1001 ft-lbs.

Next you need to figure out the stabilizing moment. To do this, you need to know how far it is from the toe of the footing to each of the three forces acting on the footing. You measure this horizontally. From the toe of the footing to the center of the wall is 1'7" plus ½ the thickness of the wall, or 1'11". From the toe of the footing to the center of the footing is half of 2'9". From the toe of the footing to the center of the soil behind the wall is 2'9" minus ½ the thickness of the soil, or 2'6". Then convert these horizontal distances to feet, multiply each weight by its respective horizontal distance from the toe of the footing, and add the three products. Add these products together and you have the total stabilizing, or clockwise, moment. The stabilizing moment is:

= (502.5 lbs × 1.92 ft) + (412.5 lbs × 1.375 ft) + (250 lbs × 2.5 ft)
= 964.8 ft-lbs + 567.2 ft-lbs + 625 ft-lbs
= 2157 ft-lbs

Figure 6-28 Typical basement block wall

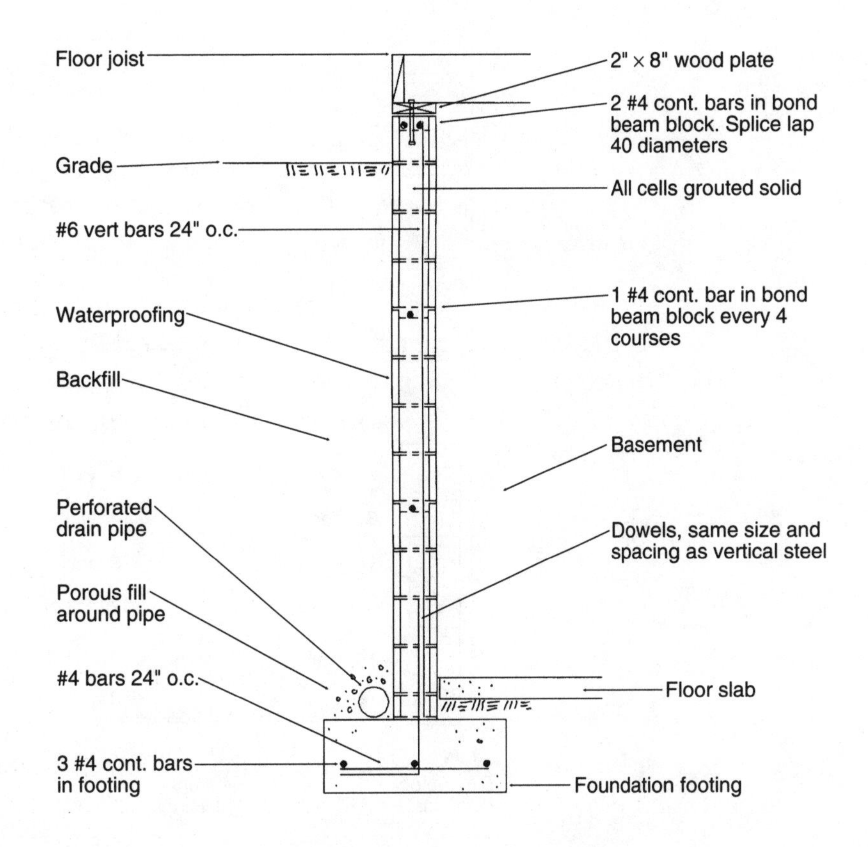

The ratio of the total stabilizing moment to overturning moment, 2157 / 1001, is greater than 1.5 which means that the trial design is satisfactory.

Most building departments say you must get an engineer's approval for retaining walls over 4 feet high. You may need an engineer's signature on the plans if the grade at the top of the wall is sloped, or if there's another structure next to the top of the wall, causing a surcharge. The normal load on a retaining wall is the level retained earth behind the wall. A surcharge on a retaining wall is any vertical load in addition to the retained earth. The surcharge may be a sloped surface, a nearby building foundation or vehicular traffic.

Basement walls are another form of retaining wall. Figure 6-28 shows a typical concrete block basement wall under a dwelling. This wall serves two purposes — it retains the earth and it supports the building wall. When ground water level is above the basement floor level, a drain pipe should be installed near the footing of the wall. The drain pipe may be a perforated plastic pipe or a clay pipe without mortar at the joints. The exterior surface of the wall should be made waterproof.

Pavers

To make a typical brick paved walk, first put down a 3- to 6-inch layer of coarse granular base. Then lay a 1-inch-thick bedding course that's level and to grade. Next, set the pavers on the bedding course with ⅛-inch-wide joints. Finally, fill the joints with sand.

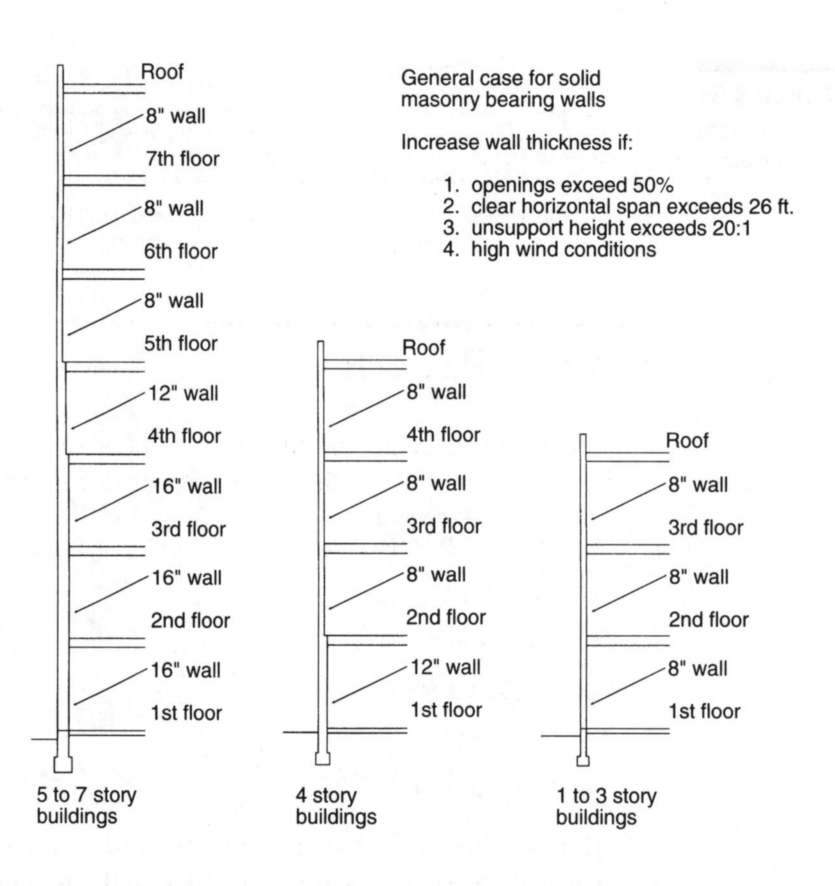

Figure 6-29
Solid brick walls

Single- and Multistory Building Walls

A typical one-story solid brick wall is 8 inches thick. A reinforced grouted brick wall is about 8¾ inches thick. Filler nonbearing walls are about the same thickness. Solid brick walls for one-story buildings less than 9 feet high may be 6 inches thick.

Generally, the thickness of a brick bearing wall depends on the number of stories above the wall. A solid masonry wall may be 8 inches thick if it's not over three stories, or 35 feet, in height. Figure 6-29 shows a rule of thumb for wall thicknesses for multistory buildings in nonseismic areas.

Openings

Surround all openings in reinforced masonry walls with steel reinforcing bars to prevent cracking. As a rule of thumb, you should put one #4 bar at each side of an opening more than 24 inches in either dimension. Extend these bars 24 inches beyond the edges of the opening. The bars around the openings are in addition to those otherwise required for earthquake requirements.

Masonry walls need expansion joints to help keep them from cracking when they expand or contract with temperature changes. Generally, walls up to 200 feet in length don't need expansion joints. Longer walls need a joint every 200 feet.

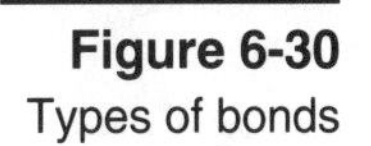
Figure 6-30 Types of bonds

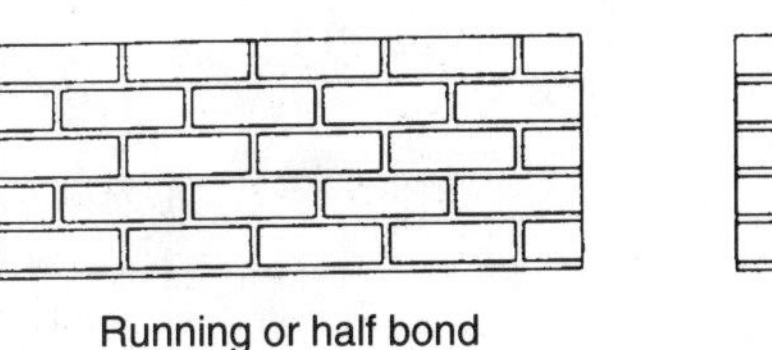
Running or half bond

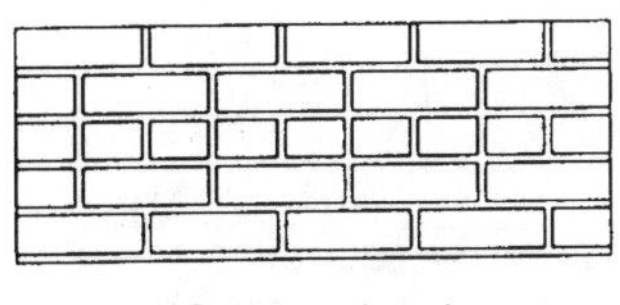
Common bond

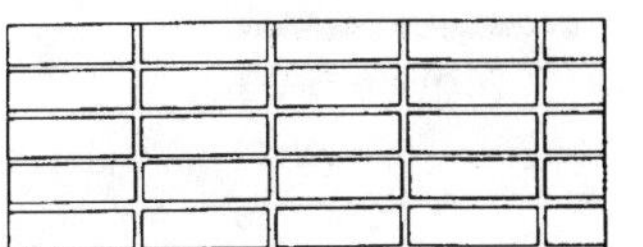

Stacked bond

Bond Pattern

The way bricks are laid in a wall is called the bond pattern. The following bond patterns were brought to the United States from Europe:

- running bond
- one-third bond
- common bond
- stacked bond
- Flemish bond
- English bond

Running bond, as shown on Figure 6-30, is where the masonry unit is laid so that the vertical mortar joints are centered over the unit below. This pattern is sometimes called "center bond" or "half bond."

One-third bond is where the brick unit is laid so that the wall corners show alternate header and stretcher courses. The stretchers continue from this point. Vertical mortar joints occur over one-third the length of the stretchers.

Stacked bond, as shown on Figure 6-30, is laid so the vertical mortar joints are in a continuous vertical alignment.

English bond consists of standard-size bricks with alternate courses of stretchers and headers. Stretchers continue from 2- and 6-inch corners, headers continue from 4-inch corners.

Flemish bond consists of courses of alternating stretcher, header, stretcher, header, etc., with headers centered over stretchers in alternate courses.

Piers, Pilasters, and Columns

The sizes of piers and pilasters you need will depend on the size of the masonry units you're using. When you use clay bricks, the width of a pilaster is the sum of the length and width of units plus the thickness of the mortar joints. For example, if you use modular bricks, the width of a pilaster is approximately 15 inches ($11\frac{3}{8}$ inches + $3\frac{3}{8}$ inches + $\frac{3}{8}$ inch), or 27 inches ($11\frac{3}{8}$ inches + $11\frac{3}{8}$ inches + $3\frac{3}{8}$ inches + $\frac{3}{8}$ inch + $\frac{3}{8}$ inch).

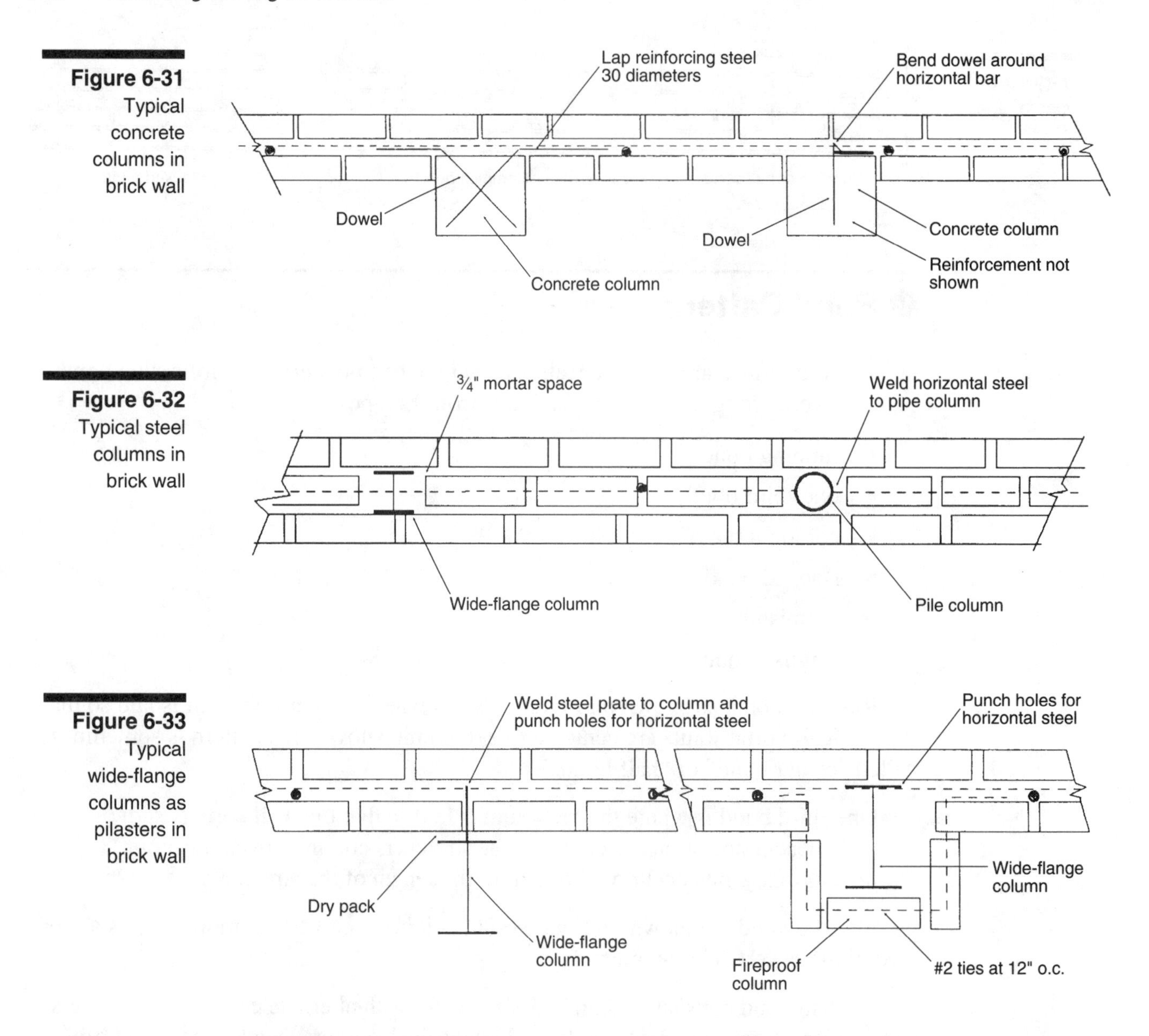

Figure 6-31 Typical concrete columns in brick wall

Figure 6-32 Typical steel columns in brick wall

Figure 6-33 Typical wide-flange columns as pilasters in brick wall

You can use cast-in-place concrete columns which are integral with a brick wall to support girders. Tie the concrete column to the wall with steel dowels. Figure 6-31 shows two concrete columns encased in a brick wall.

Sometimes you can use a small wide-flange column or a steel pipe column in the center of a 12-inch brick wall to support a girder load. This is done when the load exceeds the capacity of the wall and you don't want to thicken the wall with a pilaster. Weld short horizontal bars to the pipe column to tie it to the wall. Figure 6-32 shows a typical wide-flange column and a pipe column encased in a brick wall.

If you have a heavy vertical load on a brick masonry wall and a pipe column isn't strong enough, use a larger wide-flange steel column to carry the load. Figure 6-33 shows typical wide-flange columns as pilasters in a masonry wall. You can leave the steel members exposed or cover them with brick for fireproofing. Use welded rods to fasten a steel column securely to a brick wall.

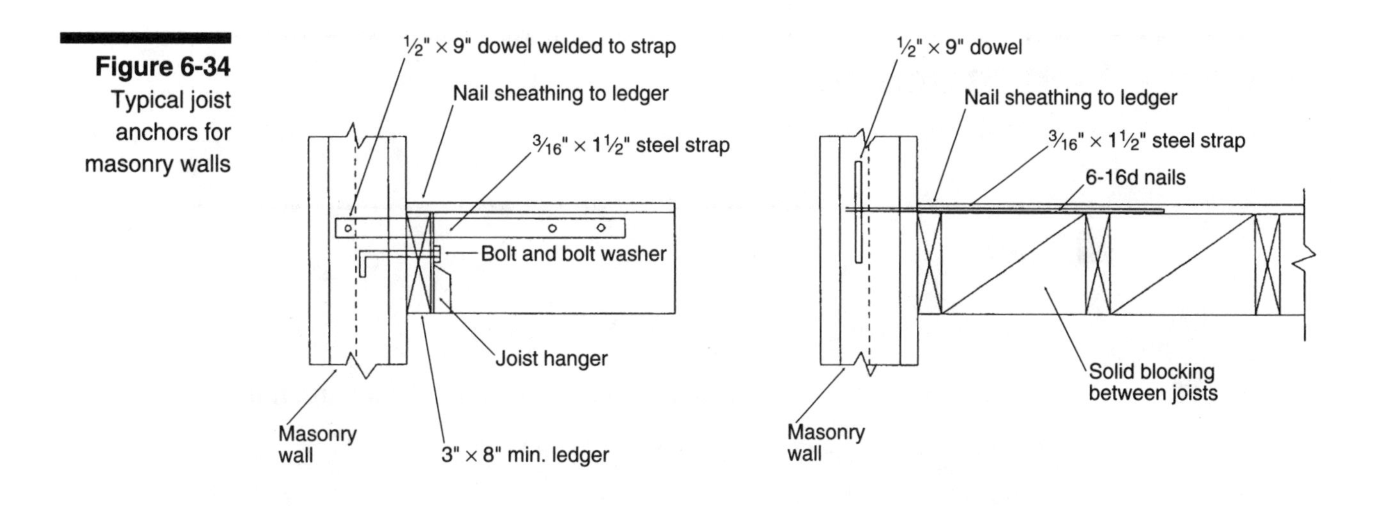

Figure 6-34 Typical joist anchors for masonry walls

Connections

In most buildings you must attach floor and ceiling joists to masonry walls. This requires connections to carry the floor and roof load. In seismic areas, you must provide anchors from the masonry wall to the floor and roof to keep the wall from falling away from the building. Figure 6-34 shows various types of wall anchors. A typical anchor strap is $\frac{3}{16} \times 1\frac{1}{2} \times$ 24-inch steel bar. One end has a hole for a $\frac{3}{4} \times$ 8-inch steel rod to be welded to the strap and encased in the wall. The other end has holes for nails or bolts for attachment to the joists or rafters.

Another common type of connection is the wood ledger. These are usually 3 × 8-inch minimum and are bolted to the masonry wall. The ledger supports the ends of the joists or rafters.

Footings

Put footings for brick masonry walls on firm, undisturbed soil of adequate loadbearing capacity. Set the bottom of the footing below the frost line. Unless local codes require otherwise, make the footings for small buildings twice as wide as the thickness of the wall they support. A footing is usually half as thick as it is wide.

Floor and Roof Support

Brick walls are often designed with a cast-in-place concrete beam which supports a concrete floor or roof slab. The integral concrete beam can also support an exterior nonbearing filler wall as well as the floor. Another option is to use steel channels or wide-flange beams to support floors, roofs, and filler walls. You can leave these steel members exposed or set them in a masonry wall.

Masonry Construction

Tools

The basic tools of a bricklayer, and their uses, are:

- wheelbarrow, shovel, and mortar hoe — used to handle mortar
- mortar box — used to mix mortar by hand
- mortar board to carry mortar
- mason's level (a wood or metal board, 36 or 48 inches long) — used to plumb and level walls
- square — used to make right angles or lay out corners
- straightedge — used as an extension to the mason's level
- trowels — used to mix, pick up, and spread mortar, and to tap units into the mortar bed
- hammer — used to split and rough-break bricks
- bolster, or broadedge chisel — used to cut brick
- jointer — used to make rounded, flat, and pointed mortar joints
- plumb bob, chalk, cotton cord — used to keep masonry work plumb and in alignment.

Scaffolding

Scaffolds are used to support bricklayers and their materials during construction. You'll need a scaffold when the bricklayers have completed all the work they can reach by standing on the floor or ground. Be very careful when you build scaffolding. Use rough lumber or prefabricated tubular steel for scaffolding. Here are some types of scaffolds:

- trestle scaffold
- foot scaffold
- putlog scaffold
- outrigger scaffold

You use a trestle scaffold when masonry work can be done from inside of a wall and the wall is over 4½ feet high. This type of scaffold consists of five 2 × 10-inch planks supported by trestles spaced 8 feet apart. The trestles are 60 inches long, 48 to 54 inches high and have legs and cross-beams made of 2 × 4s. Braces are 1 × 4s.

A foot scaffold is used when the mason requires less than 18 inches of increased height. This scaffold is made of 2 × 10-inch planks resting on brick supports.

A putlog scaffold is made of 4 × 4-inch posts with 2 × 12 × 12-inch wood footings. The posts are spaced 8 feet apart and placed about 5 inches from the wall. A 1 × 8-inch ledger is nailed to the top of the posts. The putlog is a 3 × 4-inch piece of lumber which is supported at one end by the ledger, and at the other end by an opening in the wall. The putlog supports five 2 × 12-inch planks, which form the platform. The putlogs and planks aren't nailed to their supports, the posts are tied to the wall by stays.

An outrigger scaffold is made of 2 × 10-inch planks supported by a wooden beam.

A rule of thumb for selecting wood planks for scaffolding is to use 2 × 10- or 2 × 12-inch planks with a minimum 1500 psi bending strength. Normally, a plank can be used to span a distance in feet that's less than or equal to the width of the plank in inches. Therefore, you can use a 10-inch plank to span up to 10 feet. You can use a 12-inch plank to span up to 12 feet.

Masonry Work

Soak brick masonry units in water before you put mortar on them so they don't suck the moisture out of the mortar. You don't need to soak concrete masonry units.

In rainy weather, do all masonry work under cover. When you're not actually working on exposed masonry, protect it with a nonstaining waterproof cover. When you start working on it again, clean any loose mortar off the top surface. In dry weather, moisten the tops of the walls under construction before you begin working on them again.

When the air temperature is 80 degrees F or more, put mortar in its final position within 2¼ hours after you mix it. If the air temperature is less than 80 degrees, you should have it in place in 3¼ hours.

All masonry work should be plumb, level, and square, except where otherwise shown on the plans. Adjust each masonry unit to its final position while the mortar is still soft and plastic. If a unit gets disturbed after the mortar has stiffened, remove and reset it with fresh mortar.

Figure 6-35 shows the popular patterns for facing and finishing mortar joints. To get the most strength and weather resistance from mortar joints, you need to tool them. Tooling is compressing and shaping the face of a mortar joint with a special tool other than a trowel. Take special care to make all joints weathertight and clean.

Concave and "V" joints are best, because the tooling works the mortar tight and produces a good weather joint. Water will tend to drip off the top edge of the joint instead of running into the joint. A deep concave joint produces a sharp shadow similar to a rake joint, but a thin layer of mortar may spall off.

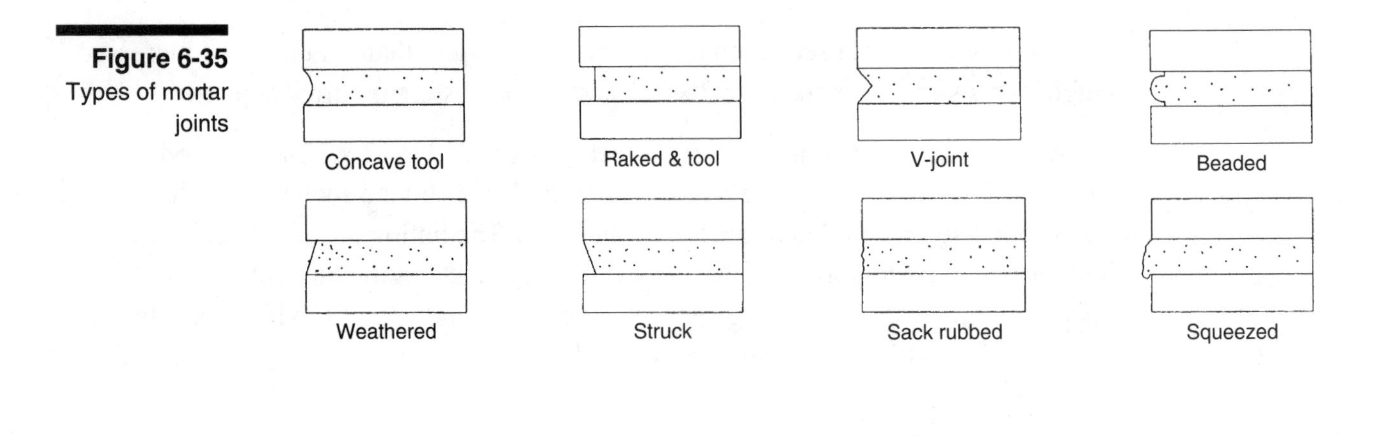

Figure 6-35 Types of mortar joints

Flush joints are used when a wall will be painted and joints covered. It's difficult to make this joint weathertight because the mortar may shrink away from masonry surface. You can make flush joints by rubbing a carpet-faced wood float against the wall.

Raked joints are used mainly to accentuate the joint line. They provide poor weatherproofing. Squeezed joints, also called "weeping" joints are used for a rustic architectural effect. But they don't make a good weathertight joint. Neither do struck joints, which let water run into the joints.

Tolerances in Masonry Work

The maximum deviation of a masonry wall or column from plumb or vertical plane should be less than ¼ inch in 10 feet, ⅜ inch in a story height or ½ inch in 40 feet or more. The maximum difference in the location of masonry walls from that shown on the plans should be less than ½ inch in any bay or 20 feet, or ¾ inch in 40 feet. Wall thicknesses must be minus ¼ inch or plus ½ inch of the dimensions shown on the drawings.

Unit Masonry Construction Checklist

The checklist on the facing page lists the items a building inspector most typically checks for on masonry work. Add to the checklist any items that your experience shows are also important and make copies as you need them to use on your jobs. A simple reminder sheet like this can save the cost and delay of a reinspection.

Unit Masonry Construction Checklist

General

- ☐ Is the horizontal and vertical alignment of the foundation in line and plumb within a 1-inch tolerance?
- ☐ Is continuous inspection required?
- ☐ Is wall thickness correct?
- ☐ Is the foundation aggregate exposed before starting masonry work?
- ☐ Is additional reinforcement around wall openings correct in size and number?
- ☐ Is the foundation square in accordance with approved tolerances?
- ☐ Is the horizontal alignment of all sills and lintels correct and in accordance with approved plans?
- ☐ Are structural pilasters correctly located in accordance with approved plans?
- ☐ Is the mortar to be used with steel reinforcing? If so, is the mortar free of any and all salts (no accelerating admixtures have been used)?
- ☐ Is the top surface of the concrete foundation clean and free of laitance?
- ☐ Is called inspection required?
- ☐ Is bond beam size correct?
- ☐ Is steel reinforcement correct for kind, grade, size, location and clearance?
- ☐ Are joist anchor and anchor bolt connections correct in type, number, size and location?
- ☐ Is the drainage provided at the wall's base sufficient for the site and in accordance with approved plans?
- ☐ Is the masonry correctly aligned, vertically and horizontally, with wall opening frames?
- ☐ Are weep holes provided and correctly located if required in approved plans?
- ☐ Is mortar "sticky" to better adhere to masonry?

Solid unit masonry

- ☐ Is the type and quality of brick in accordance with approved plans?
- ☐ Were bricks presoaked but surface dry before being mortared?
- ☐ Is reused brick in compliance with the code?

Hollow-core unit masonry

- ☐ Is the masonry being mortared dry?
- ☐ Are vertical cells clear of overhanging mortar?
- ☐ Are there cleanout openings at the bottom of cells for grout pours over 4 feet high?
- ☐ Are vertical cells continuous?
- ☐ Are reinforcing bars correct in size, number and location?
- ☐ Are cleanout cells sealed?

CHAPTER 7

PLUMBING

Good plumbing design must consider the cost of installation, operation, and maintenance. If the plumbing system is for a spec-built home, the contractor may want only the minimum needed to meet code requirements. If it's for a custom-built home, the contractor probably wants a plumbing system that's more water- and energy-efficient, and easier to maintain and extend in the future. Adding extra shutoff valves, access openings, and headers can save a lot of time and trouble later. The owner of a custom-built home is more likely to review plans and specifications closely and take an active part in the design.

Designing a Plumbing System

Every plumbing system should meet the following requirements:

- supply pure and wholesome water
- prevent unsafe water from flowing back into the plumbing system
- supply water, in sufficient volume and pressure, to fixtures
- supply and store hot water
- include adequate cleanouts to keep solids from clogging the drainage system
- include water seals to keep noxious gases out of building interiors
- be durable enough to last the life of the structure

You can design a plumbing layout using a handbook, rule-of-thumb methods, manual calculations, or a computer. If you're working on a large project, you'll probably want to use a computer for the design and drafting part. You can use word processing software for the specifications, and spreadsheet software for flow analysis and bills of materials. You can make piping drawings with computer-aided drawing (CAD) software.

Plumbing design isn't an exact science. You have to make some rough estimates and do some guessing. For example, water pressure in public mains usually rises and falls throughout the day. Water use during the day will depend on what the occupants of the structure are doing. That will vary from day to day, month to month, and year to year. Finally, your plumbing design has to comply with changes in your local plumbing code, and make use of approved materials.

Plumbing Code

The first plumbing code in the United States was printed in 1928. Codes have been revised and improved regularly since then. The present *UPC* (*Uniform Plumbing Code*) sets standards for the following applications:

- quality of materials
- drainage
- vents
- indirect waste piping
- traps and interceptors
- joints and connections
- plumbing fixtures
- water distribution
- sewers and private sewage disposal systems
- fuel gas
- water heaters and vents
- backflow protection
- fire protection
- food establishments
- trailer parks
- swimming pools
- rainwater drainage

Permits

You usually need a plumbing permit to install, repair, or change any plumbing, sewer, water, gas, or swimming pool piping. You also need a permit for water treatment devices and fixtures. The fee for a plumbing permit is usually based on the number and type of plumbing fixtures. Here are some of the plumbing permit fees listed in the *1991 Uniform Plumbing Code:*

Each plumbing permit	$20.00
Each supplemental permit	$10.00
Each plumbing fixture on one trap	$7.00
Each building sewer	$15.00
Each drain on a rainwater system	$7.00
Each cesspool (where permitted)	$25.00
Each private sewage disposal system	$40.00
Each water heater and/or vent	$7.00
Each gas piping system with 1 to 5 outlets	$5.00
Each additional gas outlet	$1.00
Each installation, repair, alteration of water piping	$7.00

To file an application for a plumbing permit you need to provide:

- job site address
- legal description of the job site including lot number, block, and tract
- sketch of the plot plan
- owner's name, address, and phone number
- contractor's name, address, phone number, and license number
- architect's name, address, phone number, and license number
- engineer's name, address, phone number, and license number
- description of building use
- class of work — new, addition, alteration, or repair
- description of work

Plans and Specifications

The basic plumbing plans for a multi-family dwelling should show cold water, hot water, sanitary waste, and gas piping in the form of:

- Schematic riser diagrams (see Figures 7-1A and 7-1B)
- Schematic isometric diagrams (see Figure 7-2)

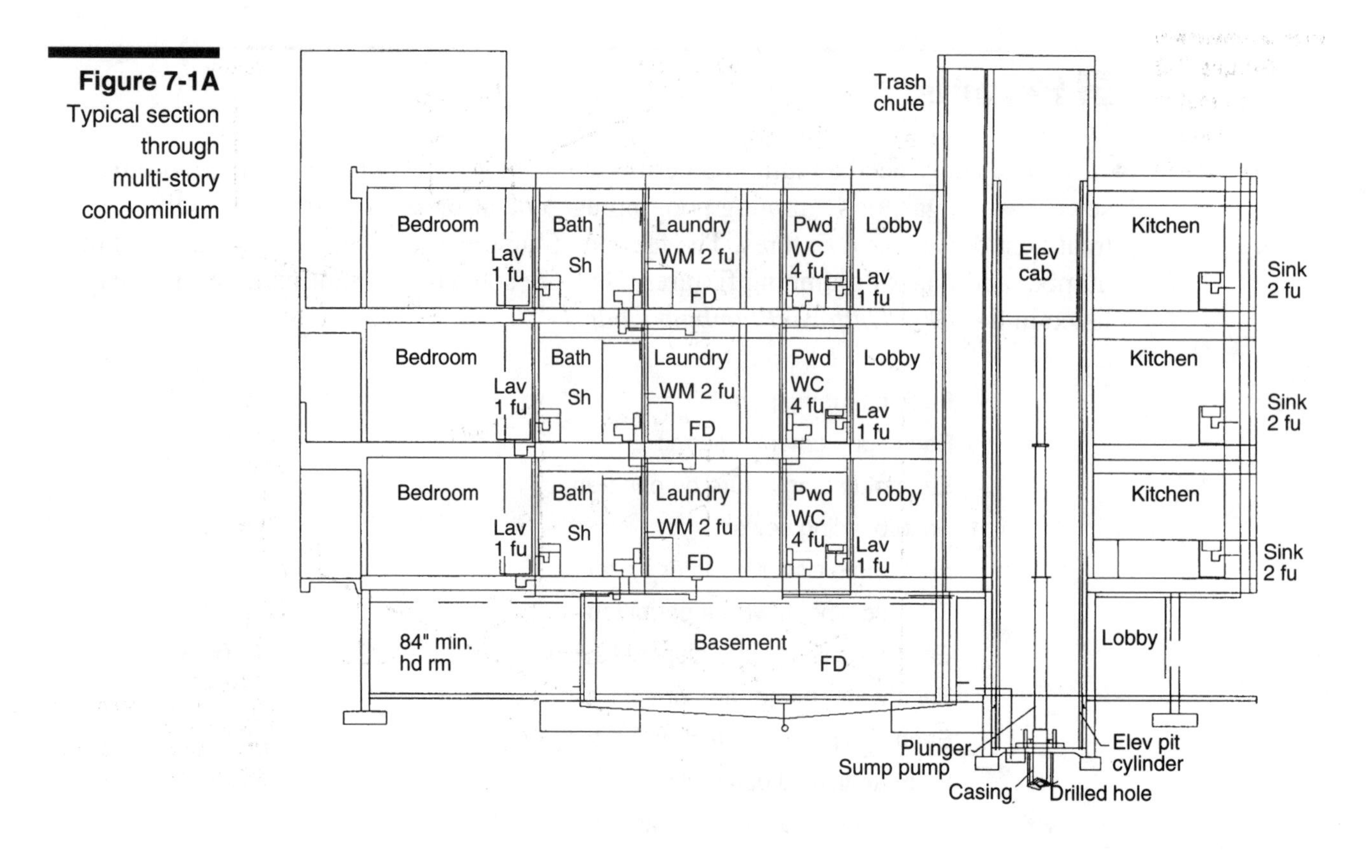

Figure 7-1A Typical section through multi-story condominium

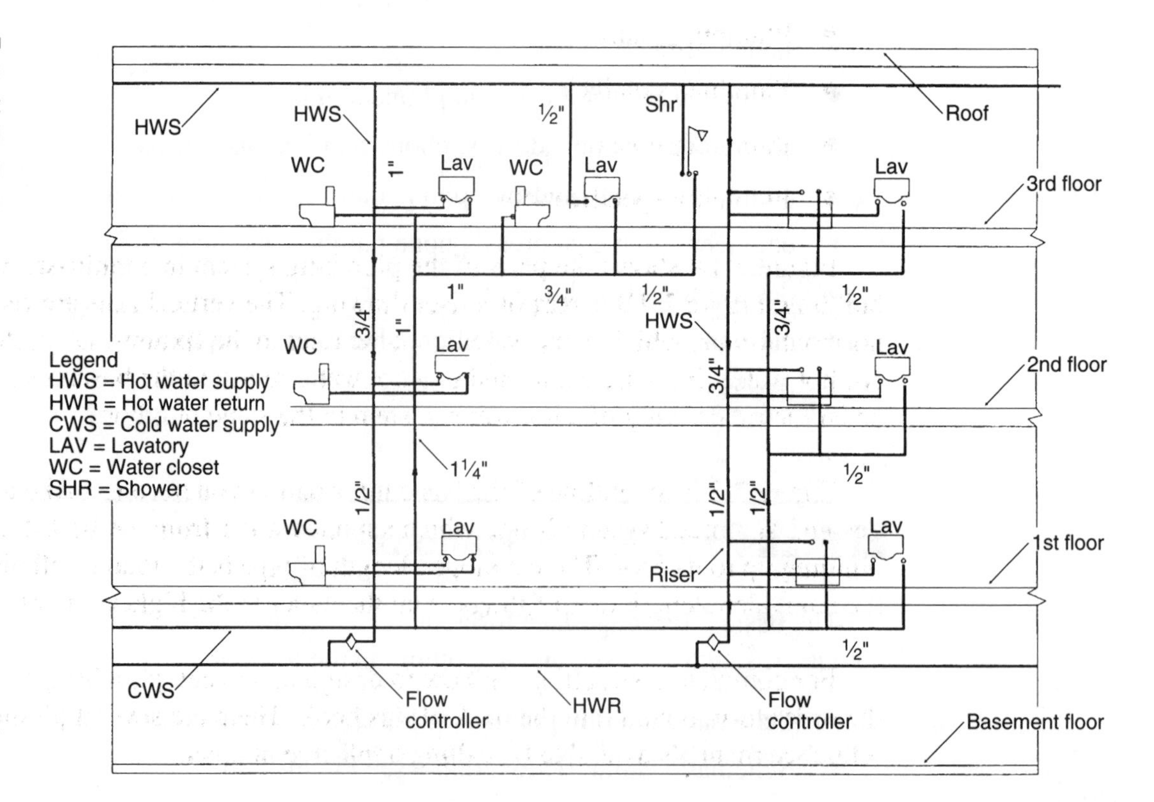

Figure 7-1B Typical hot water and cold water piping elevation

Figure 7-2
Isometric drawing of water system

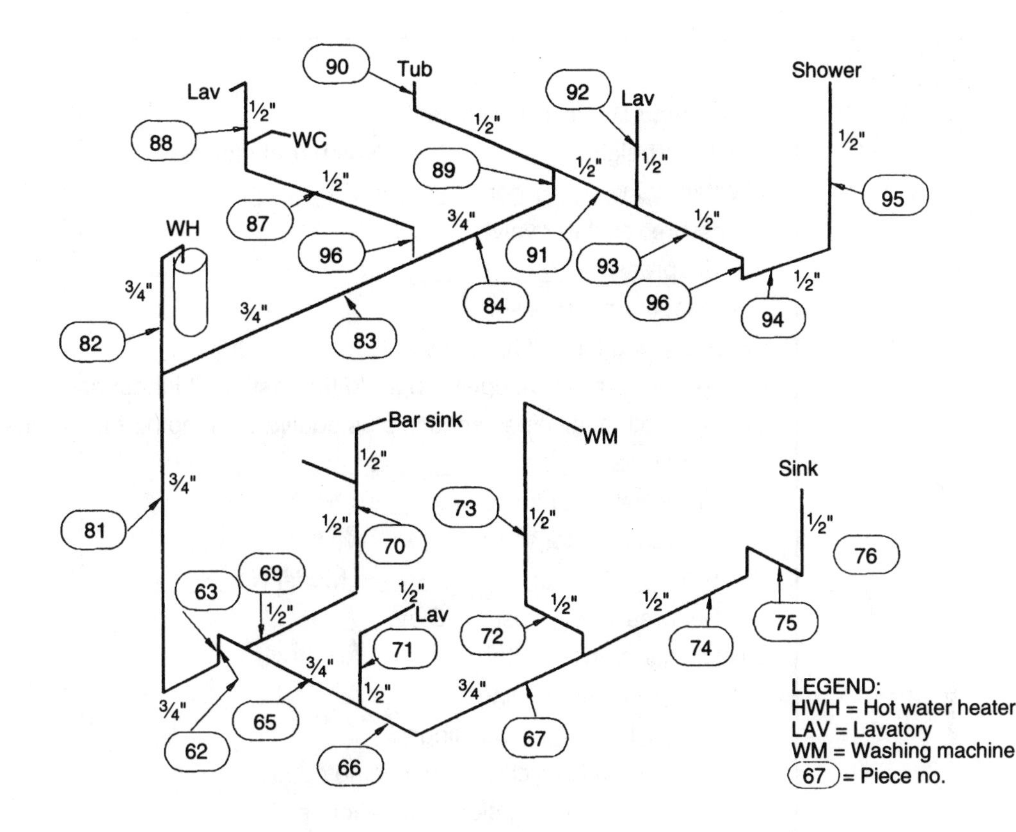

- Plumbing plans
- Plumbing details
- Plumbing notes
- Plumbing specifications

Figure 7-1A shows the parts of the plumbing system in a multi-story residential building. Figure 7-1B is part of a riser drawing. The vertical runs are risers and the horizontal runs, which carry water from the riser to the fixtures, are called headers. All hot water risers are connected to a hot water return in the basement. Figure 7-2 is an isometric drawing of a hot water system in the same building.

Figure 7-3 is an outline of the basic information you need to make a plumbing design. An upfeed system is one which supplies water from the first floor of a building up to its roof. The developed length of pipe is the total of all pipe runs plus the equivalent length of all fittings from the meter to the highest outlet.

For complete instructions on how to design and install plumbing systems, check the order form bound into the back of this book. There are several plumbing reference manuals available by calling a toll-free number.

Figure 7-3
Information for plumbing design

Plumbing Design Data Sheet
Building height:__________ Number of stories:___________
Water main:_____ gpm:______psi to_____psi ______ft elev.
Pressure regulator valve:
Set pressure_____psi_______ft
Highest fire hose valve:_____ ft
Highest fixture on upfeed system:_________ft
Total measured developed length to farthest and highest domestic fixture outlet from:
(total length of all pipe sections plus equivalent length of all fittings)
Street main:______________ft
Pressure regulator valve:____ft
Pressure regulator valve:
Size:____ Make:_____________Model No:________
Set pressure:___psi Elev:____ ft
Total fixture units (fu):
______fu per building
______fu for ____ buildings = ____
_____ gpm from chart in plumbing code
_____ gpm for irrigation, recreation, etc.
Average ____ gpm per building

Fixture Units

A fixture unit is a measure of the amount of water that goes into or out of a plumbing fixture for a particular unit of time. Figure 7-4 shows the plumbing code values for total fixture units.

The *fu* values for fixtures in public buildings are greater than those for private buildings. The probability of every fixture in a building being used at the same time is remote. But there should be enough water to supply any fixture assuming the most probable number of fixtures in use at the same time. To figure this involves considering:

- the frequency of use of each type of plumbing fixture
- the discharge rate of each plumbing fixture
- the type of building occupancy

Figure 7-4
Plumbing code values for total fixture units

Fixture name	Private fixture units	Public fixture units
Bar sink	1	2
Bathtub	2	4
Clothes washer	2	
Hose bib	3	5
Laundry tray	2	4
Lavatory	1	2
Lawn sprinkler	1	1
Shower	2	4
Sink/dishwasher	2	4
Water closet (tank)	3	5
Water closet (flush valve)	6	10
Service sink	2	4

Figure 7-5
Connections to fixtures

Fixture	Hot water	Cold water	Soil or waste	Vent
Toilet		3⁄8"	3" × 4"	2"
Lavatory	3⁄8"	3⁄8"	1 1⁄4"	1 1⁄4"
Bathtub	1⁄2"	1⁄2"	1 1⁄2"	1 1⁄4"
Sink	1⁄2"	1⁄2"	1 1⁄2"	1 1⁄4"
Laundry	1⁄2"	1⁄2"	1 1⁄2"	1 1⁄4"
Sink/tray	1⁄2"	1⁄2"	1 1⁄2"	1 1⁄4"
Shower	1⁄2"	1⁄2"	2"	1 1⁄4"

Figure 7-5 lists the sizes of typical water supply and waste connections to plumbing fixtures. The size of the supply pipe to each type of fixture is related to the fu rating for that fixture. Fixture units can be roughly converted to gallons per minute intermittent flow by:

1 fu = up to 7.5 gpm

2 fu = 8-15 gpm

4 fu = 16-23 gpm

5 fu = 24-30 gpm

6 fu = 31-50 gpm

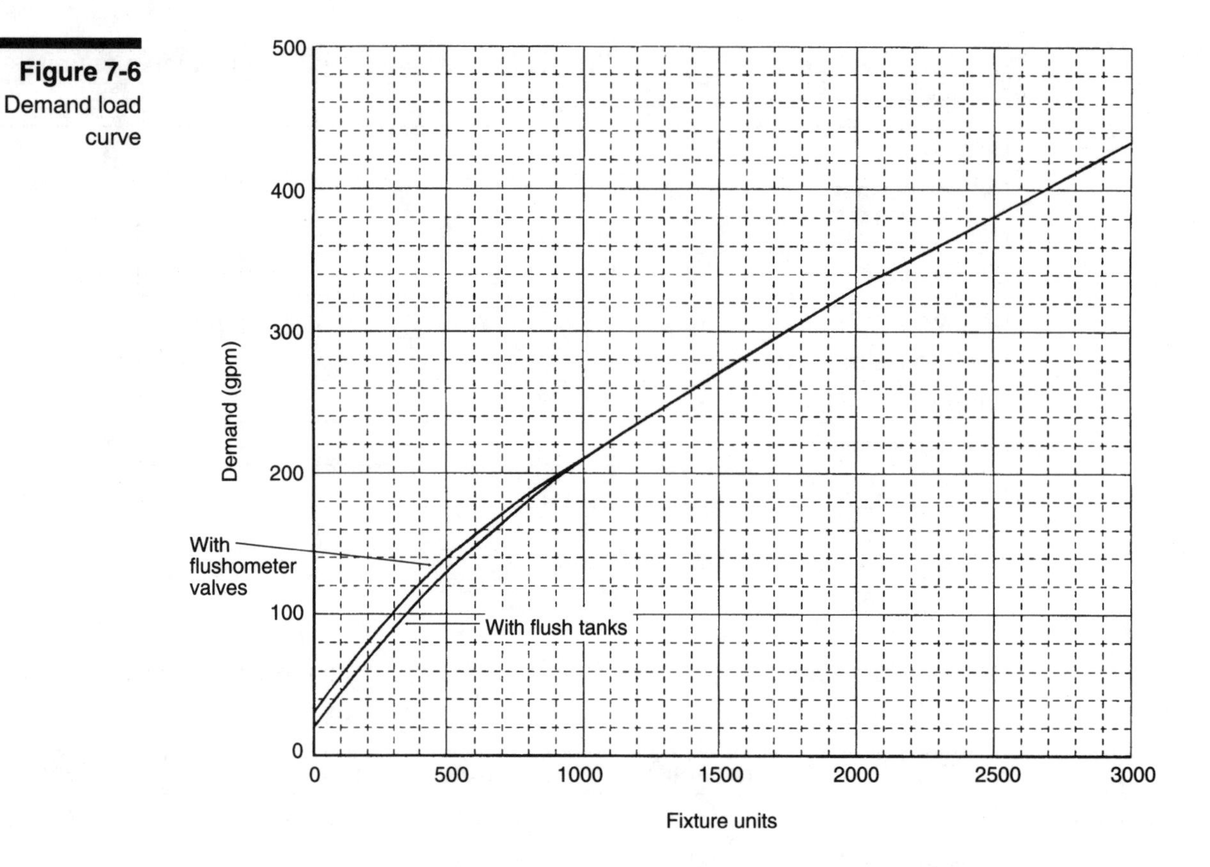

Figure 7-6 Demand load curve

The approximate water demand of buildings with up to 3000 fixture units is shown in Figure 7-6 by the curves called "with flush valves" and "with flush tanks." They show that there's a slightly higher water demand up to 1000 fu when most water closets are fitted with flush valves rather than flush tanks. Flush valves use more water than flush tanks.

Plumbing codes have tables that list fixture units, pipe sizes, water pressure ranges of 30 psi to 60 psi, and lengths of pipe up to 200 feet. As a rule, the approximate number of fixture units that can be supplied by pipe of the following sizes are:

Pipe size in inches	Number of fixture units
1/2	7
3/4	22
1	42
1 1/4	96
1 1/2	240

How to Pick a Pipe Size

Here's a step-by-step method you can use to figure out the different sizes of pipe you need for a job:

1) Calculate the total number of fixture units (fu) in the building. A fixture unit is a rating which stands for the amount of water required by a plumbing fixture. Fixture unit ratings for common fixtures in private and public buildings are given in Figure 7-4.

2) Find the developed length of water supply pipe from the meter to the cold or hot water outlet that is farthest from the meter.

3) Figure out the difference in elevation between the water meter and highest fixture or outlet.

4) Ask the water company what the lowest, highest, and average water pressure is at the street main. Design the water distribution system for the minimum pressure available.

5) Size the water service pipe and the water meter according to Table 10-2 in the *Uniform Plumbing Code* using the following steps:

   From the available pressure at the source of water supply, subtract 0.5 psi per foot of difference in height between the source of supply and highest water outlet in the building. For example, if the city pressure is 70 psi and the highest outlet is 40 feet above the water supply pipe, subtract (0.5 x 40) 20 psi from 70 psi, to get 50 psi.

   Look in the "Pressure Range" group where this pressure will fall. The pressure ranges are 30 to 45 psi, 46 to 60 psi, and over 60 psi. So for 50 psi, you use the 46 to 60 psi group.

   Select the "Length" column which is equal to or longer than the required length. The length column ranges from 40 to 1000 feet. Let's say you've got 200 feet of pipe.

   Follow down the column to the fixture unit value equal to, or greater than, the total number of fixtures in the building. Let's say you have 30 fixture units.

   The two left-hand columns will give the size of meter and street service. Our example falls between 25 and 44 fixture units, so you would use a 1¼-inch service supply and a 1-inch meter.

You can use this procedure to size any particular section of a piping distribution system, except you use the length of pipe, number of fu, and pressure in that section only.

When you design a plumbing system you should find out about the building site and water conditions. Here are some of the questions you should ask, and how to find the answers:

- What type of piping material is used in the neighboring buildings?
  Check the public records of previously-issued plumbing permits.
- Is the water extremely corrosive?
  Request a chemical analysis and commentary from the water company.
- If the water is corrosive, what's been done to keep pipe from corroding prematurely?
  Depending on just how corrosive the water is, there are several methods of prevention: using copper tubing; using plastic piping; installing magnesium capsules; and introducing direct current into the piping system.
- What is the expected service life of steel pipe versus non-corrosive pipe?
  Check the repair plumbing permits on record at the building department to find out how often the different types of pipe had to be replaced.
- What is the cost difference between the two?
  Check the prevailing cost of alternate materials and labor.

Sizing Water Piping for a Multi-Dwelling Building

Here's a suggested method for sizing a water supply pipe for a multi-dwelling building which has 30 fixture units:

1) Find the elevation of the highest fixture above the street main, in feet. For example, let's say the highest fixture is 30 feet above the street main.
2) Multiply this height, in feet, by 0.43 to convert it to psi. This gives you 13 psi.
3) Select the required pressure at the highest fixture, in psi. Let's say you want at least 15 psi at the highest fixture as permitted by the plumbing code.
4) Compute the loss of pressure due to the height of the fixture that's the highest up from the meter. This is 30 times 0.43, or 13 psi.
5) Ask the water company what the average daily service pressure at the main is. Let's say they tell you it's 75 psi.
6) Subtract the sum of the minimum required pressure at the highest fixture (15 psi), and the loss of pressure due to the height that fixture is above the meter (13 psi) from the average daily service pressure. This gives you 75 - 15 - 13, or 47 psi. This is the available water pressure at the highest outlet.
7) Now you need to subtract the friction loss at the meter. This will depend on the flow rate, or demand load, which in turn, depends on the total number of fixture units in the system. First check Chart A-2 Estimate Curves for Demand Load in

Appendix A of the *UPC* to find out the flow rate for 30 fixture units. The answer is 20 gpm. Then check Chart A-1 in the same appendix for that flow rate and, let's say, a 1-inch meter. This point indicates a pressure loss of 2.5 psi. Therefore, the available water pressure after the meter is 47.0 - 2.5, or 44.5 psi.

8) Figure out the developed length of pipe from the water meter to the highest fixture. Let's say that you have 50 feet of pipe, and fittings of another 5 feet, giving a developed pipe length of 55 feet.

9) Divide the available water pressure after the meter by the developed length of pipe. That's 44.5 psi divided by 55 feet. Then the allowable friction loss in the pipe is 0.81 psi per foot.

10) Multiply the friction loss per foot (0.81 psi) by 100 to get average permissible friction loss for 100 feet of pipe.

11) Now check Chart A-4 in the *UPC* for a 20 gpm flow and 81 psi friction loss, to find what size pipe you can use. The chart shows that a 1-inch Type M copper pipe will give you a friction loss of about 80 psi at 20 gpm. So, you can use that size pipe to supply adequate pressure to the highest fixture.

Hydraulics

Unfortunately, plumbing design, hydraulics, and physics use different terms to mean the same thing. The terms in the plumbing code are usually simpler and less technical than those in hydraulics and physics. The plumbing code also uses charts and graphs instead of formulas. Hydraulic terms are more complex because they apply to different kinds of fluids and units. Water flow can be given in gallons per minute, cubic feet per second, or pounds per minute. Sometimes, hydraulic terminology is used to explain plumbing terms.

Here are the terms used in plumbing codes for water distribution:

- Flow, Rate of Flow, or Demand in gpm
- Pressure, in feet of water column or psi
- Pressure Loss, or Pressure Drop, in psi
- Fixture and Drainage units, in arbitrary numbers which can be converted to approximate gpm
- Friction Loss, in psi per 100 feet of pipe
- Grade of Pipe or Pipe Slope in fractions of an inch per foot
- Water Velocity, in fps
- Equivalent Length of pipe, in feet per type of fitting
- Pipe Roughness, as smooth, fairly smooth, fairly rough, or rough

For gas distribution, the terms are:

- Gas Demand or Demand, in cfh or Btus per hour
- Gas Pressure, in inches of water column or psi. (Medium gas pressure is over 14 inches water column, but less than 5 psi.)

For rainfall, the plumbing terms are:

- Rainfall Intensity, in inches per hour
- Gutter Slope, in fractions of an inch per foot

Common terms that hydraulic engineers use are:

- Pressure or Force on a unit area in psi
- Static Pressure, or Head, in feet of a column of water, or psi. (One foot column of water equals 0.4335 psi.)
- Gas Pressure, in inches of column of water. (One inch column of water equals 0.03613 psi.)
- Velocity Head or pressure caused by the flow of water, in feet of water column or psi
- Energy or Work in foot-pounds or watt hours
- Pressure Energy or capacity of water under pressure to do work, in feet or height of water column in feet
- Potential Energy or Head or capacity of water at higher elevation to do work, in height of water column in feet or psi
- Mechanical Energy or capacity of a machine to do work, such as a pump, in height of water column in feet or psi
- Velocity Energy or Velocity head which is work created by the flow of water, in height of water column in feet or psi
- Total Energy which is the sum of Potential Energy, Pressure Energy, Velocity Energy, and Mechanical Energy, in height of water column in feet or psi
- Water Flow in cfs, gpm, or lbs per minute. (One pound of water equals 0.1198 gallons.)
- Gravitational factor (g) which is the rate of acceleration of a falling body due to gravity or 32 ft/sec^2
- Friction factor (f) which is a measure of the roughness on the interior surface of a pipe expressed as a decimal number such as 0.015
- Grade of Pipe or Pipe Slope, in decimal parts of a foot or percent of a foot per foot

Here's a very brief and simplified description of hydraulics as related to plumbing. If you're a little rusty on physics and algebra you may find the formulas difficult. They will give you a general idea of hydraulic engineering though.

A plumbing system gets energy from:

- the height of a column of water in the system

- the pressure of the water in the system
- the velocity of the water in the system

All of these are expressed in feet. If you reduce the height of a column of water, the energy in the system is less. The friction of water moving inside a pipe also reduces the energy in a plumbing system. The total energy in a plumbing system is called Total Head, or H. The pressure of the water in a system is called the Pressure Head, or p/w. The velocity of the water is called the Velocity Head, or $V^2/2g$.

Energy can be changed from one form to another. For example, if a pipe is connected to a large elevated tank and the outlet of the pipe is closed, the fluid in the pipe is under pressure, and the Pressure Head equals the Total Head, or $p/w = H$. If a valve at the pipe's outlet is opened and water flows through the pipe, the velocity of the water is related to the Total Head. Here's the formula:

$$V = \sqrt{2 \times g \times H}$$

The quantity of water, Q, that flows through a pipe depends on the cross-sectional area, A, of the pipe, or:

$$Q = V \times A$$

If there's no friction in the pipe, the Velocity Head plus the Pressure Head will equal the Total Head. The formula for this is

$$\text{Total Head} = \frac{p}{w} + \frac{V^2}{2g}$$

But in most pipes, the friction of the water moving along the pipe cuts down the Total Head. This loss is expressed as

$$\left(f \times \frac{l}{d}\right) \times \left(\frac{V^2}{2g}\right)$$

where

f = a decimal number which is the friction factor of the pipe

l = length of pipe in feet

d = the inside diameter of the pipe in feet

p = the pressure of the water in psf

V = the velocity of the water in fps

g = 32 ft/sec/sec

Figure 7-7 shows how the velocity of the flow from an outlet depends on the height of the water above the outlet.

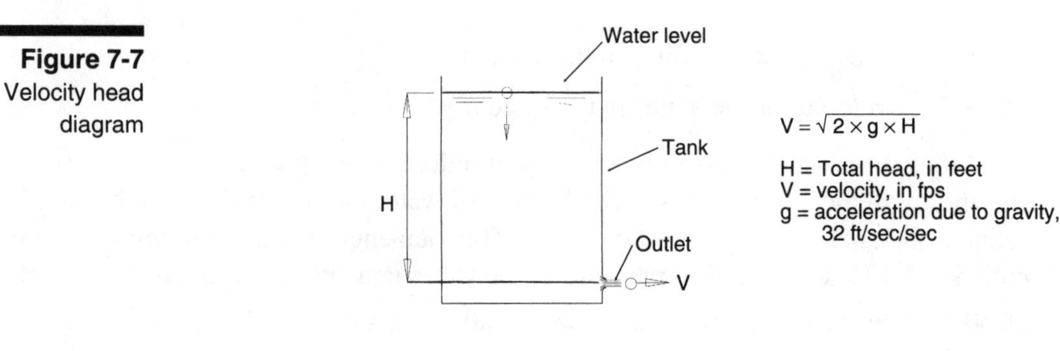

Figure 7-7 Velocity head diagram

Water Supply

Service

The drinking (domestic) water for a building usually flows from the water main in the street to a water meter at the property line. Between the meter and the house there should be a full opening valve that shuts off water to the house and yard. If the pressure in the public water main is more than 80 psi, add a regulator to reduce the pressure to less than 80 psi. The minimum water pressure at the outlet that's farthest from the water meter should be 15 psi.

In multiple-unit housing projects, you can usually use copper, galvanized steel, or polyvinyl chloride (PVC) pipe to pipe from the meter. Use a ¾-inch or 1-inch service pipe to get water from the yard pipe to each dwelling unit. A yard pipe distributes water throughout a complex and service pipes supply water to each building. Make sure the service pipe is in solid ground below the frost line.

In cold areas, pipes can freeze and burst. Make sure that all piping carrying fluids is protected from frost. Insulate pipes in exterior walls against freezing. Install pipes in unheated areas at a slope so they can be drained before the cold season.

Here are some possible causes of insufficient water supply:

- scale buildup in the pipe
- pressure regulator set incorrectly
- reduced city water pressure
- non-functioning balancing valves, flow controllers, or other devices designed to equalize the water flow
- control devices plugged by sediment in the water

Pressure drop

Too much pressure drop reduces water flow. The pressure drop per 100 feet of Schedule 40 steel pipe due to friction at various flow rates is shown in Figure 7-8.

Figure 7-8
Pressure drop per 100 feet of Schedule 40 steel pipe due to friction at various flow rates

Pipe size	Flow, gpm	Velocity, fps	Pressure drop per 100 feet, psi
2 inch	35	3.35	1.10
1½ inch	35	5.51	3.83
1¼ inch	35	7.51	8.25
1 inch	20	7.44	11.05
¾ inch	15	9.02	22.00

Noisy piping

Water hammer is the banging noise you hear in some piping systems when the water is shut off suddenly. It's more common and more severe when water is flowing very fast through the pipes. To avoid water hammer, keep the maximum velocity of water through a pipe under 8 fps for interior pipes, and 10 fps for exterior pipes. The velocity of flow is found by the formula V = Q/A, as just described.

Piping can vibrate when water flows through it. This vibration is transmitted to the building where the pipe is attached to the framing. Vibration is worse in thin wall Type M copper pipe than in the heavier Type L or K copper pipes, which are more rigid. For comparison, the wall thickness and weight of three types of copper pipe are:

- Type M ½ inch pipe has a 0.028 inch wall and weighs 0.204 plf
- Type L ½ inch pipe has a 0.040 inch wall and weighs 0.285 plf
- Type K ½ inch pipe has a 0.049 inch wall and weighs 0.344 plf

Water Distribution

The typical water system of a multiple-family building has:

- cold water headers, including valves and pumps. A cold water header is a horizontal pipe which supplies a number of fixtures with water.
- risers, to feed units in upper stories
- water heater and hot water storage tank
- hot water loop, to carry hot water continuously from the storage tank to all portions of a building
- distribution hot water piping from the hot water loop to each plumbing fixture that uses hot water
- distribution piping from the cold water headers to the plumbing fixtures
- accessible shut-off valves at each plumbing fixture, appliance, and water heater.

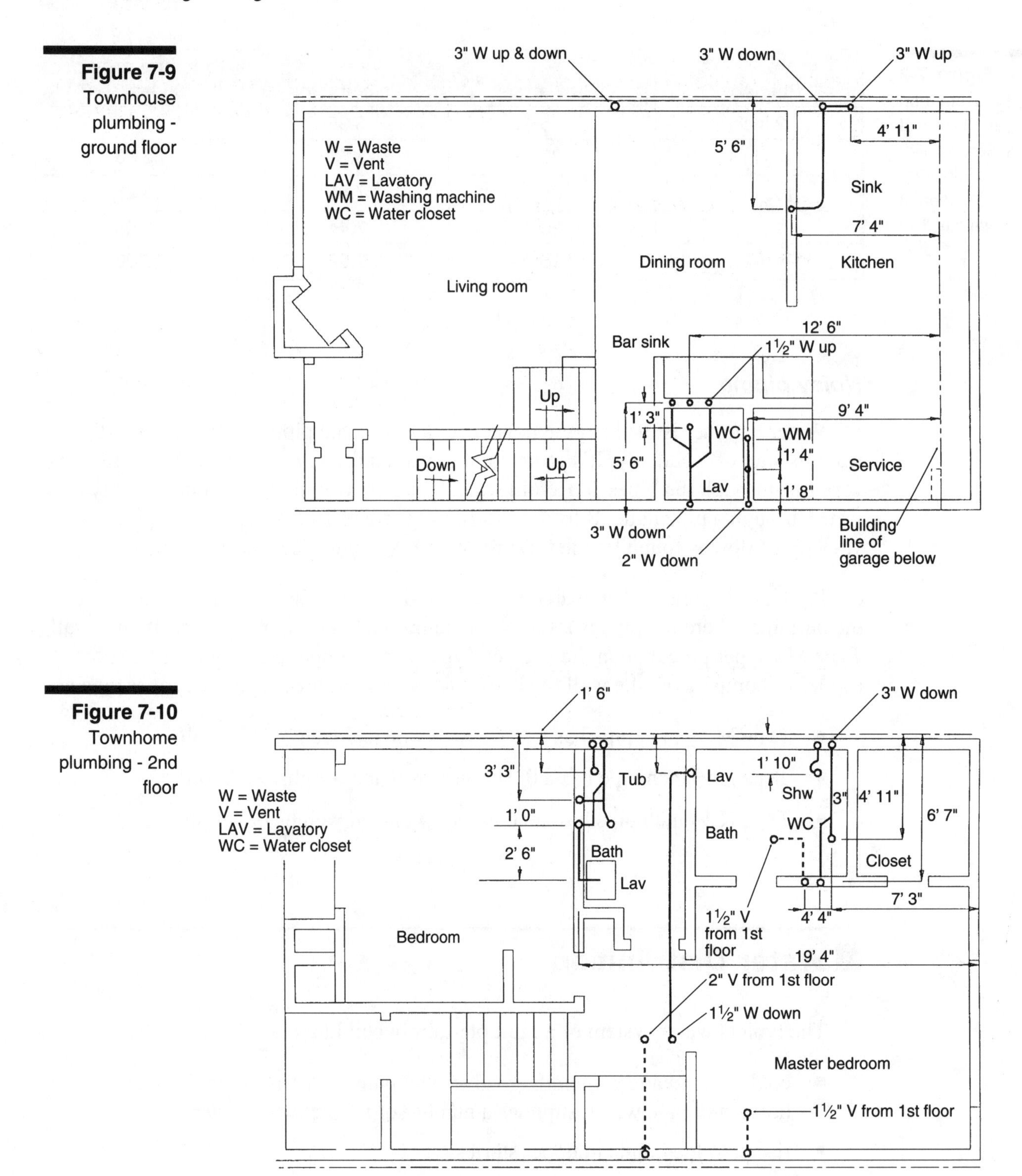

Figure 7-9 Townhouse plumbing - ground floor

Figure 7-10 Townhome plumbing - 2nd floor

Figures 7-9, 7-10, and 7-11 show a simplified plumbing layout for a typical townhouse unit. The drawings show the cold and hot water distribution and the waste system. Figure 7-9 shows the plumbing at the ground floor level. Figure 7-10 shows the second floor plumbing and Figure 7-11 shows the waste plumbing in the garage level. You should prepare a material list which shows the piece number, size, type of pipe, and fittings for each section of pipe. Here's an example of part of the material list for plumbing this townhouse:

HOT WATER - PLAN "A" UNIT #15

Piece No.	Size	Pipe: galv.	Fittings: galv.
61	3/4"	2'-3"	Ell & coupl
62	3/4"	1'-7½"	Ell
63	3/4"	SH NIP	Ell
64	3/4"	SH NIP	3/4 × ½ Tee
67	3/4"	2'-7"	3/4 × ½ Tee
69	½"	3'-6"	Ell
70	½"	1'-8½"	Ell
71	½"	1'-8½"	Ell
72	½"	1'-3"	Ell
73	½"	4'-0"	Ell
74	½"	10'-5"	Ell
75	½"	6" NIP	Ell
76	½"	1'-8½"	Ell

TOP OUT HOT WATER - PLAN "A" UNIT #15

Piece No.	Size	Pipe: galv.	Fittings: galv.
81	3/4"	6'-7"	3/4 Tee
82	3/4"	5'-11"	Ell
83	3/4"	12'-8½"	3/4 × ½ Tee
84	3/4"	5'-0"	Ell
85	3/4"	SH NIP	½ × ½ × 3/4 Tee
87	½"	4'-7"	Ell
88	½"	1'-8"	Ell
89	½"	4'-7"	Ell
90	½"	2'-3"	No ftg
91	½"	0'-8"	Tee
92	½"	1'-8"	Ell
93	½"	6'-0"	Ell
94	½"	1'-7½"	Ell
95	½"	4'-3"	No ftg
96	½"	(2 ea.) SH NIP	Ells

Ell = Elbow SH NIP = Short nipple ftg = Fitting

Hot Water Usage

When a homeowner draws hot water from the hot water heater, cold water flows into the system and lowers the temperature of the hot water still in the tank. When he draws off enough water to lower the temperature of the water in the tank below a

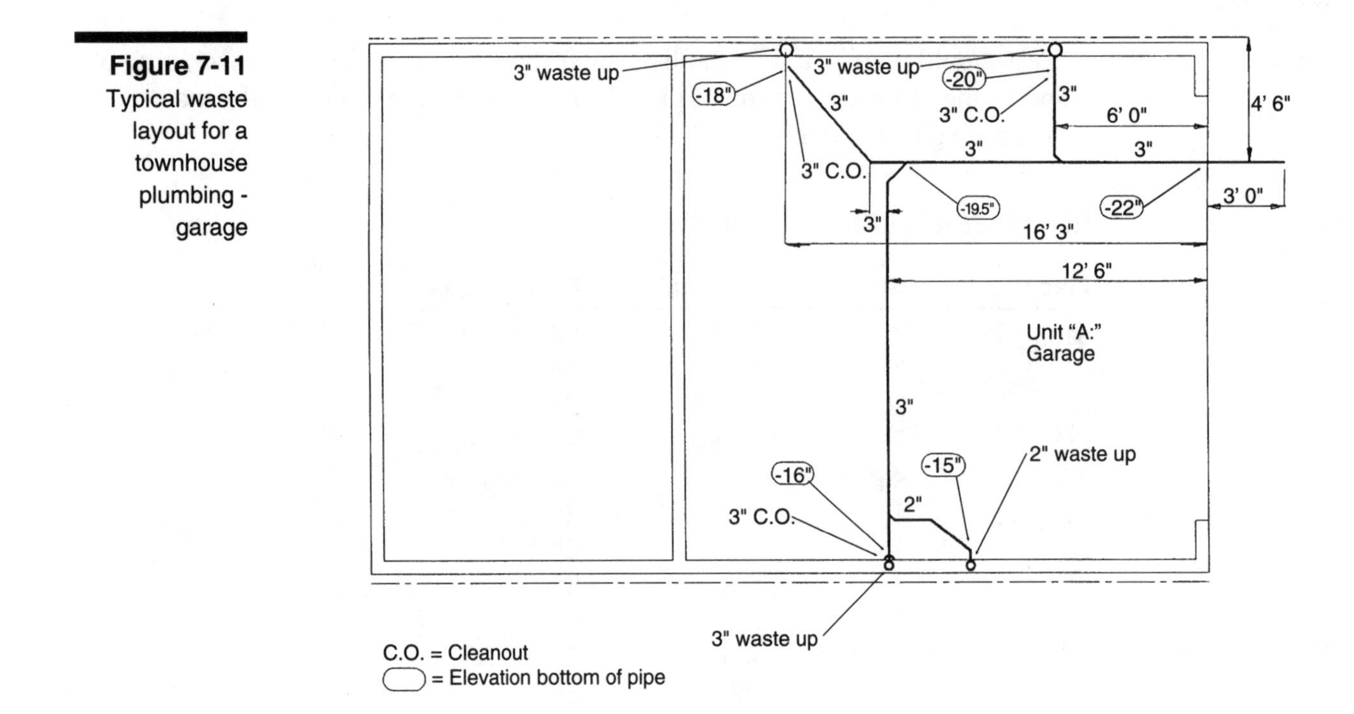

Figure 7-11 Typical waste layout for a townhouse plumbing - garage

preset temperature, the burners come on to heat the stored water. As long as the homeowner doesn't draw hot water faster than it can be replaced, he'll be happy with the system. Your task is to figure just what size heater will do the job.

The size of a water heater is based on its thermal input. Its output is measured in Btus per hour (Btuh). A Btu, or British thermal unit, is the amount of heat required to raise the temperature of 1 pound of water 1 degree Fahrenheit. To convert the output of water heater in Btuh to gallons, divide the output in Btuh by 8.33. The actual recovery capacity of a water heater is:

$$\text{Recovery (gph)} = \frac{\text{input Btuh} \times \text{eff.}}{8.33 \times \text{temperature rise}}$$

where

eff. is determined by the water heater manufacturer based on their laboratory tests.

For example, a 20,000 Btuh heater may produce 28 gph of hot water with a 60 degree F temperature rise, or 16.1 gph with a 100 degree F rise. However, a 100,000 Btuh heater may produce 140 gph with a 60 degree F temperature rise, or 84 gph with a 100 degree F rise. The difference is in efficiency. Figure 7-12 lists typical recovery rates in gph for hot water heaters, with efficiencies included.

Figure 7-12
GPH recovery at indicated temperature rise

Input Btuh	60 degrees gph	70 degrees gph	80 degrees gph	90 degrees gph	100 degrees gph
18,000	28.2	21.6	18.9	16.8	15.1
20,000	28.0	24.0	21.0	18.7	16.1
25,000	35.0	30.0	26.2	23.3	21.0
30,000	42.0	36.0	31.5	28.0	25.2
33,000	46.2	39.6	34.7	30.2	27.7
35,000	49.0	42.0	36.8	32.7	29.4
40,000	56.0	47.6	42.7	37.0	33.6
43,000	60.2	51.6	45.2	40.2	36.1
50,000	70.0	60.0	52.5	46.7	42.0
60,000	84.0	72.0	63.1	56.0	50.4
70,000	98.0	84.0	73.5	65.1	58.8
80,000	108.0	96.0	84.0	74.4	67.2
90,000	126.0	108.0	94.5	83.7	75.6
100,000	140.0	120.0	105.0	93.0	84.0

Here's a quick way to select what size water heater a single-family residence requires. Suppose a family of four lives in the house, which has two full bathrooms, an automatic dishwasher, and a washing machine. At the two hour peak load from 7:00 am to 9:00 am, the family could use:

2 persons at 20 gallons per person	40 gallons
2 persons at 5 gallons per person	10 gallons
Second full bath	10 gallons
Automatic dishwasher	10 gallons
Automatic clothes washer	20 gallons
Total	90 gallons

However, demand for hot water is proportional to the number of people using it at the same time, not the number of people using it whenever. The 90-gallon demand is over a 2-hour period, which is equivalent to 45 gallons per hour. So a water heater with a recovery rate of 45 gallons per hour would meet the requirements. Don't include the storage capacity of the tank when you figure this because you should assume the water stored in the tank is already used up.

To design a hot water system for a large building, base the demand on:

- the bathing (shower) load
- the type of shower heads
- the peak demand period you calculated the shower load on

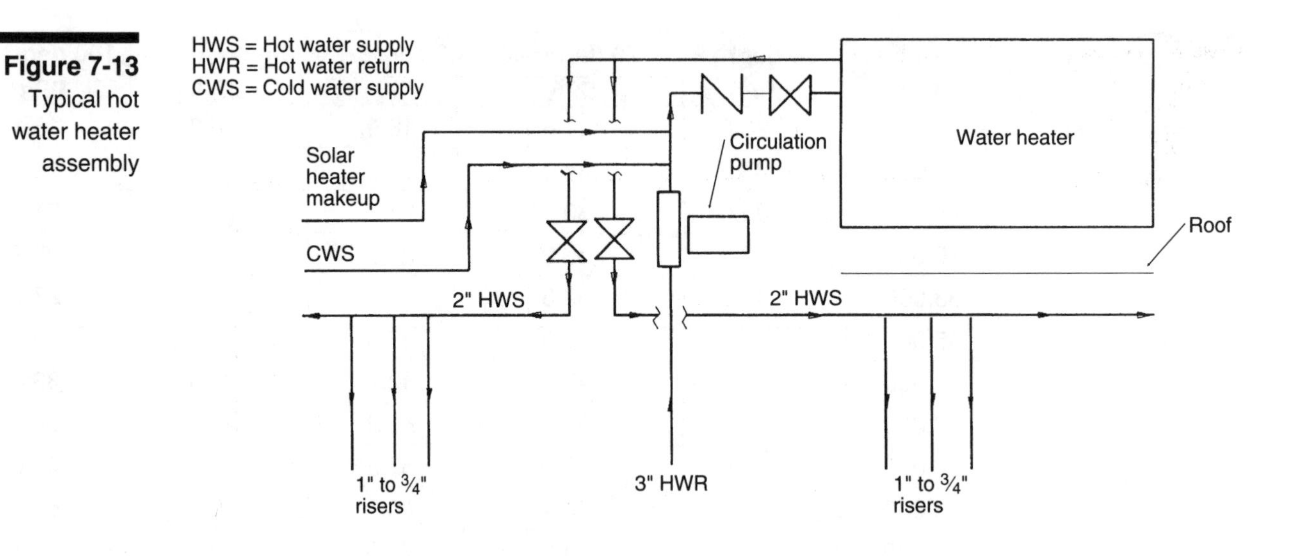

Figure 7-13 Typical hot water heater assembly

Most water heater manufacturers provide simple tables for sizing their water heaters. These are based on the type of dwelling, number of bedrooms and dwelling units served by the water heater. The peak demand may vary between 1 to 3 hours depending on the occupancy. The usual peak period for multiple-unit dwellings is 2 hours. The average occupancy is 2½ persons per unit.

Showers use the most hot water. The maximum flow from a shower is about 4.5 gpm at 105 degrees F. Assuming a mixture of 3 gpm at 140 degrees F and 1.5 gpm at 40 degrees F, for a duration of five minutes, a single shower requires 15 gallons of 140-degree F water.

A typical three-story condominium building with 12 two-bedroom units has a single hot water heater assembly on the roof. Assume that the capacities of the water heater and storage tank are:

- 330,000 Btuh input
- 264,000 Btuh output
- recovery rate of 396 gph with an 80-degree F temperature rise. If the makeup water is at 60 degrees F, this heater would produce 396 gph at (60 + 80) 140 degrees F. The temperature of the makeup water will depend on the temperature of the locale, so it could be near freezing or as high as 80 degrees F.
- vertical storage hot water tank, 24-inch diameter by 48 inches high, with 115 gallon capacity.

Here's a list of a typical hot water heater assembly. Some of the items are shown in Figures 7-13 and 7-14, which show two common types of piping arrangements for a hot water heater in a multiple-unit dwelling:

- cold water supply, CWS
- hot water supply, HWS
- hot water return, HWR

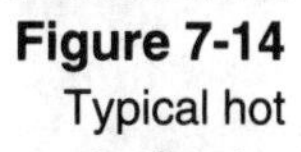

Figure 7-14 Typical hot water heater assembly

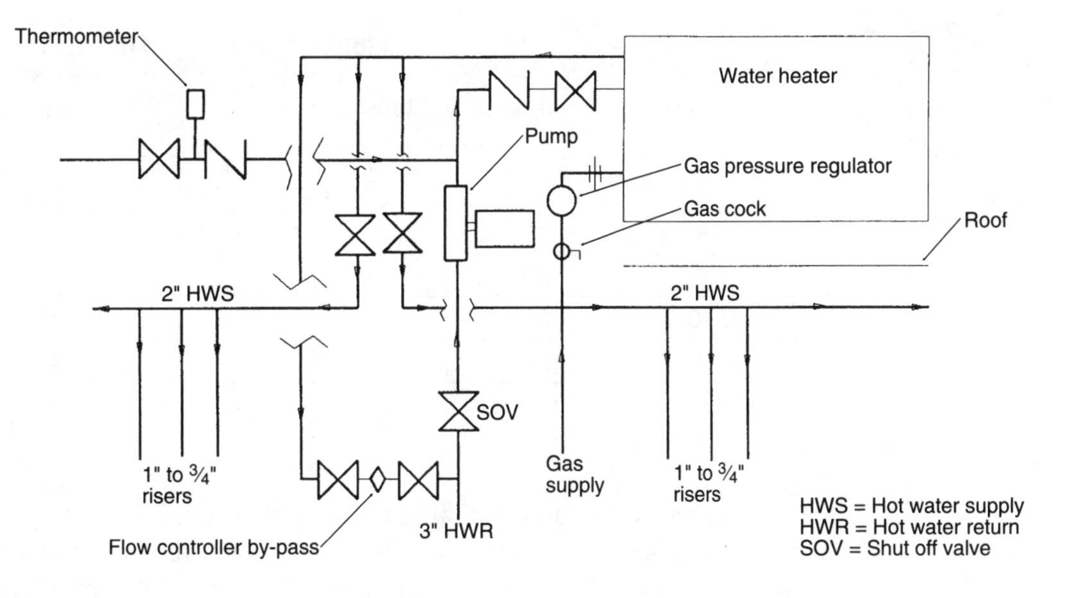

- fuel gas, FG
- boiler circulating pump
- hot water loop circulating pump
- flow controller switch
- gas pressure regulator
- pressure relief valves (part of water heater)
- thermometer
- thermostatic mixing valves
- shut off valves, (SOV)
- energy shut off valves, shuts off gas at 210 degrees F
- aquastat
- header and return header

Most water heaters are labeled to show their recovery rate, input rating, output rating, recovery efficiency, and temperature rise. The hot water storage tank is usually labeled with its capacity in gallons. The booster, or circulating pump, also has a nameplate describing its characteristics, such as horsepower, capacity in gallons per minute (gpm), head in feet of water, voltage, phase, and amperage. A head of water is the height of the fluid, in feet. It's used to measure pressure. Therefore, a 10-foot height, or head, of water has a pressure of 624 pounds per square foot or (624/144) 4.33 psi.

Usually, the hot water loop starts at the water heater assembly on the roof of a building. Hot water supply pipe goes down to the ceiling of the first floor, loops horizontally to all units, and then rises back to the roof. Most hot water distribution systems are made of Type L copper tubing. Be sure to insulate exposed piping on the roof. You don't have to insulate hot water piping that's inside insulated walls.

Figure 7-15
Hot water requirements for 3 hour period

No. units	Occupants	Gallons required for 100° temperature rise	Gallons required for 80° temperature rise	Storage tank capacity
1-3	7	124	112	50
4	10	165	138	60
5-6	15	247	207	72
7-8	20	330	275	85
9-10	25	412	345	100
11-15	37	562	470	113

The hot water loop in a building not only provides hot water near every fixture, it also stores hot water. A tankless hot water heater relies on the storage capacity of the hot water loop. Here's the capacity, in gallons, for 100 feet of copper pipe of various sizes:

Pipe size	Volume per 100 feet, in gallons
½ inch	1.21
⅝ inch	1.81
¾ inch	2.51
1 inch	4.28
1¼ inch	6.52
1½ inch	9.25
2 inch	16.06
2½ inch	24.78

Storage tanks

A hot water storage tank, also called an expansion tank, is usually made of steel, and lined with glass, plastic, and zinc. A tank should carry an ASME label describing the pressure and capacity it's designed for. It should be insulated. A pipe tank is a hot water storage tank which has a valve you can open to drain it to a floor sink. Make sure the floor or roof where the tank rests is strong enough to carry the weight of the tank and the water in it. Place the tank so that manholes, inspection cover, name plates, and drain valves are accessible.

It's a good idea to use a storage tank that's equipped with magnesium or zinc anode rods to reduce deterioration and scaling on the piping, heater tubes, and the tank itself. You should connect the return line from the circulating loop close to the bottom of the tank. Be sure to start the hot water supply line to the loop at the top of the tank.

Figure 7-15 lists hot water requirements for a 3 hour period for multiple-dwelling buildings with various numbers of units.

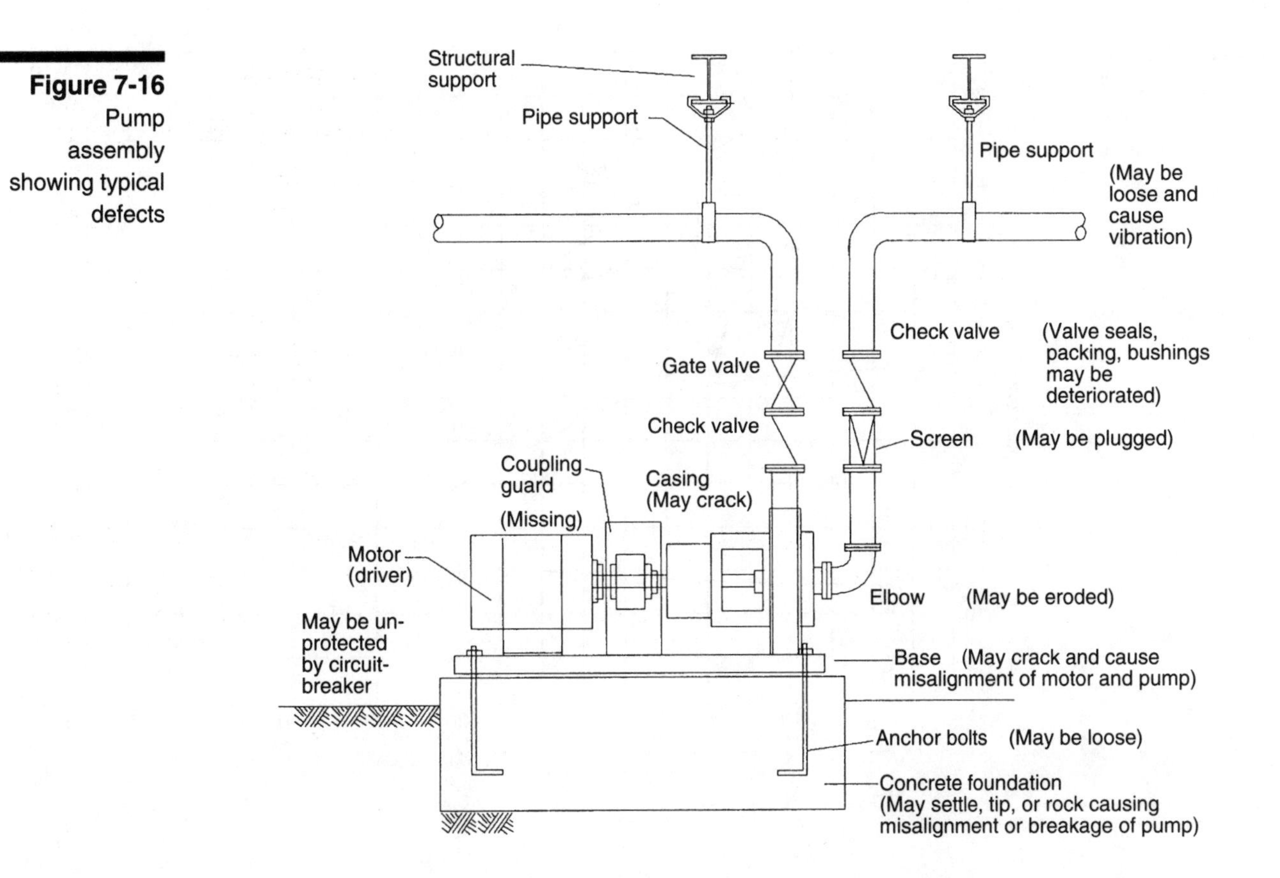

Figure 7-16 Pump assembly showing typical defects

Pumps

You may need circulating pumps between the water heater and the storage tank, in the building loop, or both. To figure out if a circulating pump is the right size, you need to know its model number, rpm, horsepower, pipe size and length, and flow rate. Figure 7-16 shows the major parts of a pump assembly, and common installation faults to watch for. The manufacturer of a pump usually provides a curve which shows the delivery of the pump in gpm and the delivery against the head, in feet. There may be several curves on the same chart, each curve showing a different capacity pump or a different size motor. Figure 7-17 shows delivery vs. head for a $\frac{1}{3}$, $\frac{1}{2}$, and $\frac{3}{4}$ hp motor.

Here's how you can use Figure 7-17 to pick a pump and motor for a particular delivery, say 30 gpm against a particular head, say a 40-foot head. The intersection of the 30 gpm line and 40-foot head line is above the $\frac{1}{3}$ hp curve and under the $\frac{1}{2}$ hp pump curve. Therefore, select the $\frac{1}{2}$ hp pump which has an efficiency rating of about 63 percent.

If the building hot water loop is over 60 feet in length, the suction connection to the tank may not be adequate because the friction loss is too great. In this case, it's a good idea to put a separate circulating pump in the return line to the boiler.

Pump design uses these units:

- Foot-pound (ft-lb), which is a unit of work.

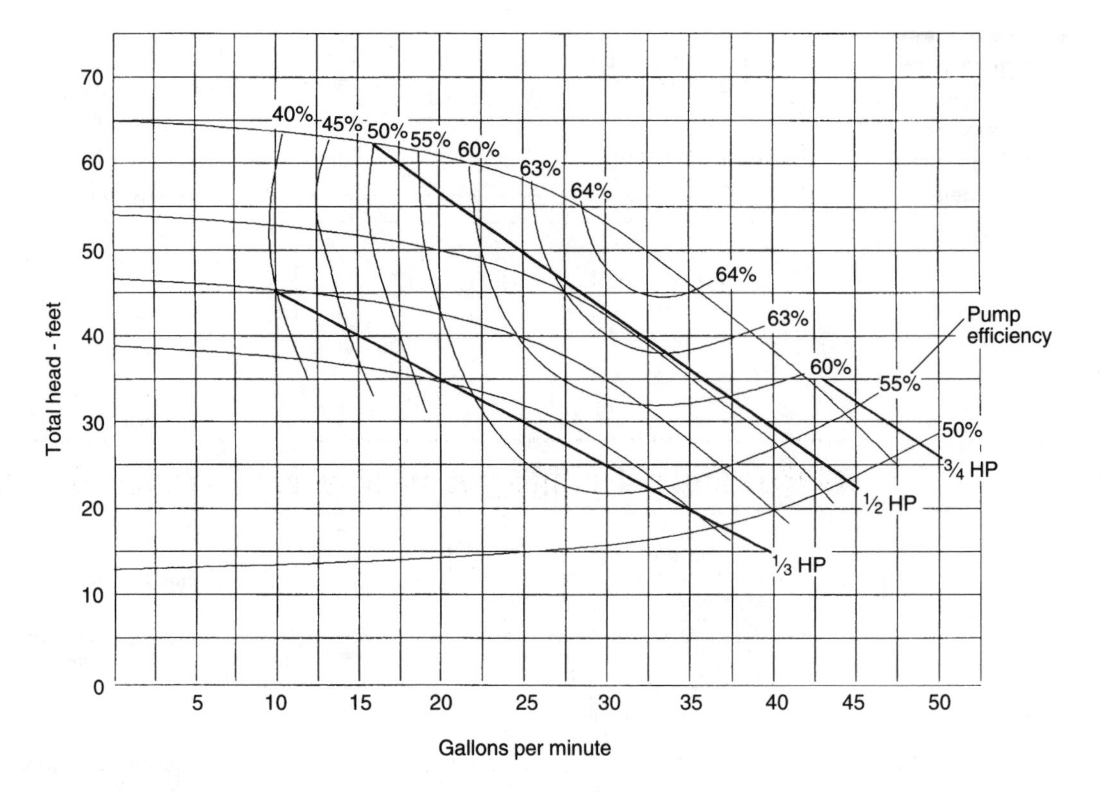

Figure 7-17 Typical pump curve

- Horsepower (hp), which is 33,000 ft-lbs per minute, or 0.746 kilowatts.
- Watt-hour (w-hr) which is a small unit of electrical work.
- Kilowatt-hour (kwh) which is a large unit of electrical work equal to 1000 watt-hours.

The cost of pumping water through a circulating system is 0.189 times the power cost per kwh, times head, divided by the product of pump efficiency and motor efficiency, or:

$$\text{Cost \$} = \frac{0.189 \times \text{\$ per kwh} \times \text{ft head}}{\text{pump efficiency} \times \text{motor efficiency}}$$

For example, suppose a pump delivers 100 gpm against a 50 foot head, power costs 10 cents per kwh, it has 75 percent pump efficiency, and 85 percent motor efficiency, then the cost is:

$$\frac{0.189 \times 0.10 \times 50}{0.75 \times 0.85 \times 60} = \text{\$0.025 per minute}$$

You need the 60 here to convert the cost from dollars per hour to dollars per minute.

Here are some other useful formulas for pumps:

$$\frac{\text{Kilowatt hours used}}{\text{by a pump motor}} = \frac{\text{number}}{\text{of hours}} \times 0.746 \times \frac{\text{horsepower rating of the pump}}{\text{pump efficiency}}$$

$$\text{Pump efficiency} = \text{gpm} \times \frac{\text{head in feet}}{3960} \times \text{brake horsepower of the pump's motor}$$

Brake horsepower, commonly used by pump manufacturers, is the horsepower absorbed by a brake, or the work required to stop the pump's motor.

$$\text{Head in feet} = 3960 \times \text{pump efficiency} \times \frac{\text{brake horsepower of the pump}}{\text{gpm}}$$

Possible Problems with a Hot Water System

Some things that can go wrong in a hot water system are:

- No hot water because the central water heater, circulating pump, or temperature control valve, which monitors the amount of makeup water, isn't working properly.
- Less and less hot water over a long period due to scale buildup in the water heater or piping.
- Varying-temperature hot water in different living units, which may be caused by heat loss between the beginning and end of the hot water loop due to insufficient insulation. Some units may be too far from the water heater. Units closer to the beginning of the loop get scalding hot water while those near the end of the loop get only lukewarm water.

Comfort Heating and Cooling

A building's comfort cooling and heating system, which is normally installed by a mechanical contractor, also uses water. Its main parts are:

- chiller, condenser, space water heater, pumps, mains, and associated items usually located on the roof
- chilled water loop from the chiller to the lowest part of the building
- hot water loop from the water heater to the lowest part of the building
- distribution supply, return chilled water, and hot water lines that feed the cooling and heating units in each dwelling unit.

In summary, pick a hot water generating system to accommodate the peak period bathing load using the recovery rate of the water heater and the capacity of the storage tank. Ideally, the system should store water at or near 140 degrees F. Water at higher temperatures can cause scalding.

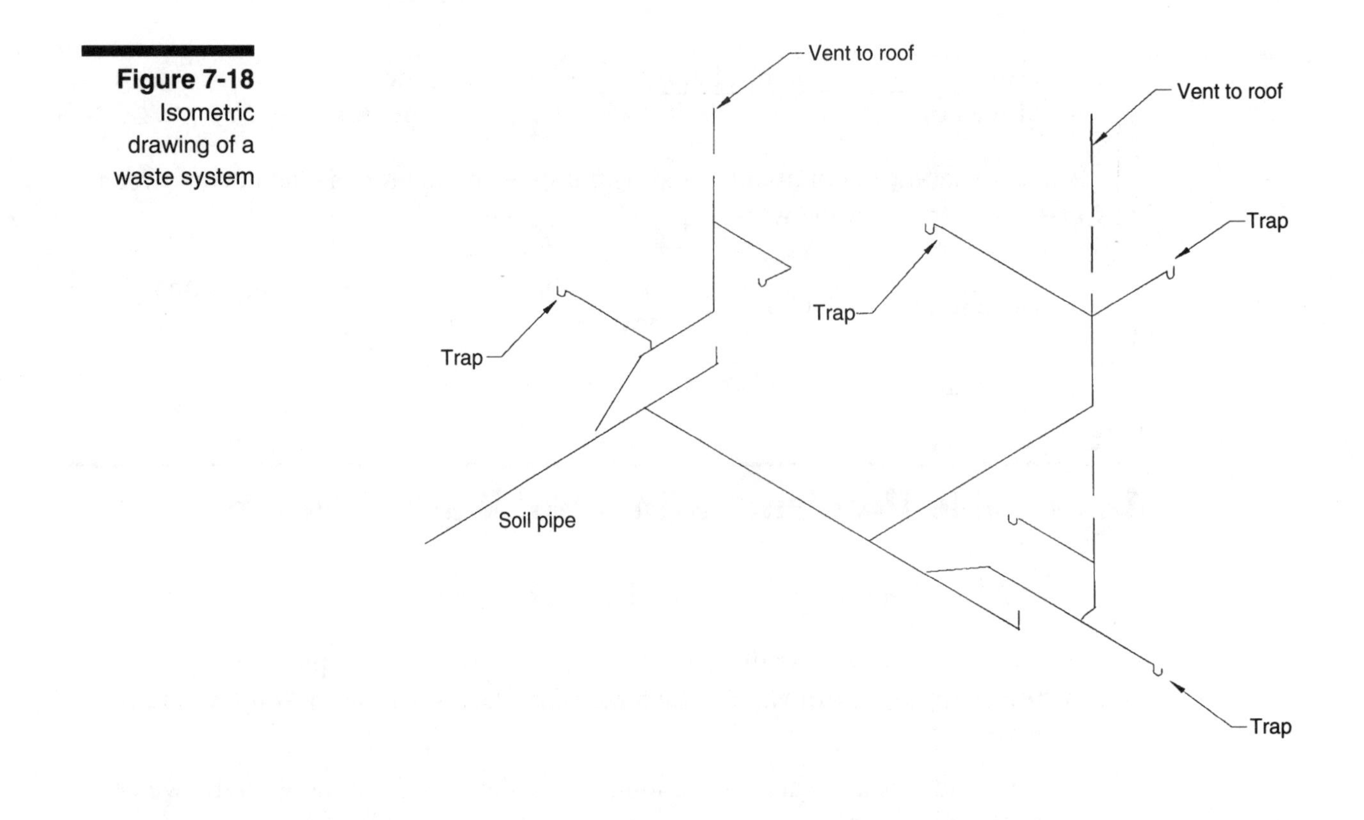

Figure 7-18 Isometric drawing of a waste system

Waste Systems

A well-designed waste system should provide adequate flow of wastes, protect against cross-connection, and keep noxious fumes from escaping. Drainage and vent piping may be cast iron, galvanized steel, copper, or brass. Buildings up to three stories in height may use ABS or PVC-DWV plastic pipe. Figure 7-18 is an isometric drawing of a waste system. The major parts of a waste system are:

- soil pipes to carry the discharge from toilets and urinals
- waste pipes to carry only liquid waste that's free of fecal matter
- traps which have special fittings, to make a liquid seal against back passage of air
- vent pipes to circulate air in the system to prevent siphoning or back pressure
- cleanouts, with removable plugs, so the piping can be cleaned

Most waste lines and sewers are designed to operate half full under gravity flow conditions. One of the formal hydraulic formulas used to find discharge or velocity in gravity flow in storm drains and larger pipes in feet/second is called the *Manning Formula.* This is beyond the scope of this book, but you can get information on it from reference books on hydraulic design.

For plumbing design, you can use the simple tables in the *Uniform Plumbing Code.* The chart in Figure 7-19 shows the approximate rate of flow based on a half-full waste pipe at various slopes.

Figure 7-19 Rate of flow of half-full waste pipes

Design size, inches	Slope, inches per foot	Flow, gpm
2	1⁄4	8.2
3	1⁄4	25.0
2	1⁄8	6.0
3	1⁄8	17.0
4	1⁄4	47.0
5	1⁄4	100.0
4	1⁄8	39.0
5	1⁄8	70.0
4	1⁄16	27.0
5	1⁄16	50.0

Figure 7-20 Sizing drain pipes by fixture units

Item	Fixture units	Drain size
Toilet	4 fu	3-inch drain
Tub	2 fu	1½ inch drain
Shower	2 fu	2-inch drain
Sink/dishwasher	2 fu	1½ inch drain
Lavatory	1 fu	1¼ inch drain
Washing machine	2 fu	2-inch drain
Bar sink	1 fu	1½ inch drain
Floor drain	2 fu	2-inch drain

Figure 7-21 Maximum number of fixture units for piping at ¼" per foot slope

Pipe size	Fixture units	Pipe size	Fixture units
1¼ inch	1 fu	4-inch	216 fu
1½ inch	2 fu	5-inch	428 fu
2-inch	8 fu	6-inch	720 fu
2½ inch	14 fu	8-inch	2640 fu
3-inch	35 fu	10-inch	4680 fu

Use the number of drainage fixture units discharging into a pipe to figure out how big it needs to be. Some values are shown in Figure 7-20. Assuming a slope of ¼ inch per foot, the maximum number of fixture units a pipe can carry is shown in Figure 7-21.

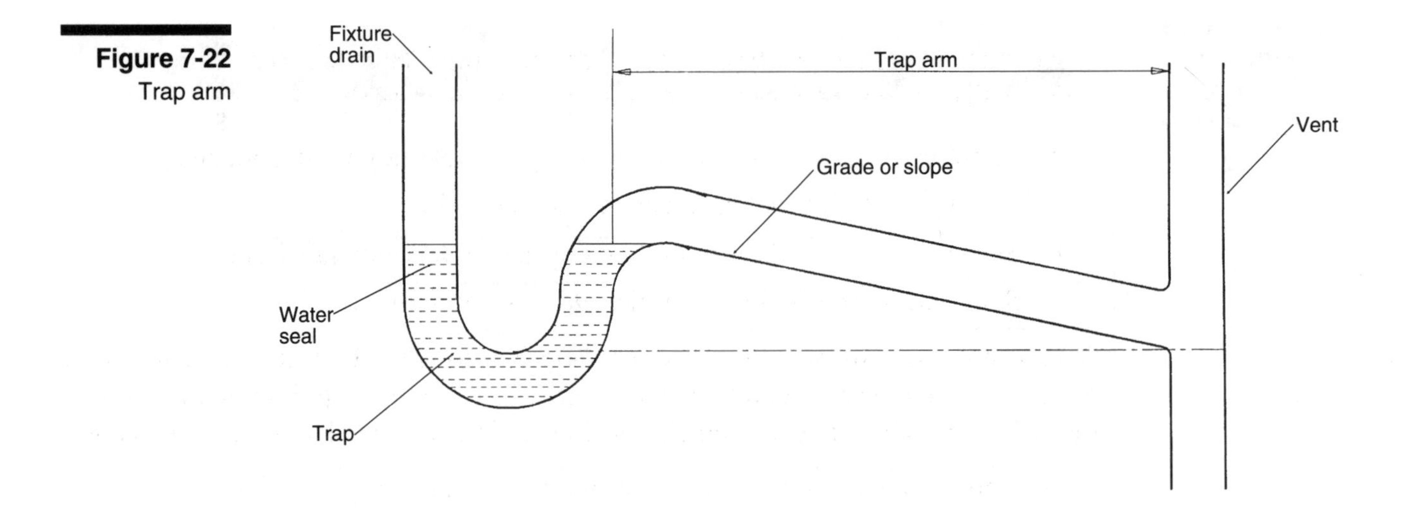

Figure 7-22
Trap arm

Table 4-3 of the *Uniform Plumbing Code* provides the basis for sizing drainage and vent piping. This is based on the number and type of plumbing fixtures, and length, size and slope of drain pipes. For example, a 6-inch drain set at a ¼-inch-per-foot slope can handle waste from 720 fixture units.

Traps and Vents

The U-shaped trap on a plumbing fixture collects water at the low point of the U. This liquid seal keeps odors, gases, (and even rats) from escaping into the building from the sewer system. For health reasons, fixture traps are very important.

Each trap should be close to a vent. The maximum grade and distance between the trap and a vent pipe should be:

Size of pipe	Max. grade, per ft	Max. distance
1¼ inch	1¼ inch	5 feet
1½ inch	1½ inch	6 feet
2 inch	2 inch	8 feet
3 inch	3 inch	12 feet
4 inch	4 inch	16 feet

The drain pipe connecting the trap with the vent is called the trap arm. Figure 7-22 shows the relationship between the trap and its arm.

Make the slope of the horizontal soil and waste piping at least ¼ inch per foot throughout the waste system, although the plumbing code permits slopes of ⅛ inch per foot for larger pipe sizes, if you have special approval.

You can use galvanized steel for vent piping if there is at least a 6-inch clearance between the pipe and the ground surface. Otherwise, use cast iron pipe or approved plastic pipe. Make the horizontal vents level or sloped to drain back to the drainage pipe they serve.

Air should circulate freely through all drainage pipe. Here are some common problems that can happen if you don't put the vents in properly:

- Plumbing fixtures drain so slowly that you think they're plugged up.
- Toilets don't clear out all waste in a single flush.
- Sewer gases come into the building because the liquid trap failed.
- Soap suds come up through toilets, tubs, or floor drains.

To keep soap suds out of pipes, clothes washers located above a basement should not be connected to shower, tub or floor drains. Suds can backup through these low drains. Install a separate waste pipe for a clothes washer located above a basement.

Common problems relating to waste and vent piping are:

- Putting a vent too far from a trap so the water seal is drawn out by vacuum
- Putting a trap too far from the waste pipe so the flow from the fixture slows down
- Using a trap arm without enough slope so the flow from the fixture slows down

Cleanouts

You need a cleanout at the upper end of each run of waste pipe unless the run is less than 5 feet, serves only a sink, and is sloped more than 18 degrees. There is less chance of a solid buildup at this extreme slope. Steep slopes are given in degrees rather than fraction of inch per foot.

Put additional cleanouts where the total change of direction is more than 135 degrees (one 90 degree and one 45 degree elbow). Make sure all cleanouts are accessible. You'll need 12 inches of clearance in front of a cleanout for 2-inch pipe, 18 inches for 2½-inch pipe, and 30 inches for all horizontal cleanouts. Put a cleanout in the sewer pipe at the connection to the building drain. You need a cleanout for each 100 feet of sewer pipe. Extend cleanouts under floor slabs to the floor level and cover them with a brass cap.

Roof Drains

Size roof drains by the area drained and the intensity of rainfall in the area. Check the tables in Appendix D of the *Uniform Plumbing Code* to find out what size drain pipes you should use.

Unless you slope the roof to drain over the roof edges, you should install a roof drain at each low point of a roof. Roof drains keep rainwater from accumulating in ponds, which could overload the framing.

Equip roof drains with strainers to keep them from getting clogged with leaves and debris. The minimum inlet area of a strainer should be 1½ times the area of the roof drain pipe it's in. Always put an overflow drain or scupper beside a roof drain in case it gets clogged. A scupper is a rectangular opening in a roof parapet. The overflow drain should be at least as big as the roof drain. The area of an overflow scupper should be at least three times as big as the area of the roof drain next to it. The flow line of an overflow drain or scupper should be 2 inches above the low point of the adjacent roof. Make sure you install a separate drain pipe for an overflow drain.

Exterior vertical drain pipes, also called leaders or downspouts, are usually made of sheet metal and installed by sheet metal contractors. Interior drains are made of cast iron or plastic. Table D-1 of the *UPC* shows that a 3-inch vertical roof drain is adequate for 8800 square feet and 1 inch of rainfall; 4400 square feet for 2 inches of rainfall; and 2930 square feet for 3 inches of rainfall.

Here are the steps you take to figure out what size roof drainage system you need:

- Determine the maximum rainfall for the local area, in inches per hour
- Figure out the horizontal areas to be drained and the areas of walls which drain to horizontal areas, in square feet
- Select drain size from Table D-1 of the *UPC,* based on rainfall and total area
- Select slope of horizontal rainwater piping (⅛, ¼, or ½ inch per foot)
- Use Table D-2 to select the size of horizontal rainwater pipe based on rainfall (in inches per hour), slope of pipe, and drainage area (in square feet).
- Select slope of gutters (1/16, ⅛, ¼, or ½ inch per foot)
- Use Table D-3 to select size of gutter based on rainfall, slope of gutter, and drainage area.

Subterranean Drains

Subterranean drainage piping, or dewatering pipe, has perforated pipe or bell and spigot clay pipe with open joints. Usually you put these pipes around the exterior basement walls.

The building code requires you to provide drainage for surface water around buildings, but it doesn't give you particular specifications. How you do this will depend on the designer and soil conditions. Standard practice for subterranean drainage around basement walls is:

- Install a perforated 3 or 4 inch clay or plastic pipe at the same level as the bottom of the foundation and about 2 feet from the foundation wall.
- Surround the pipe with a pocket of crushed rock or gravel to make it easier for ground water to enter the pipe.
- Cover the gravel pocket with building paper to prevent silt from penetrating gravel and pipe.

- Slope pipe at least ⅛ inch per foot to daylight if possible, or to a seepage pit.
- Install a sump pump in a pit to discharge the collected ground water.
- Apply waterproof coating to the outside surface of the wall and foundation.

Private Sewage Systems

If no public sewage system is available, you have to put in a private sewage system. This will usually include a septic tank, leaching field or pit, or cesspool. A cesspool is a brick-lined pit which holds raw sewage. The solids remain, but the liquids seep out through the walls of the pit. The *UPC* considers a cesspool as a temporary expedient if a public sewer will be constructed in less than two years and if the soil and ground water conditions are suitable for cesspool disposal. The *UPC* may approve a cesspool as a overflow facility to an existing cesspool.

Septic Tanks

Septic tanks may be made of concrete, steel, or various types of plastic. All the surfaces of a concrete or steel septic tank should be coated with a protective bituminous material. The coating may consist of air-blown and steam-blown asphalt, conforming to ASTM C-309, Type IV - ASTM D-41, and FHA requirements. A septic tank has 2 or more compartments that sewage flows through. The sewage is kept in contact with the air for at least 24 hours. During this time, bacteria converts some of the organic matter into methane and carbon dioxide. Periodically, the settled solids must be pumped out of a septic tank. Figure 7-23 shows a typical 750 gallon precast concrete septic tank.

To figure out what size septic tank you need for a building, use the number of bedrooms, apartments, or fixture units in the building. According to the *UPC*, the recommended capacity of a septic tank for an individual residence is 750 gallons (or 15 fixture units) for a 1 or 2 bedroom residence; 1000 gallons (or 20 fixture units) for 3 bedrooms; and 1200 gallons (or 25 fixture units) for 4 bedrooms.

Don't put a septic tank less than 5 feet from an exterior wall of a house. Put the leaching pit at least 8 feet from a building or property line.

Planning a Leaching Field or Pit

A leaching field may be a bed or a trench. See Figure 7-24 for sections through a typical leaching field and leaching pit. How big it needs to be will depend on the capacity of the septic tank it's used with and the type of soil it's in.

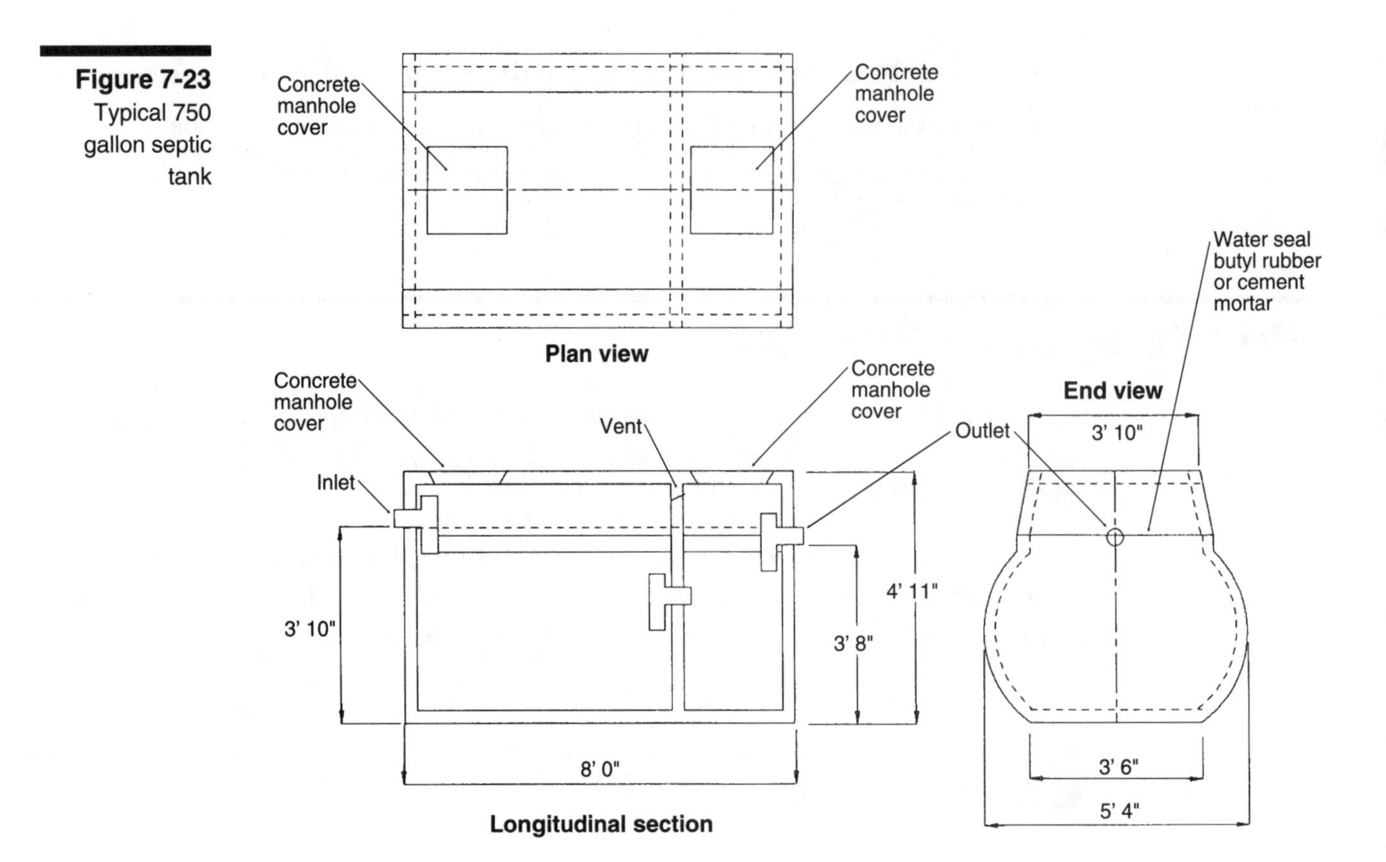

Figure 7-23 Typical 750 gallon septic tank

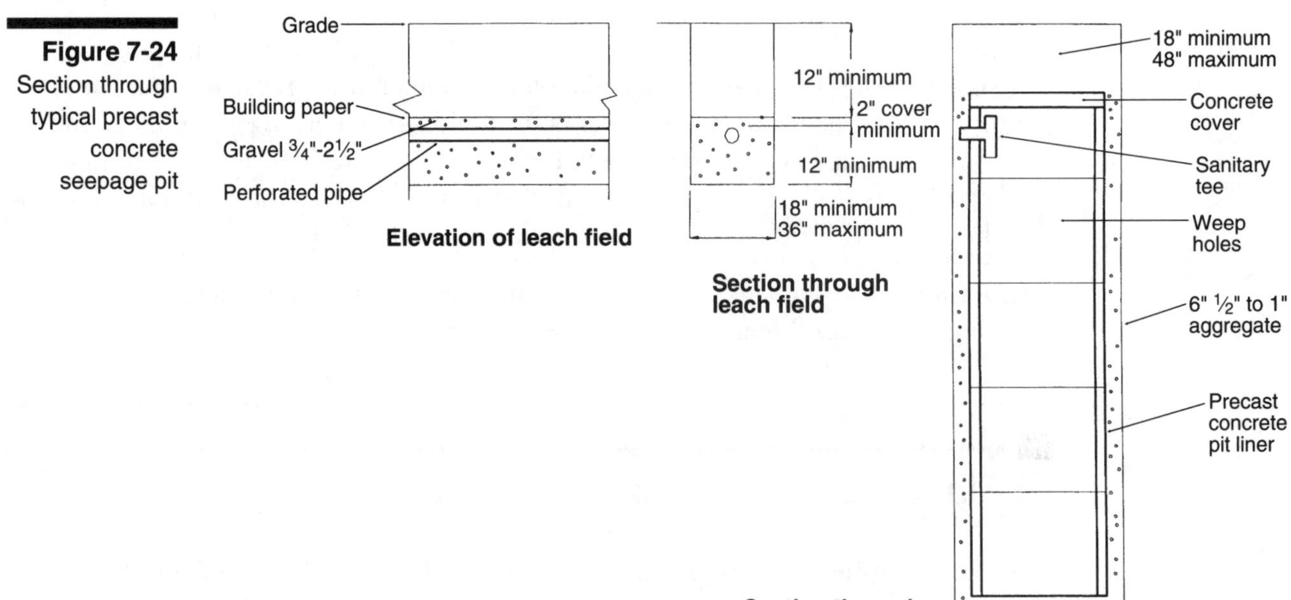

Figure 7-24 Section through typical precast concrete seepage pit

The maximum absorption capacities of five typical soils, in gallons per square foot of leaching area for a 24-hour period, are:

Type of soil	Gallons/square foot/24 hours
coarse sand or gravel	5
fine sand	4
sandy loam or sandy clay	2.5
clay with a lot of sand or gravel	1.1
clay with little sand or gravel	0.83

Let's suppose you have a 1200-gallon septic tank in sandy loam soil. Then the leaching field would have to absorb 1200 gallons of water in 24 hours. Sandy soil can absorb 2.5 gallons per square foot in 24 hours, so you'd need at least 1200/2.5, or 480 square feet of leaching area.

If the soil you're working in isn't one of these, the plumbing department may require that you do a percolation test on the soil.

To make a drainage trench, use 4-inch minimum field tiles with ½-inch open joints. Cover the pipe with at least 2 inches of crushed rock and add a layer of straw or paper. Cover the gravel and paper with at least 12 inches of earth. Make the trench at least 18 inches wide, and at least 12 inches below the bottom of the pipe. Fill the space with crushed rock.

Gas Supply Systems

The plumbing code defines gas piping as any pipe or fitting used to carry fuel gas, except service piping and connections that are less than 6 feet long. Service piping is piping between the street gas main and the gas piping system inlet into a building.

Most plumbing contracts require the plumber to install low pressure gas piping within a building. Low pressure gas is under 14 inches of water column. The service pipe is installed, controlled, and maintained by the service gas supplier.

Gas supply to the meter and regulator usually is at medium pressure. Medium gas pressure is pressure over a 14-inch water column but not more than 5 psi. The regulator reduces the pressure to low pressure (under 14-inch water column) for distribution.

Distribution Materials

Fuel gas pipes, 2 inches and smaller, are usually threaded Schedule 40 steel with screwed malleable iron fittings. Pipes 2½ inches and larger are Schedule 40 pipe with butt welded fittings. Schedule 40 pipe (standard weight pipe), should be acceptable for 125 psi gas pressure.

Figure 7-25
Pipe capacity

Pipe diameter (inches)	10 feet of pipe (cu ft/hr)	20 feet of pipe (cu ft/hr)	40 feet of pipe (cu ft/hr)	60 feet of pipe (cu ft/hr)	80 feet of pipe (cu ft/hr)	100 feet of pipe (cu ft/hr)
½	174	119	82	66	56	50
¾	363	249	171	138	118	104
1	684	470	323	259	222	197
1¼	1,404	965	663	532	456	404
1½	2,103	1,445	993	798	683	605
2	4,050	2,784	1,913	1,536	1,315	1,165

Steel pipe for fuel gas may be either galvanized on non-galvanized. In some areas, the gas company may insist on non-galvanized, or black iron pipe, because of the quality of the fuel gas. In these localities, the fuel gas contains chemicals that can cause a reaction with the zinc coating of galvanized pipe.

Put approved coating or wrapping around all underground ferrous piping you install outside of a building's walls. The coating or wrapping should be approved by the gas company or the American Gas Association (AGA). You can only use field wrapping on short connections and fittings that are stripped for threading. Put at least 12 inches of earth (or equivalent protection) over the pipe. Mark protective-coated pipe at least every 2 feet. The field wrapping should be marked with the manufacturer's logo and the ASTM number for identification.

Gas Demand

The amount of gas supplied to a building depends on the type and number of gas appliances in the building. Each appliance has a rated maximum demand in Btuh. For example, a 50 gallon water heater may use 50,000 Btuh, a gas range 65,000 Btuh, and a clothes dryer 35,000 Btuh. To convert the total amount of Btuhs to cubic feet of gas, you need to know the heating value of natural gas. The plumbing code assumes a value of 1100 Btu per cubic foot of gas.

The amount of gas a pipe can carry is based on the diameter and length of the pipe. Table 12-3 in the *UPC* lists the capacity of different diameter gas pipes at various lengths. Let's use part of this table, which is shown in Figure 7-25, to figure what size pipe you would need to supply 175,000 Btuh to a house that's 20 feet from the gas meter on the property. If each cubic foot of gas contains 1,100 Btuh, you'll need a pipe that can supply 200,000/1,100, or 159 cubic feet of gas per hour. The table shows that 20 feet of ¾-inch pipe will deliver 249 cubic feet of gas per hour. So the ¾-inch pipe will do the job.

All steel underground gas pipes should have a protective coating. This may be hot coat wrapping with asphalt or tar, tape coating, or fusion-bonded epoxy.

Figure 7-26
Color code for copper pipes

Type of pipe	Color
K	Green
L	Blue
M	Red
DWV	Yellow

Plumbing Materials

Pipes and Fittings

Water pipes are the "life lines" of a building. They supply water both for domestic use (washing and cooking) and for the heating in certain HVAC systems. Older buildings used galvanized steel piping. Newer buildings usually use copper piping.

Most pipes and fittings are marked to identify who made them, their size, grade, and material. For example, copper pipe has colored stripes to show if it's Type K, L, M, or DWV. Figure 7-26 lists the colors.

Pipe materials are classified as ferrous, non-ferrous, and non-metallic. The most common types of ferrous pipe are:

- steel A53 black, used for natural gas
- steel A53 galvanized, used for hot and cold water and vents
- cast iron screwed A126, used for waste and drains
- cast iron A74 soil, used for toilet and urinal discharge
- cast iron hubless, used for sanitary systems.

Metallic non-ferrous pipes are generally:

- copper water tube, including Type K, L, and M, hard and soft temper, used for hot and cold water
- copper drainage tube (DWV), used for drains, wastes, and vents.

Non-metallic pipes include many types of plastic, cement, and clay pipes:

- acrylonitrile-butadiene-styrene (ABS), used for domestic sewage
- asbestos cement (AC) sewer, also used for domestic sewage
- chlorinated polyvinyl chloride (CPVC) used for water lines
- concrete drain, used for rainwater drainage
- concrete sewer, used for domestic waste
- bituminous fiber, used for domestic sewage

Figure 7-27
Linear thermal expansion, inches per 100 feet

Temperature, degrees F	Steel pipe	Copper pipe
0	0.00	0.00
20	0.15	0.26
40	0.30	0.45
60	0.46	0.65
80	0.61	0.87
100	0.77	1.10
120	0.92	1.35
140	1.08	1.57
160	1.24	1.77
180	1.40	2.00

- polyethylene (PE), used for domestic cold water supply and yard piping
- polyvinyl chloride (PVC), used for natural gas yard piping
- polyvinyl chloride drainage, waste, and vents (PVC/DWV), used for domestic sewage, building drain, waste and vent systems
- vitrified clay (VC), used for private and public sewers.

Old galvanized pipe may deteriorate from corrosion or get clogged by excessive buildup of scale and sludge. You can help control this buildup by installing a water softener. It's a good idea to use copper pipe because it's less affected by corrosion and scale buildup than galvanized steel pipe.

A hot water system will leak more often than a cold water system because hot water is usually moving. Leaks can also occur due to corrosion or electrolysis between different metals, such as if you connect galvanized steel pipe directly to copper pipe, or support copper pipe with steel hangers. If you use the wrong type of flux when you solder the socket joints, copper pipe may corrode. Follow the requirements of IAPMO IS 3-89 in the *Uniform Plumbing Code* for installation of copper plumbing tube, pipe, fittings, and type of flux. If you don't deburr and ream the cut ends of pipe when you solder them, they may leak. Joints that are under high stresses due to thermal expansion and inflexible pipe may also leak. Figure 7-27 shows the relative rates of thermal expansion between steel and copper pipe.

Pipe Sizes

The nominal size of a pipe is its approximate outside diameter because fittings must fit the outside of the pipe. The diameter will be the same for various weights of the same size pipe. As the wall thickness of a pipe gets bigger, its inside area gets smaller. So, heavier weight pipe carries less water than lighter

Figure 7-28 Wall thickness of Schedule 40 and Schedule 80 pipe (inches)

Nominal size	O.D.	Sch40 wall	Sch80 wall
3/8	.675	.091	.126
1/2	.840	.109	.147
3/4	1.050	.113	.154
1	1.315	.133	.176
1 1/4	1.660	.140	.191
1 1/2	1.900	.145	.200
2	2.375	.154	.218
2 1/2	2.875	.203	.276
3	3.500	.216	.300

thin-walled pipes. Figure 7-28 is a list which shows the difference in wall thicknesses in pipes up to 3 inches in diameter between Schedule 40 and Schedule 80 weight. All values are in inches.

Corrosion

Pipe may rust because of:

- stray electric current from grounded electrical appliances
- acidic water
- bacterial content of water
- contact between dissimilar metals, such as steel and copper, or new pipe and old pipe.

A plumbing system is like a battery. The metal pipe is the anode and the water is the oxygen-bearing solution. The metal and oxygen try to reach equilibrium. To reach equilibrium, metal ions must leave the pipe and go into solution. This phenomenon is called *oxidation* or *corrosion.* It makes pips and cracks in the pipe wall. Some metals, like copper, oxidize very slowly. Magnesium, zinc, and iron corrode faster. When you coat the pipe with zinc it's called galvanizing. With a galvanized pipe, the zinc goes in solution first and this protects the iron until all the zinc is in solution.

Another way to slow down corrosion is to place a piece of magnesium or zinc in the plumbing system. For example, you can put a zinc ball in a pool strainer to slow down the rate the pool piping corrodes at. To protect the pipe in a hot water storage tank, put zinc capsules in the tank. You can also put direct electrical current into a piping system to keep the metal ions from leaving the pipe walls and going into solution.

Ground all water pipes to cut down on electrolysis in steel pipe. You should ground interior metal water piping that may become energized from a defective electrical appliance that uses the pipe for a ground. To do this, bond the pipe to the electrical service equipment enclosure, or to the grounded conductor at the service panel. In multiple-occupancy buildings, bond the interior metal water piping system for each dwelling unit to the panelboard or switchboard supplying the unit.

Another method of grounding is to attach to a metal underground water pipe that's in direct contact with the earth. This pipe should be 10 feet or more long and electrically continuous. The pipe may be supplemented by an additional electrode, such as a copper rod, plates, or other underground metal structures.

Copper tubing tends to be corrosion-resistant, but under certain conditions it can develop pinhole leaks. In some localities the water contains chemicals that attack copper pipe.

Steel pipe rusts, but heavy-wall steel pipe, such as Schedule 60 or 80, resists corrosion better and lasts longer. Zinc coating, or galvanizing, will protect pipe. The thicker the zinc coating, the longer the protection lasts.

If the water in your community is corrosive, install a water softening unit to protect piping in the building. The material used in the softening tank depends on what elements are in the water. Caustic agents such as calcium hydroxide or potassium hydroxide tend to neutralize acidic water.

Pipes, fittings, and their supports that are made of different materials can cause corrosion by electrolysis. These materials should be separated by non-conductive material, such as a dielectric fitting. For example, copper pipe supported by steel pipe hangers should be isolated with a non-conductive material. The potential for electrochemical corrosion of two different materials depends on their relative location in the galvanic series of metals.

Figure 7-29 lists metals from the least noble, or anodic, to the most noble, or cathodic, metals. The least noble metal is a sacrificial metal because its atoms flow to the more noble metal. When you connect a less noble material like aluminum or steel to a more noble material like copper, you'll get corrosion. Separate these materials by a dielectric material to prevent the flow of electrons from the steel to the copper.

Copper Water Tubing

Copper water pipe, or water tube, generally comes in three grades, Type K, L, and M. Type K, the heaviest, is usually used for underground installation within structures or where the water in the pipe with be at high pressures and temperatures. Type L is normally used above ground in residential buildings. You can use Type M, which has the thinnest walls, in residential buildings where low cost is important and the temperature and pressure are not expected to be too high. Common practice is to use Type M with ordinary water in above ground installation when the quality of the water is the basis of selection. But use Type L where water conditions are more

Figure 7-29
Galvanic series of metals

Anodic to Cathodic Metals	
Magnesium	Nickel
Zinc	Brass
Aluminum	Copper
Steel or iron	Bronze
Cast iron	Copper-nickel
Chrome iron	Stainless steel 304, 316, and 410
Nickel iron	Silver
Lead tin solder	Graphite
Lead	Gold
Tin	Platinum

severe, as Type K is about 50 percent stronger than Type M copper tubing. Although most plumbing departments let you use Type M copper tubing, a heavier wall pipe that's stronger, more durable, less noisy, and more stable is better. Standard practice says that any pipe 2 inches or smaller that you use above ground should be Type L, with wrought copper fittings. Larger pipe may be Type L with cast brass and silver brazed fittings. Generally, water pipe above ground within a building should be Type L copper tubing or heavier.

To prepare copper fittings:

- Cut the pipe to length and remove any burrs.
- Clean the outside end of the pipe and inside end of the socket with medium-grade sandpaper or emery cloth.
- Wipe away all emery dust and metal particles.
- Apply flux and heat.
- Feed solder to the joint completely.

Toilets

Toilets are made of vitreous china. The exposed surfaces are coated with ceramic glaze that's fused to the body. This coating makes the fixture resistant to corrosion and discoloration. A properly-operating toilet should:

- flush quietly, completely, and efficiently
- shut off water flow after the flushing action
- refill the flush tank to the proper depth
- shut off the water from the supply line

There are several types of toilet bowls:

- Washdown bowl discharges into trap at the front of the bowl. The trap passageway is 2 inches and the water seal is 3 inches.
- Reverse trap bowl discharges into a trap at the rear of the bowl
- Siphon jet similar to the reverse trap, but with a larger water surface area
- Flush valve toilets in which the flush valve seat is 1 inch or more above the rim of the bowl so that it'll close even if the toilet trap is clogged

The toilet supply should be a ⅜-inch straight valve or ½-inch angle valve. Undersized or kinked supply lines can restrict flow into the tank and prevent complete flushing. Most toilets need at least 35 to 40 psi working pressure at the fixture to work properly. Flushing takes about 9 to 15 seconds. Refilling the tank takes about 1 to 2 minutes. On the average, one flush takes:

- volume of discharge 3.8 gal
- period of discharge 12.8 sec
- average rate of flow 17.8 gpm
- peak rate of flow 33.4 gpm

Use this information when you compare the cost and quality of different models of toilets.

Valves

Piping systems should include shutoff valves at each fixture, pump, heater, and tank. Valves are also needed to regulate and control the system. This is very important in the hot water circulating system.

Here are some of the places you should install a gate valve:

- outside a building to shut off the water supply to the building
- outside each dwelling unit in a multi-dwelling building. Then you only have to shut off the water to one unit instead of all of them.
- on storage tanks and hot water heaters so they can be drained

Put a stop valve on all piping which supplies water to appliances and fixtures that require maintenance. Use a check or backwater valve to prevent backflow. Use a globe valve to control the amount of flow to a water supply system.

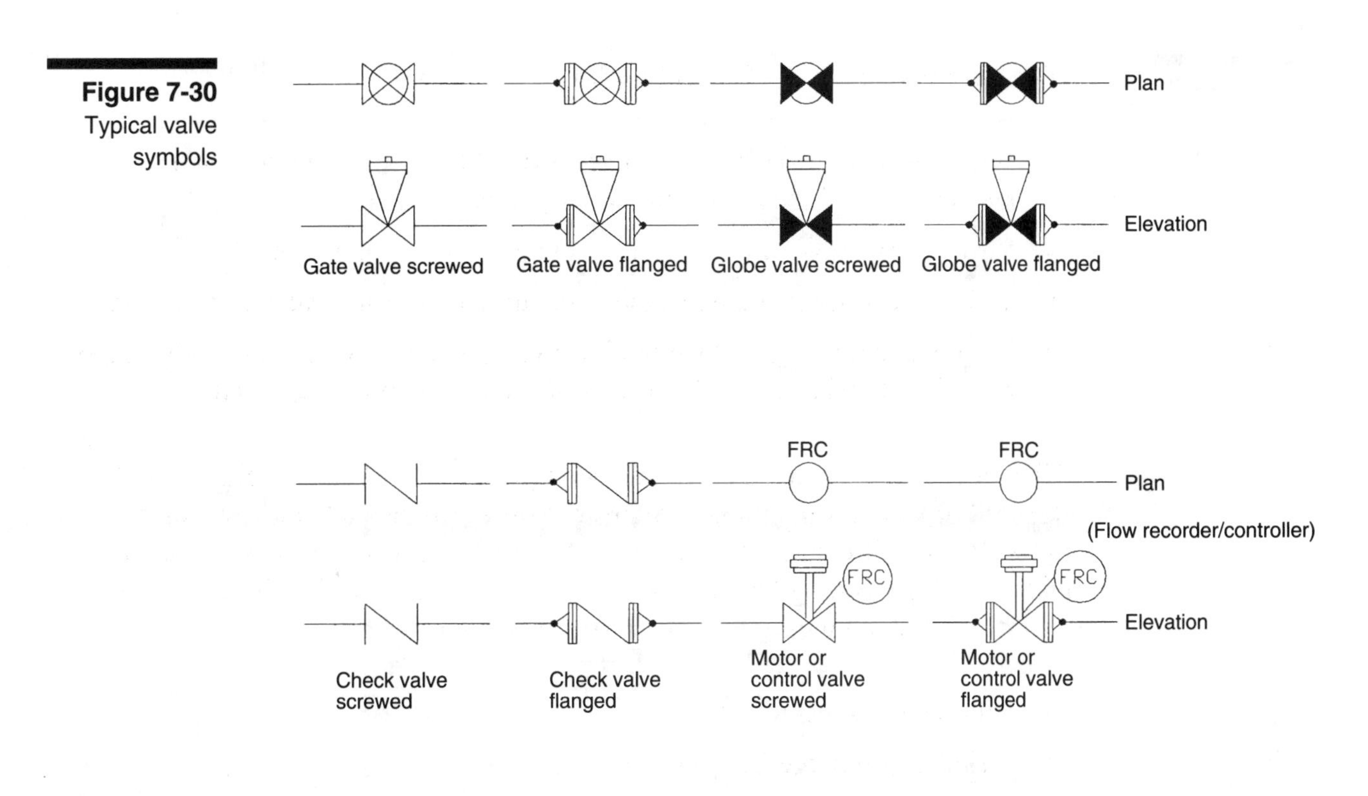

Figure 7-30 Typical valve symbols

Plumbing Documents

Here's a list of the documents you'll usually need for plumbing work:

- Architectural plans, to locate all plumbing fixtures.
- Plumbing plans, to identify plumbing fixtures, fu demand at all inlets and outlets for cold water, hot water, sanitary waste, standpipes, gas, and storm drain piping. Figures 7-30 and 7-31 show commonly-accepted drafting symbols for plumbing drawings.
- Schematic riser drawings, to identify risers and laterals for all plumbing fixtures, demand at all inlets and outlets for cold water, hot water, sanitary waste, standpipes, gas, and storm drain piping.
- Schematic isometric drawings, to identify all risers, laterals and details for all the plumbing fixtures, demand at all inlets and outlets for cold water, hot water, sanitary waste, stand pipes, gas, and storm drain piping.
- Plumbing details, to illustrate all water heaters, piping fittings, valves and other devices relating to the hot water assemblies, including the manufacturer, model, Btu input, Btu output, relief valve expansion tank, storage tank capacity and size, recirculating pump capacity and size, mixing valve, relief valve, thermometer and thermostatic mixing valve for each building of a project.
- Plumbing specifications, usually incorporated with the architectural or mechanical specifications. These documents also include general conditions, special conditions, or supplementary conditions. General conditions describe conditions required on every individual construction project. They cover insurance, ownership of drawings, royalties, any patented work, claims for

Figure 7-31 Typical pipe fitting symbols

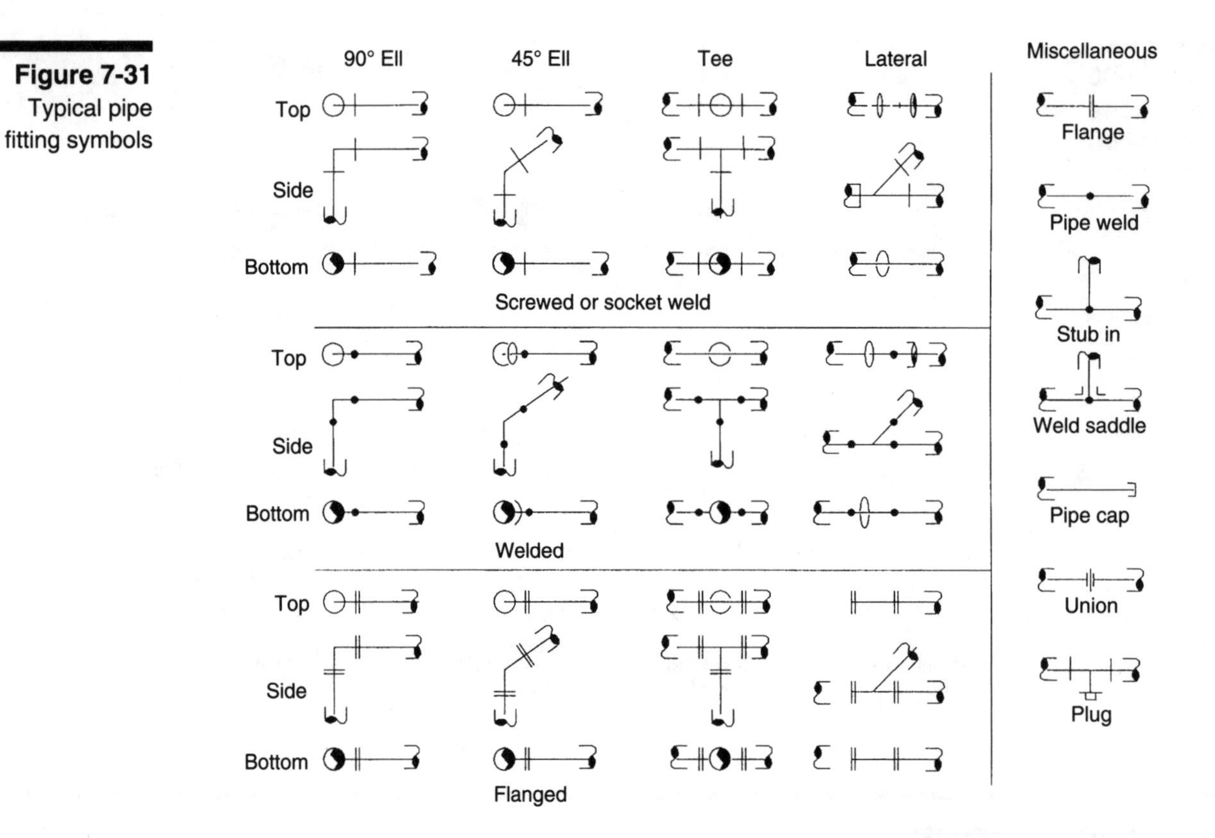

extra cost, termination of contract, and payments. Special or supplementary conditions deal with specific items that apply only to the project under consideration.

- Requests to bid for plumbing by the general contractor, which lists drawings and specifications to be included in a bid.
- Addenda to plumbing specifications, to identify all changes in the bid documents.
- Piping and fitting take-off sheets, which list each pipe and fitting of the cold water system by size, weight, length, material, and number. Purchase these items from the pipe supplier, or on major jobs, from pipe and fitting manufacturers.
- Take-off sheets for hot water, sanitary waste, vents, storm drains, building drains, and building sewer for each building of the project. Also, the material for the water service that includes the water meter, backflow preventers, main shutoff valves, and pressure regulators. Figure 7-32 shows the required pipe and fittings for a typical fixture.
- Valves and specialties take-off sheets, which include all miscellaneous plumbing devices required to control, support, insulate, and isolate the plumbing system. These also may include water-hammer silencers, pressure regulators, flow regulators and controllers, air gap devices, insulation, isolators and pipe clamps, shut-off valves, and flow restrictors.

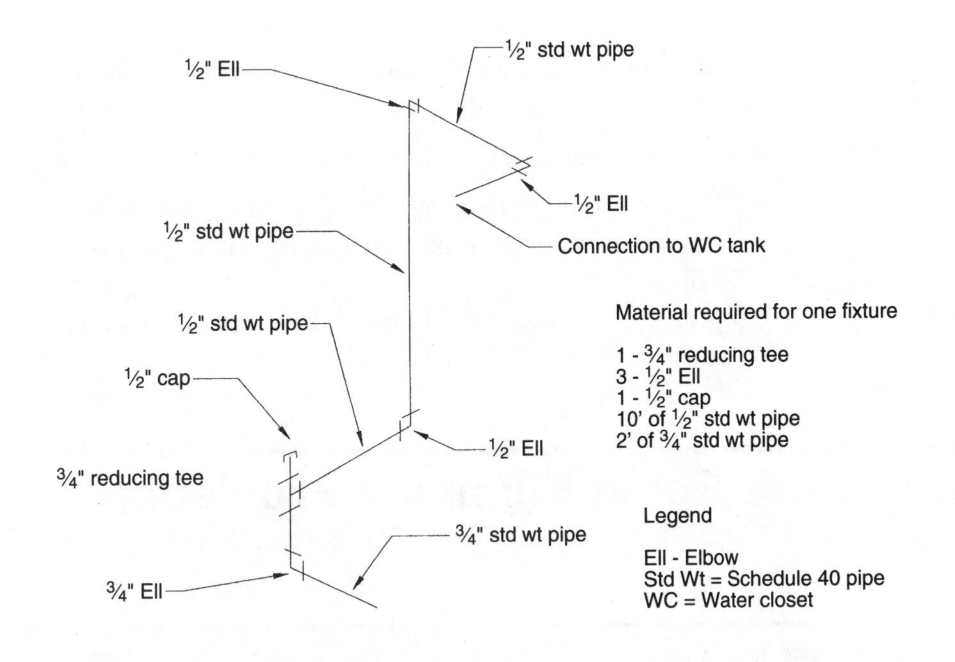

Figure 7-32 Isometric piping detail showing pipe and fittings for one fixture

- Proposals for plumbing work, which list drawings and specifications included in the bid.
- Subcontract agreement for plumbing work, which lists drawings and specifications included in a subcontract.
- Building permits or building permit applications, which identify the owner, architect and engineer; buildings involved; dates of issuance; and special approvals of zoning, health, and fire departments. There may be several permits if the project is constructed in phases.
- Bulletins, change orders, field orders and revisions, which are issued by the general contractor or architect to identify all changes made during the progress of the project.
- Plumbing permits or plumbing permit applications, which identify the mechanical engineer, contractor or authorized agent, buildings involved, number and types of plumbing fixtures, or items the permits cover. After a plumbing system is complete and passes inspection, the building department issues a certificate of approval to the person who holds the plumbing permit for the job. A copy of this certificate is kept on file in the building department.
- Record of tests, including drainage tests, waste tests, vent tests, building sewer tests, shower receptor tests, water piping tests, yard piping tests, low pressure air tests, rough gas piping tests, final gas piping tests, and PVC natural gas piping tests.
- Purchase orders for all the items listed in the take-off sheets.

- Manufacturers' manuals and data sheets for all plumbing fixtures and equipment used on the project.
- All correspondence, memos, and notes on meetings relating to plumbing work.
- Plumbing superintendent's log or field records, which list the hot and cold water, gas, sanitary waste, and storm water systems located in the buildings. This log also records daily the personnel on the job, portions of work in progress, weather conditions, and any unusual event that affects the work.

How to Cure Some Plumbing Problems

Toilets

- Toilets do not flush because of insufficient water supply.
  Check supply pipe for size or kinks.
- Soap suds coming out of toilets because of proximity of the toilet drain to the clothes washer drain.
  Use a separate waste line from a clothes washer.
- Toilets with bad seals which continue to run and do not flush properly.
  Replace seal.
- Toilet continues to run because float valve leaks.
  Replace float valve.
- Leak at base of toilet because of poor wax seal.
  Reset toilet.
- Bubbling in toilet.
  Clear drain pipe of obstructions which cause bubbling.

Noisy Plumbing

- Risers strapped to framing without isolators.
  Install isolators.
- Water hammer silencers not installed in the system.
  These are normally installed ahead of the inlet to hot and cold water connections of sinks, washing machines, and other appliances.

- Sound transmission through walls because pipe is not insulated from wall framing.
  Pipe should be held in place with clamps containing resilient material to isolate pipe from framing. At least 1/4 inch should separate the pipe from the structure.

- Complaints of noise from neighboring units due to running water or flushing toilets.
  Wrap all waste and water piping in party walls with sound insulation where it penetrates the floor/ceiling assembly. (Party walls are common walls between dwelling units.)

- Complaints of running water noises to upper or lower dwelling units.
  Strap risers to framing with isolators.

- Noise from vibrating pipes.
  Replace noisy lightweight Type M copper pipe with heavy rigid Type K or L pipes.

Waste Backup

- Backup in lower sinks when upper clothes washers operate because foam in the drain pipe obstructs flow.
- Backups in first floor laundry rooms because drain lines don't slope enough.
- Soap suds coming out of toilets and backup in sinks when clothes washers on higher floors are operating because the drain pipe from the clothes washer isn't big enough.
- Check valves ineffective because debris and paper in waste water fouls the valves. Check valves permit water to flow in one direction only.
- Drain lines plugged because there aren't enough cleanouts. There should be a cleanout at the end of a line and after aggregate turns of 135 degrees.

Low Water Pressure

- One cause of low water pressure is plugged restrictors in the hot and cold water lines. Normally, all fixtures except bath tubs are supplied with flow controllers to minimize sound when water passes through openings. Another cause is defective pressure regulators.

Leaks

- Poor soldering of joints
- Threaded connection not tight enough

Inspection and Tests

You should inspect and test plumbing as you install it. Make a rough inspection before you cover up any piping. Test the drainage and waste system, vents, sewers, shower receptors, water piping, rough gas piping, and final gas connections.

You can test a drainage and vent system by either a water test using a minimum of a 10-foot head of water, or an air test under 5 psi pressure. The water test can be applied to the entire system or by sections. The water or pressurized air is maintained for no less than 15 minutes before inspection starts.

Test the waste and sewer system by filling the entire system from the lowest point to the highest point with water. If you test the system in sections, fill each section with at least a 10-foot head of water.

Test the hot and cold water system with water at the working pressure of the system or by an air test set under 50 psi pressure. The working pressure is the maximum pressure of the city water main, or what you set the pressure regulator at.

Test the fuel gas system with air, carbon dioxide, or nitrogen pressurized to 6 inches of mercury measured by a manometer or slope gauge.

The following are suggested step-by-step procedures for testing:

You can do an air test on the drainage system this way:

- Attach an air compressor testing apparatus to any suitable opening.
- Tightly close all inlets and outlets to the system.
- Force air into the system until there is a uniform pressure of 5 pounds per square inch on the entire system or the section under test.
- Maintain the air pressure on the system or section under test for at least 15 minutes. Use soap and water at the joints to help you find any leaks.
- If there is no loss of pressure or any leaks, the system is satisfactory.

You can test a drainage and vent system using a smoke test or a peppermint test. To do a smoke test:

- Fill all traps with water.
- Introduce a pungent, thick smoke into the system.
- When the smoke appears at the stack openings on the roof, close these openings

- Apply a pressure equal to a 1-inch water column. A 1-foot high column of water has a pressure of 62.4 psf, or (62.4 / 144) 0.43 psi. A 1-inch column is $\frac{1}{12}$th of 0.43 psi, or 0.036 psi.
- Maintain this pressure for at least 15 minutes.
- Start inspection.

To do the peppermint test, put 2 ounces of oil of peppermint into each line or stack. If there are any leaks, you'll smell them.

Gas piping should withstand a 10 psi air pressure for 15 minutes without leaking.

The holder of the plumbing permit should provide the equipment, labor, and materials needed for testing. Tests should be conducted in the presence of a representative of the building department. When the tests are completed satisfactorily a certificate of approval is issued by the building department.

Proper inspection can prevent future complaints, such as:

- scalding hot water.
  Check temperature controls to see that they're set properly. The temperature controls include temperature relief valves, pressure relief valves, and energy shut off devices.
- lack of adequate hot water.
  Check the hot water distribution pipes to make sure they're the correct size.
- fluctuating hot water supply.
  Check to see that thermostatic controls are set between 110 to 140 degrees F.
- noisy plumbing.
  Check to make sure all pipes are isolated from the building structure, that pipe insulation and water-hammer arresters are installed, and correct pipe size and schedule are used.
- leaking pipes.
  Check to make sure threaded joints are correctly caulked and embedded in fittings, that copper joints are completely soldered, and bolts in pipe flanges are tight.
- poor drainage from plumbing fixtures.
  Check pipe size and slope on all gravity piping.
- waste water backup and clogged lines.
  Check for proper number and placement of cleanouts.

It's fairly easy to check plumbing in a multi-story apartment building which has a basement, because the water, gas supply, and drain lines are usually suspended from the basement ceiling. You can see the pipe material, pipe size, connections, supports, and slopes. And you can measure the circumference of a pipe to find its outside diameter. The outside diameter of any pipe is its circumference divided by pi, or 3.14. The following table lists the nominal size, actual outside diameter, and circumferences of steel pipes. Note that the outside diameter is slightly greater than the nominal pipe size.

Nominal size inches	Outside diameter inches	Circumference inches
½	0.840	2.638
¾	1.050	3.297
1	1.315	4.129
1¼	1.660	5.212
1½	1.900	5.966
2	2.375	7.458
2½	2.875	9.028
3	3.500	10.99
3½	4.000	12.56
4	4.500	14.13
5	5.563	17.47
6	6.625	20.80
8	8.625	27.08
10	10.750	33.76
12	12.750	40.04

Copper and plastic pipe are usually stamped with their size so you don't need tables of sizes for them.

Use a carpenter's level and a ruler to measure slopes of drain pipes. If you have a 4-foot long level, read the fall in the pipe in inches and divide by 4. This will give you the slope in fractions of an inch in one foot.

CHAPTER 8

Heating, Ventilating & Air Conditioning

The *Uniform Mechanical Code* is the principal code regulating HVAC work. Other codes that include requirements for HVAC work are the *Uniform Building Code (UBC)*, the *Uniform Building Code Standards (UBCS)*, the *Uniform Plumbing Code (UPC)*, and the *National Electrical Code (NEC)*. Local zoning codes also affect HVAC work.

Here are some of the typical requirements of these codes:

- Equipment supported on the ground must be on a concrete slab that's at least 3 inches above ground level.
- There must be shutoff valves in the gas piping outside each gas-burning appliance, and before each union connection.
- The shutoff valve for an appliance must be accessible and within 3 feet of the appliance.
- Safety devices for electrical service to HVAC units must conform to the *NEC*.
- Some zoning codes prohibit locating equipment in front, side, or rear yards.

Heating, ventilating, and air conditioning (HVAC) work usually includes:

- Fabrication, installation, maintenance and repair of warm-air heating systems and appliances
- Ventilating systems with blowers and plenum chambers
- Air-conditioning systems with cooling units
- Ducts, registers, flues, humidifiers, filters, and thermostatic controls for any of these systems
- Equipment that uses solar energy

You must have the appropriate mechanical permits to install, erect, repair, relocate, replace, or add to any HVAC equipment. Usually you don't need a permit for portable units. You need plumbing permits for gas piping to heating units, and electrical permits for electric service to HVAC units. To apply for a mechanical permit, you need to give this information:

- owner's name and job address
- installer's name, address, and license number
- number and horsepower of comfort cooling system
- number of air handling units
- number of smoke dampers, appliance vents, fire dampers, smoke detectors, and evaporative coolers

There are two inspections for mechanical work: rough and final. The first one, the rough mechanical inspection, is required on all equipment that will be covered up by construction. The other, the final mechanical inspection, must be done after the building is completed but before it's occupied. This is the last chance you have to make sure all piping, ductwork, registers, grilles, fuel-supply lines, and appliances are in place and can operate properly. You may have to take the ceiling tiles out to make sure the ductwork is correctly installed.

Mechanical plans and specifications are usually required for buildings that have an area of 15,000 square feet or more, including the basement. In addition, buildings with environmental heating systems with a Btuh input capacity greater than 350,000, or absorption cooling systems with over 350,000 Btuh input also require mechanical plans and specifications.

You should include calculations, diagrams, and other backup data on your mechanical plans and specifications for a building. Make sure the plans are drawn to scale and show clearly the location, nature, and extent of the proposed work. These documents are usually prepared by mechanical contractors, but some states require that drawings be prepared by a licensed architect or engineer. Generally, mechanical plans should have:

- an exhaust and ventilation system showing location, size, occupancy, and nature and extent of work proposed.
- sensible heat sources, including type of equipment, heat output, usage factor, and desired room temperature. Sensible heat sources include solar heat gain through windows, roofs and walls, and heat from people, appliances, and air infiltration.
- latent (or latent and sensible) heat sources, listing the quantity of equipment, breakdown of sensible and latent load, usage factor, and desired room temperature. Latent heat is the amount of heat that's absorbed or evolved when a substance changes its state without changing its temperature — vaporizing water, for example. It's an effect of moisture given off by people, appliances, and infiltration of air.

Figure 8-1 shows some common symbols used on HVAC plans. These symbols are based on standards of the Sheet Metal Air Conditioning National Association (SMACNA). They're generally accepted by the HVAC industry, but some designers don't use them. If you see a symbol on HVAC plans that you don't understand, always request an explanation from the designer. Mechanical plans should always include a legend to describe the meaning of each symbol.

Heating Systems

Gravity Systems

Floor furnaces have burners, a fire box, and a cast iron heat exchanger. The heat exchanger heats the air in the immediate area. This type of unit is usually located under the floor in a central hallway. Figure 8-2 shows a simple floor furnace in a house. Warm air generated from the furnace moves to adjoining rooms through open doors. The floor register is a sturdy grille you can walk on.

Make sure you leave enough room around a floor furnace so someone can inspect, repair, or replace it if necessary. Allow a work area that's at least 24 by 18 inches. Leave at least a 6-inch clearance from the bottom of the furnace to the ground, at least a 30-inch clearance on the side of the unit that has the controls, and at least a 12-inch clearance on the other three sides.

To supply warm air to both sides of a wall, install a dual floor furnace under the wall. Use a damper to direct the heat to one or both sides. Don't put a floor furnace where a door can open over, or against, its air outlet. Never install a register less than 6 inches away from any inside room corner.

Put a gravity heater in a basement where the ducts can carry warm air to floor outlets throughout the building, as shown in Figure 8-3. This heating system needs a basement with at least 7 feet of headroom. Gas fumes exit through a vent running through the roof to the outside. When you put a fan on this unit, you have to put a return air duct from the heated space back to the furnace to force the colder air back to the heater to be reheated.

The vented wall furnace in Figure 8-4 is another simple and inexpensive type of gravity heater. You can install this furnace between 4- or 6-inch wall studs. The louvered side of the unit is exposed in the room or area you want to heat.

This unit has a burner and heat exchanger. A Type BW oval double-wall vent is attached to the top of the heater. This vent extends through the attic space, and ends above the roof line. To prevent a fire where the vent penetrates the ceiling and roof, leave about a 2-inch separation from the vent and any combustible material. Check your local building code for the actual separation distance it requires. Above the roof line, make the vent watertight with a storm collar, flashing, and vent cap. Be sure the vent rises at least 1 foot above the roof surface and 4 feet from any vertical wall.

Figure 8-1
Ventilation and air conditioning symbols

Symbol meaning	Symbol	Symbol meaning	Symbol
Fan & motor with belt guard & flexible connections		Gooseneck hood (cowl)	
Ventilating unit (type as specified)		Back draft damper	BDD
Unit heater (downblast)		Automatic air dampers, motor operated	SEC ELEV
Unit heater (horizontal)		Access door (AD) Access panel (AP)	AD
Unit heater (centrifugal fan) plan		Damper as specified	
Thermostat	T	Acoustical lining	
Power or gravity roof ventilator-exhaust (ERV)		Sound trap	ST
Power or gravity roof ventilator-intake (SRV)		Fire damper and sleeve	FD
Power or gravity roof ventilator-louvered		Flexible duct	
Mixing plenum	MB	Exhaust or return air inlet ceiling	12 × 20 GR 700 cfm
Louvers & screen	36 × 24 L	Supply outlet, ceiling, round, direction of flow	20 Ø 700 cfm
Direction of flow		Supply outlet, ceiling, square, direction of flow	12 × 12 700 cfm
Splitter damper		Supply outlet wall	12 × 20 GR 700 cfm
Duct section (exhaust or return)	E or R 12 × 20	Supply grille (SG)	12 × 20 SG 700 cfm
Duct section (supply)	S 12 × 30	Return grille - note at ceiling or floor	12 × 20 RG 700 cfm
Duct (side shown x side not shown)	12 × 20	Exhaust grille (EG)	12 × 20 EG 700 cfm
Inclined drop, in respect to air flow, top flat	D	Exhaust or return inlet - wall (EG or RG)	12 × 20 G 700 cfm
Inclined rise, in respect to air flow, bottom flat	R	Door grille	D G 12 × 6
Turning vanes		Note: Grilles with volume control are designated as - SR, RR or ER	
Volume damper	VP		

Figure 8-2
Floor furnace

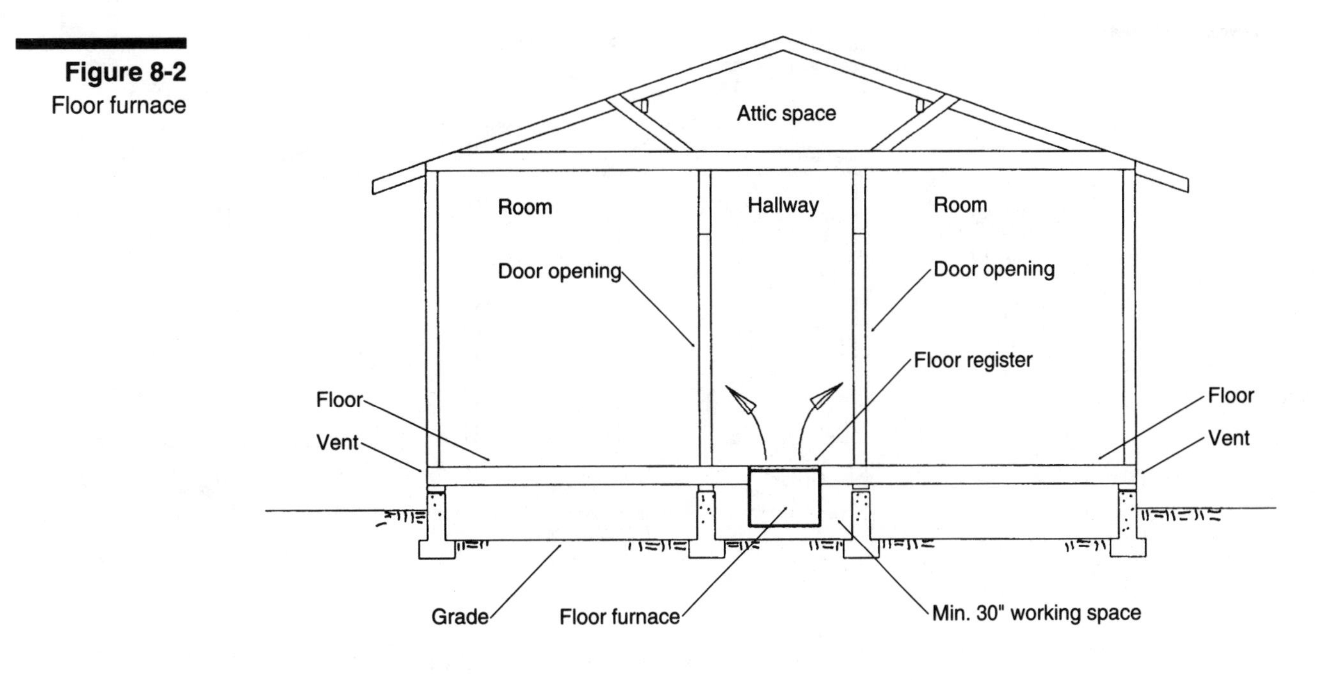

Figure 8-3
Gravity warm air heating system

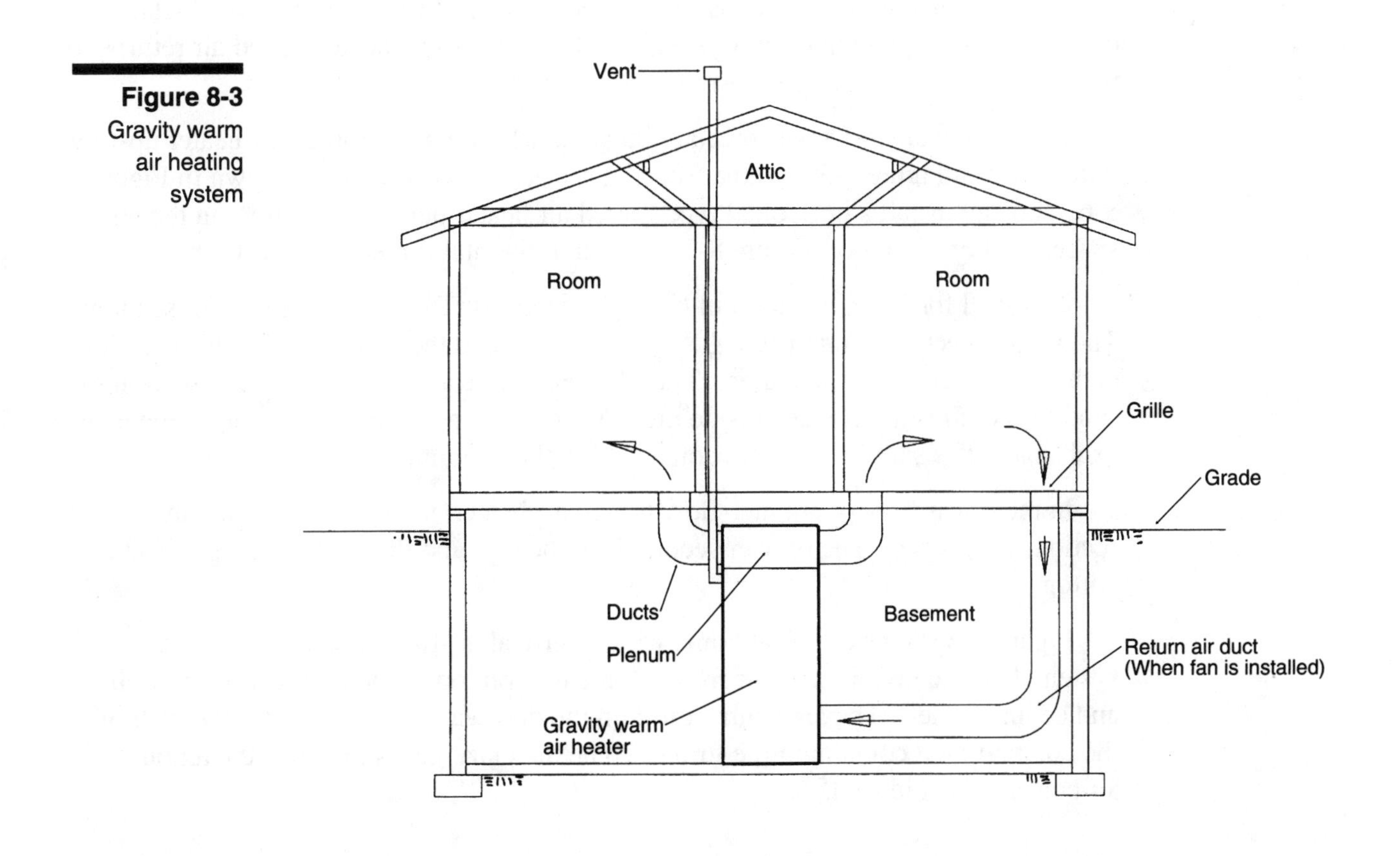

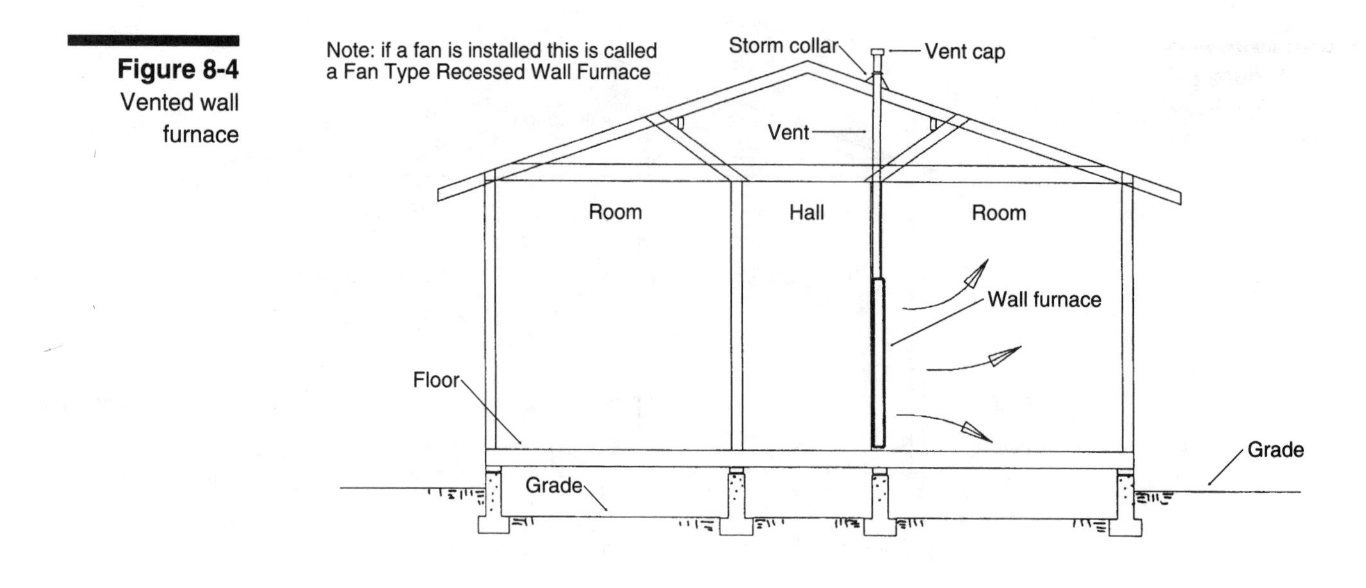

Figure 8-4 Vented wall furnace

Don't put a door where it will open within 12 inches of any air inlet or outlet of a wall furnace. Don't attach a duct to a wall furnace.

Forced Air Heating Systems

The main part of a forced air heating system is the central heating furnace, which includes a burner, fire box, heat exchanger, and a fan. Forced air furnaces also have a filter, a plenum, controls, and possibly an electrostatic air cleaner. Supply ducts connected to the plenum carry warm air to the rooms to be heated. Used air returns to the furnace through a register or return-air ducts.

There are many types of forced air heaters, including the forced air heater up-flow unit shown in Figure 8-5, and the forced air heater down-flow type shown in Figure 8-6. You can install a horizontal type forced air heater below the ceiling, in the attic space, or above the roof. Figure 8-7 shows a horizontal forced air heater.

A typical forced air heater unit has three sections. The first is the blower section. The second section contains the gas or oil burners and the controls. The third section is the firebox and heat exchanger. There's a vent pipe attached to the third section to send the combustion fumes outside the building. The plenum is at the top of the heat exchanger. The supply ducts are connected to the plenum.

In the down-flow or counter-flow forced air heater, the position of the three sections is reversed. The blower section is at the top, and the heat exchanger is at the bottom.

Figure 8-8 shows a typical home with a vertical up-flow forced air furnace system. Ducts convey warm air from the plenum on top of the unit to various wall and ceiling outlets. The air returns through the doorways to a register in the wall of the compartment containing the furnace. The duct size gets smaller as the amount of air conveyed is reduced.

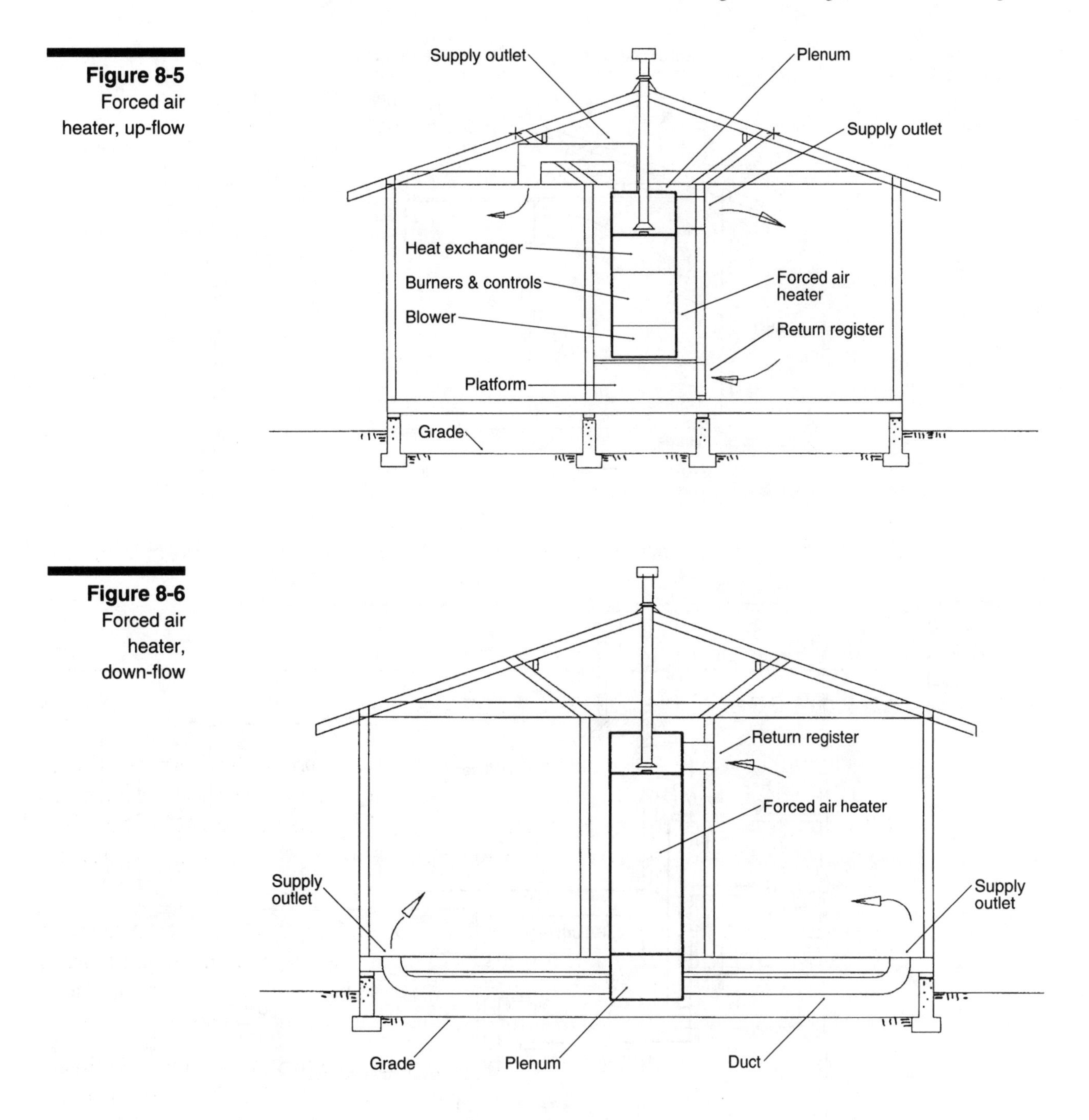

Figure 8-5 Forced air heater, up-flow

Figure 8-6 Forced air heater, down-flow

Usually you put this type of furnace in a compartment that has at least twice as much floor area as the unit. Mount it on a raised platform to let air return into the blower section. Don't put it in a bedroom, bathroom, closet, or confined space that you can only get into by going through the room. Put it in a closet off a corridor or entry way.

If there's a cooling coil section above the heat exchanger, the unit is a combination heater/cooler that can either heat or cool a building. The same ductwork carries both heated and cooled air. Always put the cooling coils downstream of the heating section to prevent condensation in the heating element.

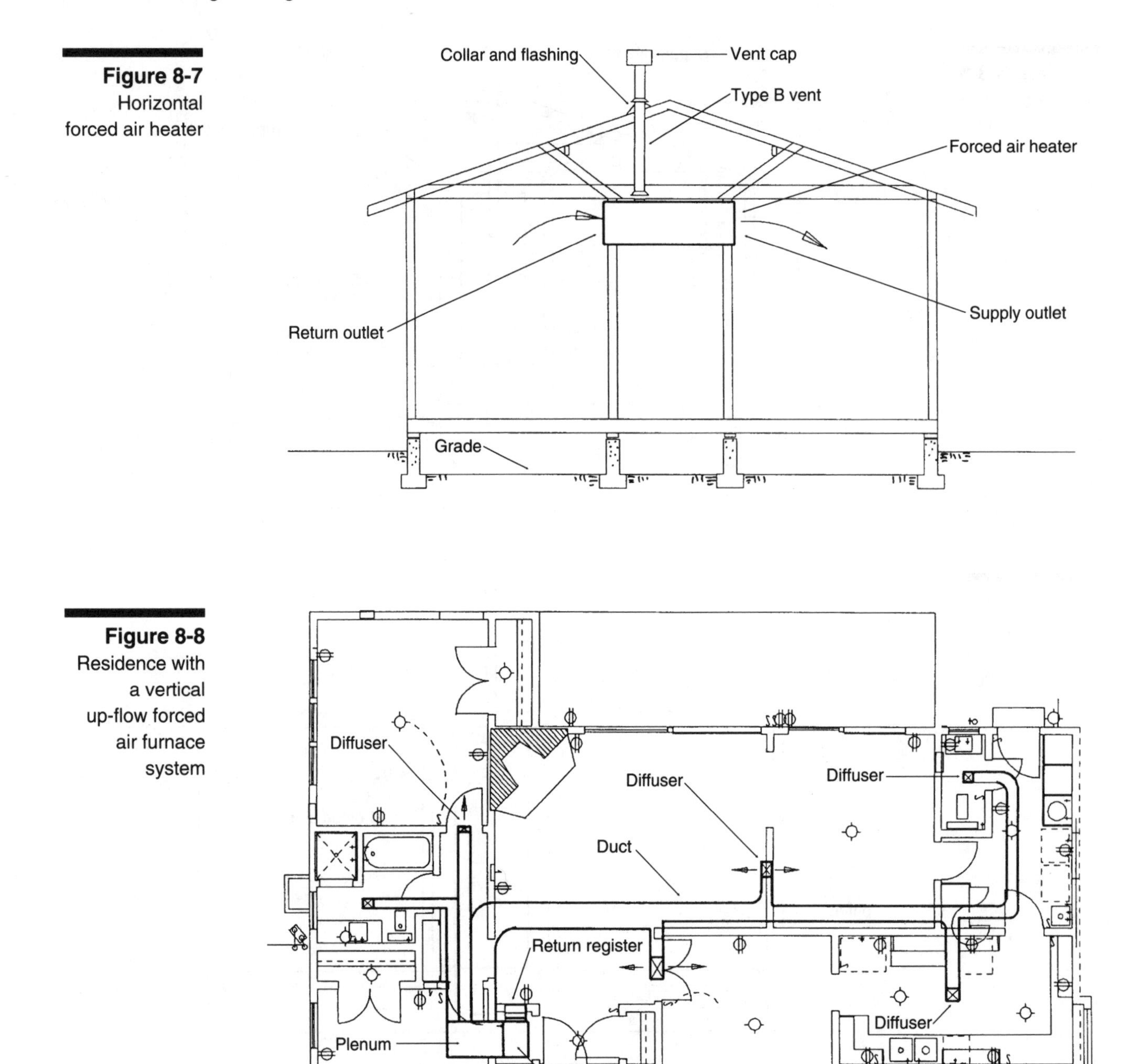

Figure 8-7 Horizontal forced air heater

Figure 8-8 Residence with a vertical up-flow forced air furnace system

A typical horizontal heating/cooling rooftop unit is shown in Figure 8-9. The unit consists of a return air plenum, filter unit, heat exchanger, cooling unit, air handling section and outer housing. This unit has a watertight pan under the cooling unit to collect dripping condensation. The entire housing is supported by vibration isolators. Some units use hot water coils as the heat exchanger in place of gas burners and a fire box.

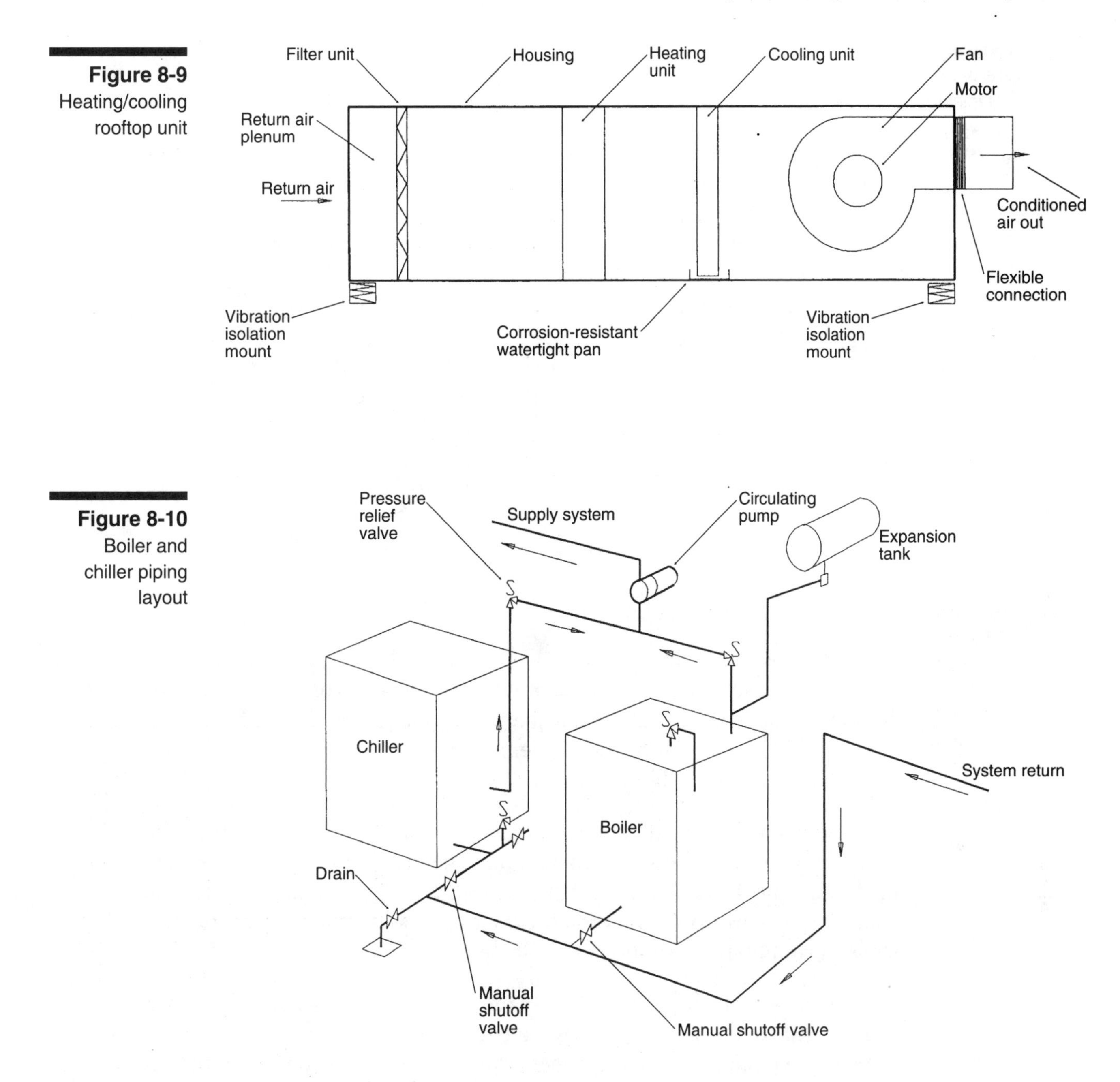

Figure 8-9 Heating/cooling rooftop unit

Figure 8-10 Boiler and chiller piping layout

Radiant Heating Systems

Radiant heating systems use hot water, steam, or electricity to heat radiators, pipe coils, or resistance cables so they will radiate heat. This is called *thermal radiation.*

The major pieces of equipment in a hot water radiant system are the boiler, circulating pump, expansion tank, supply and return piping, and control devices. Figure 8-10 shows the piping around the chiller and boiler in a heating/cooling system.

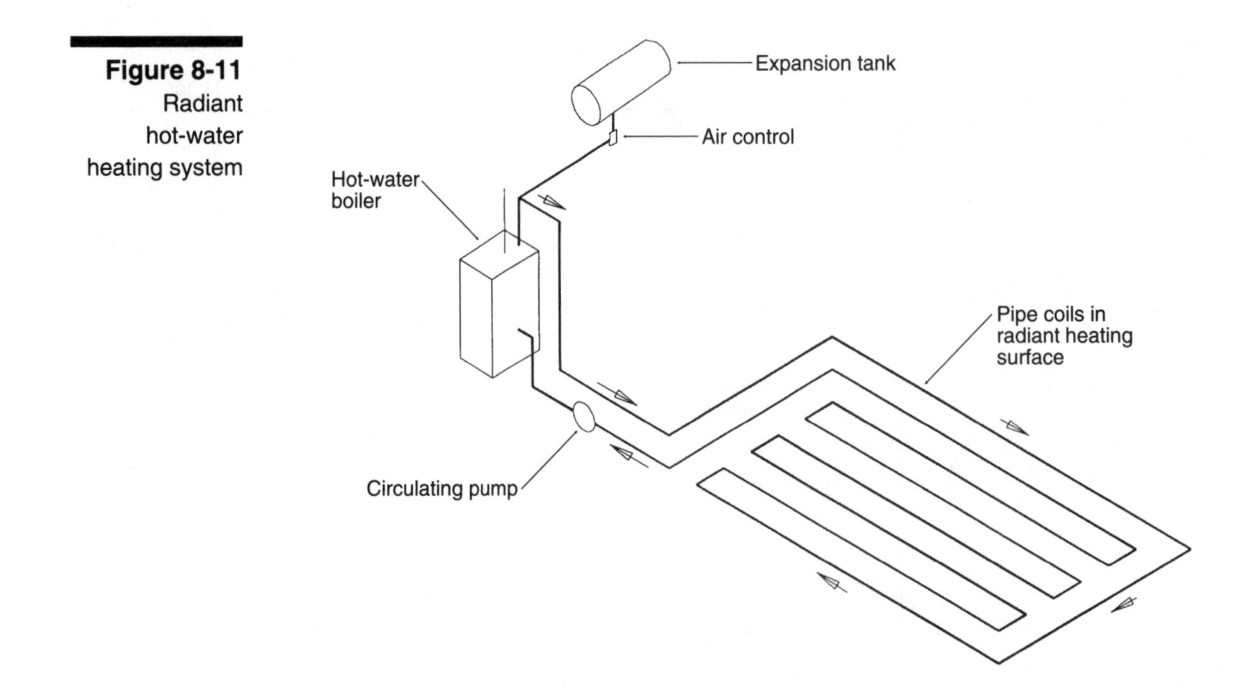

Figure 8-11 Radiant hot-water heating system

A hot water or hydronic system heats water in boilers, and pumps it through pipes to radiators, baseboard convectors, or tubing in the floor or ceiling. Radiators and convectors are usually near the outside wall of the rooms. Figure 8-11 shows one type of hot water radiant heating system. The expansion tank removes water vapor and air from the hot water piping.

Systems with radiators or convectors are generally mono-flow, two-pipe direct return, or two-pipe reverse return systems. The mono-flow system costs less than the others because it uses less piping. The two-pipe direct system can have water flow balance problems, so it's better to use the two-pipe reverse return system. Figure 8-12 shows a simplified illustration of the three piping systems.

Figure 8-13 shows a single-family residence which has a radiant hot water heating and cooling system that uses a boiler for heating and a chiller for cooling. Usually the piping is made of copper tubing and it runs, in coils, through the floor, as shown in Figure 8-14. But you can also put the pipe coils in the ceiling.

The maximum heat emission for floor panels is usually 50 Btuh per square foot for all habitable rooms. The maximum heat emission is 60 Btuh per square foot for bathrooms, entrance halls, and areas within 18 inches of exterior walls. The maximum design surface temperature for floor panels is 85 degrees F.

You should put thermal insulation on each boiler to make sure that heat loss is less than 0.5 Btuh per square foot per degree F. Some common types of insulation are magnesia blocks, hydrous calcium silicate blocks, fiberglass, and insulating plaster. The manufacturers of these products provide tables and charts which list the conductivity in Btu per inch per square foot per degree F per hour of the materials. Use these tables and charts to pick the type and thickness of insulation you need for a job.

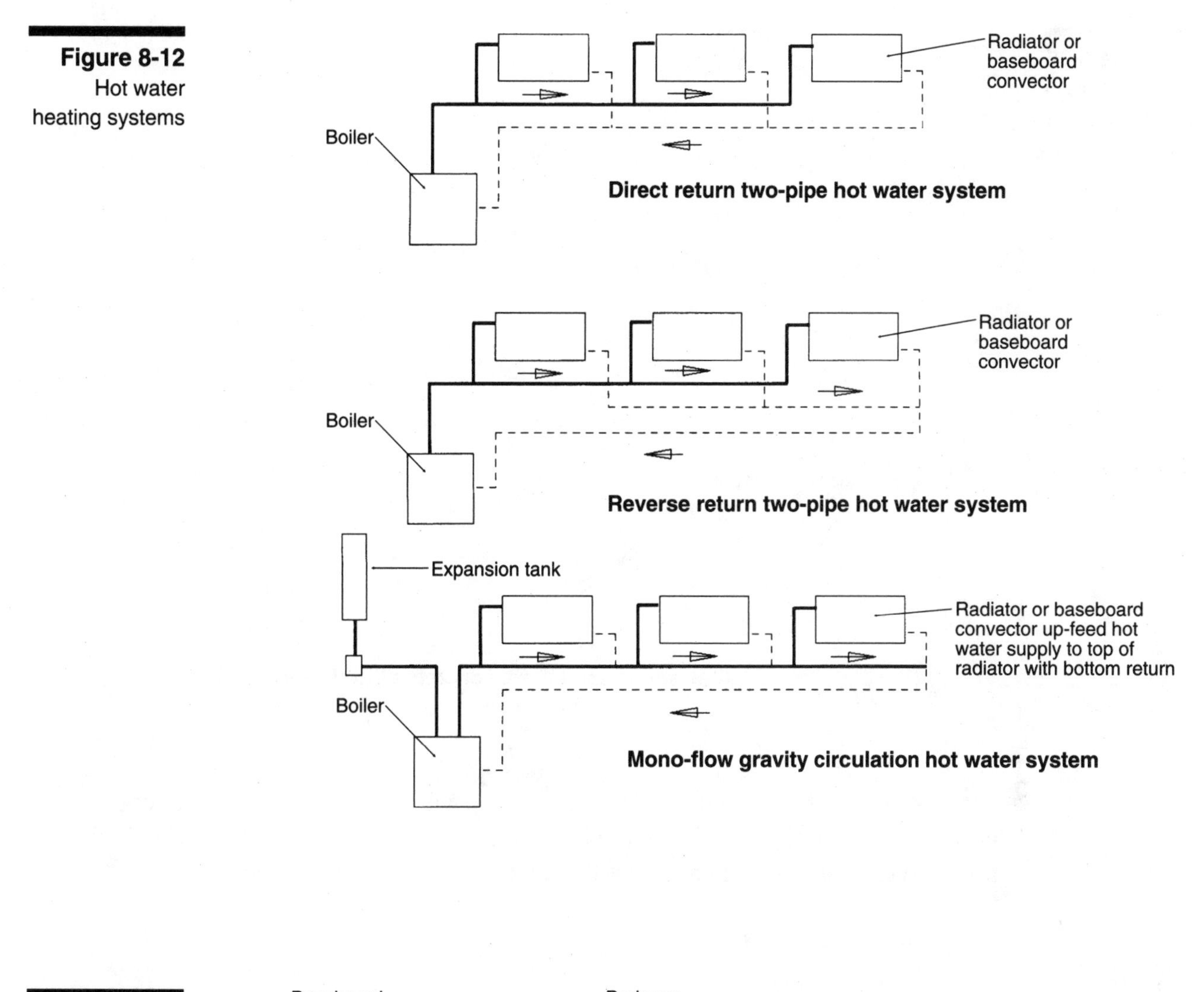

Figure 8-12
Hot water heating systems

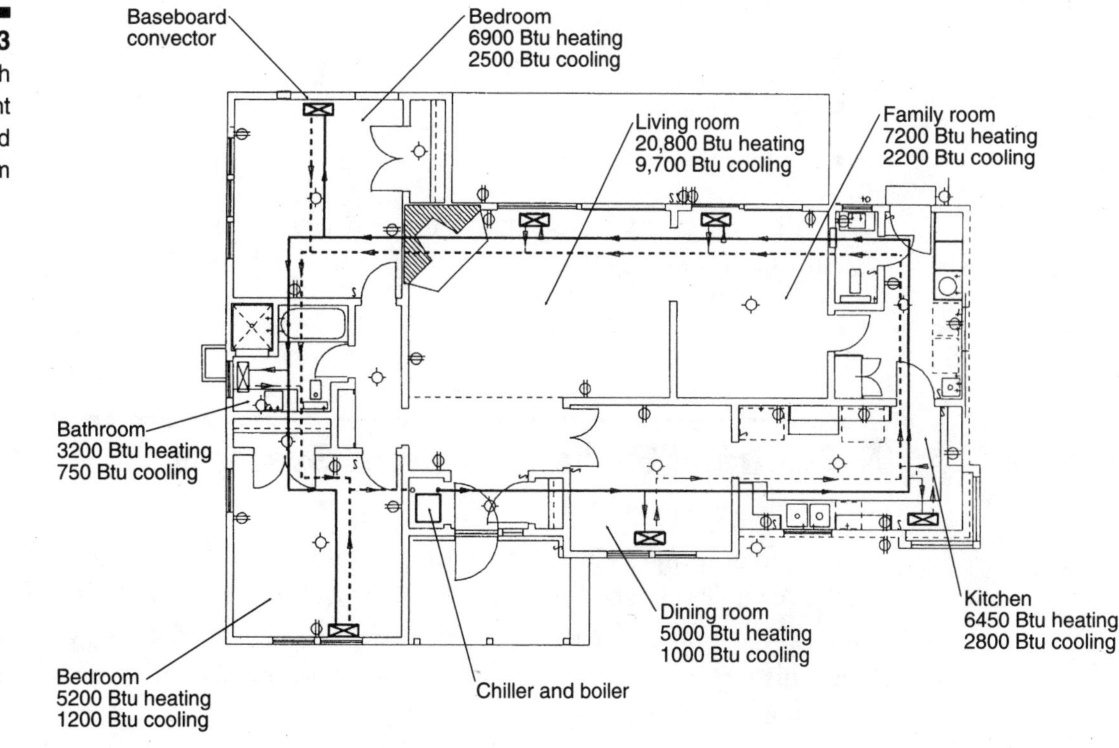

Figure 8-13
Residence with a radiant hot/chilled water system

Figure 8-14
Heating coils in concrete floor

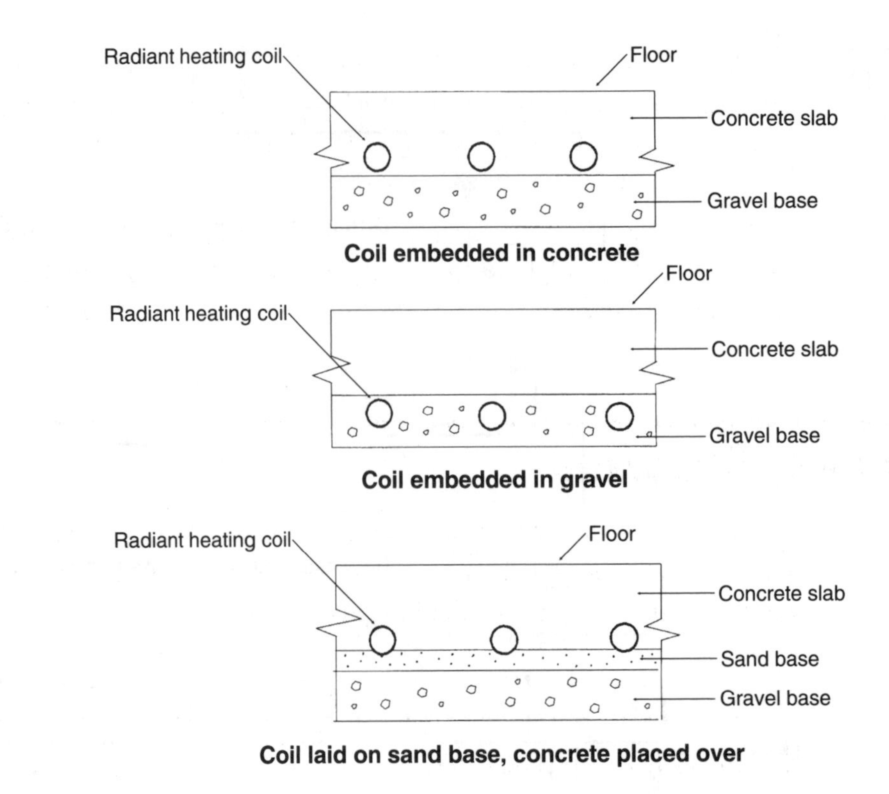

Connect the boiler to the cold water supply with a shutoff valve, drain cock, and hose connection. Put pressure-reducing valves between the shutoff valve and the boiler. Each boiler should arrive at the job site with a label that identifies:

- manufacturer's name
- type of boiler
- size
- Btuh input
- Underwriters Laboratories, Inc. listing

Don't connect any boiler, steam, or condensate drain directly to a drainage system. Don't send any water above 140 degrees F into a drainage system.

Electrical Radiant Heating Systems

A popular form of electrical heating system called the radiant heat ceiling has resistance cables attached to gypsum lath or wallboard in the ceiling. The cables are spaced about 6 inches apart and covered with ¼-inch to ⅜-inch gypsum plaster or filler. This type of system provides fast and uniform heat emission. It usually costs less than a mechanical system.

The panels should be thermostatically controlled. The maximum surface temperature should not exceed 150 degrees F. Put a control thermostat in each heated room or space. Obviously, you don't cut a panel, and you don't fasten lighting fixtures into the active area of a panel.

Use heating units and control devices listed by the Underwriters Laboratories, Inc. (UL) and install them to comply with the *NEC*. The labeled voltage of the equipment should be within 5 percent of the service voltage provided.

Ventilation Systems

The most basic way to ventilate a room is with windows. The area of windows which open should be at least 1/20th of the floor area of the room. But you can also ventilate a room by mechanical means.

Gravity Systems

Gravity ventilation is used mainly for unoccupied areas such as attics, machinery rooms, and similar types of spaces which can use roof ventilators and wall openings to provide ventilation. Louvers protect wall openings against rain.

Roof-mounted gravity ventilators may be the stationary, pivoting wind directional, or rotating-turbine type. Whatever type ventilator you use, it should have a minimum wind velocity of 4 mph. When you pick a unit, you should consider the mounting height, which is the distance from the air intake of the building to the ventilator on the roof. Ventilators are more effective when they are higher, when the temperature difference between the outside and inside air is greater, and when the wind velocity increases. A change from 5 mph to 10 mph wind would increase the exhaust capacity of a gravity ventilator by about 50 percent, because the rotating vanes on a ventilator make a low pressure area which draws air up from a building. The faster the wind, the faster the ventilator draws the air out.

Here's a formula you can use to figure out how many gravity roof ventilators you need. It comes from the ventilator manufacturer's manual:

$$\text{Number of single ventilators required} = \frac{V}{R \times F \times B}$$

where:

V = volume of building in cubic feet

R = desired rate of air change in minutes per change

F = height and temperature factor

B = base capacity of the ventilator

Let's go through an example. Suppose you want to ventilate an 850,000 cubic foot building with an air change every 5 minutes. The ventilator height above intake is 30 feet and average temperature difference between the inside and outside air is 15 degrees F.

Now let's suppose that you've found a ventilator you think might work. First you need to check the manufacturer's information on the ventilator for its base capacity, and throat and curb area. The throat area is the cross-sectional area of the ventilator duct. The curb area is the area of the opening in the roof. Make sure the area of the intake openings to the building are equal to, or greater than, the total curb areas of the ventilators. Suppose the ventilator has a base capacity of 17,000 cfm. The height and temperature factor for 15 degrees F and 30 foot height is 0.91. The manufacturer usually publishes the height and temperature factor. Then:

$$\frac{850{,}000}{5 \times 0.91 \times 17{,}000} = 10.99$$

So, you'll need eleven ventilators.

The following table shows a very general guide for recommended periods of air change for different types of buildings. Every ventilation situation is special so you have to consider each one individually.

Type of Building or Area	Minutes per Air Change
Attics	2 - 4
Auditoriums	4 - 15
Bakeries	1 - 5
Banks	3 - 10
Bars	1 - 5
Beauty parlors	4 - 10
Boiler rooms	1 - 4
Churches	4 - 15
Classrooms	3 - 8
Corridors	3 - 8
Dining rooms	1 - 5
Factories, light manufacturing	1 - 2
Garages	½ - 1
Kitchens	2 - 8
Machine shops	1 - 6
Offices	1 - 2
Recreation rooms	1 - 2
Residences	2 - 10
Restaurants	4 - 6
Retail stores	5 - 30
Toilets	10 - 15
Warehouses	5 - 10

If you use a gravity ventilation system, remember that something has to replace the air you take out. Provide some way for air to get back into the area you're ventilating.

Forced Air Ventilation Systems

If you use mechanical ventilation, plan to provide at least two air changes per hour for each habitable room. One fifth of the air should come from the outside. Bathrooms, water closet compartments, laundry rooms, and similar rooms should have at least five air changes per hour. This is figured by multiplying the floor area in square feet by the height of the room in feet to obtain the volume in cubic feet. Then multiply the volume by the number of air changes per hour required to get the cubic feet per hour. Then divide the cfh by 60 to get cubic feet per minute of makeup air.

A fan exhaust ducted directly to the outside should provide one air change in 12 minutes. A mechanical ventilating supply or exhaust system should deliver approximately 7 cubic feet per minute (cfm) of outside air per person. As a rule-of-thumb, use the number of square feet per occupant for each use as described in Table 33-A of the *Uniform Building Code*. This is in the column titled "Occupant Load Factor." For example, if a classroom is 400 square feet and the Occupant Load Factor is 20, only 20 people can legally occupy the room at the same time. Here are some of the common occupant load factors for different types of uses:

Use	Occupant Load Factor
Assembly areas, without fixed seats	7
Conference and dining rooms	15
Classrooms	20
Dwellings	300
Garages	200
Hotels and apartments	200
Warehouses	500

Fan capacity in cfm is based on the number of air changes needed to ventilate a space, and the volume of the space in cubic feet. Fan capacity in cfm is:

$$Fc = V \times N/60$$

where

V = volume of the space in cubic feet
N = number of air changes per hour

When selecting a fan, you need to consider the resistance the ducting adds to air movement. The outward force of air within a duct is called static pressure. This can be due to friction of air moving through ducts, around elbows, or through filters. A fan loses its efficiency as the static pressure increases. Therefore, fan manufacturers list the capacity of each model fan in cfm against a specific static pressure.

Roof exhausters are roof ventilators with electrically-powered fans. Fans may be either direct-driven or belt-driven. Some roof fans have automatic dampers that close when the fan stops to keep warm air in during cold weather.

A wall fan has a metal housing that fits conveniently between studs in a wall. It's also called a side wall propeller fan. It may be direct-driven or belt-driven. Wall fans also can be fitted with automatic dampers to prevent backdraft.

Special Ventilating Situations

You'll need ventilation in the following special areas and rooms:

- Ventilate enclosed attics by openings or mechanical means.
- Ventilate underfloor areas by mechanical means or by openings in the exterior foundations.
- Provide at least 25 cfm ventilation, or 2 cfm per square foot of area, in bathrooms and toilet compartments. To do this you can undercut the bathroom doors, or install air intake grilles.
- Ventilate air in the kitchen by a fan in a grilled opening in the ceiling, in the wall near the kitchen range, or through a kitchen range hood. The fan should discharge directly to outside air. The recommended ventilation rate for a cooking area is 150 cfm.
- Ventilate boiler rooms so the air temperature stays below 120 degrees F. Control the exhaust fan with a thermostatic switch and a manual switch.
- Ventilate elevator machinery rooms so the room temperature isn't more than 10 degrees F above the outdoor temperature. Avoid overheating the electrical equipment. Set a thermostatic switch to start the exhaust fan at 90 degrees and stop it at 50 degrees F, or as specified by the elevator installer. Don't forget that electric motors and pumps generate heat in a room.
- Vent each elevator shaft or hoistway independently to the outside. The area of the vent should be at least 3.5 percent of the area of the elevator shaft.
- Ventilate rubbish rooms and chutes to the outside air.

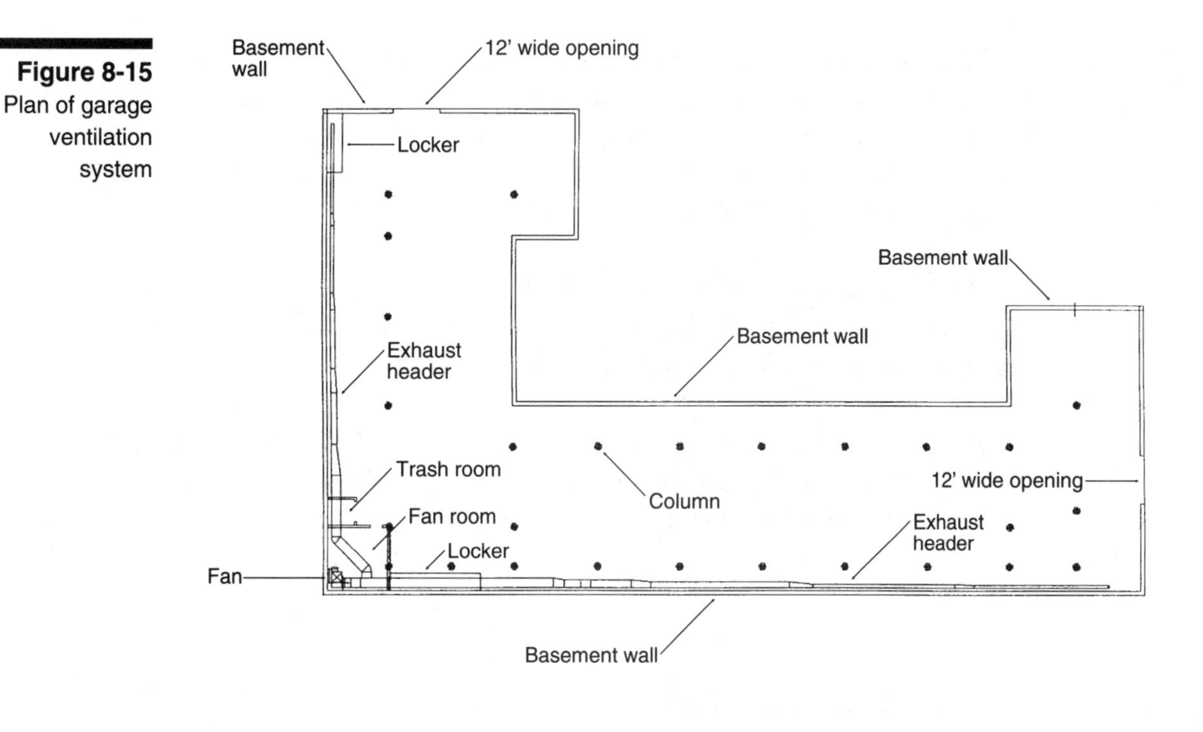

Figure 8-15 Plan of garage ventilation system

Parking Garage Ventilation

Parking garages must be effectively ventilated because of the carbon monoxide gas exhausted by vehicles. Figure 8-15 shows a typical system for an L-shaped garage. The exhaust fan at the corner of the garage induces a vacuum in the two headers. Vertical ducts extend from the headers to within 18 inches of the garage floor to remove the heavier exhaust gases.

Here are some requirements for ventilating parking garages:

- Garage ventilation should exhaust 1.5 cfm per square foot of floor area, or 14,000 cfm for each operating vehicle in the garage.
- Use carbon monoxide sensing devices in garages with a maximum concentration of 50 parts per million (ppm) during an 8-hour period. There are a number of gas sensing devices available. One type will set off a remote alarm when the maximum concentration of carbon monoxide is detected. Other types will display the concentration on a meter, or they'll change colors.
- You don't need ventilators in a garage if its walls have permanent openings equal to 2.5 percent of its floor area.
- You need a ventilating system which gives one complete change of air every 15 minutes in all enclosed parking garages.
- In a parking garage, put the exhaust ventilation inlets of a mechanical exhaust system within 18 inches of the floor level.
- Protect every duct in a garage from vehicular traffic by putting a heavy steel plate or pipe around the duct at bumper height.

Figure 8-16
Evaporative cooler at grade

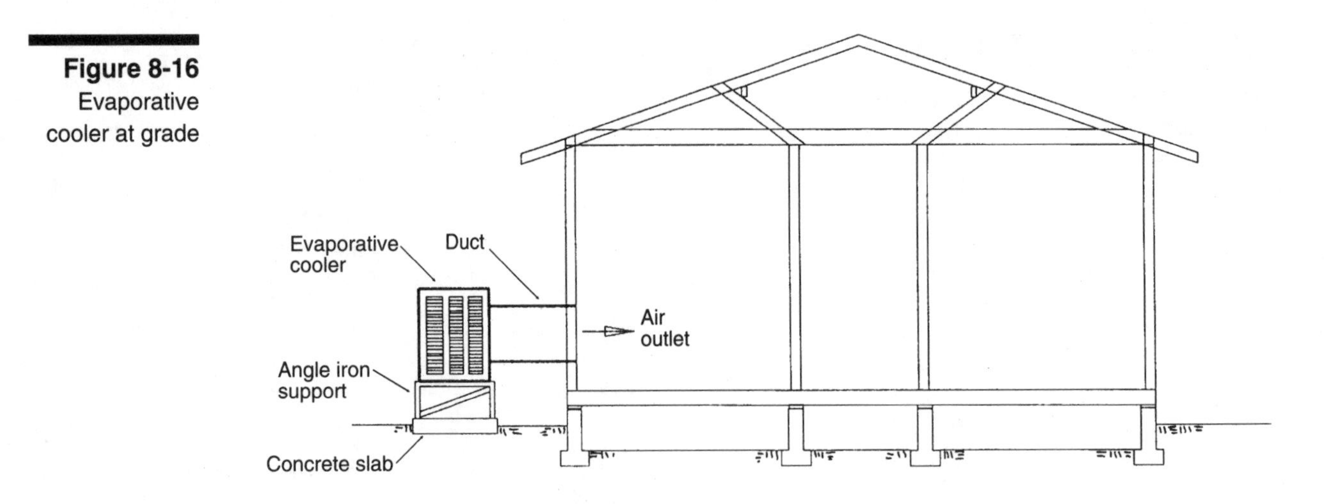

- Don't space ventilation duct openings more than 50 feet apart around the perimeter of the garage.
- Make sure any exhaust outlets that carry noxious gases end outside the building, and at least 10 feet from any adjacent building, adjacent property line, or air intake opening into any building. The end of the exhaust must also be at least 10 feet above the adjoining grade level.
- You must extend every exhaust opening at least 2 feet above the nearest roof surface within a 10 foot radial distance.

Here are some recommendations on dampers for a ventilating system:

- You can put control dampers that are operated by electric or pneumatic actuators within ducts.
- Put automatic fire dampers on the ducts that go through fire separation walls. The dampers should have a fusible link or other heat actuating device so they close automatically. The fusible link should operate at 165 degrees F.
- Make sure any ventilating system you install is designed to keep heat, flame, and smoke from spreading rapidly through a building.

Cooling and Air Conditioning Systems

Evaporative Cooling Units

The simplest type of cooling unit is the evaporative cooler. It's mainly used in areas with somewhat dry climates. The cooler evaporates water to cool the flow of air. Figure 8-16 shows such a unit mounted outside a house. A watertight duct carries cooled air to the interior of the building. Figure 8-17 shows a unit that's installed on the roof. If you install a unit this way, make sure the roof is strong enough to support

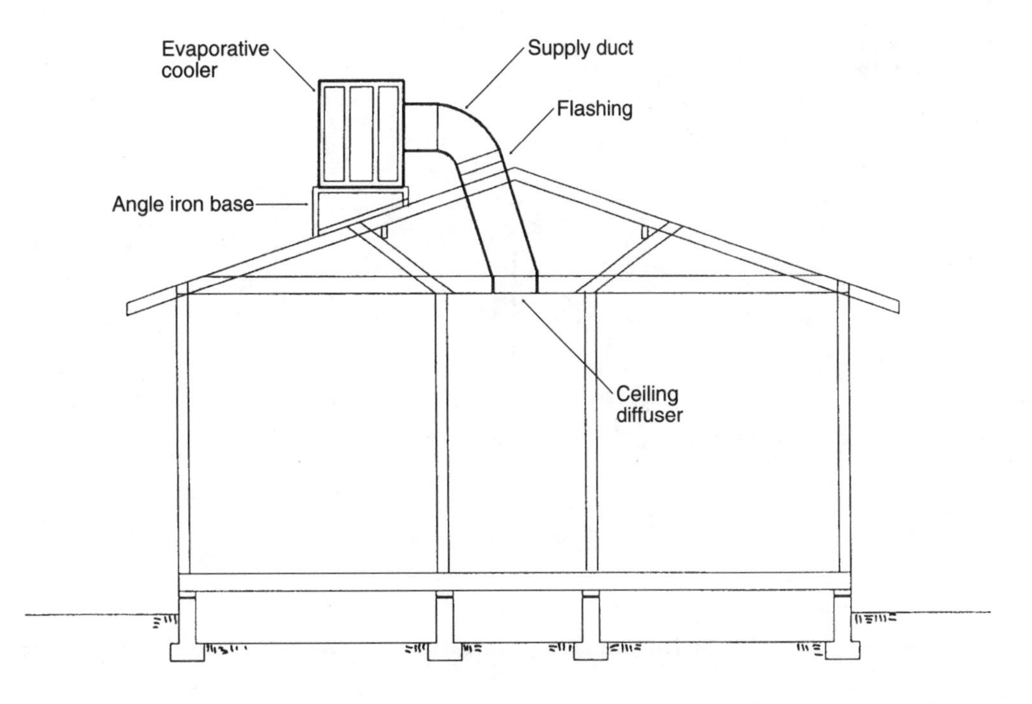

Figure 8-17 Evaporative cooler on roof

the weight of the unit. Make sure you level the cooler when you install it because it has a pan that holds water which must be level. You'll need electric power, water supply, and overflow drainage to operate an evaporative cooler.

Mechanical Air Conditioning Systems

The simplest form of mechanical cooling is a through-the-wall air conditioning unit. Typically these units can handle between 250 and 350 cfm, and about 10,000 Btuh cooling load. Air conditioning units are rated in Btuh cooling load or tons of refrigeration. A standard ton of refrigeration is equal to 200 Btus per minute, or 12,000 Btus per hour. Wall-mounted air conditioners are made in various combinations; only cooling, cooling with electrical heat, heat pump heating and cooling, and heat pump with cooling alone.

A refrigeration system is classified according to the type of refrigerant it uses. Group 1 refrigerants are non-toxic and non-combustible. They are used to cool residential buildings. Sealed absorption systems use Group 2 refrigerants. They are used to cool commercial and industrial buildings.

Figure 8-18 shows a residential split-package air conditioning unit which supplies air to several rooms. Usually you install this system with the air inlet in the center hall of a residence. To reduce compressor noise and promote air circulation around the condenser coils, set the condenser on a pad away from the building. Use refrigerant piping to connect the evaporator on the roof with the condenser outside.

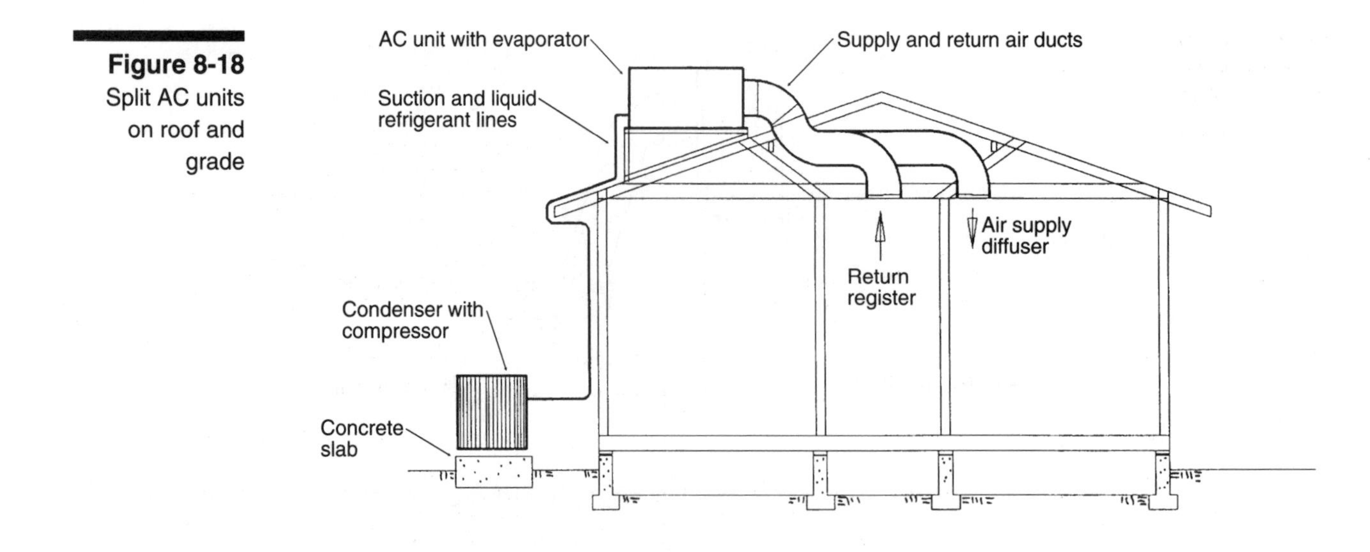

Figure 8-18 Split AC units on roof and grade

All factory-built air conditioning systems should have a label showing:

- Manufacturer's name
- Model number
- Amount and type of refrigerant
- Factory test pressure
- Normal Btuh or power input rating
- Minimum Btuh
- Cooling capacity in Btuh
- Type of fuel or energy
- Symbol of approval agency
- Instructions for starting, operating, and shutdown

Heat Pumps

A heat pump is like an air conditioner that runs two ways. By reversing the flow of the refrigerant between the evaporator and the condenser, it can supply either warm or cool air to the interior of the building. Figure 8-19 shows a simplified flow diagram of a heat pump. The dark lines show how the refrigerant flows during cooling or heating.

When a heat pump is used for cooling, it pumps the warm inside air across the pump coils which have a refrigerant in them. The refrigerant absorbs the heat, and is then pumped to the outside coils. There, a fan cools the coils so they can absorb more warm air. Figure 8-20 shows a complete heat pump package with a plenum, evaporator, and condenser section.

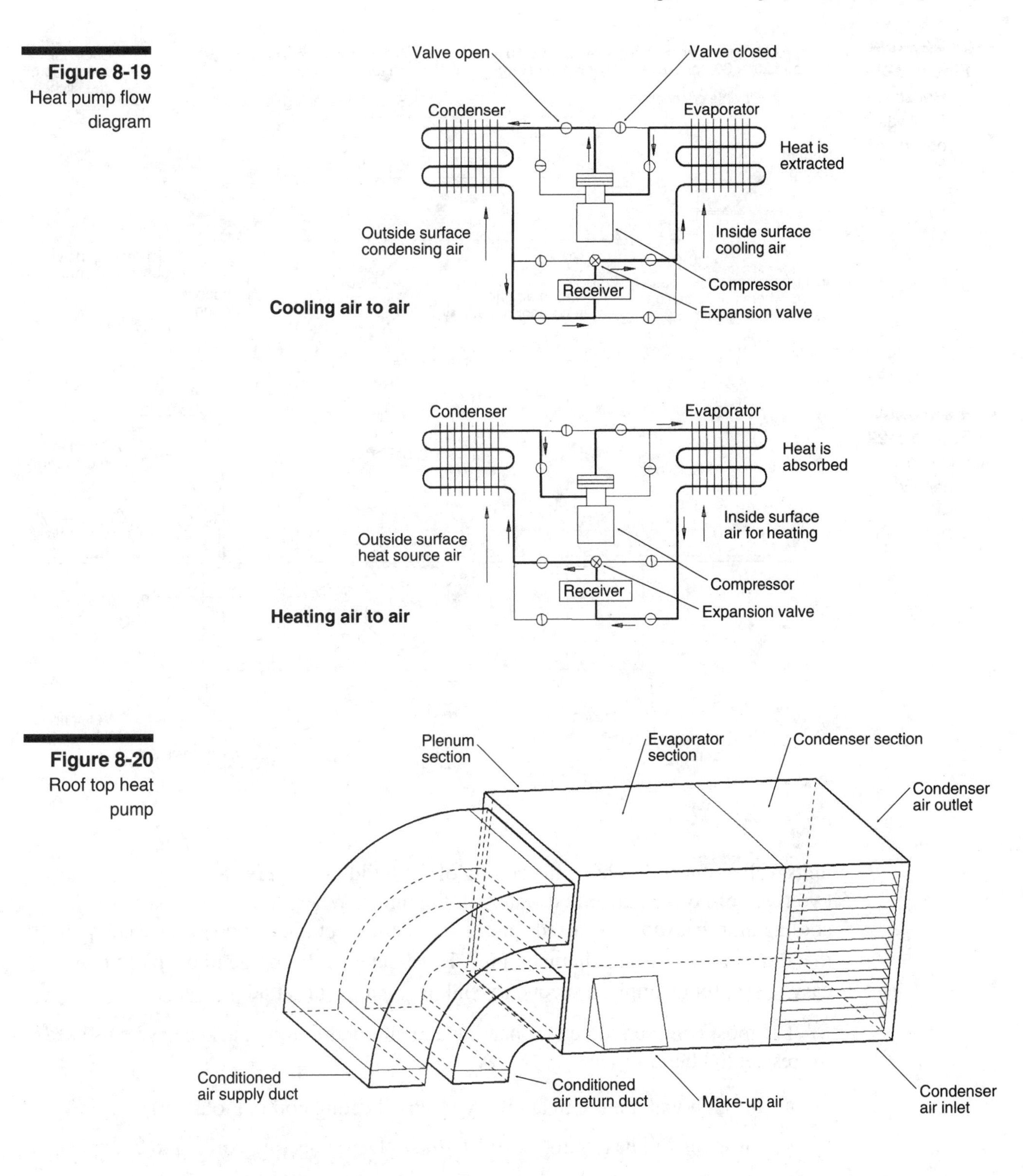

Figure 8-19 Heat pump flow diagram

Figure 8-20 Roof top heat pump

Chilled Water Systems

This type of system is commonly used in multi-story residential buildings. It has a chiller, condenser, boiler, pumps, and other equipment. It's usually located on the roof of a building or in an equipment room. The system usually has multiple supply risers and a single return riser. The hot and cold water are fed to cooling and heating

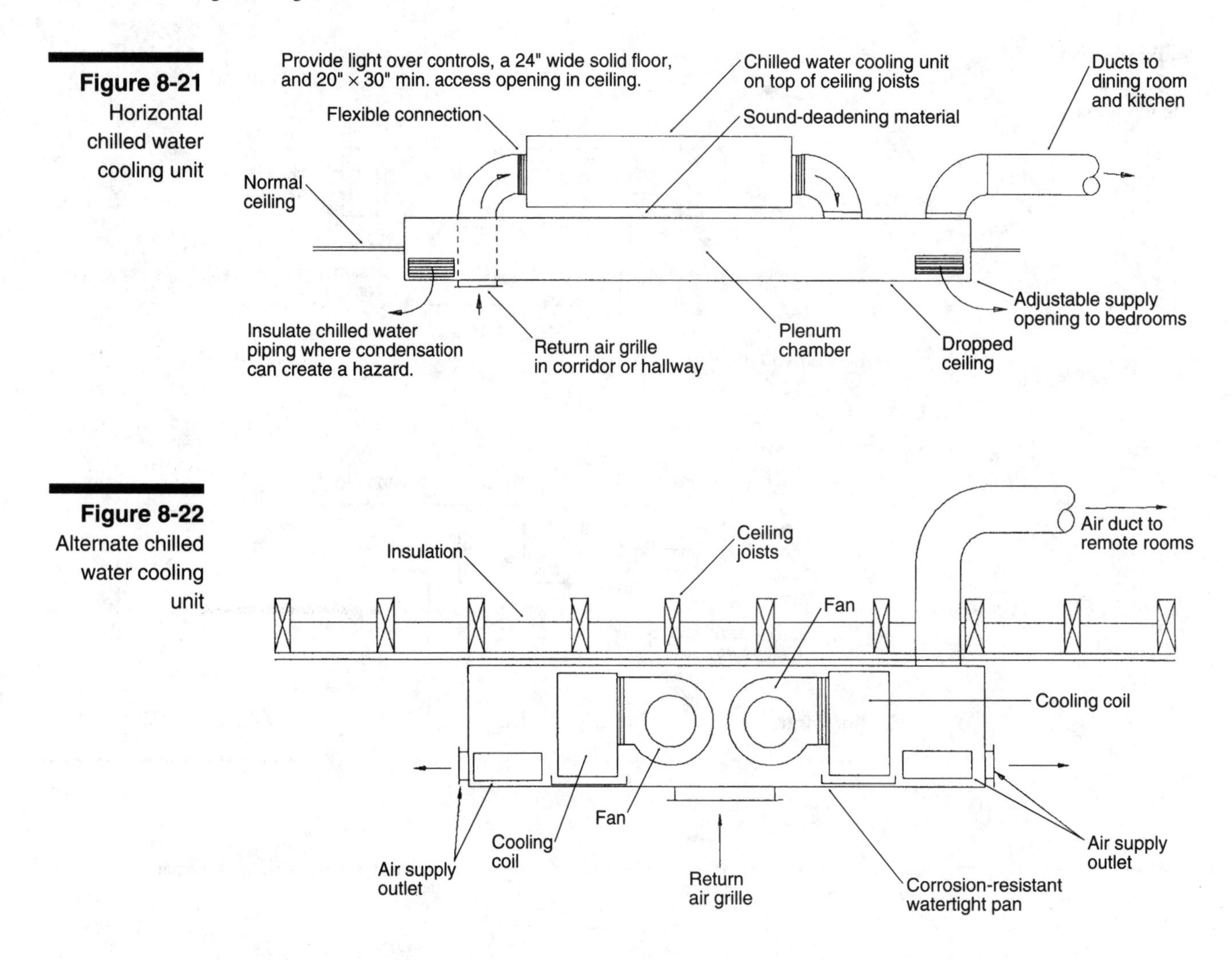

Figure 8-21 Horizontal chilled water cooling unit

Figure 8-22 Alternate chilled water cooling unit

units, called fan coils, above the ceiling of the building. Figures 8-21 and 8-22 show two types of horizontal chilled water cooling units. When you install this type of cooling unit in an attic or furred space in the ceiling, put in a condensate catch pan that drains into a ¾-inch diameter pipe. Have the pipe discharge into a place that's easy to see, for example, a secondary sink in a garage or laundry room.

The most common forced air heating and air conditioning package systems used for residential buildings are:

- through-wall unit with auxiliary electric heating coil (¾ to 2 ton)
- through-wall heat pump with auxiliary electric heating coil (¾ to 2 ton)
- electric furnace with top mounted evaporator and remote air-cooled electrically-operated condensing unit (2 to 10 ton)
- rooftop combination heating/cooling unit powered by electricity or gas (2 to 15 ton)
- water-cooled draw-through unit
- gas furnace with top mounted evaporator and remote air-cooled electrically-operated condensing unit (2 to 10 ton)

- water-cooled multi-zone unit with electric duct heaters (15 to 50 ton)
- heat pump (3⁄4 to 2 ton)

The HVAC packages usually used in commercial and industrial buildings are:

- rooftop multi-zone unit with combination heating/cooling (10 to 50 ton)
- multi-zone two pipe all-electric system with chillers and electric duct heaters, hot water boiler, pumps, and expansion tanks (50 to 1000 tons)
- low pressure variable air volume system with chilled water passing through air handling unit (50 to 200 tons)
- high pressure variable air volume reheat all-electric system with electric duct coils. This system includes centrifugal chillers, cooling towers, pumps, and expansion tank (200 to 1000 tons).

Ductwork and Piping

Duct and Fitting Materials

The major components in duct design are gauge and construction, corner closures, hangers, elbows, vaned elbows, tee connections, access doors, fire dampers, and air intakes. Most elbows are 45 and 90 degrees, or reducing type. Other duct fittings are offsets, single branch take-off, caps, register box, square-to-round transitions, and many combinations of the above.

The galvanized sheet steel you use in ducts should be of lock forming quality (LFQ). It should have a galvanized coating on both sides of the sheet equal to at least 1¼ ounces of zinc per square foot. All 30 × 30-inch ducts should be cross broken for reinforcing the flat sheet. Flat sheets are cross broken by machine in a sheet metal shop. Sheet metal ducts should conform to the following gauges:

Maximum Dimension	Steel Gauge and Thickness
4 to 30 inches	24 (0.024 inches)
30 to 60 inches	22 (0.030 inches)
60 to 90 inches	20 (0.036 inches)

Install tightly-constructed metal ducts rigidly to eliminate vibration. Rectangular ducts should be made with standing seams and braced with angle iron to stiffen the sheet and avoid vibration. It's a good idea to lower the velocity in the main duct and the remote branches gradually to distribute the air uniformly. This also cuts down the friction in the smaller ducts, where it would otherwise be the greatest. To lower the air velocity in a duct, increase the size of the duct.

Pressure losses in duct systems are caused by the velocity of the air flow, the number of elbows, and the friction of air against the sides of the duct, heating coils, air filters, air washers, dampers, and deflectors.

Solder, weld, or otherwise seal all joints and seams to make them tight. You can also rivet or spot weld them. Double lock, make a 1-inch lap, or rivet the longitudinal lap joints or seams with 3-inch or less rivet centers.

Make the lap joints so the outlet of one length fits into the inlet end of the next length in the direction of air flow. Lap the joint at least 1¼ inches. Don't put the rivets more than 4½ inches apart on center. Use at least four rivets in any lap joint. You can use an equal number of spot welds in place of rivets.

A typical round flexible duct is made of a vinyl-coated fabric which is reinforced with spiral wound wire. Some flexible ducts also have fiberglass insulation in them. Round ducts can also be made of spiral fabricated steel bands.

Diffusers and Registers

Diffusers are outlet fittings specially designed to uniformly distribute air to rooms and spaces. Some diffusers have round concentric louvers, and others have rectangular adjustable louvers.

Registers are inlet fittings that return air from conditioned space to the return air duct. They usually have a protective screen or grid to keep debris from being sucked into the duct.

Flues and Vents

Connect all gas-fired equipment to a suitable flue or gas vent. The products of combustion can be hazardous if they discharge into a space where there's equipment. A flue should be structurally safe, durable, smoke tight, non-combustible, and able to withstand the action of flue gases without softening, cracking, corroding, or spalling. Separate flues from combustible framing according to Table No. 9-C of the *Uniform Mechanical Code.*

Roof Penetrations

It's very important to flash the roof penetrations. To make an effective watertight seal around a duct that goes through a roof, put a curb around the opening. Put a cant strip around the curb from the roof level so that the roofing doesn't have to bend more than 45 degrees. Attach metal flashing to the duct wall so it overlaps the curb and cant. Be sure any joints in the metal duct that are exposed to the weather are watertight.

Protect any outlet of an exhaust duct or inlet of a fresh air duct against rain and insects. One way to do this is to equip an exhaust outlet with a backdraft damper. Another is to position the opening downward at 45 degrees off the horizontal. Any horizontal outlet you use should have louvers.

Filters

You need to get the dust and soot out of the fresh air that comes into a system. This is particularly true if the intake is close to a street or alley. There are several compact air filters that are good for this. They go on the inlet side of the system. Don't forget they'll add some resistance to the system that you need to allow for when you decide what size fan to use in the system. Check the filter manufacturer's catalogs for a listing of static pressure for each size fan. See the section on Forced Air Ventilation Systems for more information on static pressure.

Insulation

The most common types of duct insulation are made of mineral fiber, rock, slag, or glass blankets. The insulation is usually clad with Kraft paper or aluminum foil. The thermal conductivity, or k-value, varies from 0.24 to 0.30. Insulate warm air ducts to reduce heat loss. Insulate cool air ducts to keep warm air from entering a duct and prevent condensation on its outside surface. In extremely cold locations, heat or temper cold air coming into a system.

Duct liners, made of fiberglass or rock wool, are similar to exterior duct insulation, but you install them on the interior surface of the duct. They cut down on the noise generated by air turbulence and mechanical equipment and the booming and cracking sounds caused by sheet metal contraction and expansion. Sound insulation also restricts cross-talk (sound transmitted from one room to another through the ducts) within the duct system.

Duct Hangers

Usually you use hangers to support ductwork. A hanging system has three parts; the upper attachment to the building structure, the hanger itself, and the lower attachment to the duct. When you support the ducts from a concrete slab or beam, you can use concrete inserts or power-actuated drive pins for the upper attachments. Insert expansion bolts into drilled holes or drive pins into concrete with a pneumatic gun.

In steel frame buildings, you can use C-clamps or welded studs to support ductwork. The hangers may be steel straps or hanger rods. Rods are usually ¼ inch in diameter and threaded for adjustment by couplings. You can attach vertical ducts to walls using steel straps or angles. Fasten these to the wall the same way as the hangers.

Controls

Any HVAC equipment that can develop hazardous pressures or temperatures must be equipped with controls to safely relieve such pressures and temperatures. For example, you need to:

- Cut off the fuel supply when there's a failure or interruption of flame or ignition.
- Prevent or reduce the input of additional heat when a predetermined pressure or temperature is exceeded.
- Cut off the fuel supply when the water level in a steam boiler drops below a predetermined level.
- Protect the heating system against excessive line pressure by installing relief valves.
- Install thermostats to regulate heating and cooling machines and moderate temperature in machinery rooms.

Thermostats

Some thermostats contain strips of two different metals, usually brass and iron, which expand and contract at different rates so the strip bends and triggers a switch. A rod thermostat contains a steel rod in a brass tube. The tube expands and contracts more than the rod, so it can trigger a gas valve.

Other Controls

Other controls used to moderate air flow are:

- dampers which control air flow in ducts
- flow control valves which provide zone control for hydronic heating and cooling systems

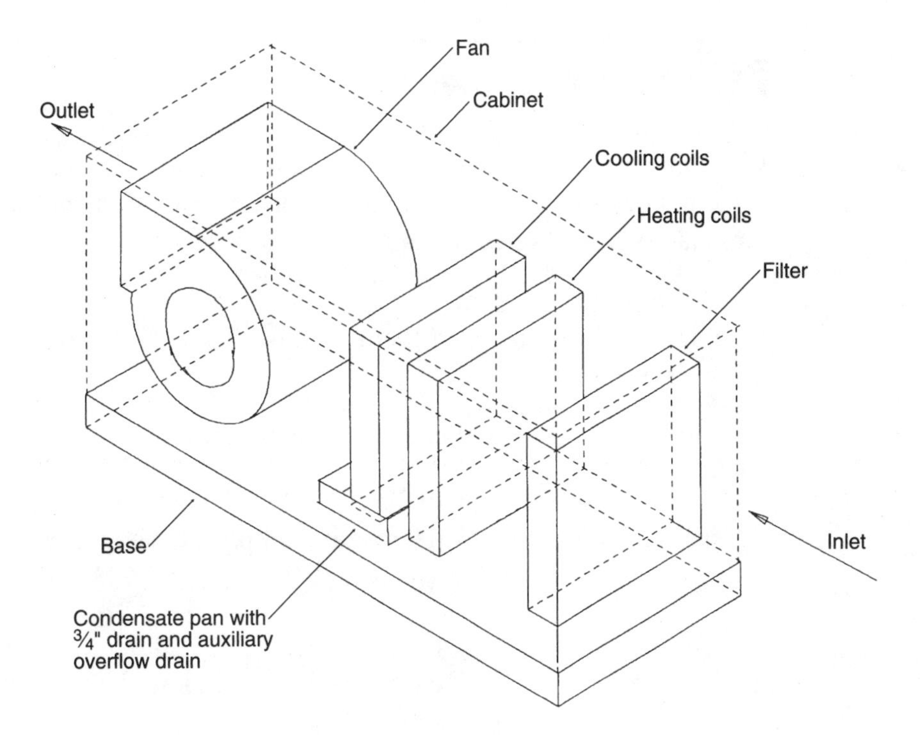

Figure 8-23 Combination heating/cooling unit

- check valves which protect against gravity circulation and prevent unwanted circulation in multiple zone systems
- pressure regulators which reduce pressure of gas to burners
- steam traps which collect condensate in steam piping

Package HVAC Units

Most HVAC manufacturers make modular units. Typical modules have duct coils, filters, a fan assembly, furnace, heat pump, condenser, and an evaporator. These units may be assembled vertically or horizontally.

Air Handling Units

A typical air handling unit has a housing, blower, motor, filter, and base which you install on the roof or within a building. If you add heating or cooling coils to the air stream, you convert the equipment to a heating or cooling unit. Figure 8-23 shows the parts of a heating/cooling air handling unit. Using a burner and heat exchanger instead of heating coils converts the unit to a fuel-burning heating unit.

Fans and Blowers

A fan or blower draws in air from one system and forces it through another system. For example, it brings in outside air, and exhausts used or impure air. It also circulates warmed or cooled air.

It's important that you select motors and fans that are powerful enough to move the quantity of air you want them to move. Here's how to pick a fan that will move air in a system:

1. Measure the length of each straight duct run.
2. Add the equivalent length of each elbow in the system by checking in a duct design manual to find out which factor to multiply the elbow diameter by to get the equivalent length. Then you add the equivalent of each transition, coil, and filter section.
3. List the total length of each size duct including the equivalent lengths of elbows, transitions, and coils and filter.
4. Calculate the air velocity in each size duct by dividing the volume of air by the cross-sectional area of the duct.
5. Use an air friction chart to find the static pressure for each size duct.
6. Add the static pressure of all ducts which are connected to the fan to find the total static pressure.
7. Using a fan manufacturer's catalog, select a fan which provides the required volume of air at the total static pressure you figured.

When you install a motor or fan, make sure there's enough space around it so it can be serviced and maintained.

To control vibrations and fan noise, use flexible connections to ducts and spring or neoprene isolator mounts. Cut the operating pressure of a fan down to under 1.5 inch water column to make it run more quietly.

Fans are either directly connected to a motor, or belt driven. Direct-drive fans are more efficient because there's less energy loss from friction of belts and pulleys. They also take up less space because you can mount the motor within the blower housing.

The only way to adjust a direct-driven fan is to use a variable speed motor. On a belt-driven fan, you can adjust fan speed by changing pulley sizes. But be sure to protect all belts and pulleys with a guard.

Fans and blowers have these properties and characteristics:

- Fan size is usually given as a letter and number that represents the wheel diameter in inches.

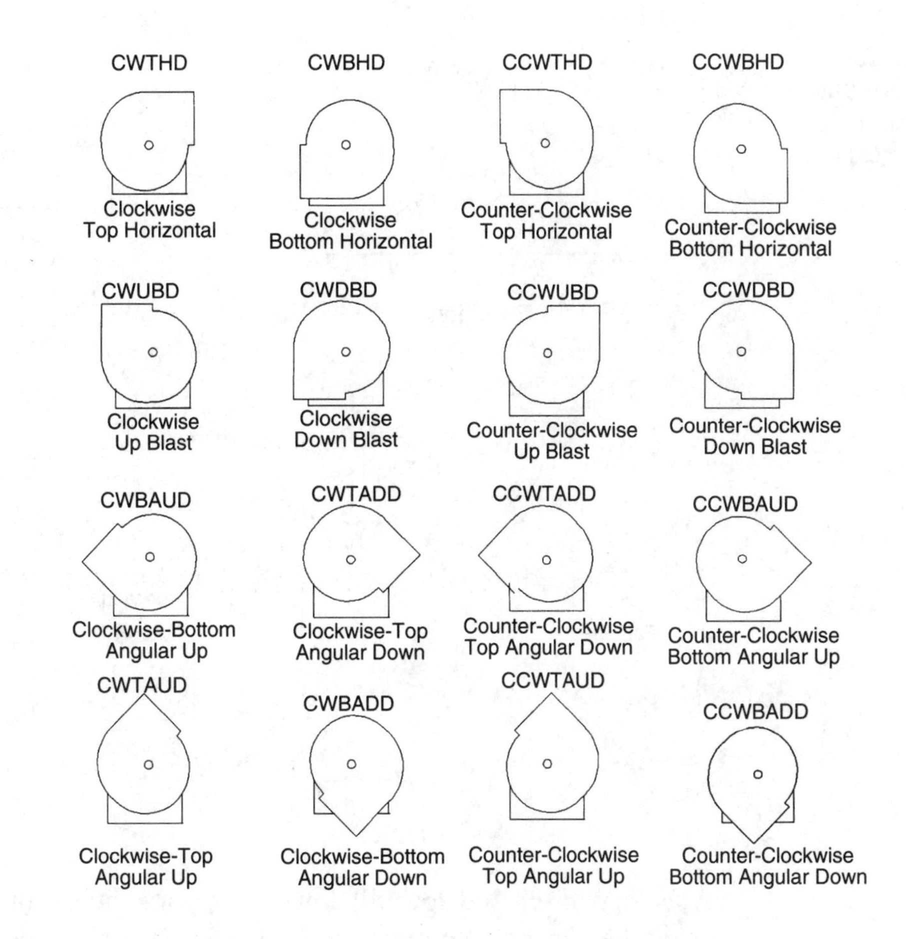

Figure 8-24 Standard designation for blower rotation and discharge

- Tip speed is given in fpm and depends on the wheel diameter and rpm. Tip speed is the velocity of the outer edge of the fan blades. It affects the amount of noise the fan makes.
- Inlet size is given in inches of diameter or square feet of opening.
- Outlet size is given in inches of diameter and square feet of opening.
- Air flow is given in cfm and varies according to the outlet velocity given in fpm.
- Outlet velocity is given in fpm and depends on the rpm of the wheel.
- The motor of a belt-driven fan can be front, rear, inside, or outside the fan shaft.

The way a blower turns and discharges air depends on the drive of the fan when it's resting on the floor, as shown in Figure 8-24. On a single inlet fan, the drive side is on the side opposite the inlet. The abbreviation for each type of rotation and discharge position is shown above the figure. The fan and motor are both rotated by a shaft connected to a pulley. The two pulleys are connected by a drive belt. The motor pulley drives the fan pulley.

Figure 8-25
Positions of fan discharge ducts

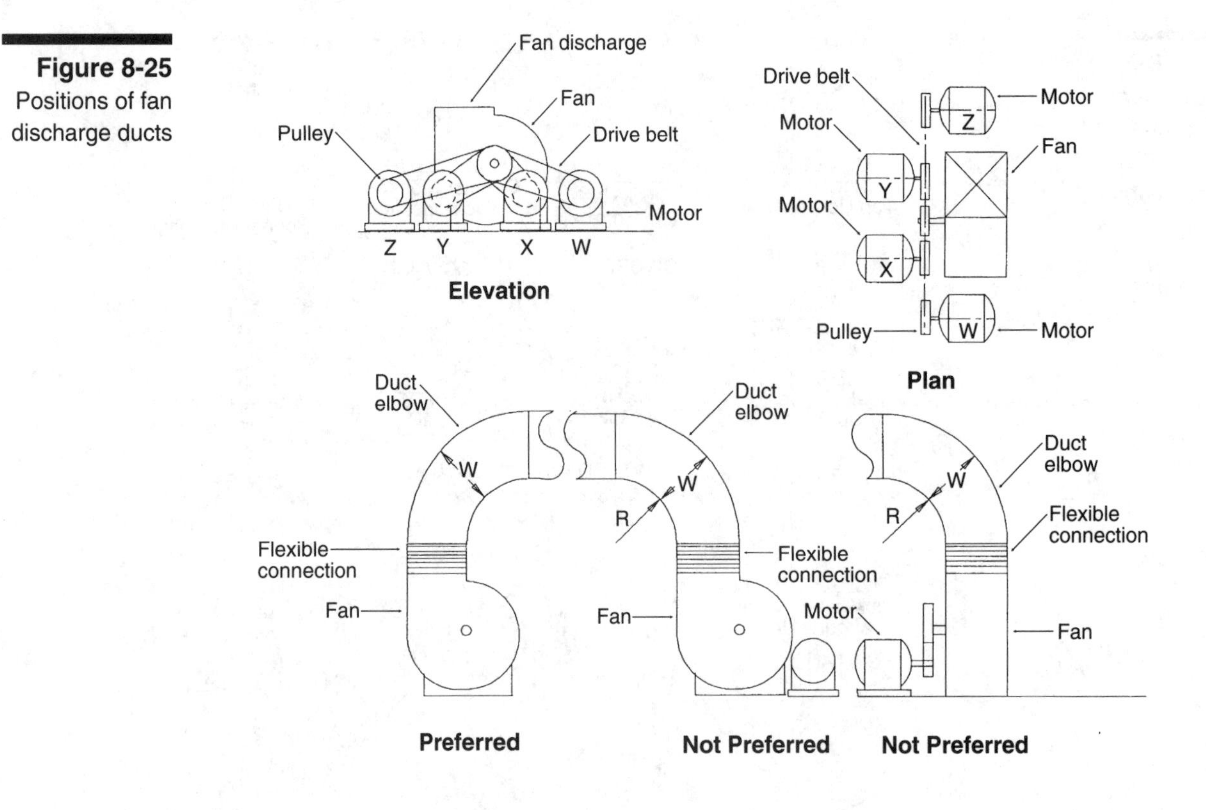

A fan may discharge upward, downward, horizontally, or diagonally. The mechanical engineer or contractor selects the best position depending on the available space, duct layout, and area for maintenance. The orientation of the fan and motor are important in specifying and installing the equipment correctly.

For belt-driven fans, some manufacturers assign a letter to each motor position relative to the fan. They do this so they can select the proper size drive belt and prepare a certified drawing the builder can use when he installs the fan. The letters W, X, Y, and Z shown on the top of Figure 8-25 show four motor positions. As the fan and motor are both rotated by shaft-mounted pulleys, the pulleys must be in line to accommodate the drive belt. In the X position, the drive belt is between the fan and motor, which is away from the fan discharge. Position Y is similar to Position X except the motor is near the fan discharge. In Position W, the motor and fan are on the same side of the belt, but the motor is on the opposite side of the fan discharge. Position Z is similar to Position W except the motor is near the fan discharge. You select the best position depending on the space available for the motor and access for maintenance. It's better to position the motor so the upper part of the drive belt is tight and the lower part is slightly slack.

Steam and Hot Water Boilers

Each steam boiler you install should be approved by the Code of Low Pressure Heating Boilers. The following list shows other required features.

- an ASME label
- a safety valve with an ASME steam rated Btu capacity not less than the gross output rating of the boiler
- a steam gauge, and water gauge glass with gauge cocks and drain
- a low water cutoff, when the boiler is automatically fired
- thermal insulation to cut down the rate of heat flow from the surfaces of the heat-producing equipment into the building
- a cold water supply with a shutoff valve, drain cock, and hose connection
- pressure reducing valves installed between the shutoff valve and the boiler

A hot water boiler heats air, not water. It should have a pressure relief valve with a rated Btu capacity of not less than the gross output rating of the boiler.

Chillers

A typical water chiller is the reciprocating type, which has chilled water going to the cooling coils at 44 degrees F, and returning to the chiller at 54 degrees F. The condenser which removes the heat from the system operates at an ambient temperature of 95 degrees. If the ambient temperature is higher than 95 degrees, the system is less efficient. The package contains a compressor, motor, cooler, condenser, internal piping, wiring, starter, and controls.

Pumps can circulate hot or chilled water. They can pump up to 150 gpm, against a 35 foot head, with a 1½ hp motor, through a ¾ to 3-inch pipe. The most commonly-used pump is the in-line booster pump.

Engineering

Designing an HVAC engineering system involves understanding thermodynamics, fluid mechanics, weather conditions, and several other things. These are all rather complex and would take many pages to explain, so the following section summarizes the fundamentals and gives you the commonly-used rules-of-thumb.

Heating

Some general design rules for heating equipment are:

- Heating equipment is sized by its heating capacity measured in Btuh.

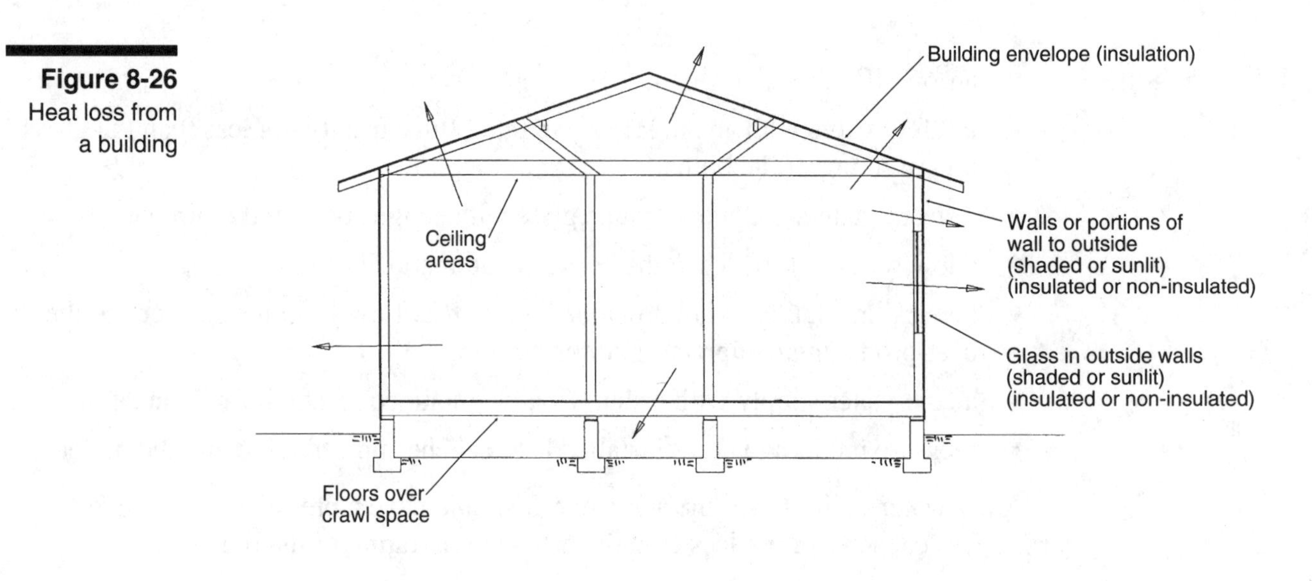

Figure 8-26 Heat loss from a building

- When floor furnaces or vented recessed heaters are installed in spaces other than the room being heated, the distance from the heater to the center of the space should be less than 18 feet, or 20 feet for kitchens. Measure this distance through no more than one interconnecting door.
- Fans in heaters should have a minimum capacity of 9 cfm per 1000 Btuh output, or enough capacity to limit the temperature rise to 100 degrees F at the rated heater capacity.

Generally, inside design conditions shouldn't be higher than 80 degrees F DB and 50 percent RH. Exceptions to this rule are those areas where design DB temperature exceeds 100 degrees F. In this case, a 20-degree F differential between outside and inside design temperature may be used.

To figure out the size of a heating system for a residential building, you have to find out the amount of heat loss in Btuh from the building envelope. Figure 8-26 shows some common paths of heat loss. To find heat loss or gain from a specific building, you'll need to know:

- site conditions: typical design basis for residential buildings is summer dry bulb (DB) temperature of 74 to 76 degrees F and relative humidity (RH) of 50 to 45 percent. Winter DB temperature is 74 to 76 degrees F and RH 35 to 30 percent.
- heat transfer coefficients and surface areas of the east, west, and south windows, exterior doors, walls, ceiling, roof, and floors
- infiltration in cfm
- number of building occupants
- sensible heat from window infiltration, glass transmission, walls, ceilings, roof, lights, motors, appliances
- latent heat from infiltration, occupants
- duct loss
- heat generating motors

As a rule-of-thumb, you can just figure out heat loss by multiplying the net square foot area of the building envelope by a factor representing a 20, 40, or 55 degree F temperature differential. The differential is the desired difference between outside and inside air temperature. You also multiply air infiltration (in cfm) by this factor to get total heat loss in Btuh for a specific building or space. Here are some examples of these factors:

Item	Factor 20°	Factor 40°	Factor 55°
Walls, or portions of walls, exposed to the outside			
Not insulated and shaded	7	14	18
Not insulated and sunlit	5	10	13
Insulated	3	5	7
Glass in outside walls			
North or shaded	22	45	62
Sunlit and not shaded	21	42	61
Sunlit with inside shades	20	40	60
Ceiling areas under attic			
No insulation	10	18	24
With insulation	3	5	6
Ceiling areas under flat roof or beam ceiling			
No insulation	10	20	25
With insulation	6	10	12
Floor			
Over crawl space	5	9	12
Over finished rooms	0	0	0
Slab	0	6	10
Outside air, in cfm, based on building volume/60	22	43	60

The capacity of the heating equipment shouldn't be less than the total heat loss under design indoor and outdoor conditions.

Here's a simple example of the heat loss factors used above. Assuming the house shown in Figure 8-8 and a temperature difference of 40 degrees F, the calculations are as follows:

Item	Heat Loss, Btuh
Walls or portions of walls exposed to the outside	
Not insulated and shaded	450 sf × 14 = 6300
Insulated	1030 sf × 5 = 5150
Glass in outside walls	
North or shaded	190 sf × 45 = 8550
Sunlit with inside shades	225 sf × 40 = 9000
Ceiling areas under an attic	
With insulation	1430 sf × 5 = 7150
Floor over crawl space	1430 sf × 9 = 12870
Outside air, in cfm, based on building volume/60	490 cfm × 43 = 21070
Total heat loss in Btuh	= 70,900 Btuh

Therefore, a heating load of at least 70,900 Btuh is required.

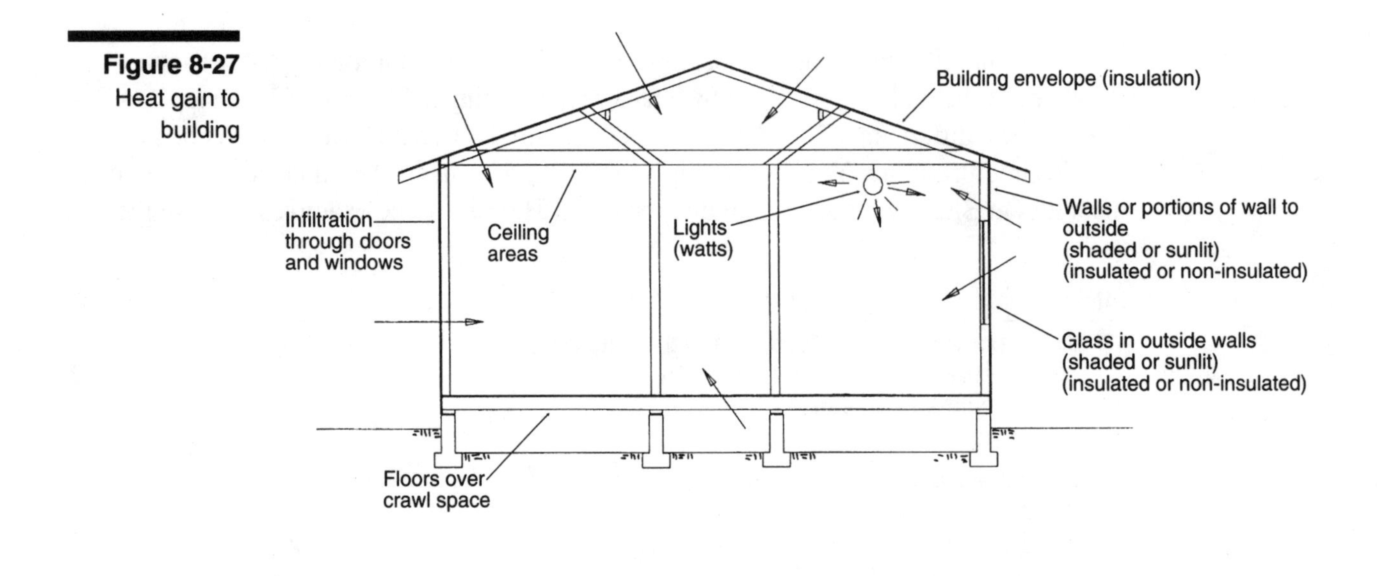

Figure 8-27 Heat gain to building

Air Conditioning

Size cooling or air conditioning systems for residential buildings just about the same way you size systems for heating. You need to calculate the amount of heat gain from the outside into the building envelope. Figure 8-27 shows the heat paths into a building.

Begin by calculating the net square footage of the building components. To find the Btuh heat gain, multiply these areas by a factor representing 20, 25, or 30 degree F temperature differential between the inside and outside of the building. Similarly, you multiply the amount of air infiltration in cfm, the number of people, and total wattage of lighting by the listed factors to obtain the heat gain in the building. When you multiply by these factors, you determine the total amount of heat gain in the building.

	Factor		
Item	**20°**	**25°**	**30°**
Walls, or portions of walls, exposed to the outside			
Not insulated and shaded	6	7	8
Not insulated and sunlit	9	11	14
Insulated	3	4	4
Glass in outside walls			
North or shaded	22	28	34
Sunlit and not shaded	120	130	150
Sunlit with inside shades	40	50	60
Sunlit with awnings or shade screens	30	35	40
Ceiling areas under attic			
No insulation	13	15	16
With insulation	5	6	7
Ceiling areas under flat roof or beam ceiling			
No insulation	18	20	22
With insulation	7	8	9

Item	Factor 20°	25°	30°
Floor			
Over crawl space	4	5	6
Over finished rooms	5	6	7
Slab	0	0	0
Outside air, in cfm, based on building volume/60	22	27	33
Number of people	200	200	200
Lights, in watts	3.4	3.4	3.4

The sum of these items is the net heat gain to the building or space in Btuh. The capacity of the cooling equipment shouldn't be less than the total heat gain under design indoor and outdoor conditions.

Here's an example for heat gain using the same house with a temperature differential of 30°, four occupants and 600 watts of lighting:

Item	Heat Loss, Btuh
Walls or portions of walls exposed to the outside	
Not insulated and shaded	450 sf × 8 = 3600
Insulated	1030 sf × 4 = 4120
Glass in outside walls	
North or shaded	190 sf × 34 = 6460
Sunlit with inside shades	225 sf × 40 = 9000
Ceiling areas under an attic	
With insulation	1430 sf × 7 = 10010
Floor over crawl space	1430 sf × 6 = 8580
Outside air, in cfm, based on building volume/60	190 cfm × 33 = 6270
Number of people	4 each × 200 = 800
Lights in watts	600 watts × 3.4 = 2040
Total heat loss in Btuh	= 50,880 Btuh

Therefore, you need a cooling load of at least 50880 Btuh, or 4 tons.

Another rule-of-thumb method for sizing HVAC capacity for single- and multi-family residential buildings is:

- single-family: 550 to 650 square feet per ton for cooling and 50 to 70 Btu per square foot for heating
- multi-family: 425 to 575 square feet per ton for cooling and 50 to 60 Btu per square foot for heating

However, you have to adjust these factors for extremely hot or cold climatic conditions.

Another simple rule for selecting evaporative coolers uses the zones specified below, the floor area of the building, and capacity of the cooler in cfm. The general description of the zones is:

- Zone 1: Desert — arid climate
- Zone 2: Plains — semi-arid climate
- Zone 3: River valleys — humid climate
- Zone 4: Coastal plains — high humid climate

Here's a list of the capacity of a cooler in cfm and the area of the space being cooled in square feet for each zone.

Capacity cfm	Zone			
	1	2	3	4
3000	750	565	420	315
4000	1000	750	560	420
4500	1125	845	635	475
5600	1375	1030	775	580
6600	1625	1220	915	685

If you need a detailed analysis of heat loss and gain, you have to calculate the thermal insulating or conducting properties of the building envelope. Figure 8-28 shows a sample analysis of heat transfer through a wood stud and plaster wall. The R-values are based on tests by ASHRAE (American Society of Heating, Refrigerating and Air Conditioning Engineers). This involves:

- Thermal resistance, (R), which is the measure of the resistance of a material or building component to the passage of heat in degrees F/square feet/Btuh

Figure 8-28 Typical heat transfer coefficients

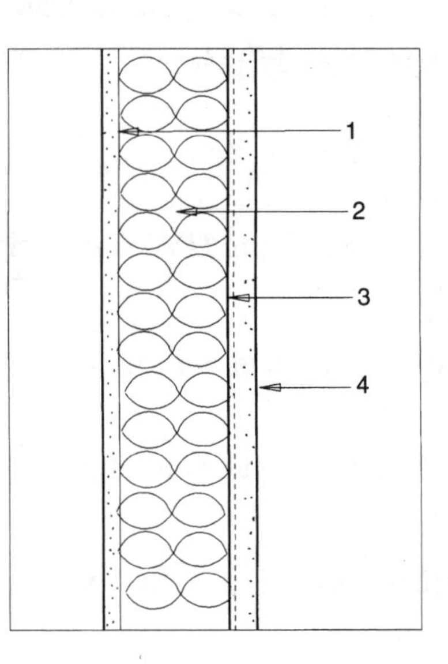

Building component	R-Value
1. 0.5" gypsum board	0.45
2. 3.5" mineral fiber insulation	11.00
3. Building paper	0.05
4. Wire reinforced stucco 0.875"	0.18
Inside surface air film	0.68
Outside surface air film	0.17
Total thermal resistance (Rt)	12.53
1/Rt = overall heat transfer coefficient (U-value)	0.08

- R-values of the materials per inch of thickness is (1/k), and per thickness as (1/C). R-value = (thickness of the material) × (1/k). Thermal resistances of building and insulating materials are listed in the ASHRAE Handbook of Fundamentals in both values of (1/k) and (1/C).
- Overall coefficient of heat transfer, or U-value, which is the flow rate through a given construction assembly, air to air, in Btuh/square foot/degree F temperature difference.
- The sum of the R-values of the individual components of the assembly gives you the total thermal resistance, Rt, of the wall assembly. The reciprocal of the Rt is called the Heat Transfer coefficient, or U-value. Multiply the U-value by the area of the wall in square feet to get the total heat gain.

In designing a cooling system you also have to know how much heat you lose along the edge of a concrete floor slab. Here's the formula you use to calculate heat loss through an uninsulated concrete slab floor where outside temperature is from 20 to 16 degrees F:

$$H = F \times P$$

where:

H = Heat loss in Btuh

F = Heat loss coefficient per linear foot of exposed edge. Heat loss coefficient varies from 32 for a 40-degree temperature differential, to 73 for a 90-degree differential. Assuming a 50-degree differential, the coefficient is about 40.

P = Perimeter of floor in linear feet

Heat Gain from Electric Motors

Electric motors also create heat. You can calculate the heat gain from the total input wattage of the electric motors. The wattage of each motor is:

$$\text{Watt} = \frac{\text{Brake hp} \times 746}{\text{Motor efficiency}}$$

The manufacturer of a motor rates it by the loading conditions which occur on the motor. For example, the efficiency of a 1 hp 220v motor at ½ load is 74.5 percent, at ¾ load it's 78.2 percent, and at full load it's 78.8 percent. Match the motor to the work it needs to do.

One watt generates about a 3.4 Btuh heat gain to the interior of a building. For example, the heat gain of typical motors of fans or pumps in air conditioned spaces is:

Nameplate or Brake hp	Btuh
1	3220
1.5	4770
2	6380
3	9450
5	15600

Ventilating and Ducts

To make air flow through a duct, the pressure at the two ends of the duct must be different. This is done by a fan that can move air at a greater static pressure than the resistance in the duct system plus the velocity pressure. Static pressure is the resistance caused by the friction of the air flow through the ducts, elbows, filters, dampers and other devices in the ductwork.

Air flow in ducts produces both static pressure and velocity pressure. The velocity pressure is the energy needed to move the air. The units used for static pressure and velocity pressure are inches of water column. The algebraic sum of the static (SP) and velocity pressure (VP) is the total pressure (TP). Figure 8-29 shows a typical supply system where the pressure in the duct is above atmospheric pressure,

Figure 8-29 Static and velocity pressures in duct

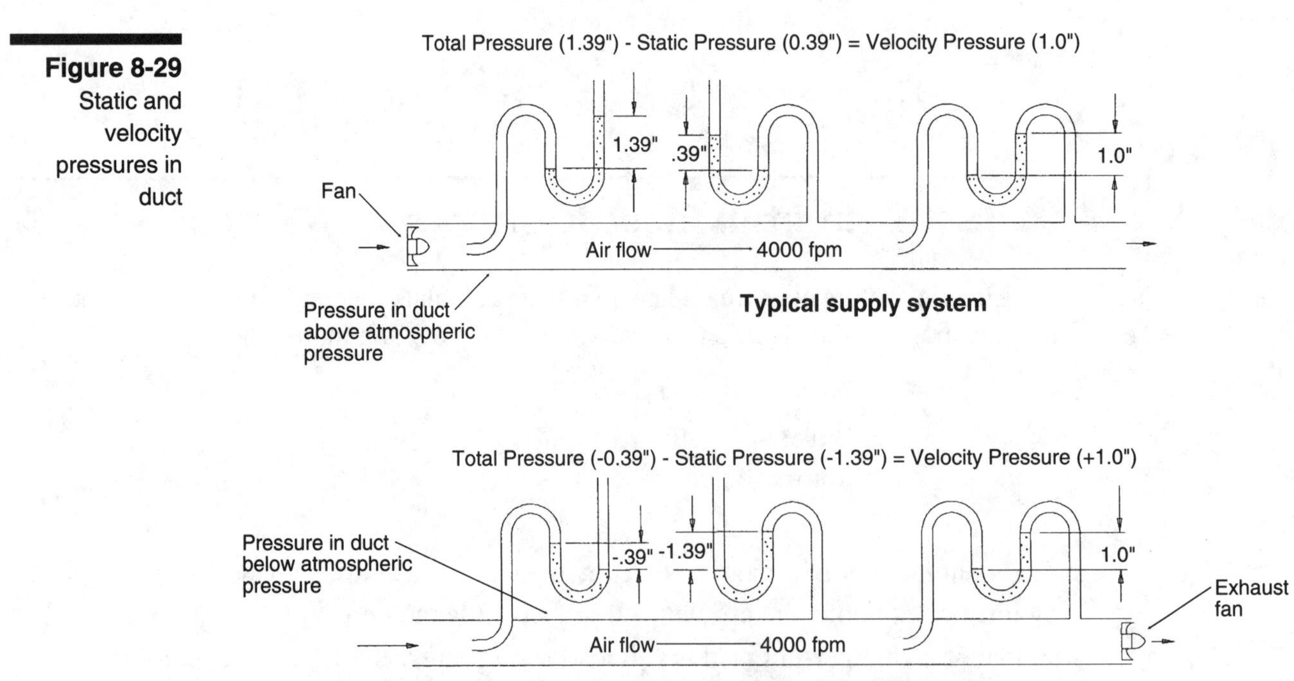

and a typical exhaust system where the pressure in the duct is below atmospheric pressure. The total pressure (TP) in the duct minus static pressure (SP) equals the velocity pressure (VP).

Another basic equation for determining total pressure in a duct system is:

SP1 + VP1 = SP2 + VP2 + Losses

where:

SP = static pressure at points 1 and 2
VP = velocity pressure at points 1 and 2
Losses = resistance in duct system.

Total static loss is the sum of the air flow resistance through the following elements:

- straight lined or unlined ducts
- regain or loss due to duct transition
- fittings, such as elbow and tee connections
- branch takeoffs
- obstructions
- fire dampers
- regulating dampers
- takeoff neck for air terminal device
- air terminal device
- outside air intake louvers
- air filters
- heating coils
- cooling coils
- fan entrances, and
- static pressure loss when the fan outlet velocity is lower than duct velocity

Duct systems are classified in three groups:

- Low pressure ductwork, which is under 2 inch water pressure
- Medium pressure ductwork, which is 2 to 6 inch water pressure
- High pressure, which is over 10 inch water pressure

To find the friction loss of any system, you need to know the velocity and size of the ducts in the system. You can find the static pressure of different size ducts in HVAC manuals, but here are a couple of examples:

Diameter, Round Ducts	Static pressure, inches WC
8"	0.0012/foot
10"	0.0017/foot

You also need to add an allowance for elbows, bends and other fittings. To figure this allowance, use the following table to arrive at the equivalent linear feet of duct to add for each type of fitting:

Round 45 elbows	5 duct diameters
Round 90 elbows	10 duct diameters
Rectangular elbows	8 duct diameters
Plenum outlet	35 duct diameters
90 wall box	35 duct diameters

Here are some common rules to consider when you design a ventilating system:

- The recommended amount of outside air that should come in, per person, for apartments is 20 cfm. The minimum is 10 cfm.
- Ventilation air should be calculated as follows:

  cfm/person × ____ persons = cfm

  or

  cfm/sq ft × ____ sq ft = cfm

- Infiltration air (cfm) in a room or space = height × length × width × air changes/60
- Fan capacity (cfm) = room volume (cf) × air changes per hour/60
- Recommended air changes per hour is 12 for bathrooms and 15 for kitchens.
- Make the neck length of air outlets at least 4 times the duct diameter.
- The ideal duct is square or round.
- Rectangular ducts shouldn't be more than 4 times as wide as they are deep.
- Avoid sharp obstructions in the duct because they increase static pressure.
- Streamline any obstructions inside ducts.
- Put air outlets at least 12 inches below the ceiling to prevent smudge produced by recycled dust-laden air from accumulating.
- Put anti-smudging rings on round ceiling outlets.
- Make the transition sections for housing filters, washers, coolers, and heaters less than 15 degrees from the slope of the duct centerline.
- Make the air movement inside conditioned air space between 20 and 50 fpm.
- Don't obstruct more than 20 percent of the cross-sectional area of a duct.
- Make the slope of transition in the duct walls on the upstream side of the transition section 15 degrees, and 30 degrees on the downstream side.
- Increments in ductwork sizes should be in one dimension only and in increments of 2 inches, such as 6 × 6, 6 × 8, 8 × 8, or 8 × 10 inches.
- Don't use a duct that's less than 6 × 6 inches.
- The throat radius shouldn't be less than three-quarters of the radius of the centerline of the elbow. The throat radius is the curve of the inside wall of a 90 degree elbow. If you can't do this, use turning vanes. Turning vanes are

curved blades you install in the interior of 90-degree elbows to smooth out the air flow. See turning vanes on Figure 8-1. Don't use a miter elbow unless you also use turning vanes.

- Put access doors at filters, cooling coils, heaters, sound absorbers, volume and splitter dampers, and fire dampers.
- The maximum temperature which combustible material can be exposed to is 175 degrees F. Combustible material includes wood, compressed paper, and plant fibers, whether flame-proofed, plastered or not.
- Use ducts that are made of sheet steel or aluminum, except underground, where you can use asbestos-cement, concrete, clay, or ceramic.
- Make the inside radius equal to the depth of the duct on standard duct elbows.
- Make vaned elbows with vanes 2¼ inches apart, curved to a 4½-inch radius, enclosing 90 degrees.
- Use reach-through, hinged, sliding, or removable access doors. They may be insulated or non-insulated.

When you pick the size of a duct or opening, you need to consider the amount and velocity of the air you want to move through the duct or opening. You also need to consider the effect of the air velocity on people in the air conditioned space. The ideal velocity for human comfort is 25 fpm with a maximum velocity of 50 fpm when people are seated and 75 fpm, standing. At 15 to 20 fpm, people will complain of stagnant air. These velocities should be measured where the air leaves the outlet, which is at a point about 6½ feet above the floor. A smaller duct increases velocity. A larger duct reduces air velocity, making it quieter, but costs more.

	Preferred velocities in supply ducts where quietness is important (fpm)	Preferred velocities in supply ducts where quietness is not important (fpm)
Main ducts	1000 to 1200	1200 to 1800
Branch ducts	500 to 800	800 to 1000
Branch risers	500 to 700	800

	Preferred velocities in exhaust ducts where quietness is important (fpm)	Preferred velocities in exhaust ducts where quietness is not important (fpm)
Main ducts	1000 to 1500	1300 to 2000
Branch ducts	800 to 1200	1000 to 1500
Branch risers	600 to 800	1000

The checklist on the next page is based on items typically checked by building inspectors. Make copies to use on your next job to reduce the risk of failing inspection.

HVAC Checklist

Cooling and Air Conditioning Systems

- ☐ Shipping blocks removed from compressor
- ☐ Blower wheel is in alignment and tight
- ☐ Unit operates quietly
- ☐ Safety controls are installed and operational
- ☐ Compressor hold-down bolts are tight
- ☐ Unit is level and properly supported
- ☐ All appliances have the proper approval labels
- ☐ Equipment has adequate accessibility

Ductwork (checked by visual inspection)

- ☐ Cross-breaking or beading
- ☐ Corner closures
- ☐ Tee connections
- ☐ 90 degree straight with vanes
- ☐ Radius tap-in on taper
- ☐ Fire dampers operate automatically
- ☐ Ducts don't leak
- ☐ Air flow is balanced
- ☐ Longitudinal seams
- ☐ Hangers
- ☐ 90 degree tap-in
- ☐ Radius tap-in
- ☐ 45 degree tap-in
- ☐ Fire dampers are labeled "FIRE DAMPER"
- ☐ Hot and cold air ducts are insulated

Vents

- ☐ Vent connector and draft hood for furnace are properly installed
- ☐ Construction around furnace is fire resistant and has required clearances
- ☐ Gas-vent caps terminate at least 1 foot above the roof surface and within 10 feet of the vent

Fans and Blowers

- ☐ Flexible duct connects the fan to the ducts
- ☐ Exposed flywheels, fans, pulleys, belts, and moving machinery have guards made of expanded metal or split or one-piece sheet metal
- ☐ Equipment has adequate accessibility

Piping

- ☐ Piping doesn't vibrate
- ☐ Condensate drainage pipes from air-cooling coils and overflow from evaporative coolers are at least ¾ inch
- ☐ Copper tubing has correct markings for service pressures
- ☐ Condensate waste pipes slope ⅛ inch per foot
- ☐ Condensate drains discharge to floor sinks

Electrical

- ☐ Fuses are properly sized
- ☐ Smoke detectors are installed in main air-return ducts in front of the outside-air inlet
- ☐ High voltage wire is proper size
- ☐ When it's activated, the smoke detector automatically shuts down the air-moving equipment

G L O S S A R Y

ABS pipe: acrylonitrile-butadiene-styrene pipe. Used for sewer pipe, waste pipe, building drains and fittings.

Absorption: 1. *masonry.* the amount of water, expressed as a percentage of total weight, absorbed by a masonry unit immersed in water. **2.** *refrigeration.* the process by which a gas, evolved in the evaporator, is taken up by some other material (absorber).

Acre: a unit of land containing 43,560 square feet, or a parcel of land about 208.71 feet square.

Admixtures: materials added to grout or mortar to increase or retard setting. Also used in making concrete, for air entrainment, water reduction, retarding and accelerating curing, increasing workability, and damp proofing.

Adsorption system: a process that dehydrates air by bringing it into contact with some other material (the adsorbent), for example activated carbon (charcoal).

Air conditioning system: equipment used to control the temperature, humidity, cleanliness, and distribution of air. Examples include the following systems: central, split, absorption, adsorption, evaporative cooling and heat pump.

Air filter (or air cleaner): a device that removes very small particles from the air. There are two types: dry filters, which strain dirt out using a cloth, felt or paper screen; and viscous (or wet) filters, which trap dirt on the surface of the fluid or trap gases by absorbing them.

Air gap (or air break): a physical separation between a water inlet and the flood level of a plumbing fixture. It prevents contamination of water supply.

Allowable stress increase: the increase in the stress permitted on a member, based on the length of time that the load causes the stress to act on the member, expressed as a percentage. The shorter the duration of the load, the higher the percent increase in the allowable stress.

Alloy pipe: steel pipe with one or more elements other than carbon added. This gives the pipe more strength and resistance to corrosion than plain carbon steel pipe.

Angle stop: a shutoff valve at the supply pipe of a plumbing fixture in which the inlet and outlet of the valve is at 90 degrees to the supply pipe.

Anodic reaction: ions released from a less noble metal into water carrying one or more positive charges. The metal sacrifices itself by corroding and deteriorating.

Arc welding (or fusion welding): a non-pressure welding process. The welding heat comes from an electric arc between the base metal and an electrode, or between two electrodes.

Automatic dry sprinkler system (or dry pipe system): a sprinkler system that uses air pressure in the system. When the sprinkler head automatically opens from the heat of a fire, the air escapes and allows water to flow through to the open sprinkler heads.

Automatic sprinkler system (or wet pipe system): a sprinkler system that has water under pressure in the pipes. That allows water to be discharged immediately when a sprinkler head opens automatically from the heat of a fire.

Axial force: a push (compression) or pull (tension) acting along the length of a structural member, usually expressed in pounds. For example, a load applied to the center of a structural member such as a column, strut or tie rod.

Axial stress: the load or force at the ends of a member, divided by the cross sectional area of the member (P/A), usually expressed in pounds per square inch (psi).

Backflow: the flow of water or other liquids, mixtures, or substances into the distributing pipes of potable water from any source other than the intended source.

Backflow preventer (or backup valve): a device installed in a pipe to prevent drainage from backing up from the main sewer in the street to a lower level, or a device to prevent backflow into a potable water system.

Backing strip: material (metal, asbestos, or carbon) used for backing up the root of a weld or the most remote point in the space filled by a weld material.

Base lines (surveying): the principal lines running east and west in the Rectangular System. Also see *U.S. Rectangular System.*

Batter: recessing or sloping of a wall in successive courses. The opposite of *corbel.*

Beam: a horizontal or inclined load-carrying structural member, supported on two or more points. Joists, girders, rafters, and purlins are beams.

Bearing: 1. *surveying.* an angular deviation of a line relative to the north-south line. This is measured in degrees, minutes, and seconds. **2.** *building.* the pressure between steel surfaces connected or held together by high strength bolts.

Bearing wall: a wall that supports a vertical load other than its own weight.

Bed joint: the bed of mortar that units of masonry are laid on.

Bench mark: a permanent landmark at a known position and altitude, or elevation. When used with a surveyor's level, a bench mark is the primary reference point for determining the elevation of other points.

Bending moment: a measure of the bending effect due to an externally-applied load acting on the member.

Bending stress: the force per square inch of area acting at a point along the length of the member resulting from the bending moment applied at that point, usually expressed in pounds per square inch (psi).

Black pipe: a type of ungalvanized steel pipe, normally used for gas lines. Not recommended for water lines due to its rapid corrosion.

Board foot: a unit of measure equivalent to a board 1 foot square and 1 inch thick.

Boards: lumber that is 2 (or more) inches wide and 1½ inches or less thick. Also see *Strips.*

Boiler, high pressure: a boiler that furnishes steam at pressures above 125 psi, or hot water at temperatures above 250 degrees F.

Bond: 1. *adhesion.* tensile strength between a masonry unit and the mortar or grout holding it. **2.** *masonry.* the manner in which masonry units are laid up in a wall, as English cross, Flemish, running, stack, and third or quarter bond.

Bond beam: courses of reinforced masonry units at floor and roof levels designed to resist horizontal forces.

Bridging: the blocking used between floor joists and rafters to keep the members from twisting.

British thermal unit (Btu): the amount of heat required to raise the temperature of 1 pound of water 1° F at sea level.

Built-up member: a single structural component made from several pieces fastened together.

Bulkhead: a partition built into wall forms to terminate each pour of concrete.

Bull header: a brick laid perpendicular to the plane of a cavity masonry wall to tie the two wythes together.

Called inspection: periodic inspection made by the building department inspector at certain stages of construction.

Camber: a predetermined curve set into a steel beam during fabrication to make up for the sag that will occur when the beam is loaded.

Cathodic reaction: the flow of electrons between dissimilar metals that causes corrosion in the less noble of the two metals.

Caulking: an elastic material used to seal joints between different materials and between building components.

Cavity wall: a wall built of masonry units having a continuous center air space. You can't use this type of wall in earthquake prone areas.

Celsius (or centigrade): a unit of temperature measurement named for Swedish physicist Anders Celsius, based on the freezing (0 degrees C) and boiling (100 degrees C) temperature of water. To convert Celsius to Fahrenheit, multiply the degrees C by 1.8 and add 32.

Cesspool: a lined pit that receives waterborne waste and permits the liquids to seep through its sides and bottom into the surrounding soil.

Chain: an early land surveying tool used in the days before metal measuring tapes. A chain is made up of 100 interlocking metal rods. One chain equals 66 feet, and 80 chains equal 1 mile. If tract maps are based on the original sections, or parts of a section, the measurements are often listed in units of chains and links.

Chair: a support of an individual reinforcing bar.

Chase: a continuous recess in a wall for pipes and ducts.

Checks (lumber): splits or cracks in a board, usually caused by overly rapid drying or seasoning. A separation of the wood naturally occurring across or through the annual growth rings.

Chord: the structural member on the top or bottom of a truss, normally called top chord or bottom chord.

Cleanout: 1. *masonry.* a grouted opening at the bottom of cells or walls used to remove debris or any obstruction from the wall. **2.** *plumbing.* a fitting in a drainage system that permits access for clearing obstructions from the pipe.

Closer: the last masonry unit laid in a course.

Combined stress: the combination of axial and bending stresses acting on a member simultaneously, expressed in pounds per square inch (psi).

Common vent: a vertical vent pipe that's connected to the traps of two fixture branches on the same level.

Compression: the force exerted on a structural member that has a compressive or pushing effect on the member and its end connections.

Compressor (air conditioning): a machine for compressing a refrigerant to change its state from a gas to a liquid.

Concentrated load: a superimposed load centered at a given point.

Condenser (air conditioning): a vessel or arrangement of piping that liquifies a vaporized refrigerant by cooling it.

Conductance: the flow of thermal energy in matter as a result of molecular collisions. For example, if one end of a metal bar is held in a flame, the heat will be conducted along the bar.

Conductor: 1. *plumbing.* a vertical pipe used to drain water from a roof or gutter. Also called roof leader. **2.** *electrical.* a copper or aluminum wire or bar that conducts electricity.

Corbel: a shelf or ledge formed by successive courses of masonry that stick out from the face of a wall.

Coupling (plumbing): a threaded fitting used to connect ends of pipes in a common line.

Course: 1. *masonry.* a single continuous horizontal layer of masonry units. **2.** *surveying.* the direction of a line. It's given with respect to north or south. The same line may be described as 45 degrees east of north (N 45° E) or 45 degrees west of south (S 45° W).

Cross connection: any connection between a potable water supply system and any plumbing fixture or device that non-potable, used, unclean, polluted or contaminated water, or any other substance, can get into.

Cross-grain: grain that is not parallel to the axis of the member. This characteristic is undesirable as it lessens the strength of the wood.

Curing: a chemical reaction occurring in cement that solidifies concrete.

Curtain wall (or filler wall): a nonload-bearing wall built between exterior columns and beams.

Dead load: any permanent roof load, such as the weight of the roof framing, roofing, ceiling, etc.

Deflection: the amount of deformation of a member; the displacement of a structural member under load.

Deformation (or strain): any change in shape, including: shortening, lengthening, twisting, buckling or expanding.

Design loads: the dead and live loads that a structural member is designed to support.

Developed pipe length: the total length of pipe, including the equivalent pipe length of all fittings and valves.

Dewatering pipe (or subterranean drain pipe): perforated pipe installed around buildings to drain away ground water.

Double-headed nail: a special nail used in formwork that can be driven in tightly, but is easily removed when stripping forms.

Drainage system (or piping): all piping within public or private premises, that conveys sewage or other liquid wastes to a legal point of disposal. Does not include the mains of a public sewer system, sewage treatment, or disposal plant.

Dry bulb temperature: the temperature of the atmosphere as indicated by an ordinary thermometer.

Dry rot: deterioration of wood caused by fermentation and chemical breakdown when it's attacked by fungus.

Duct system: all ducts, fittings, plenums, and fans that are put together to make a continuous passageway for distributing air. Low pressure: a system operating at a water-column pressure under 2 inches. Medium pressure: between 2 and 6 inches water column pressure. High pressure: over 6 inches water column pressure.

DWV system (or drain, waste, and vent system): a plumbing system that includes: all drain piping from sinks, lavatories, tubs, showers, clothes and dishwashers; all waste piping from toilets and urinals; all vent piping associated with both the drain piping and the waste piping.

Dynamic head, total: the vertical distance between source of supply and point of discharge when pumping at required capacity plus velocity head, friction, entrance and exit losses.

Easement: an interest in the land that belongs to another person. This may entitle the owner of the interest to certain use of the land, such as the right to cross over the land (affirmative easement). It may prohibit the owner of the land from doing something to the land, such as constructing a tall building (negative easement).

Efflorescence: a white crystalline deposit of water-soluble salts on the surface of a masonry wall.

Elastic action: the state in which a strain accompanied by a stress vanishes when there's no stress on the system. Plastic or inelastic action occurs when there is a residual strain after the stress is removed.

Elbow (or ell): a fitting used to connect two pipes at a 90-degree angle.

Elevations: height, usually measured against sea level. Surveyors usually set elevations in hard surfaces such as concrete pavements, and state them in hundredths of a foot. Elevations of rough grading are normally stated in tenths of a foot.

Entry loss: a loss of pressure caused by air flowing into a duct, measured in inches of water.

Equivalent fluid pressure: the lateral pressure exerted by backfill on a retaining wall.

Erection drawings: plans that show all separate pieces of a steel structure, including subassemblies, with assigned shipping and erection numbers noted in their correct position on the finished framework.

Evaporative cooler: a device used for reducing the temperature of air by evaporating water in the air stream.

Expansion joint: a designed crack in a concrete wall or slab to allow for thermal expansion and contraction.

Facing: material used as an integral finished surface of a masonry wall.

Factor of safety: the allowable unit stress based on judgment of a competent authority. The margin of safety is based on the risk involved, consistent quality of material and control of the loading conditions.

Ferrous pipe: pipe made of steel, wrought steel, wrought iron and iron.

Fire wall: a fire-resistant masonry wall that goes from the foundation to the roof, and partitions a building.

Fixture: an installed receptacle, device or appliance that is supplied with water, or that receives liquids and/or discharges liquids or liquidborne wastes, either directly or indirectly into a drainage system.

Fixture branch: the portion of a drainage system from the fixture trap to the vent.

Fixture supply pipe: a water supply pipe connecting the fixture with the fixture branch. This is the pipe that brings water from behind the wall to the plumbing fixture.

Fixture unit (fu): a quantity that represents the amount of water required by different kinds of plumbing fixtures.

Flame cutting (or gas cutting): a method of cutting steel using an excess of oxygen with acetylene gas. The gases are fed through a blowpipe and ignited at the tip. The heated steel burns in the presence of oxygen, producing a narrow gash.

Flexural strength: resistance to bending stresses.

Flow regulator: a device designed to deliver a constant flow of water over a wide pressure drop.

Flux: a chemical material used to clean and prepare the surface of copper tube connections for soldering.

Foot-pound: a measurement of work equal to the work done by a force of 1 pound acting for 1 foot in the direction of the force.

Friction head (or friction loss): the loss of pressure measured in feet of water caused by the flow of water in a pipe.

Friction-type high strength bolt: a bolt that clamps connected parts together using very high pressure. Shearing forces are resisted by the friction between the connected parts.

Galvanic action: the interchange of atoms between two dissimilar metals in contact with water. The anodic metal corrodes while the cathodic metal is protected. Example: anodic metal: magnesium, cathodic metal: iron. Result: the magnesium corrodes, protecting the iron.

Galvanized pipe: pipe coated with zinc inside and out by the hot dip system, used in both hot and cold water distribution systems.

Gas-metal arc welding: a method similar to submerged arc welding except that a stream of inert gas shields the arc.

Gauge: a method of measuring the thickness of wire or sheet metal, or the spacing of bolt holes in a structural member.

Girder: a major horizontal member used to carry a series of beams or a large load.

Girt: a horizontal member used to support wall siding.

Grade: an elevation of slope of the ground. In piping, the slope or fall in a pipe line expressed as inches or a fraction of an inch per foot.

Grades of lumber: the classification of lumber according to strength and utility.

Green (lumber): freshly sawn or unseasoned wood.

Hardwood: lumber made from broad-leafed deciduous trees, heavy and close-grained. Used mainly for flooring.

Head joint: a vertical mortar joint at the ends of a masonry unit.

Head loss: the loss of pressure in a fluid given in equivalent feet of water.

Head of water: the height of a column of water given in feet.

Header: 1. *framing.* a short joist that other joists are framed to. **2.** *masonry.* a masonry unit laid flat with its greatest dimension at right angles to the wall. **3.** *plumbing.* a horizontal pipe that supplies fixtures with water or gas.

Heart (lumber): the hard wood in the center portion of a tree, surrounded by softer sapwood.

Heat gain: the rise of temperature in water measured in degrees per pound.

Heel: the part of a roof truss where the top and bottom chords meet.

Hood: an air-intake device connected to a mechanical exhaust system to collect vapors, smoke, dust, steam, heat, or odors.

Horizontal branch piping: the part of a drain pipe that receives discharge from one or more fixture drains.

Horsepower: a measurement of work given in foot-pounds per minute.

House drain (or building drain): part of the lowest piping of a drainage system that receives all soil, waste and drainage systems inside the walls of a building and discharges it to the building sewer.

House sewer: part of the drainage system starting 2 feet from an outside foundation wall to the main sewer or septic tank.

Hub (surveying): a wooden stake, usually made of 2" x 2" x 12" long hardwood, pointed at one end and painted white at the top.

Humidity, absolute: pounds of water vapor per cubic foot of air.

Hydrated lime: quicklime that you add water to, on site, to make lime putty.

Hydrology: the science of the distribution and circulation of water as it occurs in the atmosphere, on the ground surface, and underground.

Hydrostatic pressure: the pressure exerted by a head of water, or the pressure exerted by a mechanical device, such as a pump, equivalent to the head of water, in psi or feet.

I-beam: a common name for the American Standard Beam because of its resemblance to the capital letter "I."

Inch water: pressure exerted by a column of water 1 inch high.

Industrial waste: any liquid or waterborne waste from industrial or commercial processes, except domestic sewage.

Initial point (surveying): the intersection between base line and principal meridian in the U.S. Rectangular System.

Interceptor (or clarifier): a device designed and installed to separate and hold hazardous or undesirable matter from normal wastes, and to permit normal sewage or liquid wastes to go into the disposal terminal by gravity.

Isometric drawing (plumbing): a schematic drawing of a piping system with horizontal lines drawn at 30 degrees.

Jack beam: a horizontal member used to support another beam or truss and eliminate a post.

Jack truss: a truss used to support another truss and eliminate a post.

Key: a protrusion in concrete that interlocks with a keyway.

Keyway: a depression made in concrete to receive a key, forming an interlocking joint.

Kiln dried lumber: wood seasoned in a special chamber by artificial heat.

Kiln run: unsorted brick from one kiln with a random mixture of size and color.

Knee brace: a corner brace placed at an angle.

Laitance: material that accumulates on the surface of fresh concrete from using too much water in the concrete or overworking the concrete.

Lead and tack (L&T): a means of marking a survey point in concrete by chiseling a hole, filling it with lead, and setting a brass tack.

Legal description: a written description necessary for an owner to establish, maintain, and transfer the right to occupancy and use. The description may be by government subdivision, metes and bounds, or reference to a tract map.

Licensed surveyor (LS): a surveyor who is registered to practice land surveying under the state regulations.

Lift (masonry): a continuous pour of concrete, usually into wall forms.

Lime putty: quicklime mixed with water to make a paste.

Lintel (or cap, header): a horizontal structural member placed over an opening in a wall, such as a door or window, to carry superimposed loads.

Liquid waste: the discharge (except fecal matter) from any fixture or appliance into a plumbing system.

Live load: any roof load that is not of a permanent nature, such as snow, wind, people and temporary construction loads.

Main soil (or waste) vent: the principal artery of a venting system that vent branches connect to.

Manual shield-metal arc welding (or manual, hand, or stick welding): an electric arc is produced between a coated metal electrode and the steel components to be welded. The arc heats the base metal and the electrode so they melt together and form a molten pool on the surface of the work.

Mechanical joint (ducting): a joint between metal parts made with a positive holding mechanical connection. For example, duct joined together with sheet metal screws.

Mechanical refrigerating system: a system in which the gas in the evaporator is compressed by mechanical means.

Metes and bounds: a method of naming irregularly-shaped parcels. The boundary of the land is described by courses, distances, and fixed monuments at the corners of angles. Metes mean measurements and bounds mean boundaries.

Modular masonry unit: a masonry unit whose nominal dimension is based on the 4-inch module.

Moment of inertia: a property of a structural shape. Also, a measurement of a structural member's ability to resist changing shape that's used to indicate the member's strength. This is used in calculating the bending stress.

Monuments: fixed permanent marks used by land surveyors to locate the boundary of parcels.

Negative reinforcement: steel bars that resist tensile stress at the top of a beam.

Nominal dimension: **1.** *lumber.* the rough size of lumber before it is finished and surfaced. A dry *nominal* 2 x 4 is about 1½ x 3½ inches. **2.** *masonry.* the actual size of a masonry unit plus the thickness of the mortar joint.

North: True north and magnetic north are rarely used in any description of land subdivision. North in most legal descriptions is based on a previously-recorded document. Plot plans often indicate an arbitrary north arrow to make it easier to indicate building sides as north, south, east, and west. These are sometimes called *plant north.*

Pan form: a metal form used for casting concrete joists.

Parapet wall: part of an exterior wall above the roof line.

PE pipe: polyethylene pipe, used in cold water distribution systems.

Pier: a vertical masonry column with a height less than 4 times its least width.

Pilaster: a portion of a masonry wall that projects from its face and acts as a vertical beam or column.

Pile: a long pole driven into the ground to make a firm foundation.

Pipe chase: a vertical shaft in a building built to enclose pipes.

Pitch: **1.** *lumber.* an accumulation of resinous material. **2.** *roofing and stairs.* a measurement of overall slope or angle (off horizontal), measured and expressed as inches of rise per horizontal foot. For example, a roof with a 6/12 pitch gains 6 inches height per foot. **3.** *roofing.* the resinous material used for flashing and to cement roofing felts together. **4.** *bolt spacing.* the distance separating each bolt in a line of bolts.

Plank: a wide piece of sawed timber, usually 1½ to 4½ inches thick and 6 inches or more wide.

Plasticizing agent: a material that makes mortar or grout more plastic and easier to work with.

Plenum: a chamber with one or more ducts that are connected to form part of the air conditioning system. Also a pressure equalizing chamber.

Plumbing fixtures: devices such as bathtubs, toilets, lavatories and sinks that receive or supply water, or receive liquid or waterborne waste to discharge into a drainage system.

Plyform: sheathing or form lining made out of a plastic-coated plywood. Concrete won't adhere to this material, making the forms easy to strip from the hardened concrete.

Pointing: putting mortar into the joints of a masonry unit after it's laid.

Positive reinforcement: steel bars that resist tensile stresses at the bottom of a beam.

Post-tensioned beam: a concrete beam in which the tendons are stretched *after* the concrete has cured.

Pre-tensioned beam: a concrete beam in which the tendons are stretched *before* the concrete has cured.

Prefabricated forms: factory-made concrete forms, usually made in modules.

Private sewer: a sewer with an independent sewage disposal system that's not connected to a public sewer.

Public sewer: the sewage collection system maintained and controlled by public agencies.

Pumice: a porous, volcanic rock used as aggregate to make lightweight concrete.

Pump curve: a diagram showing the characteristics of a pump, based on its horsepower, rotation and head.

Purlin: a horizontal member that acts as a beam and supports common rafters or ceiling joists.

PVC pipe: polyvinyl chloride pipe, used in cold water distribution systems.

Rafter: a sloping beam from the ridge of a roof to the eaves, supporting the roof.

Ranges and townships: a grid pattern of blocks of 36 square miles that form a rectangular system. Columns running parallel to, and east and west of, the principal meridian are called ranges, identified as R1E, R1W, etc. Rows running parallel to, and north and south of, the base line are called townships, identified as T1N, T1S, etc. See Figure 2-33 in Chapter 2.

Reshore: to remove the form and shores together, and replace each post immediately with a new post that's wedged into place to support the concrete.

Refrigerant: a substance, usually a fluorocarbon, used to produce refrigeration by its expansion and evaporation. It becomes a liquid and gives off heat when compressed, and when it absorbs heat from the air it becomes a gas.

Reglet: a built-in slot in the concrete wall for installing metal flashing.

Reinforced grouted masonry: brick masonry that has steel reinforcing bars in its grout space.

Relative humidity: the ratio of the actual pressure of the water vapor in a space to the saturated pressure of pure water at the same temperature.

Relief valve: a spring-loaded or dead-weight valve arranged to provide an automatic relief in case of excess pressure.

Relief vent (or revent): a pipe that provides air circulation between the drainage system and the plumbing system vent.

Resistance welding (or fusion or heat welding): a welding process involving the resistance to electric current between two parts to be connected. When the metal reaches a high temperature and pressure is applied, the two components are united in the contact area.

Retempering: adding water to mortar and remixing after the original mixture.

Ridge: the highest point of a roof, described as a horizontal line running the full length of a building.

Riser diagram: a drawing showing the vertical piping on a plumbing system.

Root (welding): the most remote point in the space a weld material fills.

Rowlock: a brick laid on its face edge with its long dimension at right angles to the wall face.

S-DRY: a grade indicating that the moisture content of wood does not exceed 19 percent.

Sailor: a brick laid on its end so that its greatest dimension is vertical and its wider face exposed.

Sanitary sewer: a drain pipe that conveys waterborne organic matter.

Sapwood: the outer layers of growth on trees between the bark and heartwood, exclusive of the bark, that contains the living elements (sap). Usually lighter in color than heartwood.

Saw-sized lumber: lumber that's uniformly sawn to the net size for surfaced lumber.

Schedule pipe: an identification method for telling the thickness of the pipe wall for a specific steel pipe diameter. Schedules are normally 40, 80, and 120. Schedule number = 1000 x P / S, where P = internal pressure in psi and S = allowable fiber stress of steel in psi. Pipes are also classified as standard, extra strong, and double extra strong.

Screed: a temporary guide for maintaining uniform thickness of a concrete slab.

Screw jack: a threaded steel rod used at the top or bottom of a shore for adjusting the length of the shore.

Seamless pipe: steel pipe with no welded joints.

Section modulus: a way to find the strength of a structural member when the member's cross section is known, expressed as the structural member's moment of inertia divided by the distance between its neutral axis and outer edge, or I/c.

Section: 1/36 of a township. Each section is approximately 1 mile square, and contains approximately 640 acres, or 10 square chains. Sections are numbered consecutively from the northeast corner to the southeast corner from 1 to 36, as shown on Figure 2-34 in Chapter 2. Sections may be further divided into one half, one fourth, and so on.

Seepage pit: a lined underground excavation that gets discharge through a drain pipe from a septic tank. Effluent seeps through the pit's bottom and sides.

Seismic load: an assumed lateral load, caused by an earthquake, acting in a horizontal direction on a structural frame.

Septic tank: a watertight receptacle that receives all or part of the discharge of a drainage system and digests organic matter through a period of detention, then allows liquids to go into the soil outside the tank through a system of open joints, piping or a seepage pit.

Service pressure: the maximum pressure, excluding test pressure, that a system or component may be subjected to, including any surge pressures that may develop in the system.

Setback line: a line established by law as the minimum distance from the street line to the face of the building.

Sewage system: a system of drains and sewers that carry waterborne waste from inhabited areas to a collecting point for treatment and disposal.

Shake: **1.** *lumber.* a defect (usually a split or crack) in wood, the result of damage during growth or unequal shrinkage during drying. **2.** *roofing.* wood shingle hand-split from a timber section.

Shale: a laminated clay-like rock.

Shape: the general term applied to steel rolled sections that have at least one of their cross-sectional dimensions greater than 3 inches.

Shear connectors: steel studs welded to the top flange of a steel beam supporting a concrete slab. The studs make the two parts, steel beam and concrete slab, an integral whole.

Shear stress: the stress that tends to keep two adjoining planes of a body from sliding on each other when two equal parallel forces act on them in opposite directions.

Shiner: a brick laid on its edge so its greatest dimension is horizontal and parallel to the face, and its wider side exposed.

Shipping mark (or erection mark): a system of marking various fabricated structural steel pieces that tells the type of the piece and its location on the detail sheet.

Shipping piece: a steel beam or column with its connection attachment or complete assembly of members, such as a truss or cross frame.

Shoe plate: the bottom plate of a wall form that provides a level plane.

Shotcrete: a type of pneumatically-applied concrete.

Slag (or cinder): a vitrified waste material from steel production or from a volcano eruption.

Sleeve (pipe): a form for casting a round opening in a concrete wall or slab to put pipe into.

Slip form (or sliding form): a movable form that is raised vertically, or moved horizontally, as concrete is placed.

Slopes and grades (or pitches): the direction of a line relative to a horizontal plane. The measurement may be in inches per foot (used by carpenters and plumbers), or decimal parts of a foot per foot, or percent of a foot per foot (used by civil engineers and sewer contractors).

Smoke damper: a damper that seals off smoke automatically.

Snap tie: crimped wire that is temporarily attached to wales to maintain space between opposite sides of a wall form.

Softwood: lumber from trees with a needle or scale-like leaf.

Soil pipe: any pipe that carries discharge from water closets, urinals, or similar fixtures to the building drain or sewer.

Soil stack: the vertical section of a pipe that receives discharge from water closets and carries it to the building drain.

Solar heating system: a means of heating water by radiation from the sun.

Soldered joint: a joint made by joining metal parts with metallic alloys that melt at a temperature between 400 and 800 degrees F.

Soldier: a brick laid on its end so that its greatest dimension is vertical and its edge exposed.

Spandrel wall: part of a panel wall above the top of a window in one story, and below the window sill in the story above.

Special inspection: required continuous inspection by a deputy inspector who is retained by the owner to check higher permitted stresses in masonry.

Split: the separation of wood due to the tearing apart of the wood cells.

Stack: a vertical pipe that carries waste and serves common fixtures in a multistory building.

Stack vent: an extension of a soil or waste stack that continues through the building roof.

Static head: the vertical distance between the source of water supply to a structure and the highest outlet in the structure.

Station (surveying): a length of 100 feet along a transit line.

Stiffener: a piece of steel plate solidly welded to the web and flange of a beam to stiffen both against buckling.

Stop valve: a valve to control the water supply, usually to a single fixture.

Story pole: a pole marked with each course to measure vertical heights during construction.

Straightening: an operation to remove any distortion in a steel member caused by welding or rolling.

Stress-grade: lumber that has an assigned working stress.

Stretcher: a masonry unit laid with its longest dimension horizontal and parallel to the face of the wall, with its edge exposed.

Stringer: a long horizontal timber that supports a floor or deck sheathing

Stripping: the disassembling of forming and shoring, usually for reuse.

Strips: boards less than 6 inches wide.

Strongback: a structural member used to support a wale.

Struck joint: a mortar joint that is finished smooth with a trowel.

Structural lumber: lumber that is more than 2 inches thick; used where working stresses are required.

Stud: a vertical wood or metal member that supports the sheathing in wall forming.

Subdivision: a tract of land that is divided into a number of parcels that are offered for sale.

Submerged arc welding: a method in which a bare wire electrode is used instead of a coated-wire electrode. The flux is supplied separately in granular form. Loose flux is placed over the joint to be welded, and the electrode wire is pushed through the flux. As the arc is established, part of the flux melts to form a slab shield that coats the molten metal.

Surcharge: any additional load applied to the top of a backfill behind a retaining wall, such as sloping surface, vehicular traffic, or an adjacent building.

Survey markers: pipes, lead and tack, wood hubs, and brass monuments used to mark points of a survey.

Temper (metal): a particular degree of hardness and elasticity produced in steel by heating the steel and then quenching it in a cold liquid.

Template: a guide or pattern. A thin metal sheet used to guide flame cutting, center punching, or position of anchor bolts. Also, an assembly used to hold anchor bolts or column forms in position.

Tensile strength: resistance to tension forces.

Tensile stress: a pulling force over a unit area, such as a square inch. Tensile stress is also called intensity of stress, stress per unit of the cross-sectional area, or unit stress. A "stress" is the tensile force divided by the tensile area.

Tension: the force exerted on a structural member that has the effect of either pulling apart or elongating the structural member in question.

Test prism: a cylinder or cube-shaped specimen of mortar or grout used for compression testing.

Thermodynamics: the science relating to the rate of heat flow to temperature differences and material properties.

Tie beam (or collar beam): a beam that ties roof rafters together.

Tie wire: iron wires used to hold opposing wall forms in position.

Tier (or withe, wythe): each vertical masonry thickness.

Tilt-up: a precast concrete panel that is lifted and set in place by a crane.

Timber: lumber 5 inches or more in the least dimension, and all lumber classified as beams, stringers, posts, caps, sill, girders, and purlins.

Tooling: shaping a mortar joint with a special tool.

Top chord: an inclined or horizontal member that establishes the upper edge of a roof truss, usually carrying compression and bending stress.

Total dynamic head: the vertical distance between source of supply and point of discharge when pumping at required capacity plus velocity head, friction, entrance and exit losses.

Transit line: a survey line used as a basis for measurement of roadways or pipe lines.

Truss: a rigid framework of wood or steel members arranged in triangles. Usually used in roof or bridge construction.

Tuberculation: a condition that creates hemispherical lumps (tubercles) on the interior surface of a pipe due to corrosive material in the water passing through the pipe.

Ultimate stress: the maximum stress that a material can stand before it breaks apart.

Uniform load: a load that is equally distributed over a given length, usually given in pounds per linear foot (plf).

Unit stress: the amount of stress on 1 square inch of area of a material.

United States Geological Survey (U.S.G.S.): an agency of the federal government that surveys and sets monuments of the U.S. Rectangular System.

United States Rectangular System: a method for surveying public lands by surveying 6-mile-square townships, containing 36 one-mile-square sections.

Velocity head: the kinetic energy developed in water flow expressed in height of water column in feet. $H = V^2/2g$ where V = velocity of water in fps, and g = 32.16 ft/sec^2.

Veneer: **1.** *masonry.* facing attached to a backing but not structurally bonded and therefore unable to carry vertical loads. **2.** *wood.* a thin outer layer of a select wood bonded on to a less-attractive wood, used in paneling and furniture.

Vent: pipe and fittings that carry flue gases to the outside atmosphere. Type B: gas vent for appliances that burn only gas. Type BW: gas vent for gas-fired vented wall furnaces. Type L: gas vent for oil-burning and gas-burning appliances.

Vent header: a single pipe connected to the main vent stack that receives the connection of two or more vent pipes.

Vent stack: the vertical part of a vent pipe that provides air circulation to the drainage system.

Vent system: all the vent pipes of a building that provide free air flow to and from a drainage system.

Ventilation system: equipment that supplies or removes air by mechanical means.

Venting collar: the outlet of an appliance that connects to a vent system.

Vitrified: a clay unit that has been exposed to high temperatures in a kiln.

Volume damper: a device that slows or directs air or gas flow in a duct.

Waffle form: a form for casting two-way concrete joists.

Wale (or waler): a horizontal member running on the outside of concrete forms, or a reinforcement for the studs on wall forms.

Water stop: an elastic strip placed in a concrete joint to prevent penetration of water.

Water/cement ratio: the number of gallons of water per sack of cement.

Webs: the diagonal and vertical members that join the top and bottom chords of a roof truss to form triangular patterns that give truss action, usually carrying tension or compression stresses, but no bending.

Wet bulb temperature: the temperature at which a liquid evaporating in the air can bring the air to saturation at the same temperature.

Wet standpipe: an in-structure fire protection water source with constant water pressure in the line; generally intended to be used, with a connected hose, by building occupants for small fires.

Wide-flange shapes: a term used for both beams and columns but often covering the H, B, and CB shapes in which the flange is wider than in the I-beam.

Working stress: the unit stress that experiment has shown to be safe in a material, while maintaining a proper degree of security against structural failure.

Wrought iron: wrought steel and wrought iron pipe made by forming furnace-welded pipe from skelp (flat plates) or seamless pipe from billets (unrolled blocks of iron ingots).

Wythe (or tier): each vertical masonry thickness.

Yard hydrant: a hydrant located in a private water distribution system, normally with a fire department pumper connection.

Yield point (or yield strength): the unit stress at which a material first exhibits an increase in strain with no increase in stress.

Yield stress: the amount of stress that will result in permanent deformation.

Yoke: a wood or metal frame that holds rectangular column forms in place.

INDEX

A

B

C

D

E

Q

R

S

T

U

V

W

X, Y, Z

Other Practical References

Concrete Construction & Estimating

Explains how to estimate the quantity of labor and materials needed, plan the job, erect fiberglass, steel, or prefabricated forms, install shores and scaffolding, handle the concrete into place, set joints, finish and cure the concrete. Full of practical reference data, cost estimates, and examples. **571 pages, 5½ x 8½, $20.50**

Masonry Estimating

Step-by-step instructions for estimating nearly any type of masonry work. Shows how to prepare material take-offs, figure labor and material costs, add a realistic allowance for contingency, calculate overhead correctly, and build competitive profit into your bids. **352 pages, 8½ x 11, $26.50**

Wood-Frame House Construction

Step-by-step construction details, from the layout of the outer walls, excavation and formwork, to finish carpentry and painting. Contains all new, clear illustrations and explanations updated for construction in the '90s. Everything you need to know about framing, roofing, siding, interior finishings, floor covering and stairs — your complete book of wood-frame homebuilding. **320 pages, 8½ x 11, $19.75. Revised edition**

Estimating & Bidding for Builders & Remodelers

New and more profitable ways to estimate and bid any type of construction. This award-winning book shows how to take off labor and material, select the most profitable jobs for your company, estimate with a computer (**FREE** estimating disk enclosed), fine-tune your markup, and learn from your competition. *Includes Estimate Writer, an estimating program with a 30,000-item database on a 5¼" high density disk when you buy the book.* (If your computer can't use high density disks, add $10 for *Estimate Writer* on extra 5¼" 360K disks or 3½" 720K double density disks.) **272 pages, 8½ x 11, $29.75**

Rough Framing Carpentry

If you'd like to make good money working outdoors as a framer, this is the book for you. Here you'll find shortcuts to laying out studs; speed cutting blocks, trimmers and plates by eye; quickly building and blocking rake walls; installing ceiling backing, ceiling joists, and truss joists; cutting and assembling hip trusses and California fills; arches and drop ceilings — all with production line procedures that save you time and help you make more money. Over 100 on-the-job photos of how to do it right and what can go wrong. **304 pages, 8½ x 11, $26.50**

How to Succeed With Your Own Construction Business

Everything you need to start your own construction business: setting up the paperwork, finding the work, advertising, using contracts, dealing with lenders, estimating, scheduling, finding and keeping good employees, keeping the books, and coping with success. If you're considering starting your own construction business, all the knowledge, tips, and blank forms you need are here. **336 pages, 8½ x 11, $19.50**

Manual of Professional Remodeling

The practical manual of professional remodeling that shows how to evaluate a job so you avoid 30-minute jobs that take all day, what to fix and what to leave alone, and what to watch for in dealing with subcontractors. Includes how to calculate space requirements; repair structural defects; remodel kitchens, baths, walls, ceilings, doors, windows, floors and roofs; install fireplaces and chimneys (including built-ins), skylights, and exterior siding. Includes blank forms, checklists, sample contracts, and proposals you can copy and use. **400 pages, 8½ x 11, $23.75**

Spec Builder's Guide

Shows how to plan and build a home, control construction costs, and sell to get a decent return on the time and money you've invested. Includes professional tips to ensure success as a spec builder: how government statistics help you judge the housing market, cutting costs at every opportunity without sacrificing quality, and taking advantage of construction cycles. Includes checklists, diagrams, charts, figures, and estimating tables. **448 pages, 8½ x 11, $27.00**

Daily Job Log

This handy builder's calendar will help you stay on schedule, on budget and on top of what's happening at each of your jobs for the whole year. Here you'll keep a daily log of appointments, change orders, deliveries, submittal deadlines, job visits, weather delays and due dates -- everything that affects your work schedule -- in one handy place. There's a calendar page for each day, each with a humorous construction cartoon to help start your morning right, as well as monthly summary pages, estimating forms, cost reference data, and advice from a lawyer on how the data you enter in this log can serve as protection against unfounded claims by others and as the foundation for admissible evidence should you have a case end up in court. **352 pages, 8½ x 11, $24.50**

Contractor's Guide to the Building Code Revised

This completely revised edition explains in plain English exactly what the Uniform Building Code requires. Based on the most recent code, it covers many changes made since then. Also covers the Uniform Mechanical Code and the Uniform Plumbing Code. Shows how to design and construct residential and light commercial buildings that'll pass inspection the first time. Suggests how to work with an inspector to minimize construction costs, what common building shortcuts are likely to be cited, and where exceptions are granted. **544 pages, 5½ x 8½, $28.00**

Residential Electrical Design Revised

If you've ever had to draw up an electrical plan for an addition, or add corrections to an existing plan, you know how complicated it can get. And how many electrical plans -- no matter how well designed -- fit the reality of what the homeowner wants? Here you'll find everything you need to know about blueprints, what the NEC requires, how to size electric service, calculate and size loads and conductors, install ground-fault circuit interrupters, ground service entrances, and recommended wiring methods. It covers branch circuit layout, how to analyze existing lighting layouts and install outdoor lighting, methods for remote-control switching, residential HVAC systems and controls, and more. **256 pages, 8½ x 11, $22.50**

Paint Contractor's Manual

How to start and run a profitable paint contracting company: getting set up and organized to handle volume work, avoiding mistakes, squeezing top production from your crews and the most value from your advertising dollar. Shows how to estimate all prep and painting. Loaded with manhour estimates, sample forms, contracts, charts, tables and examples you can use. **224 pages, 8½ x 11, $19.25**

Excavation & Grading Handbook Revised

Explains how to handle all excavation, grading, compaction, paving and pipeline work: setting cut and fill stakes (with bubble and laser levels), working in rock, unsuitable material or mud, passing compaction tests, trenching around utility lines, setting grade pins and string line, removing or laying asphaltic concrete, widening roads, cutting channels, installing water, sewer, and drainage pipe. This is the completely revised edition of the popular guide used by over 25,000 excavation contractors. **384 pages, 5½ x 8½, $22.75**

Craftsman Book Company
6058 Corte del Cedro
P. O. Box 6500
Carlsbad, CA 92018

In a hurry?

We accept phone orders charged to your MasterCard, Visa or American Express

Call 1-800-829-8123
FAX (619) 438-0398

Name (Please print) ______________________
Company ______________________
Address ______________________
City/State/Zip ______________________

☐ Send check or money order and we pay postage
If you prefer, use your ☐ Visa, ☐ MasterCard or ☐ Am Exp
Card # ______________________
Expiration date ______________ Initials ______

10-Day Money Back Guarantee

- ☐ 22.00 Basic Plumbing with Illustrations
- ☐ 15.00 Building Layout
- ☐ 20.50 Concrete Const. & Estimating
- ☐ 39.75 Construction Forms & Contracts with **FREE** $3\frac{1}{2}$" high density ***Forms & Contracts*** disk.*
- ☐ 19.25 Construction Surveying & Layout
- ☐ 28.00 Contractor's Guide to the Building Code Revised
- ☐ 24.50 Daily Job Log
- ☐ 29.75 Estimating & Bidding for Builders & Remodelers with **FREE** $5\frac{1}{4}$" high density ***Estimate Writer*** disk.*
- ☐ 17.00 Estimating Home Building Cost
- ☐ 22.75 Excavation & Grading HndbkRev
- ☐ 19.50 How to Succeed With Your Own Construction Business
- ☐ 23.75 Manual of Professional Remod.
- ☐ 26.50 Masonry Estimating
- ☐ 31.50 NationalConstruction Estimator with **FREE** $5\frac{1}{4}$" high density ***Const. Estimate Writer*** disk.*
- ☐ 32.50 National Repair & Remodeling Estimator with **FREE** $5\frac{1}{4}$" high density ***Repair & Remodeling Estimate Writer*** disk.*
- ☐ 19.25 Paint Contractor's Manual
- ☐ 22.50 Residential Electrical Design Revised
- ☐ 19.00 Roofers Handbook
- ☐ 26.50 Rough Framing Carpentry
- ☐ 27.00 Spec Builder's Guide
- ☐ 19.75 Wood-Frame House Const.
- ☐ 34.00 Basic Engineering for Builders
- ☐ **FREE** Full Color Catalog

****If you need ☐ $5\frac{1}{4}$" or ☐ $3\frac{1}{2}$" double density disks add $10 extra.***

Craftsman Book Company
6058 Corte del Cedro
P. O. Box 6500
Carlsbad, CA 92018

In a hurry?

We accept phone orders charged to your MasterCard, Visa or American Express

Call 1-800-829-8123
FAX (619) 438-0398

Name (Please print) ______________________
Company ______________________
Address ______________________
City/State/Zip ______________________

☐ Send check or money order and we pay postage
If you prefer, use your ☐ Visa, ☐ MasterCard or ☐ Am Exp
Card # ______________________
Expiration date ______________ Initials ______

10-Day Money Back Guarantee

- ☐ 22.00 Basic Plumbing with Illustrations
- ☐ 15.00 Building Layout
- ☐ 20.50 Concrete Const. & Estimating
- ☐ 39.75 Construction Forms & Contracts with **FREE** $3\frac{1}{2}$" high density ***Forms & Contracts*** disk.*
- ☐ 19.25 Construction Surveying & Layout
- ☐ 28.00 Contractor's Guide to the Building Code Revised
- ☐ 24.50 Daily Job Log
- ☐ 29.75 Estimating & Bidding for Builders & Remodelers with **FREE** $5\frac{1}{4}$" high density ***Estimate Writer*** disk.*
- ☐ 17.00 Estimating Home Building Cost
- ☐ 22.75 Excavation & Grading Hdbk Rev
- ☐ 19.50 How to Succeed With Your Own Construction Business
- ☐ 23.75 Manual of Professional Remod.
- ☐ 26.50 Masonry Estimating
- ☐ 31.50 National Construction Estimator with **FREE** $5\frac{1}{4}$" high density ***Const. Estimate Writer*** disk.*
- ☐ 32.50 National Repair & Remodeling Estimator with **FREE** $5\frac{1}{4}$" high density ***Repair & Remodeling Estimate Writer*** disk.*
- ☐ 19.25 Paint Contractor's Manual
- ☐ 22.50 Residential Electrical Design Revised
- ☐ 19.00 Roofers Handbook
- ☐ 26.50 Rough Framing Carpentry
- ☐ 27.00 Spec Builder's Guide
- ☐ 19.75 Wood-Frame House Const.
- ☐ 34.00 Basic Engineering for Builders
- ☐ **FREE** Full Color Catalog

****If you need ☐ $5\frac{1}{4}$" or ☐ $3\frac{1}{2}$" double density disks add $10 extra.***

Mail This Card Today For a Free Full Color Catalog

Over 100 books, videos, and audios at your fingertips with information that can save you time and money. Here you'll find information on carpentry, contracting, estimating, remodeling, electrical work, and plumbing.

All items come with an unconditional 10-day money-back guarantee. If they don't save you money, mail them back for a full refund.

Name (please print) ______________________
Company ______________________
Address ______________________
City/State/Zip ______________________

Craftsman Book Company, 6058 Corte del Cedro, P. O. Box 6500, Carlsbad, CA 92018

NO POSTAGE
NECESSARY
IF MAILED
IN THE
UNITED STATES

BUSINESS REPLY MAIL
FIRST CLASS MAIL PERMIT NO. 271 CARLSBAD CA

POSTAGE WILL BE PAID BY ADDRESSEE

CRAFTSMAN BOOK COMPANY
PO BOX 6500
CARLSBAD, CA 92018-9974

NO POSTAGE
NECESSARY
IF MAILED
IN THE
UNITED STATES

BUSINESS REPLY MAIL
FIRST CLASS MAIL PERMIT NO. 271 CARLSBAD CA

POSTAGE WILL BE PAID BY ADDRESSEE

CRAFTSMAN BOOK COMPANY
PO BOX 6500
CARLSBAD, CA 92018-9974

NO POSTAGE
NECESSARY
IF MAILED
IN THE
UNITED STATES

BUSINESS REPLY MAIL
FIRST CLASS MAIL PERMIT NO. 271 CARLSBAD CA

POSTAGE WILL BE PAID BY ADDRESSEE

CRAFTSMAN BOOK COMPANY
PO BOX 6500
CARLSBAD, CA 92018-9974